Also by Dick Russell

The Man Who Knew Too Much
Black Genius: And the American Experience

DICK RUSSELL

MAPS BY EBEN GIVEN

Simon & Schuster

New York London Toronto Sydney Singapore

Eye of the Whale

Epic Passage from

Baja to Siberia

SIMON & SCHUSTER
Rockefeller Center
1230 Avenue of the Americas
New York, NY 10020

Book design by Ellen R. Sasahara

Manufactured in the United States of America

10 9 8 7 6 5 4 3 2 1

Library of Congress Cataloging-in-Publication Data
Russell, Dick.
Eye of the whale ; epic passage from Baja to Siberia /
[Dick Russell].
p. cm.
Includes bibliographical references (p.).
1. Gray whale—Migration. I. Title.
QL737.C425 R87 2001
599.5'2217742—dc21 2001020572

ISBN 0-684-86608-0

To Mookie, my muse,
and Dexter José Arugula

Canst thou draw out leviathan with an hook? . . .
I will not conceal his parts, nor his power, nor
his comely proportion. . . .
He maketh a path to shine after him. . . .
He beholdeth all high things: he is king over all the
children of pride.

—Job 41:1, 4, 24, 26

Contents

The Eye of the Whale

Homero Aridjis

(Genesis 1:21)

To Betty

(Tr. Richie Guerin, Jessie Benton, Devora Wise, Jeremy Greenwood)

And God created the great whales
there in Laguna San Ignacio,
and each creature that moves
in the shadowy thighs of the water.

He created dolphin and sea lion,
blue heron and green turtle,
white pelican, golden eagle,
and the double-crested cormorant.

And God said unto the whales:
"Be fruitful and multiply
in act of love that may be
seen from the surface

only through a bubble,
or a fin, slanted,
the female is taken below
by the long prehensile penis;

for there is no splendor greater than the gray
when the light turns it to silver.
Its bottomless breath
is an exhalation."

And God saw that it was good,
that the whales made love
and played with their young
in the magical lagoon.

And God said:
"Seven whales together
make a procession.
One hundred make a dawn."

El Ojo de la Ballena

Homero Aridjis

(Genesis 1:21)

A Betty

Y Dios creó las grandes ballenas
allá en Laguna San Ignacio,
y cada criatura que se mueve
en los muslos sombreados del agua.

Y creó al delfín y al lobo marino
a la garza azul y a la tortuga verde,
al pelicano blanco, al águila real
Y al cormorán de doble cresta.

Y Dios dijo a las ballenas:
"Fructificad y multiplicaos
en actos de amor que sean
visibles desde la superficie

sólo por una burbuja,
por una aleta ladeada,
asida la hembra debajo
por el largo pene prensil;

que no hay mayor esplendor del gris
que cuando la luz lo platea.
Su respiración profunda
es una exhalación."

Y Dios vio que era bueno
que las ballenas se amaran
y jugaran con sus crías
en la laguna mágica

Y Dios dijo:
"Siete ballenas juntas
hacen una procesión.
Cien hacen un amanecer."

And the whales came out
to catch a glimpse of God
between the dancing furrows of the waters.
And God was seen through the eye of a whale.

And the whales filled
the oceans of the earth.
And it was the afternoon and the morning
of the fifth day.

—After a trip to Laguna San Ignacio,
Mexico City, March 1, 1999

Y las ballenas salieron
a atisbar a Dios entre
las estrias danzantes de las aguas.
Y Dios fue visto por el ojo de una ballena.

Y las ballenas llenaron
los mares de la tierra.
Y fue la tarde y la mañana
del quinto día.

—Después de un viaje a Laguna San Ignacio,
México D.F., 1 de marzo de 1999

Part One

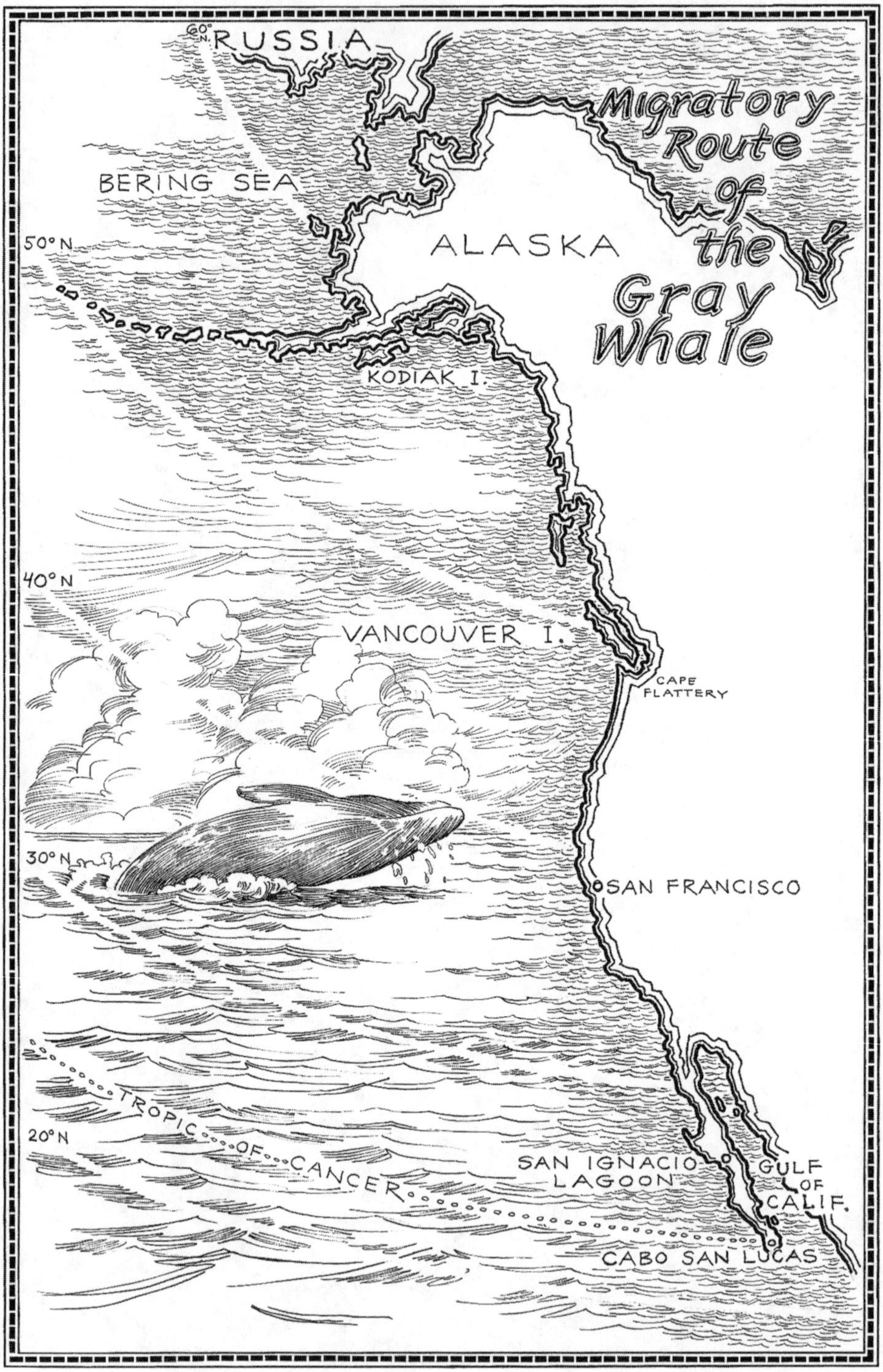

Migratory Route of the Gray Whale
RUSSIA
60° N.
BERING SEA
ALASKA
50° N
KODIAK I.
40° N
VANCOUVER I.
CAPE FLATTERY
30° N
SAN FRANCISCO
TROPIC OF CANCER
20° N
SAN IGNACIO LAGOON
GULF OF CALIF.
CABO SAN LUCAS

Prologue

"Whales of Passage"

As the whale is great, so to cherish it can be proof of our greatness. Meanwhile, and for a little longer, the great whale glides through the sea, feeling its vibrations and reading its meaning by senses it has gained through eons of time. Had the whale been created only to deepen our sense of wonder, that were enough, for it is imagination that makes us human.

—Dr. Victor B. Scheffer, former director, U.S. Fish and Wildlife Service, *The Year of the Whale*, 1969

Halfway down the remote Pacific coastline of Mexico's Baja California peninsula, 435 miles south of the U.S. border at San Diego, the San Ignacio Lagoon (Laguna San Ignacio) winds its way inland along the southern boundary of the Vizcaíno Desert. Millennia ago the lagoon was formed when rising sea levels breached the sand dunes and flooded portions of the desert. Today the surrounding region is as dry as the Sahara. The lagoon's sixty square miles fall within a stark landscape of intertidal mudflats and red and white mangrove estuaries, sand dunes, and salt mounds. Miles of natural salt flats extend to the north and west, created gradually by repeated saltwater flooding over the centuries. Three mountain ranges crest near the lagoon's narrow, often shallow reaches.

There is no electricity or plumbing, no invasive human development at all, around the San Ignacio Lagoon. Because of a lack of rainfall and any immediate source of freshwater, the area contains few permanent residents. Almost four hundred people live in six fishing communities along the shoreline. For water, propane fuel, and primary foodstuffs, they must travel either to nearby ranches or for several hours along a rutted dirt road to the oasis town of San Ignacio.

Perhaps this very isolation is why the Eastern Pacific's California gray whales *(Eschrichtius robustus)* have been coming here for thousands, perhaps hundreds of thousands, of years—to breed, bear, and nurse their young in the lagoon's warm, calm waters.

In March 1998 I traveled for the first time to Laguna Ballenas, as the Mexicans once designated it. "Lagoon of Whales." I was on assignment for *The Amicus Journal*, a quarterly magazine published by the Natural Resources Defense Council. The story focused on a proposed joint venture by Japan's Mitsubishi International Corporation and Mexico's Ministry of Trade. Together they wanted to build the world's largest salt factory inside the last pristine lagoon habitat of the gray whale.

I had heard about what marine biologists call "the friendly gray whale phenomenon" at the San Ignacio Lagoon. Here, with increasing frequency over the past quarter century, mother gray whales have been approaching small boatloads of visitors and "introducing" their newborn calves. Often the pairs will linger alongside and allow themselves to be petted. They permit you to place a hand in their mouths. Or even plant a kiss on their massive foreheads.

What makes this phenomenon all the more astonishing is that, in this same lagoon, hundreds of gray whales were slaughtered during the second half of the nineteenth century. They fell prey to an American whaling fleet led in by Charles Melville Scammon, the first Western man known to have found a safe way into the lagoon from the Pacific. To Scammon and other whalers, the California gray was the most dangerous quarry of all—especially when protecting its young. The gray whale's ferocity prompted a San Francisco newspaper to state in 1863: "As many men are lost in catching them as in all the other whaling grounds put together." In those years American whalers called the gray whale the devil-fish. The same appellation, *koko-kujira* or "devil-fish," was bestowed by Japanese whalers upon its cousins in the Western Pacific.

Today, as the descendants make an entirely different type of contact with human beings, some go so far as to say that the whalemen's devil-fish has become an angel.

I never anticipated how my first encounters with the gray whales on the waters of the San Ignacio Lagoon would affect my life. As a journalist I had been writing about ocean-related issues for many years. My interest began with the Atlantic striped bass. In the early 1980s this magnificent fish appeared headed for the Endangered Species List. The combined effects of overfishing along their migratory range and pollution in their Chesapeake Bay spawning grounds had proven too much. I ended up organizing a coastwide campaign of sports fishermen and conservationists, calling for stronger management of commercial fishing pressure. It worked. The dramatic surge in striped bass abundance is today considered proof of a threatened fish's ability to recover. I'd gone on to fight against ocean dumping and for stronger protection of Atlantic tunas and marlin.

During the early 1990s I fell in love with Baja California. A group of close friends and I pooled resources and built a house along the southern coast. I spent many a day marveling at whales and dolphins passing just offshore of our beach. Removed from the often frenetic pace of my home base in Boston, I found a clarity of thought, a sense of hope and possibility.

At the San Ignacio Lagoon, I would find something more. Meeting the gray whale, I met the ineffable. I came to discover that many others have had a similar transformative experience: not only whale watchers and environmentalists but marine scientists, indigenous peoples, and even the nineteenth-century whaling captain who discovered the lagoon. The gray whales would lead me on a long journey, examining the many interactions between human beings and this remarkable animal as I followed its migratory course. From the desert shores of the San Ignacio Lagoon, I would travel across California and Oregon, to the temperate rain forest along Washington's Olympic Peninsula, and on to Vancouver Island. Eventually I would make my way up the Alaskan coastline to the Bering Strait, then across to the Siberian native hunting villages of Chukotka. I would even venture to the only known outpost of the critically endangered Western Pacific gray whale, an isolated lagoon in the Russian Far East.

It is not known how, when, and where gray whales originated. The first whale ancestors lived on land about 50 million years ago, but no fossil remains of a direct gray whale progenitor have ever been found. The grays, in fact, pose an evolutionary puzzle. It was once thought that they evolved from a long-extinct whale family called the Cethotheridae. However, the single complete skeleton (and a few partial ones) so far discovered of the earliest modern gray whales dates back only 50,000 to 120,000 years. So scientists are no longer willing to hold to their earlier theory. Gray whales have been placed in their own family, Eschrichtidae.

Whales are, of course, the largest animals on the planet. Grays, while outsized by the blues, bowheads, fins, sperms, and humpbacks, make a twice-annual migration that must be regarded as one of the most spectacular achievements on the planet. The majority of a population now estimated at more than 26,000 travels from three warm lagoon areas around central Baja to Arctic feeding grounds near the Bering Strait and then back again. At a minimum they are swimming five thousand miles in each direction. If we consider that some have been reported as far north as the Canadian Beaufort Sea and others as far south as Baja's Sea of Cortés, the gray whales lay claim to making the longest annual trek of any mammal on earth.

Unlike other whales, who meander between coastal and deep-ocean waters, only the California gray follows a continuous path, generally within a few miles of shore. They swim in small groups, sometimes alone, in waters often less than thirty feet deep. Some may occasionally do opportunistic feeding along their route, and a few are even known to drop out and take up seasonal residence in food-plentiful locations along the way. Most, though, do not pause for long, if at all. In a twenty-four-hour day gray whales are capable of covering almost eighty miles. The position of the sun, the sound of the surf, and long-familiar underwater landmarks are thought to be their primary means of navigating through storms and darkness.

For an astonishing eight months between their summertime foragings in the Arctic, most gray whales will have eaten little or fasted altogether. Arriving in the chilly waters of the Bering Sea, they forage continuously along the shallow bottom near the continental shelf. Their diet consists primarily of tiny, shrimplike amphipods, of which they ingest as much as a ton per day. Gray whales have no teeth; they capture and strain their food through a fringed curtain of baleen, which hangs from the roof of the mouth.

Usually by sometime in October, as the days grow shorter and the gray whales sense that the Land of the Midnight Sun will soon be covered by a thick layer of ice, the long journey starts anew. Pregnant females lead the way south, for time is of the essence. They are followed by mature males, females with whom the males will soon mate as part of a thirteen-month cycle, and juveniles bringing up the rear. All will endure water temperatures ranging between 41 and 72 degrees Fahrenheit. A thirty-ton gray whale will expend so much energy on its return trip to the Baja lagoons it may lose fully eight tons of its blubber.

We don't know why they make this unique migration, or for how long they've pursued it. Unlike all the other species of baleen whale, the gray is never known to have occupied the southern oceans. Some scientists speculate that gray whales might have resided year-round in Baja approximately eighteen thousand years ago, when sea levels were lower and those bottom-feeding areas were richer. Then, as melting glaciers receded and sea levels rose, they had to move north seeking food. Another view holds that these same sea levels, which caused the disappearance of a low-lying land bridge across the Bering Strait, provided gray whales entrée to the extensive and fertile feeding grounds of the Bering and Chukchi Seas.

So it's conceivable that a new migratory orientation for gray whales overlapped with and paralleled a new journey by human beings. Probably between ten thousand and fourteen thousand years ago, most archaeologists believe, Paleoindian pioneers from Beringia in today's Asiatic Russia crossed over the broad plain of a thousand-mile-long exposed continental shelf into the New World. Over the centuries some of these people settled at various locations along the Pacific coast. They include such tribes as the Inuit in northern Alaska and the Tlingit in southeast Alaska, the Tse-shahts on Vancouver Island, the Makah along Washington's Olympic Peninsula, and the Chumash in California and Baja.

All these peoples came to know the gray whales who swam in close proximity to their shores. Many of them pursued the huge marine mammals from their canoes with harpoons and spears. They ate the meat, used the oil in cooking, fashioned combs and roof supports from the bones, and made chairs from the vertebrae. And each aboriginal group, in their own fashion, honored the animal.

In Siberia the Chukchi erected settlements and memorials from the bones of gray whales. The Koryak danced wearing special masks to invoke the gray whale's spirit. Before a hunt the Koryak confessed their sins and asked for forgiveness. After a whale was killed they held a communal festival. They believed that the whale had given the tribe permission to harvest it. The gray whale's soul would return to the sea, where it would tell relatives whether its reception among the Koryaks had proven hospitable. For the whales were believed to live in villages beneath the sea. They could avenge the murder of their own. Or be grateful for kindnesses received.

In Alaska the Inuit of the North Country called the gray whale *antokhak* and imitated its movements in ritual celebrations.

On Vancouver Island the Tse-shaht called the gray whale *Ee-toop* and sang to it during feast days: "I search for *Ee-toop*, bigness, largeness, the mass that moves upon the seas." Other tribes of that region believed Whale People lived in undersea dwellings as *Homo sapiens*, assuming the form of whales when they emerged.

The Makah, who sailed from nearby Vancouver Island to settle at the very northwestern edge of the contiguous United States, called the gray whale *sih-wah-wihw* (beings with itchy faces) and referred to it as "ruler of the world." Their legends tell of a giant eaglelike being called Thunderbird. The Makah believed that Thunderbird created the universe and was also the first whale hunter. Thunderbird possessed talons and wings powerful enough to lift a whale from the waters and carry it into the mountains to feast upon the flesh. The lightning and thunder arose from the tumultuous encounters between Thunderbird and the whale.

One day, according to another Makah legend, all the forest animals conceived a scheme against Thunderbird's reign. The animals built an imitation whale. They covered the outside with pitch, sealing it watertight, and headed out to sea with it. There the animals surprised Thunderbird when he attacked their whale. The god, along with several of his children, was slain.

From that time on, say the Makah, the privilege of hunting the gray whale passed to a few mortals born of godly unions. These became the Makah whaling families, revered among the tribe. The gray whale would feed the people, but only if they were prepared in spirit. Selected Makahs underwent a year of ritual cleansing to become "at one with the whale." They prayed and fasted. They rubbed their bodies with stinging nettles. They bathed in icy waters, mimicking the actions of the whale. "If a man is to do a thing that is beyond human power," Makah lore said, "he must have more than human strength for the task."

Along Southern California and into Baja, the Chumash villages fanned out along the coastlines. At the culmination of the gray whale's migration, Chumash often camped beside the San Ignacio Lagoon. There the Indians feasted upon shellfish and green turtles. Occasionally they would eat the meat of a gray whale that had died and washed ashore. But the Chumash are said not to have hunted the massive creature that came to their waters to give birth and nurture its young.

Only remnants of Chumash culture remained when Charles Melville Scammon came to the Baja lagoons during the 1850s. In later years the most renowned whaling captain of the West Coast would publish his observations of the gray whale. Scammon watched the males wait offshore as the females swam through the narrow passages. There, he wrote, "if not disturbed, they gather in large numbers, passing and repassing into or out of the estuaries, or slowly raising their colossal forms midway above the surface, falling over on their sides as if by accident, and dashing the water into foam and spray about them. At times, in calm weather, they are seen lying on the water quite motionless, keeping one position for an hour or more. At such times the sea-gulls and cormorants frequently alight upon the huge beasts."

Scammon studied the "huge beasts" closely, observing that "many of the marked habits of the California Gray are widely different from those of any other species of '*balaena.*'...The length of the female," he noted, "is from forty to forty-four feet, the fully grown varying but little in size. ... The male may average thirty-five feet in length, but varies more in size than the female."

As Scammon traveled all along their vast range, he called them "whales of passage." He witnessed in their eight-man canoes the Makah, who "consider the capture of this singular wanderer a feat worthy of the highest distinction." He witnessed the northern Eskimos, for whom the skin of a gray whale formed "an indispensable article of clothing."

Scammon listened to the gray whales at the apex of their migration in the Bering Sea, "emerging from the scattered floes, and even forcing themselves through the field of ice, rising midway above the surface, and blowing in the same attitude in which they are frequently seen in the southern lagoons; at such times the combined sound of their respirations can be heard, in a calm day, for miles across the ice and water."

Charles Melville Scammon recorded all this, in words and sketches, with the keen eye of a naturalist. Scammon's Lagoon in Baja is named after him. So is Alaska's Scammons Bay. And so are the small parasites that live on the skin of the gray whale: *Cyamus scammoni*. There is considerable irony in the last distinction. For Scammon, one of the most daring and innovative whalers of his time, bears ultimate responsibility for bringing the California gray whale to the brink of extinction in the nineteenth century. Yet his close encounters with this animal would have a profound effect upon his life—and upon our knowledge of natural history.

Only a few centuries ago gray whales roamed the Atlantic Ocean as well as the Pacific. Their remains have been found in England, Holland, and Sweden, as well as along the eastern seaboard of North America. Indeed, some biologists think that the species first evolved in the North Atlantic, then crossed into the North Pacific when temperatures became warm enough to permit an easier passage through the Arctic. Atlantic gray whales have not been seen since the advent of commercial whaling by Europeans and then Americans. Colonial records indicate that this shore-hugging species may have been among the most popular whales sought by the early New Englanders. By 1750 the Atlantic gray had disappeared from both European and American coastlines. It is the first—and so far the only—whale species known to have been driven into extinction.

A related Western Pacific gray whale (also called the Korean gray whale) once existed in large numbers as well. Today, a century after its overexploitation by Japanese and Korean whaling ships, its population has been estimated at less than one hundred, barely enough to sustain itself. These whales frequent a feeding area offshore of Piltun Lagoon, along Russia's Sakhalin Island. Their southern range remains unknown to science, although it is speculated to be somewhere around Hainan, China, at a latitude similar to that of southern Baja California.

Given what happened to their brothers and cousins, the California gray whales' survival is nothing short of miraculous. They were massacred by whaling fleets at two junctures—first during the nineteenth century and, just as their population began to recover, again during the early part of the twentieth century. By 1921 they were so rare that a single sighting merited publication in the *Journal of Mammalogy*. Two scientists reported in 1930 that it was doubtful "whether more than a few dozen individuals remained." In 1937 the California gray became the first whale designated for protection under the International Agreement for the Regulation of Whaling promulgated by the League of Nations. A decade later the grays were still thought to be all but extinct. Their recovery over the second half of the twentieth century is considered the most dramatic achieved by any species of whale. In 1995 the California gray became the first whale to be removed from the U.S. Endangered Species List. Left alone by humans, gray whales are estimated to have a life span as long as seventy-five years.

As a shore wanderer the gray whale is seen by more people, in more places, than any other type of leviathan. It's the one that started the whale-watching phenomenon during the 1950s in California in what has become a multimillion-dollar global industry. For many its twice-yearly passage engenders a feeling of both awe and companionship. I have come to see the gray whale as a kind of metaphor as we enter a new millennium, holding a mirror to a number of ecological, political, and social issues concerning our relationship to nature and particularly to the oceans.

In some ways the life and times of Charles Melville Scammon mirror the same complex themes. If this man had been merely a shrewd and daring whaler willing to risk venturing into Arctic waters and the Baja lagoons, we would remember him as an exploiter and probably even a villain. But in eventually turning away from whaling, Scammon became an explorer, writer, and the foremost expert of his era on cetaceans of all kinds. He was really our first marine mammalogist. I believe his personal metamorphosis marked the start of a shift in consciousness about the wondrous creatures who inhabit our oceans: that they are there not simply to exploit but to appreciate and to learn about.

In this book we will follow Scammon's odyssey as we traverse the gray whale's migratory path from Baja to Siberia. We will also meet Scammon's naturalist heirs, the marine scientists who devote their lives to studying the unique habits and habitats of the gray whale. We will look at the fight to preserve those habitats—at San Ignacio Lagoon and elsewhere—from the encroachments of industry. We will examine the magnetic effect this particular whale has upon individuals, sometimes even whole societies, up and down the coast. These societies include the Makah Tribe, a Native American people who recently hunted and killed a gray whale after a seventy-year hiatus, igniting a storm of protest. These societies also include the aboriginal villagers along Russia's Chukotka Peninsula, for whom hunting the gray whale has again become a matter of survival.

From the touch of a whale to the thrust of a harpoon, this is my journey as well. Over the past several years the gray whale has taken me to places, both external and internal, that I had never imagined going. Coming to trace its journey has proven a life-changing experience. Finally, this is the story of what the gray whale has taught me—about mystery and trust, about the spirit of place, about the animal kingdom, about my fellow human beings, about myself.

The journey begins with the moment I entered the domain of the gray whale and fell under its spell.

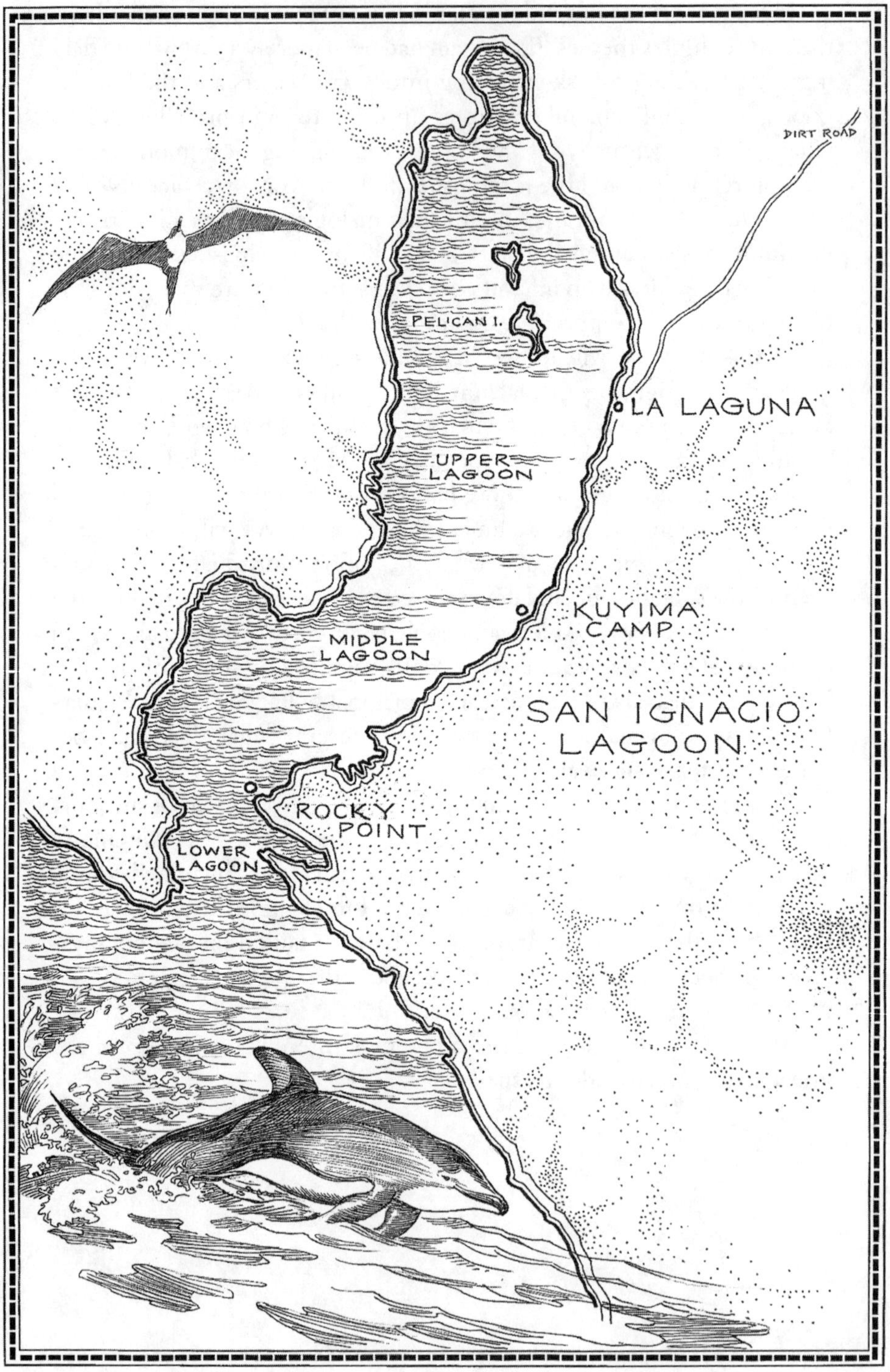
DIRT ROAD
PELICAN I.
LA LAGUNA
UPPER LAGOON
KUYIMA CAMP
MIDDLE LAGOON
SAN IGNACIO LAGOON
ROCKY POINT
LOWER LAGOON

Chapter 1

Close Encounter at San Ignacio Lagoon

The holy hangs undisturbed over the whales' huge cradle.

—Loren Eiseley

It is early January. Beyond the Mexican border the pregnant gray whale picks up her pace down the Pacific coast of Baja California, past Tijuana and Ensenada, San Quintín and El Rosario, into the Bahía de Sebastían Vizcaíno. She bypasses Laguna Ojo de Liebre, also known as Scammon's Lagoon, which is frequented by more than half of the gray whales in her present condition. She continues south for another eighty miles. Here lies San Ignacio Lagoon, protected by sandbars from the pounding surf of the Pacific. She enters through an inlet into a wide channel. She moves as quickly as possible past a number of single males looking for courtship, beyond the outcropping called Punta Piedra or Rocky Point, onward for almost ten miles to a warm, shallow, isolated sector of the Middle Lagoon. Here she prepares to give birth.

Maintaining a stationary position, slowly she raises her tail flukes vertically above the surface and then lowers them again. For about ten minutes this procedure is repeated. Then, underwater, her calf emerges, its flukes clutched to its chest. The umbilical cord, short and rigid, breaks off easily. The newborn is a dark pinkish gray color. At birth it is already fifteen feet long and weighs almost two thousand pounds. Within moments the calf swims for the surface to take its first breath. Its flippers and flukes are flaccid and rubbery, curled from being folded for twelve months inside the mother. Awkwardly its head lunges above the surface, then comes crashing down again.

At first the baby doesn't so much swim as dog-paddle, but within three hours it gains enough coordination to properly keep itself afloat. In the tranquillity of the lagoon, the two rest and move with the changing tides. The newborn nurses underwater, having, less than an hour after birth, located its mother's nipples, recessed in shallow folds on her belly along either side of her genital opening. The mother's milk is so thick that it adheres easily to the baleen curtain inside the calf's mouth. Her milk is more than 50 percent fat, about fifteen times more fat than is in cow's milk. It also contains about six times the protein content of human milk and has no sugar. For days the mother strokes the baby with her broad flippers and rubs her body against its gathering strength.

> *This species of whale manifests the greatest affection for its young, and seeks the sheltered estuaries lying under a tropical sun, as if to warm its offspring into activity and promote comfort, until grown to the size Nature demands for its first northern visit.*
>
> —CHARLES MELVILLE SCAMMON, 1874

Soon the baby will be cavorting around, even breaching and landing right on its mother—and she will deem it ready for deeper water. Traveling close to its mother's midsection, where the water passing between the pair helps pull it along, the calf will leave the nursery after a few weeks. In the Lower Lagoon, near the mouth of the channel, it will encounter the first taste of strong tides—and perhaps the first touch of a human being.

One does not pass easily into the sanctum of San Ignacio Lagoon. Of course, I did not know this when I set out near the close of the gray whale birthing season. I made advance reservations at a whale-watching campground called Kuyima. During my periodic visits to southern Baja, I generally took a plane. This time my wife, Alice, and I planned to drive south from Los Angeles in an old Dodge Travco motor home and explore the countryside along the way. Going overland with our ten-year-old nephew and two close friends would be an adventure.

Baja California is the longest and narrowest peninsula on earth. It stretches for more than eight hundred miles and generally measures no more than sixty or seventy miles across. The Pacific Ocean borders the western side, from Tijuana down to the popular resort town of Cabo San Lucas. To the east, between the peninsula and the mainland of Mexico, flows the Gulf of California, also called the Sea of Cortés. Over millions of years this vein of the Pacific formed as tectonic activity along the San Andreas Fault separated the two landmasses. While mainland Mexico was evolving complex civilizations like the Mayans and Aztecs, Baja remained primitive and isolated, peopled by a few wandering hunter-gatherer tribes.

This land route toward the winter home of the gray whales is still a rugged and arid place. The Transpeninsular Highway, which snakes for 1,050 miles between Baja's two coastlines, was completed only in 1973. Before then one either flew into small private airstrips or expected to spend a month or more traveling on dirt roads. We headed south past Ensenada and onto a two-lane paved road that wound through treacherous mountain passes into the Central Desert. Max, our young nephew, insisted on stopping to explore the granite boulder field of Cataviña, with its remarkable desert vegetation. I walked in with him among the giant cardon cactuses, spiraling upward like medieval candelabra. The tallest cacti on the planet, some reach nearly seventy feet high, weigh as much as ten tons, and are believed to be more than two hundred years old. Here, too, are the spindly boogum trees, found only in Baja. They look like inverted carrots and are named after the creatures in Lewis Carroll's *The Hunting of the Snark*.

As we climbed onto a rise, I explained to Max that Baja seemed always to have possessed a kind of enchantment. Not far from here almost five hundred prehistoric murals had been discovered tucked in the overhangs of canyons. They are popularly known as the cave paintings. In vivid red and black they portray springing mountain lions, buck deer contesting with their antlers, native shamans with elaborate headdresses and outstretched arms. Some of them, many miles of desert from the sea, depict huge whales.

Within the images of a single cave, archaeologists have noted age differences of more than three thousand years. Yet no one knows who did the cave paintings, or how old they really are. Or, indeed, how anyone could have fashioned them. No traces of scaffolding have ever been found. So how did the artists manage to cover a vertical wall twenty-five feet off the ground with image after image? Even if these ancient designers used ladders or scaffolds, how had they drawn eight-foot-high figures in the deep recesses of a cave with such precision? At the end of the seventeenth century, missionaries were told by local Indians that the murals had been created by a race of giants.

The first Westerners known to have landed on Baja were a ship's crew dispatched by Hernán Cortés, looking for an "island of pearls" that the

Spanish conquistador had heard about from the Aztec ruler Moctezuma. During the 1500s a popular chivalric romance narrative also mentioned a race of Amazon women who ruled a gold-filled island. Their queen was called Califia, the place California. The early Spanish expeditioners apparently believed that the Baja terrain resembled that of the fictional island. Indeed, Baja was widely thought to *be* an island until the end of the seventeenth century. This was the first California. The peninsula would later be called Lower California, as differentiated from its neighboring American territory. (The Spanish adjective *baja* means "geographically lower.")

In the face of unfriendly Indians and food and water shortages, Cortés's early notion of founding a colony on Baja was soon abandoned. Not for 150 years would Europeans evince any further interest. When Jesuit missionaries began arriving during the eighteenth century, one of their memoirs described the region as "nothing but rocks, cliffs, declivitous mountains, and measureless sandy wastes, broken only by impossible granite walls." Now I watched as my nephew scrambled to the top of a massive boulder to pose as "king of the mountain."

Our last gas stop was the dusty desert town of Guerrero Negro, named after a whaling ship *(Black Warrior)* that sank trying to enter its lagoon to hunt gray whales in 1859. East of here the Sierra de San Francisco mountains loom abruptly above the many miles of scrub and cactus that line the desert plain. The Vizcaíno Biosphere Reserve, which encompasses San Ignacio and two other gray whale lagoons, stretches across more than 6 million acres. It is the largest such protected area in all of Latin America. This region is the last remaining habitat for fewer than one hundred peninsular pronghorn antelope, or berrendo. These deerlike animals can endure long periods without water and resist extreme heat by regulating their body temperature through hollow hairs, which they move by way of muscular tissue beneath the skin. Sprinting to speeds of fifty-five miles an hour, they are among the fastest mammals in the world.

After two and a half days on the road, the old mission town of San Ignacio was a welcome sight. The Cochimí Indians called it Kadakaaman, "Place of the Reeds," for it is fed by an underground river. Thick groves of Arabian date palms and citrus, first planted by Jesuit missionaries about three hundred years ago, remain the primary means of livelihood for the town's nearly one thousand residents. We parked at the edge of a square shaded by huge laurel trees, behind which stands one of

Baja's most beautiful churches. It was constructed of nearly four-foot-thick lava-block walls between 1728 and 1786.

San Ignacio is a village with a population, including the suburbs, of about twenty families. The only buildings of any importance in the place are those belonging to the mission. . . . The church buildings, consisting of the church itself, and two lateral wings, one of which is prolonged into an L, are in excellent repair, and are the most imposing buildings of this class in the territory. They are very solidly built of stone with arched roofs. . . .

The residents here claim that a good port exists below here, which they call the "Laguna." I had not time to visit it, but Captain Scammon, who is familiar with every nook and corner of this coast, has doubtless described it in full in his report.

—WILLIAM H. GABB,
The Settlement and Exploration of Lower California, 1867

According to our map, the laguna was thirty-eight miles to the south. At the Kuyima camp's main office in town, I was told that the road from here on was a rough one. Just how rough I could not have envisioned. Immediately upon departing the town, we found ourselves on a bone-jarring, washboard dirt track. It was impossible to move faster than ten to fifteen miles an hour—especially in a twenty-year-old motor home. A hot sun beat down incessantly on the thick-trunked elephant trees, cardon forest, and yucca cactus dominating a horizon capped by flat-topped clay cliffs. After about two hours we spotted a section of tidal salt flats to the west. Occasional salt mounds and piles of shells dotted the roadside. There were no mileage signs, but I guessed we must be getting close.

Six P.M. A dragon-fire sunset looming. Over a particularly rutted sector, our friend David Gude tried edging the camper to the far right, where the road seemed a little smoother. The shoulder suddenly turned to soft sand. Our wheels spun. We came to a lurching halt. Our motor home was mired almost up to the front axle, and the two rear left tires were both off the ground. We were perilously close to tumbling into a five-foot-deep gully.

We'd not seen a single vehicle since leaving the oasis. We located a small trenching shovel. Brian Keating, who'd come along to take pictures of the whales, joined David in attempting to dig out enough sand to jack up the front wheel. Alice, Max, and I gathered what few rocks we could find in the desert, hoping they might supply some traction. Twenty minutes went by. The situation looked pretty hopeless. I suggested to Alice that the two of us should start walking and try to find help. It would be dark soon, but the moon was full. I figured that the Kuyima campground couldn't be more than a few miles farther. Max insisted on coming along. Carrying a small flashlight, the three of us set out into the gathering dusk.

Off to the west the lagoon's northern arm shimmered where it meets

Scammon's drawing of whalers in San Ignacio Lagoon, the Frontispiece of his 1874 book

the salt flats. There water and land appeared to melt into each other across this minimalist's landscape. There, I realized, Mitsubishi and Mexico planned to cover more than 120,000 acres with salt ponds. Across the other side of the lagoon, the volcanic granite domes of the Sierra Santa Clara mountains were barely visible through the haze. Where the lagoon commenced I could make out a long sand barrier island, Isla Pelicanos, and another that the map listed as Isla Garzas. Salt-tolerant plants—pickleweed and *meado de sapo* (toad piss)—sprouted near the lagoon shoreline. A flotilla of driftwood lay exposed across a sea of mud, accentuating the feeling of being marooned. I remembered how the Baja coastlines had been havens for pirates, primarily English who set upon Spanish galleons. One of these British brigands had landed in Lower California in 1709 with a passenger he'd rescued from a lonely Pacific island. This was Alexander Selkirk, who became the model for Defoe's Robinson Crusoe.

Mirages clouded my vision. Night descended. Probably three miles from our vehicle now, we crossed a small bridge and came to an unmarked crossroads. Having no idea which direction might lead to the Kuyima camp, we went straight. Another mile. We could see headlights approaching. I flagged down a pickup. Three burly men sat in the front seat. My heart pounding, I mustered as much Spanish as I could from long-ago college days. How far to Kuyima camp? I asked. About twenty kilometers ahead, one replied—if this is the right road. But if we kept walking, another said with a grin, there was a village a couple of miles down. They roared off.

We kept walking. For what seemed an eternity, I could see headlights in the distance behind us. By now we had walked at least five miles. Finally a car pulled to a stop. A Mexican woman was at the wheel, a man and boy in the backseat. They'd never heard of Kuyima. Hop in the front, she said, she would take us to the village.

Such as it was. At the bottom of a steep hill, the car pulled up beside a little compound consisting of two wooden houses. We disbarked. Near the far dwelling I saw a man walking toward a parked vehicle and raced over to him. *"Habla inglés?"* I asked. "Yeah," he replied. He was stocky, balding, and bearded, and looked to be in his fifties. I explained our plight.

"Well," the man said with a broad smile and something of a western drawl, "I guess if I can haul the *Titanic*, I can haul you."

He extended his hand and introduced himself as John Spencer. "You're lucky, I just got in yesterday from Idaho." He walked into a shed and returned with two Coronas and a Pepsi. "Just relax, come on in and have some dinner with us. Then we'll get you out of your fix."

He pushed open a screen door. Inside a candlelit, clay-walled dwelling that served as a combined kitchen and living room, about a dozen men stood against a wall, eating from paper plates. "These are the most wonderful people you're ever gonna meet," our host said. As he introduced three generations of the family, each gave a nod. The only one sitting down, in a folding chair, was a thin, elderly gentleman missing several front teeth. "El Jefe," John said, and bowed.

He motioned to a table where a stack of tortillas waited to be filled with meat. "Get yourselves a plate. Sorry I don't have enough to take to your friends, we weren't really expecting company." I noticed a gold earring hanging from John's right earlobe. It was a whale's tail.

Ten minutes and two delicious tacos later, we followed John and most of the others outside. The men piled into the back of a heavy-duty modified pickup with oversized wheels and a winch attached to the front bumper. PACIFIC MOVERS—HEAVY HAULERS was emblazoned along the side. Only later would I learn that John Spencer had indeed been the supervisor who built the infrastructure that tilted some two thousand tons with hydraulic jacks—as the *Titanic* sank in the last scene of 1997's Oscar-winning film.

John sped off onto a sandy track, shifting into four-wheel-drive as the lagoon suddenly surfaced alongside us. He flipped on a tape deck and began singing along to an old Creedence Clearwater tune. "I see a bad moon risin' . . . ." Then he winked at me.

"See, that road you're stuck on is actually one of the best in the world," John said. "Because if they ever pave it, the lagoon is gone. Too damn many people would start showing up." For the last twenty-five years, John continued, he'd been making an annual pilgrimage to the San Ignacio Lagoon. "I'm driven by something," he said, "something very personal. If I don't come here, my life falls apart. It's that simple. I don't know whether it's the people or the whales. Probably both."

He pointed up a steep rise toward the main road. "Think that must be your vehicle up there." The pickup jolted up the gully, swung around, and skidded to a halt just beyond our camper. The men jumped out and began setting up lights. By now David and Brian had almost managed to free the front tires. "We'd probably have had the rear ones out in a day or two," David said, as John advanced with a cable to our front bumper.

Brian shook his head in wonderment at our coterie of rescuers.

"Where did they *come* from?" I smiled and said: "Magical reality. This is right out of a Gabriel García Márquez novel."

David took the wheel again. Alice, Max, Brian, and I stood in the desert below, well out of range of the tilting Travco in case something went awry. John gunned the pickup's engine into reverse, backing to a point that appeared ominously close to another ditch across the road. The cable strained against the winch. The old camper stalled a couple of times. Then, as quickly as we'd become derailed, we were back on course.

Our rescuers waved good-bye and disappeared into the night. Slowly we proceeded toward the lagoon. We passed by the little village, darkened now, and down a winding road skirting the moonlit water. About five miles farther we arrived at Kuyima and pitched a tent where the desert embraced the sea. The Pleiades seemed to be hovering just above the horizon in a star-filled sky. I thought of the "strange beasts and fishes" I'd seen described in a nineteenth-century book about Baja: an eight-armed giant squid, a manta ray that required six men to lift it with blocks and tackles. Overhead an osprey circled silently.

Somewhere out there four of the world's seven species of sea turtles were resting in the shallows: leatherbacks, hawksbill, green turtles, olive ridleys, all endangered, come to feed on the eelgrass, the clams, the snails and sponges. Somewhere out there brown and white pelicans, egrets, and herons, and half the world's population of a seagoing goose called the brant were nesting near the turtles. Somewhere out there fishermen may have been harvesting pismo clams and the big ones they call *mano de león*, hand of the lion. Somewhere out there gray whales exhaled a heart-shaped mist.

Somewhere out there, I heard a resounding splash.

> *Armed with the only chart in existence, a century-old map, we arrived at Laguna San Ignacio. Except for mention of it in the early missionary records of the 1700s, we had discovered that almost nothing was known about "La Laguna" by anyone other than the handful of Mexican fishing families. The lagoon remained much as it did in pre-Spanish days.*
>
> —MARY LOU JONES AND DR. STEVEN SWARTZ,
> describing their 1977 entry into San Ignacio Lagoon

The chart used by the marine biologists Mary Lou Jones and Steven Swartz had been drawn by Charles Melville Scammon. It was Scammon's brother-in-law, Jared Poole—a whaling captain from San Francisco by way of the Massachusetts isle of Martha's Vineyard—who was actually the first to spy the inviting mouth of San Ignacio Lagoon. The year was 1858, and Poole could find no safe point of entry. Over the centuries the tides and prevailing winds had deposited sand, which formed barrier islands behind which rushed strong tidal flows. The sandbar stretching across the mouth of the lagoon carried little more than twelve feet of water at high tide, not enough, Poole reasoned, for a ship to pass through.

Poole knew that Scammon had been whaling in Lower California for several years, in fact had been the first to penetrate another lagoon eighty miles to the north. After an arduous journey Scammon had found it teeming with gray whales. His crew had taken forty-seven in three months, yielding a whopping seventeen hundred barrels of whale oil that brought about fifteen thousand dollars on the San Francisco market. To find yet another untapped whaling ground seemed worth whatever perils might present themselves. And if anyone could find an opening, it would be Scammon.

So it was that Scammon's *Ocean Bird*—accompanied by Poole's and five other whaling vessels—cast anchor offshore of the southern boundary of the Vizcaíno Desert. The new lagoon was in sight. Latitude 26°40' N and longitude 112°15' W, Scammon noted in his logbook. The captain quickly realized that the winds here appeared to be more dependable than at any other point he'd visited along the peninsula's coast. A strong morning land breeze, coupled with an equally good afternoon sea breeze, could probably propel even a three-hundred-ton bark through the narrow northernmost channel.

He was right. After one of the smaller tenders made it inside and found a good anchorage, the bigger vessels sailed through the currents relatively easily. It was now early in 1860, the year of Abraham Lincoln's election. Scammon would save for his scrapbooks a San Francisco newspaper article referring to him as the discoverer, "for all practical purposes," of this "inland sea." The article went on: "Though a few natives and Indians have always been aware of the facts, it evidently is unknown to geographers, for in no published maps—from the earliest periods down to the present time—is any such body of water laid down."

This lagoon, which Scammon first called simply Ballenas (Whales), wasn't nearly as large as the one he'd already discovered. San Ignacio Lagoon extended into the desert for about sixteen miles, reaching a maximum width of five miles. Not long after their arrival, Scammon and a few of his crew set out to explore the area. The captain observed flocks of gray gulls covering the shell-laden beaches of its islands. Pelicans and cormorants filled the air and surrounding waters. Hawks had built high nests of dry sticks. Around the shores huge green turtles in large numbers lay sleeping. Shoals of cowfish and porpoise, Scammon would write, "played their undulating gambols."

The men had proceeded a few miles along the banks of an arroyo when they were met by a group of Mexicans on mules. Scammon spoke Spanish. An old man, who called himself Don José and appeared to be a sort of patriarch, informed him that he had many times come down from the mountains but never seen another soul.

Nowhere along the lagoon was there any sign of wood, freshwater, or human habitation. Scammon did come upon remnants of Indian culture—shells and charcoal, stone artifacts including awls and basalt milling stones. The countryside in the immediate vicinity consisted of sandy plain or low marsh. It was nearly level and extremely barren: a few stunted mesquite trees, scatterings of a species of rush grass. To the southwest Scammon could see a long tableland that rose to a height of a thousand feet. The men passed through a forest of cardon cactus and ascended to the summit. There was nothing inland but wild, mountainous country as far as the eye could see.

Scammon observed something else along the margins of the lagoon: vast salt ponds, produced by the natural process of evaporation. "These salt deposits are of infinite extent," the San Francisco press would report in 1863, "and exist in all directions around the borders of the lake. There are also evidences of great mineral wealth in the highlands."

The other great wealth was the California gray whales—hundreds of them during the winter calving season. Whalers bestowed a number of names upon this species. One was mussel-digger, for its propensity to descend to the soft bottoms and then surface with its head smeared with dark mud. Another was hard-head, for its ability to "root the boats" with a quick upward movement, like a hog upsetting a trough. The most telling designation was devil-fish. Other types of whales generally reacted to a harpoon only once the line was taut; the grays would respond at the first piercing of the iron. One captain called gray whales "a cross 'tween a sea-serpent and an alligator." A young seaman wrote home: "No steamer's paddle wheels can cause such a watery commotion as an angry gray whale's fins and flukes."

When the parent animal is attacked, they show a power of resistance and tenacity of life that distinguish them from all other cetaceans. Many an expert whaleman has suffered in his encounters with them, and many a one has paid the penalty with his life.

—CHARLES MELVILLE SCAMMON, 1874

A fuchsia sun is rising. Several traditional Mexican fishing boats, small motorized skiffs called *pangas*, are moored to rocks just off the beach. Seagulls doze on the rails. A slight distance from our camper, several other visitors are emerging from tents. We meet them, two couples who are old friends, over breakfast. One is a middle-aged man who wrote such popular sixties folk songs as "Lemon Tree" and today runs a successful management consulting business in California. To my surprise, as he describes their first experience yesterday on the lagoon, his eyes brim with tears. "When you have an experience with these whales," he says, "you come out a different person."

What does Kuyima mean? I ask a rotund, bearded fellow named Carlos Varela Galván, who supervises the camp's operation. "Light in the darkness," he replies.

Teresa, an American woman who's married to Carlos and cooks the camp's meals, reminds us that the lagoon is a sanctuary. The Mexican government regulates the number of skiffs allowed in the water. Only twelve boats, most of them licensed to three tourist camps along the lagoon, can be out there at any given time. Various areas within the lagoon are managed separately. No boats are permitted in the southern inlet, or Upper Lagoon, where the gray whale calves are nurtured. Our group will be confined to the more open stretches of the north inlet, where the calves are older. "We have very strict rules about whale watching," Teresa adds. "If the whales come to the boat, that's their choice. The drivers are instructed that they can't go within thirty meters of a whale."

Shortly after 10:00 A.M., Carlos calls everyone—the five of us and the two couples—over to be fitted with life jackets. He's wearing shorts and a Baja bush pilot cap. He offers parting instructions. "Don't everyone jump to the same side of the *panga* if the whales come to you. You can touch them—but not the eyes, or the blowholes, the fins and the tail. These are very sensitive parts of the whale. The trip to the observation area takes around fifteen or twenty minutes. Today is very calm, and I think it will be fast."

We all shed shoes and roll up our pant legs, wading out through the mud at slack tide and crawling up onto the stern of the twenty-six-foot-long *panga*. The water is tranquil and glistening as our driver motors out, bound for an eighty-foot-deep channel in the Lower Lagoon. Visible to the north, several miles inland, is a magnificent stretch of dunes. Looking across the neck of the lagoon, I can see leathery-leafed red mangroves on the western shore. After about twenty minutes we round a tiny peninsula not far from the lagoon mouth. This is Punta Piedra, or Rocky Point. There's an American whale-watching campground here called Baja Discovery, perched above a sandstone beach. The currents moving seaward beyond the point are strong and steady. Flocks of brown cormorants fly overhead. Caspian terns with bright orange bills and black crowns flap their wings, preparing to dive at the silvery flash of mullet near the surface. Every year some eighty species of shorebirds and waterfowl arrive from northern latitudes to winter here, at the same season that the gray whales find refuge in these still waters.

All eyes scan the surface. Gray whales are said to be instantly recognizable by their lack of a dorsal fin. Instead, they possess a low hump two-thirds of the way along the back, followed by a series of knucklelike ridges that extend to the broad tail flukes. The grays are also the most heavily barnacled of whales, carrying up to a ton of these little limestone-shelled creatures on their bodies. Our guide motions to a spot about fifty yards ahead. I glimpse a massive torpedo shape, an arched back, a huge angular nose pointed down. A fan-shaped geyser of seawater erupts and then subsides with a whoosh. I imagine the plume of spray could be seen for miles. Suddenly the ten-foot-long, heart-shaped flukes arch heavenward as the whale dives, leaving behind a sparkling waterfall.

The skipper cuts back on the throttle, moving in slowly. A mother, followed closely by her baby, swims just ahead. The *panga* stays parallel and a little behind at first, letting the whales take the lead. Gradually our captain overtakes them and passes about thirty feet beyond. He puts the engine into idle. "*Las amistosas*," he says, leaning forward and pointing one finger. The friendly whales. One of our companions, an acupuncturist who'd been out the day before, suggests we all lean over the sides and "brush" the water outward with both hands. He says the whales respond to the sound.

Yes, abruptly, magically, there they are, only a few feet away from the starboard side. Inches below the surface, they appear not so much gray as whitish blue. The immensity of these creatures is overwhelming. Fully grown they reach at least thirty-five feet in length and weigh more than thirty tons—ten times the size of a large elephant. The mother dwarfs our boat. The calf is nearly one-third her size. With a mere flick of the tail, either whale could overturn us. My heart is racing. Yet, curiously, I feel no trace of fear.

The mother uses her body as a natural breakwater for her calf, seeming to coax it toward the boat. Slowly, the baby makes the rounds of each person's outstretched hands. I'd expected a feeling like hard leather, but the skin is rubbery, shiny, and amazingly soft. The whales circle the bow, advancing again down the port side. Trailing close to her calf, now the mother raises her long, tapered head to be touched. Her skin is more mottled, covered with a patchwork of sea lice and barnacles. But she, too, feels surprisingly elastic.

As the two whales reach out to commune with our human realm, my mind is devoid of thought. Only later, for days thereafter, will images reemerge for conscious reflection. For the timeless minutes that the whales are alongside, I seem almost to be holding my breath. It's a primordial feeling, ancient beyond words. A fusion of two worlds, capsizing time and space. Indeed, there is a dinosaurlike quality to their faces. Now I could swear I saw a gray whale smile by my fingertips. Tears well up behind my eyelids.

Leaving a foam of bubbles as they take simultaneous breaths, the pair dive beneath the surface and glide away with extraordinary grace. There is not a hint of clumsiness in their movements, as if they are taking extreme care not to send even a ripple toward the boat. Sensitivity feels magnified a thousandfold. We are mute.

The waiting begins. Across the main channel leading out into the Pacific, I can see a few other *pangas* being similarly approached. We must have patience. Here on San Ignacio Lagoon, the incessantly paced Information Age pales into insignificance. One becomes more accepting of the natural rhythms of time and place.

A swirl of black brant geese, like the whales come all the way from Alaska, calls from overhead. A fish leaps. The saw-toothed peaks of the Sierra Santa Clara to the west are swathed in a high-noon sun. I glimpse a silhouette on the water, which I learn is a fluke print left by a recently surfaced whale. As our *panga* closes to within about thirty feet of another boat, a mother and baby come calling again. The little one seems to realize the presence of my nephew onboard; it is eager to rub against young Max's waiting hands.

Alice leans far over the bow rail. A fountain of spray surges from the depths. It resembles a hibiscus flower opening its petals. Others have described such spouting as offensive, but these whales haven't eaten in

Alice being sprayed by a whale (Max Burnett)

months, and the smell seems fragrant. My wife turns, her clothes dripping wet but shouting with joy, a look of wonder etched across her face. Baptized.

On the bow David begins humming a lonely tune. Both whales swim right up to him, as if somehow they are responding to the plaintive sound. Their own tones, as scientists have heard them through underwater microphones, are generally low rumblings; some have been compared to the metallic ring of Caribbean steel drums. The sounds often take the form of loud clicks—not unlike those of a boy placing a finger in his mouth and popping his cheek. Such "talk," it's been learned, increases when a gray whale maneuvers toward a boat. It's also a way for a mother to call her calf.

Now, as their faces crane out of the water, one sees a kind of parrot-beaked countenance, an upper jaw that overhangs the lower. The triangular heads are bow-shaped, comprising about one-sixth of their body length. The wide, capacious mouths contain no teeth. I glimpse the hair-like bristles of the long baleen plates, with which they obtain food from the currents and sediments.

The baleen, of which the longest portion is fourteen to sixteen inches, is of a light brown or nearly white, the grain very coarse, and the hair or fringe on the bone is much heavier and not so even as that of the right whale or humpback. . . . The eye, the ball of which is at least four inches in diameter, is situated about five inches above and six inches behind the angle of the mouth.

—CHARLES MELVILLE SCAMMON, 1874

The eye. . . . Along the edge of the boat, once more the mother surges upward. She turns on her side and appears to gaze at us out of one eye. Gray whales are said to have brown eyes, but this one looks moonstone blue. It's the size of a baseball, intense and riveting. The look exchanged penetrates to my very depths. It feels as though I am being read by the whale, as though my entire life—for one endless moment—is an open book.

Mother and baby swim directly under the *panga,* resurfacing on the other side. Riding one of her powerful pectoral flippers, the calf pushes up until its head is level with the stern. I touch something fine, coarse, and cream colored. In my hand it feels like a thick mustache. It's hard to believe, but the whale wants its gums rubbed. I find myself shouting with delight. The baby wriggles, slaps the water with its fluke, emits a gentle plume of vapor, and vanishes.

Not far away another boat is being approached. Near the stern a man smiles and waves at us. It's John Spencer, bare-chested and wearing swimming trunks. He leans far over the side and cradles the head of a mother whale between his hands. Then, as if playing a game known only to the pair of them, he gently twists the whale's head until she submerges in another burst of foam. I see John tugging at his whale's-tail earring. He waves again.

After three hours on the lagoon, our skipper prepares to head back. Off to starboard two whales hover together in the turquoise sea, side by side, motionless, bidding farewell. As a single unit all nine of us aboard the *panga* stand and face them. Nobody speaks. In their presence we are utterly humbled.

The whales dive and disappear.

We are more than captivated. We are captured.

How does this phenomenon happen? Why do the gray whales greet their former predators with open arms, so to speak, at the San Ignacio Lagoon? Marine biologists have made prosaic guesses: the whales' barnacles are uncomfortable and they like to scratch against small boats. Or they are attracted to the sound of the *pangas'* engines. Or they are simply curious about our presence.

Yet even in relatively recent times gray whales in the San Ignacio Lagoon could hardly have been considered friendly. Early in 1948 the first scientific expedition to the lagoon was undertaken by Dr. Carl Hubbs of the Scripps Institution of Oceanography in San Diego. Hubbs was accompanied by the actor Errol Flynn, whose father was a marine biologist at the University of Belfast and a friend of Hubbs. The thought was that Flynn might do a documentary film, and the well-known star of swashbuckling films put up funding for a plane to transport the crew and a helicopter from which to film the whales. This was the first time a helicopter would be utilized for scientific research—and the gray whales didn't seem to care for the idea one bit.

Here's the account of what happened by Lewis Wayne Walker, another member of the team:

> On the first few flights it was obvious that the nearness of the helicopter disturbed the whales, causing them usually to seek deeper water. However, by hovering behind and to one side, we found it possible to herd the animals in any desired direction, and before long subjects to be photographed were being coaxed into shallow stretches where deep dives were impossible and where muddy trails lingered as sinuous paths against the normal blue of the lagoon. After this had been done a few times, we noticed a decided change in whale temperament. Instead of swimming along in a placid manner, some of the grays churned the water with flukes and fins until their wakes became swirling cauldrons of foam. Before such displays of angry power, the pilot invariably lifted the craft to a safe twenty-five or thirty feet.

Walker continued: "On our arrival at San Ignacio Lagoon the Mexican fishermen scoffed at the traits of meanness so often ascribed to the whales in the works of Scammon, and they rowed their *pangas* through groups of whales with impunity. The large beasts often scattered at their approach. However, after the elusive helicopter had pestered the whales for a full week, they evidently became more like the 'Devil Fish' of old."

Walker recounted that, on his last day, "cleaning up campsites and paying off debts incurred by the expedition, a boatload of excited natives hurried to the command car and told of the persistent attack by a mother *ballena.* They had been crossing the channel to reset turtle nets and barely gave a second thought to a whale that submerged after a noisy blow. Suddenly, however, they were thrown from their standing positions at the oars. . . . The initial strike was the hardest of the attack, and they felt that the boat's combination of flat bottom, extreme buoyancy, and small size was all that prevented a cave-in of the stern. The whale continued to batter and nudge as oars were used on her broad nose to push the craft shoreward. She only desisted when the water became so shallow that her wake was a ribbon of brown mud."

If the gray whales were sending a cease-and-desist message to the Hubbs-Flynn crew, it came through loud and clear. As a result of various engine problems, the helicopter had to do three forced landings in less than a week. Then, on the way back home, a small Piper Cub carrying Hubbs and a companion went down in sudden hurricane-force winds in what one newspaper called "a near miss from death."

At Scammon's Lagoon to the north, researchers faced similar reactions from the whales. In 1956 Dr. Paul Dudley White journeyed there in an attempt to record the heartbeat of a gray whale on a cardiograph. White was a cardiac specialist who had supervised President Eisenhower's recovery from a heart attack. He and Donald Douglas, manufacturer of the airplanes bearing that name, planned to hand-insert small darts with wires attached into a gray whale. In his book *Hunting the Desert Whale*, Erle Stanley Gardner, the mystery writer of *Perry Mason* fame, told what ensued:

> A whale came charging up to the boat, smashed the rudder to smithereens, knocked off the propeller, and bent the drive shaft at a forty-five-degree angle—all with one blow of his tail. Then he swam away a little distance, turned around, looked at what he had done, took a deep breath, and charged, smashing in the side of the boat.
>
> If it hadn't been for executive ability of a high order and a perfectly coordinated effort, those men would have been plunged into shark-infested waters. But as it was, they worked with speed and efficiency. They stripped off life preservers, stuffed them into the hole, took a piece of canvas, wrapped it around the outside of the boat, signaled for help, and, by frantic bailing, were able to keep afloat until a rescue boat, which had been standing by

just in case there should be any trouble, was able to come and tow them into shallow water.

In 1963 a diver trying to get photographs of the gray whales for the Scripps Institution found himself "attacked" by a large female that took four swats at him with her tail in Scammon's Lagoon. Later in the 1960s a team of divers working for Jacques Cousteau decided to chase a female gray whale for several hours in a motorized Zodiac. Finally she turned, breached, landed on top of the rubber boat, and destroyed it.

The first documented shift in the gray whales' behavior occurred at the San Ignacio Lagoon in 1976. That winter a large vessel called the *Royal Polaris* passed through the still-treacherous channel and dropped anchor in roughly the same spot as Scammon's *Ocean Bird* once had. It was carrying a group of Americans on a nature study tour sponsored by the Smithsonian Associates Travel Program. They'd sailed down from San Diego, stopping along the way for beach walks, tide-pool wades, bird-watching climbs, and shell-collecting rambles. Now they'd come for a look at California gray whales, which had rebounded by then to a population of over ten thousand.

A *New York Times* travel story, published the year before, had first brought the possibility of a lagoon voyage to millions of readers. The reporter Jack Goodman went out in a small dinghy. As he wrote, "After watching a big gray for a time and seeing a calf swimming in the shelter of her fifty-foot-long flank, I wondered how men could ever take the lives of such fascinating, obviously intelligent creatures. At times, when we pulled abreast of a whale and held a speed equal to the gray's slow gait, we could peer directly into one of the creature's widely separated eyes—and discern that it was peering back at us with more than mild curiosity. Or was it?"

Twelve people aboard the *Royal Polaris*, more than any before them, were about to receive an answer to that question. It was the early morning of February 16, 1976, and they'd gone out in two aluminum skiffs about the same size as the old wooden harpoon boats. Afterward, at the captain's request, each would set down what he or she remembered about their stunning adventure. One described a whale they dubbed Primo, which at first was feared "would capsize our skiff." But after swimming under and scratching its back, the gray lifted its head for about forty-five minutes of petting by everyone aboard.

As the phenomenon persisted, on February 29 the *Los Angeles Times* ran a page-one head shot of a gray whale nicknamed Nacho surfacing beside and nuzzling a raft. That summer, the *San Diego Union* began another front-page story with the question: "Is the California gray whale reaching détente with humans? There is increasing evidence that it is."

As I was about to find out, the first close encounter had actually occurred several years earlier than 1976. We were finishing lunch in the Kuyima camp's dining area when John Spencer arrived in his pickup. He had with him the elderly Mexican he'd identified as El Jefe the night before. "You tell me you're a journalist," John said. "Well, this is someone you should talk to. He speaks a little English. His name's actually Pachico. He was the first person to experience what you did today."

I sat down with Francisco (Pachico) Mayoral in the shade outside the thatch-roofed Kuyima *palapa*. Pachico appeared to be part Indian. His face was wizened and furrowed. He said he was about to turn sixty, but he appeared older. I could not look away from his penetrating dark eyes. Through a combination of my halting Spanish and Pachico's broken English, we managed to communicate. In February 1972 he had been out alone in his *panga*, fishing for grouper, when a gray whale surfaced alongside him. He was well aware that small boats generally kept their distance from the whales. He was surprised at first, and rather frightened. But when the animal lingered, Pachico felt himself compelled to place a hand in the water. The whale rubbed up against him, remaining almost motionless.

That was the beginning.

"The whales," he said, placing a hand over his heart, "they are my family."

He lit a cigarette. What do the *ballenas* mean to you? I asked. "I have not the words to express," Pachico continued. "The rest of my life since, I have activity with the scientists, the tourists, and the whales."

He had guided numerous marine biologists into the lagoon to study the gray whales' mysterious behavior. A few years ago, shortly before Christopher Reeve had the tragic horseback accident that left him paralyzed, it was Pachico who shepherded the actor out onto these waters for Reeve's narration of a gray whale documentary.

"The whales enter the lagoon and come to me," Pachico added. "You see, I have a position." I must have looked puzzled, trying to grasp what he was saying, and he laughed. Then he repeated: "The whales, they are my family."

Pachico glimpsed John Spencer emerging from the Kuyima dining room. "John," he added, "my very, very good friend."

Pachico got into the front seat of the pickup alongside him and disappeared in a swirl of dust.

Later I would consider the curious timing of Pachico's initial meeting with a gray whale. In 1972 the Mexican government had decreed a "Reserve and Refuge Area for Migratory Birds and Wildlife" at the San Ignacio Lagoon habitat. That same year the United Nations had voted for a resolution calling for the end of worldwide whaling, and the U.S. Congress passed the Marine Mammal Protection Act. Then, in 1973, Congress passed the Endangered Species Act, which had the gray whale on the list, where it would remain protected for the next two decades of its dramatic recovery. And during this same period gray whales first made direct overtures to a Mexican fisherman.

Their decision to make contact with our species is but one of many mysteries surrounding the gray whales. Our second morning on the lagoon, not a single one would approach us. There was an east wind blowing against the tide, creating a slight chop on the water. Later John Spencer would inform us that, at such times, the whales are apparently so aware of their capacity to inflict damage on a small craft that they remain in the near distance. Instead they offered us a fascinating show we had not witnessed the previous day. They performed what's known as spy-hopping. After a dive the whales' heads emerge vertically out of the water. They are sometimes said to be standing on their tails. They remain, as if perched, for as long as half a minute. They appear to be scanning the horizon.

Science has no explanation for this manifestation. Nor for the breaching that gray whales and a few other whale species exhibit. I watched in awe as the grays burst from the water, launching as much as three-quarters of their bodies skyward. They twist onto their backs or sides before plunging again into the sea. Sometimes they repeat this over and over, breaching as many as forty times, with about fifteen-second intervals between leaps. Is this a display of power or exuberance? A signaling system? Some kind of courtship ritual? Or is it simply pleasurable to them? It's anybody's guess.

As the easiest to observe of any whale, the gray has long been of special interest to zoologists. Between 1977 and 1984 Dr. Steven Swartz and his scientific colleague, Mary Lou Jones (later to become his wife), spent every winter at San Ignacio Lagoon. Their most memorable moments revolved around the relationship between mothers and calves, probably the most tender ever observed in a marine mammal. "While playing," Jones and Swartz have written, "the little ones climb all over their resting mothers. They swim onto her rotund back and slide off, roll across her massive tail stock, and pummel her with their leaping back-flops and belly-flops. Mothers appear very tolerant of all this and frequently join in, repeatedly lifting the calf out of the water, whereupon the calf flails itself back down with a splash."

During late February and into March, it's time for "spring training" in preparation for the long migration. When the tides are running in the Lower Lagoon, the mothers match their speed to that of the strong incoming currents as they tread water. Alongside them the calves gather strength by swimming as if on a treadmill. Sometimes, though, things don't work out as they should. On one such occasion Jones and Swartz witnessed a remarkable example of the lengths to which gray whales will go to protect their young. "From our observation tower on Punta Piedra next to a deep channel, we saw a calf thrashing as it left the channel and tried to cross a shallow sandbar. Instantly, an adult whale we took to be the calf's mother surged out of the channel and beached itself beside the calf. Seconds later another whale beached itself on the other side, sandwiching the calf between two adults. Both adults thereupon raised their heads and flukes, pivoted with the calf between them, and slid smoothly back into the channel." The following year the two marine scientists saw the same thing happen in the same place.

The spring training period is also when mothers begin nudging their calves toward humans—an apparently learned pattern which scientists have observed in individual whales that return year after year. These approaches steadily increased during the six years Jones and Swartz were at the lagoon. Only a few whales came to people in 1977 and 1978. Within five years after that scientists had chronicled over two hundred such encounters.

Jones and Swartz nicknamed some of the whales after their distinctive colors and markings: Rosebud, Pinto, Cabrillo, Peanut. One large female was christened Amazing Grace. In the two biologists' first encounter with her, "she readily adopted us along with our fourteen-foot inflatable outboard as her personal toys. She would roll under the boat, turn belly up with her flippers sticking three to four feet out of the water on either side of the craft, then lift us clear off the surface of the lagoon, perched high and dry on her chest between her massive flippers. When she tired of the bench-press technique, Grace would do the same thing with her head, lifting us out of the water and letting us slide off to swirl around her in circles, like a big rubber duck in the bathtub with a ten-ton playmate."

Experiences like this must make mythologists out of marine mammalogists.

On the way out of San Ignacio Lagoon, we stopped again at the village where I'd run into John Spencer. He was just emerging from a backyard shower. We sat down alone inside the little house. He leaned forward and asked if I had come up with any thoughts on why the whales here are the friendliest on the planet. I said I'd just be guessing. After a moment John reached to a shelf behind him, lined mostly with foodstuffs, and handed me a bottle of wine with a gray whale on its label. "Read it," he instructed.

I read aloud: "Whales Pinot Noir 1987 Idaho. San Ignacio Lagoon. Close Encounters of the First Kind. Whales possess the greatest living power on earth. Though driven to the brink of extinction, they choose to present only gentleness, tranquility and love. Through these qualities they enable humanity to visualize the possibility of world peace. When the whales are gone, there is no reason to go on living. John Spencer, 1987."

He explained that he'd made the wine himself, giving numbered bottles to friends and family.

I described the inner stillness I'd felt when the whales made contact, as if there was some sort of psychic interchange happening.

John responded: "What you're really saying is, it's being humbled. You're sitting there with something so magnificent, so close to God or whatever we have to deal with. If you don't feel it then, you're never gonna feel it. We've been asked to come into their house, and to share a moment with them. How can we be so blessed? And to be assured that nothing wrong will happen? You almost come to tears just thinking about it."

He paused a moment, then gestured toward the lagoon beyond, and went on: "I feel so safe out there. Like maybe somebody's got ahold of me. I remember one afternoon, must be quite a few years back now. Me and Luis, a guy I'd worked with for years, are dickin' around out there when our boat motor craps out on us. We're drifting. All of a sudden I look up and these three whales are right next to our boat. Mating! I'm sure we're gonna get tipped over because during mating you always stand way off, like a hundred yards. They're thrashing, and you figure they don't even know you're there. Now these whale tails are sliding right by our little boat.

"Well, I look over the side, and damned if we aren't on a reef! We're in two feet of water and then it drops straight off. Our motor is all that's holding us from going farther onto the reef. So there we are, safe and sound, three whales mating right beside us. And I've got no film in my camera!

"But I've often thought, well, we shouldn't have been there anyway. You can't capture a moment like that. It was about a thirty-minute situation. Unbelievable. Eyes. Everything. I love their eyes. Too beautiful."

He stood up. "I'll give you one clue about what's really happening out there. The best clue anybody's ever gonna give you—short of being with the guys that live with them. Not all those whales come over and play with you, right? Why do some, but not others, swim right up to your boat?"

I couldn't answer him. He said, "You think about this for a couple weeks, you'll figure it out," and extended a hand. "Now don't get too close to any gullies on your way out."

All the way back to the town of San Ignacio, I mused about the experience. Visions flooded my consciousness from other soul-rending moments in my life: Sleeping at the feet of the Sphinx and climbing the Great Pyramid at dawn. Camping at the edge of the Grand Canyon. Venturing down into Tanzania's wildlife-filled Ngorongoro Crater. This lagoon journey, as with so many of the most memorable times, was marked by an initiation. Had we not broken down in a potentially perilous situation, we might never have met John Spencer. Nor would I have been introduced to Pachico.

In the town square I stopped by the Kuyima camp offices to visit with the English-speaking director, Carlos's brother, José de Jesús Varela Galván. As I prepared to depart, reticently I asked a question my wife had urged me to raise. "Do you know Pachico Mayoral?"

José nodded.

I asked, "Do you think it's possible that Pachico somehow taught the gray whales to interact with people?"

José looked at me for a long moment. "Well," he said finally, "that is the legend."

I thought again of the stillness I had felt on the lagoon, and in the presence of Pachico. I thought of the sweetness of the air, the purity of the wilderness, the busy winging of cormorants above the spouting of the whales.

But it was not ever thus. Here, too, and not so very long ago, the sands had flowed crimson.

Chapter 2

The Whaler Who Became a Naturalist

The large bays and lagoons, where these animals once congregated, brought forth and nurtured their young, are already nearly deserted. The mammoth bones of the California gray lie bleaching on the shores of those silvery waters, and are scattered among the broken coasts from Siberia to the Gulf of California; and ere long it may be questioned whether this mammal will not be numbered among the extinct species of the Pacific.

—Charles Melville Scammon, 1874

Atop a masthead hoop in the San Ignacio Lagoon, a lookout stood watch. "Whales ahead!"

A gray whale calf surfaced in the near distance, its mother alongside. From the deck of Scammon's *Ocean Bird*, three whale boats were lowered from davits, ropes, and pulleys. Each craft held six men: the boat header, the boat steerer, and four "ship keepers"—a bowman, midship oarsman, tub oarsman, and after oarsman. Each craft was kept as light as possible, so it could move easily and quickly in the shallow lagoon waters. The equipment required was nonetheless considerable. Besides the oars, sails, and paddles for propulsion, there were two harpoons at the head of each boat, alongside three hand lances on the starboard side, designed to deliver the coup de grâce. Near the stern was the loggerhead, which controlled the line once a whale was secured.

The ploy was to move the boat between the two whales if possible, as if the intent was to take the baby. This would bring the alarmed mother near enough to kill. A small flag, known as a waif, was raised by one of the whale boats. The whalers closed in slowly, using the paddles rather than the heavier oars. The boat steerer braced himself by one leg against the clumsy cleat, a stout seat with a rounded notch. He would throw the first harpoon when they came within darting distance of about sixteen feet.

"Stand up!" the boat header ordered as the mother gray broke water. A harpoon ripped through the air. It made contact just behind the head. A foam of blood crested in the waters. The men threw the anchor. The steerer passed the harpoon to the headsman as they changed positions. "Stern all!" the headsman called out. The oarsmen worked feverishly to keep the boat astern of the wounded animal. The whale took out line and attempted to "sound" into the murky lagoon waters. The other two boats moved in close, in case she dashed her flukes against the one holding her fast.

This time, she didn't. Within an hour, affixed to two harpoons and the hand lances, the mother whale was floating lifeless at the surface.

After only two whaling seasons—yielding 8,200 barrels of oil, valued at $123,000—San Ignacio Lagoon would be virtually devoid of gray whales. Scammon would look back upon a vista crowded with whale boats crisscrossing the lagoon waters, dead whales floating with identifying flags planted in their backs, calves wandering aimlessly in search of their mothers, and a few native people scouring the shoreline for carrion.

Charles Melville Scammon "is primarily remembered as a whaler, a fact that would not please him," writes Lyndall Baker Landauer in *Scammon: Beyond the Lagoon*, a doctoral thesis published as the first volume of the Pacific Maritime History Series in 1986. "He was a whaler by necessity, not choice. . . . He was a man who . . . ventured into new territory, both physical and intellectual, and left his imprint on his times and ours."

Other than Landauer's short and scrupulously detailed volume, Scammon's protean accomplishments have been largely ignored by historians. He was a sea captain eventually elected to membership in the California Academy of Sciences. His book, *Marine Mammals of the North-Western Coast of North America, Described and Illustrated: Together with an Account of the American Whale-Fishery*, published in 1874, has been drawn upon by all subsequent zoological studies of the Pacific's whales, dolphins, porpoises, seals, and sea otters. The book's concluding section remains one of a handful of definitive nineteenth-century histories of the whaling industry, from its "origin and ancient mode" to its practice in America and Scammon's personal experiences, especially as related to "lagoon-whaling." Everywhere he went, as commander of twenty ships between 1848 and 1883, Scammon produced assiduous notes and sketches. He worked with a Darwin-like dedication to record the first complete descriptions of numerous marine mammals, foremost being the California gray whale. A century later, when scientists from the Scripps Institution of Oceanography compared Scammon's drawings and mappings of Baja California with satellite imagery, they were astonished at his near-pinpoint accuracy.

Bruce Mate, one of today's leading whale scientists, describes Scammon's contribution as "our first enlightened look at whales. Obviously because he first had a commercial interest," Mate adds, "but Scammon was also a very good observer and a pretty decent writer. So he left a legacy far beyond simply a record of how many barrels of oil were taken. In fact, his work is some of the first that pieced together the migratory habits of the gray whales. Much of what he wrote more than a hundred years ago about gray whales rings true today."

I first became aware of Scammon on my trip to the San Ignacio Lagoon, reading a Monterey Bay Aquarium book about gray whales that contained a brief biographical sketch of the captain. It noted that, during his extensive travels, Scammon "observed a vast array of animals, everything from the sea otters and whales in cold northern seas to the manatees in Florida's warm channels and springs." These included what was, for many years, "the most comprehensive [written] portrait available of the gray whale."

Even at first glance this seemed a highly unusual shift in orientation for a whaling captain. I raised this point over the phone with Steven Swartz, the marine biologist whose gray whale studies at the San Ignacio Lagoon had provided a more recent "comprehensive portrait." Swartz indicated that Scammon was indeed a fascinating individual, one whose career bore further scrutiny. A collection of Scammon's papers, he added, was available for research at the University of California's Bancroft Library in Berkeley, where his descendants had bequeathed them during the late 1940s. And, although he didn't know her whereabouts, Swartz believed that one of those descendants—the captain's granddaughter, Mildred Scammon Decker—might still be alive.

Eventually I managed to locate Mildred Decker, then ninety-one, as well as her son and grandson, living in Citrus Heights, California, a suburb of Sacramento. I traveled there to meet with them in December 1998. Mildred had known Scammon as a little girl, and her memories of him remained vivid.

•

Scammon was born in Pittston, Maine, a small community on the upper reaches of the Kennebec River, on May 28, 1825. His father, Eliakim, was a man of means and influence: Methodist preacher, postmaster, township treasurer, and later state representative. Charles was the fourth of eight children, with six brothers and an invalid sister. Several of his siblings would go on to achieve prominence in nineteenth-century America. From an early age Charles wrote poems to his sister and enjoyed reading and sketching. Yet he longed for the vastness of the oceans. When Scammon was fifteen, a Maine sea captain wrote to his father:

> *Your son wishes to go with me to sea. I do not think myself a fit person to take charge of a young man, but if he is inclined to do right, I think I may be of service to him.*
>
> *The first year he would be of no service to me; the second year he would. If he goes with me, I will take him in the cabin this voyage. It will be necessary for him to go in the forecastle if he intends to make it a business to be a sailor. I will take him for two years, send him to school when an opportunity offers; teach him navigation, and give him a good chance to study on board the ship and forty dollars a year he finding his clothes or I will furnish his clothes and give no wages. If this is agreeable to your wishes please inform me.*
>
> *Yours with respect,*
> R. MURRAY

This was *not* agreeable to Eliakim. But two years later Charles thwarted his father's desire that he attend college and shipped out as an apprentice with Captain Robert Murray. From 1842 onward Scammon would spend most of his life on one sector of ocean or another—from the tropical seas of South America to the ice-laden Arctic Ocean. Perhaps it was something in the ancestral blood. The Scammon family traced its origins to the Waldron clan of Normans, said to have settled in England with William the Conqueror after 1066. One of Scammon's ancestors captained a British ship in a seventeenth-century attack on the Barbary pirates. Others had scattered to the American colonies, first arriving in Boston around 1630.

By the time he was twenty-three, Charles Scammon had assumed his first command, as skipper of a vessel out of Bath, Maine, which traded for

Charles Scammon in younger days
(courtesy John Decker)

turpentine, resin, and peanuts with the Carolinas. That same year he married a Pittston girl named Susan Crowell Norriss. She was pregnant when he departed without her, late in August 1849, as the skipper of a merchant bark bound for the West Coast. The risky voyage was to cover over seventeen thousand nautical miles and occupy 168 days. It would take Scammon from the Atlantic into the Pacific at the nadir of South America, and around the treacherous Cape Horn.

"A place where gloomy weather prevails and storm follows storm in quick succession . . . the highest elevations ever wear a wintry garb," he would later write. At last, on February 21, 1850, Scammon arrived in San

His wife, Susan Crowell Norriss Scammon
(courtesy John Decker)

Francisco. From all along America's eastern seaboard, up to twenty ships like his were docking on the Barbary Coast every single day; in 1850 alone population swelled by some 36,000.

The motivation for the exodus was summarized in a four-page newspaper, *The Californian*, which Scammon would save to include in his scrapbooks. Dated March 15, 1848, it described a gold mine found "in the newly made raceway of the sawmill recently erected by Captain Sutter. . . . California no doubt is rich in mineral wealth, great chances here for scientific capitalists. Gold has been found in almost every part of the country."

Scammon never recorded whether gold seeking was his own original intention but, if so, he did not pursue it. He soon set sail again on trading vessels, venturing as far as Valparaíso, Chile. "The Andes . . . shade the fertile valleys . . . diversified here and there by the retiring hills which are relieved by belts of timber," he would write. But berths were generally scarce on merchant or clipper ships. Scammon recalled: "The force of circumstances compelled me to take command of a brig, bound on a sealing, sea-elephant and whaling voyage or abandon sea life, at least temporarily." He went to work for a San Francisco ship's chandler, A. L. Tubbs, who owned a number of seafaring vessels.

In 1852 Scammon traveled for the first time down the Baja coast, then known as Lower California. His quest was the sea elephant. So named for its massive, hooked nose, it was the largest of the seals, weighing more than two tons and sometimes reaching over twenty feet long. Sea elephant oil was considered "next to sperm [whale oil] for lubricating purposes," as Scammon described it. Later that same year he captained a ten-month cruise to the Gulf of Panama, Ecuador, and the Galápagos Islands in pursuit of the lucrative sperm whales.

As the gold fever subsided, many other newcomers turned to whaling. It had commenced out of San Francisco in 1851—the year *Moby-Dick* was published—with two former New Bedford vessels as the initiators. By the time Scammon started the city's fleet had increased to eight. In the preindustrial era whale oil was used mainly for lighting and, in some cases, heating or lubrication. And whaling was an enterprise of staggering proportions. Around the time Scammon entered the "fishery," a fleet of 650 ships and fifteen thousand men were engaged in whaling in the Pacific Ocean alone. Altogether some seventy thousand people derived their primary income from whaling-related business, which bore an investment of at least $70 million.

During the spring of 1854, Scammon's wife and their four-year-old son, Charles—who had yet to meet his father—sailed from Maine to join the captain for a voyage to China on a trading schooner. Their return cargo included 169 Chinese passengers. The round trip took nine months. Less than a month after they arrived home, Tubbs dispatched Scammon to Baja's Magdalena Bay, then known as Marguerita Bay. This time his target was to be the California gray whale.

Until midcentury gray whales had not been widely sought. For one thing their oil was not as plentiful and was considered inferior to that of sperm whales or humpbacks. Besides, large whaling vessels tended to avoid the near-shore waters through which the grays migrated. And nobody yet knew about any calving lagoons in Baja. But with all the ships arriving from the East Coast, shore-whaling stations quickly arose along California. So-called whalebone from the grays' baleen could be utilized for things like corset stays, umbrella ribs, and carriage whips. For the gray whale it was only a matter of time.

It was Scammon who, in the winter of 1857–58, would discover the existence of the lagoons where hundreds of gray whales came annually to give birth. It was he who would initiate the whalers' brutal slaughter of pregnant or nursing females. Without their mothers the calves could not survive to make the northern journey. During the early 1850s there may still have been upward of twenty thousand California gray whales. Within two decades probably fewer than two thousand remained.

As Wesley Marx wrote in a 1969 article for *American Heritage* magazine: "Most whalers forgot about the gray whales. But not Charles Scammon. Even as his adventurous instincts had delighted in recording the color and excitement of lagoon whaling, the reflective side of his nature had been fascinated with the whales themselves. At the same time he had been ordering his harpooners to bomb the whales and his flensers to strip the blubber, Scammon was also measuring the girth of dead whales, inspecting the contents of their stomachs, and executing precise drawings of their conformations. The Captain jotted down his detailed observations alongside log entries that recorded the number of whales struck and barrels filled."

Shortly after the outbreak of the Civil War in April 1861, Scammon enlisted in the U.S. Revenue Marine Service, the predecessor of today's Coast Guard. He did embark on two more whaling voyages before retiring permanently from the business early in 1863. Then he became a full-time Revenue Marine commander of the only official U.S. guard ship patrolling the West Coast against Confederate raids during the remainder of the Civil War.

Patriotism was much in evidence among two of Scammon's older brothers, who were the subject of several newspaper accounts in his scrapbooks. Thirteen years older than the captain, Jonathan Young Scammon was a "strong Union man" and longtime friend in Illinois of Abraham Lincoln's. Starting out there as a lawyer, Jonathan had founded the first railroad west of Lake Michigan, established Chicago's first bank and laid the groundwork for its public school system, and helped start the *Chicago American* newspaper. As the first president of the Chicago

Astronomical Society, he built the Dearborn Observatory—which had the largest refracting telescope in the world—and he was also among the founders of the Chicago Academy of Sciences. In 1865 Jonathan's son, Charles T. Scammon, would form a law partnership with the president's son, Robert T. Lincoln.

Another brother, Eliakim Parker Scammon, was a brigadier general in charge of the volunteer Twenty-third Ohio Regiment during the Civil War. Under his command at different times were two future American presidents, James Garfield and Rutherford B. Hayes. One article recounted what happened on a night in 1864, when E. P. Scammon's steamer was anchored on a river in Ohio: "Thirty-five guerrillas appeared on the opposite side, thirteen of whom crossed in a skiff and took possession of the boat, capturing Gen. Scammon and forty officers and soldiers, all of whom were asleep. The guerrillas burned the boat and paroled all the prisoners, except Scammon and three other officers. These were mounted and started for the interior. Forces have been sent in pursuit." Eliakim survived to be elected president of the Ohio Military Academy when the war ended. He went on to excel as a mathematician and teach at West Point.

The eldest Scammon sibling, Franklin, was a medical doctor who also settled in Chicago. Upon his retirement, his obituary from 1869 noted, "he was a very accomplished botanist, the West knowing no superior. His collections in this department of natural history are extensive and valuable."

Three of the brothers, including Captain Scammon, were followers of the teachings of an eighteenth-century mystic, Emanuel Swedenborg. Jonathan instituted Chicago's Swedenborgian Church of the New Jerusalem and introduced homeopathic medicine to the midwestern city through a new hospital. One of Swedenborg's most comprehensive works was *The Economy of the Animal Kingdom*, in which he described all of life as a marvelous unity, tautly structured according to a grand design. "Certain animals seem to have prudence and cunning," Swedenborg wrote, "connubial love, friendship and seeming charity, probity and benevolence, in a word, a morality the same as with men."

After the Civil War ended, Charles Scammon received a government appointment as flagship commander for the Western Union Telegraph Expedition. Its primary mission was to lay a cable that would link North America with Europe. Starting overland across British Columbia and then-Russian Alaska, the goal was to run a telegraph line underwater across the narrow Bering Strait into Siberia and ultimately all the way to St. Petersburg. This was also to be a scientific voyage, since the U.S. government was keen on surveying the potential resources of Alaska. Although the eventual success of an Atlantic-laid cable saw the Western Union Telegraph Expedition end in failure, its discoveries were instrumental in paving the way for America's purchase of Alaska from Russia in 1867 for $7.2 million. By then Scammon had cruised the Northern Pacific waters between San Francisco and Siberia for more than two years. His resulting friendship with expedition scientists dispatched from the Chicago Academy of Sciences and the Smithsonian Institution, in particular William Healey Dall, would inspire him to begin to write.

The first of Scammon's seventeen travel and natural history articles for San Francisco's *Overland Monthly* magazine began appearing in 1869. The titles include: "On the Lower California Coast," "About the Shores of Puget Sound," "The Aleutian Islands," "Seal Islands of Alaska." They appear alongside pieces by a number of new writers who'd settled in San Francisco and would go on to widespread fame. Several of the earliest stories by Mark Twain first appeared in the *Overland Monthly.* So did Jack London's Klondike stories and "The Luck of Roaring Camp" by Bret Harte, who served as the magazine's literary editor. Poems by Joaquin Miller, and John Muir's account of the wonders of Yellowstone, came to grace its pages. A kind of *New Yorker* of its day, the *Overland Monthly* offered a new western style of literature, one that brought critical acclaim from the East Coast and Europe during the six years Scammon was among the magazine's regular contributors. Charles Dickens, it was reported, eagerly anticipated each issue's arrival in London.

Scammon would draw upon many of his articles for the *Overland Monthly* in assembling his book on marine mammals, which was described thirty-five years after it appeared as "the most important contribution to the life history of these animals ever published." As an 1872 article in the *San Diego Union* put it: "No more devoted investigator of the whale and its habits ever existed, than Captain Scammon." By then he had been elected to membership in the California Academy of Sciences. "That was *very* unusual for a non-scientist, especially a whaler and ship's captain," according to the California whaling historian Alan Baldridge. "Besides obviously being a leader when it came to handling a crew, Scammon was a brilliant navigator and mapmaker. If he were a modern-day skipper, he'd be captain of a research vessel."

Not long after his book came out, Scammon took up the life of a gentleman farmer in Sebastopol, California, but he didn't officially retire from the Revenue Marine until the 1890s. He died in 1911 at the age of eighty-six, within twenty-four hours of his wife, leaving behind their three sons. Almost forty years would pass before his boxes of memorabilia emerged from an attic and into a university library. The life and times of this self-taught sailor, whaler, explorer, serviceman, scientist, historian, writer, and artist could now be resurrected.

What Scammon set down about the gray whale in the lagoons of Baja and along its migratory route offers the first perspective we have. The frontispiece of *Marine Mammals of the North-Western Coast of North America* is a Scammon drawing of the San Ignacio Lagoon. It depicts gray whales being set upon by whale boats, while larger ships wait at anchor nearby. Indeed the gray whale was the volume's dominant character, starting with a fourteen-page first chapter and appearing at numerous junctures in Scammon's tales of his whaling years. He wrote of the gray's migration patterns and its approximate distribution between twenty and seventy degrees of northern latitude. He knew that the whales congregated in the Arctic Ocean as well as the Sea of Okhotsk. He reported that females are larger than males and described the barnacle-forming parasites that cover the whales' skin. He didn't shy away from the difficult questions—the length of a female's gestation period, what the grays eat—and he honestly cited lack of available data for what he couldn't answer. It's all held up. Nothing Scammon wrote has ever been contradicted in any critical way by later marine scientists. His observed size figures for males, females, and embryonic calves are very close to the statistical norms measured in the years since.

Pieter Folkens, who is considered one of the finest contemporary illustrators of marine mammals, calls Scammon's renderings "the best of his time, compromised only because they had to be done from memory. I used some of his wonderful illustrations as examples of field sketching when I was teaching. I remember one in particular, which has rarely been better captured. It was a little pencil drawing, of the eye of a whale."

What did Scammon see when he looked into that eye? That is very much the underlying question his life portrait conjures. It is the question that, as this book progresses, we shall see unfolding in a man who could write that "the scene of slaughter was exceedingly picturesque" yet go on to describe his victims in the most evocative of terms: "This species of whale manifests the greatest affection for its young, and seeks the sheltered estuaries lying under a tropical sun, as if to warm its offspring into activity and promote comfort, until grown to the size Nature demands for its first northern visit."

In Scammon's era such observations of the natural world were not only unusual but extremely rare. At the 1876 Centennial Exhibition in Philadelphia, nature would be represented by a few cases of stuffed birds and animals. The bison and the passenger pigeon had already been hunted into extinction. Even John James Audubon generally shot the birds he took home to paint and only hinted at the need for conservation. Certainly in terms of marine life the word had scarcely entered the lexicon. Only a few voices in the wilderness, such as Henry David Thoreau in 1864, raised questions like "Can he who has discovered only some of the values of whale-bone and whale oil be said to have discovered the true use of the whale? Can he who slays the elephant for his ivory be said to have 'seen the elephant'? These are pretty and accidental uses; just as if a stronger race were to kill us in order to make buttons and flageolets of our bones."

So Scammon's metamorphosis from killer to chronicler is a mysterious one. His book, like *Moby-Dick* in its time, was a financial failure. His legacy, among those who study the great whales today, remains both a puzzle and a treasure.

Chapter 3

The Poet and the Saltworks War

This body of water was first discovered for all practical purposes, by Captain Scammon, of the whaling ship Ocean Bird, *of this port, who entered it three years ago and quickly filled up his ship there. . . . Still another resource has been discovered. Immense deposits of salt exist along the margins of the lake, produced by the process of evaporation which has been going on for ages. A Mexican has secured from his government the exclusive right to work and export this, and he had at the last accounts, several schooner loads ready for shipment.*

—Article about the San Ignacio Lagoon from a San Francisco newspaper, 1863, in one of Scammon's scrapbooks

THE same natural versatility that made the lagoon region of Baja California so hospitable to gray whales—isolation from the sea, a low frequency of rainfall with essentially no freshwater runoff, brisk and persistent winds to accelerate the evaporation process—offered ideal conditions for solar-evaporated salt. The Cochimí Indians who once lived along the lagoons were undoubtedly the first to collect the salt, as a preservative for their fish and shellfish. Later the Jesuit and Dominican padres used it for food preservation and seasoning.

The first forays of Charles Melville Scammon into the lagoons brought awareness of a product at least as valuable as whale oil, and the captain seemed well aware of the import of his discovery. Writing in 1867 for a report about the resources of the Lower California coast, Scammon described the salt fields of the Ojo de Liebre [Scammon's] Lagoon, eighty miles north of San Ignacio's, as "capable of supplying an almost unlimited quantity of excellent salt. Vessels of four hundred tons' burden can find good anchorage within five miles of where the commodity can be embarked in lighters of twenty-five to fifty tons' capacity."

Scammon's history continued:

> A year or two after the whaling commenced, vessels were dispatched from San Francisco, Upper California, for cargoes of salt; the first two, after cruising a length of time off the desired port, returned with the account that no such lagoon existed, or, if it did, no channel could be found to get into it. A third vessel was sent with a master determined to either find the place or "break something"; he lost his vessel between Black Warrior and Upper Lagoon.
>
> Subsequently the late Captain Collins, of San Francisco, a gentleman of much experience, and a skilful seaman, obtaining the most reliable information at hand, sailed for the place that seemed to baffle the efforts of his predecessors to find. In due time he arrived at the desired haven, without difficulty procured a cargo of salt, and returned to San Francisco. These voyages were followed up for a length of time, but the low price of the article compelled the proprietors to abandon the trade.

Collins's successful voyage had occurred in 1862, when two ships headed back with up to four hundred tons of salt apiece. Within two years the exports ceased—until a San Francisco merchant named C. J. Jansen installed thirty-six men at Scammon's Lagoon in 1869. They brought along cattle and mules, and utilized some previously constructed rail and pier facilities. Several barges weighing up to eighty tons would carry salt from a dock across the shoals to waiting vessels.

It was a clandestine venture. When the Mexican government found out about the American's claim, it protested vehemently and instructed Baja authorities to get rid of Jansen. In 1873, having exported some five thousand tons of salt over a four-year period, the operators quietly slipped away. When a Mexican-appointed administrator decided to leave as well, fishermen and other residents made off with a bonanza of abandoned equipment—the most attractive, perhaps, being sixty-three sixty-gallon casks of beer.

Over the next seventy-five years salt deposits between Scammon's (Ojo de Liebre) and Black Warrior (Guerrero Negro) Lagoons continued to be exploited by the Mexicans on a modest scale. By the mid-1930s another pier for salt harvesting had been installed at Scammon's. It wasn't until the mid-1950s, however, that production began in earnest. The original owner of Exportadora de Sal, S.A. de C.V. (ESSA for short) was National Bulk Carriers, owned by the reclusive American billionaire Daniel Ludwig. He had been advised that a salt-mining operation in San Francisco had dried up, meaning that West Coast paper producers faced a shortage of sodium chloride for their chlorine-bleaching process. Ludwig was always looking for new commodities to transport in his lucrative shipping business. So he sailed down the Baja coastline and bought the rights to extract salt from the adjacent lagoons.

Where coyotes and scorpions had been the predominant residents, the town of Guerrero Negro was created out of the desert. The salt process consisted of moving seawater from the Pacific Ocean into specially created onshore wetlands, then letting the salt water evaporate to form salt crystals for collection and processing. Over the ESSA company's first decade of operation, salt exports based out of Black Warrior Lagoon soared from about 50,000 tons to some 3 million tons annually. At the same time gray whales abandoned this relatively small lagoon, apparently because of dredging operations. They returned scantily only after 1967, when production was relocated to the much larger area of tidal flats around Scammon's Lagoon.

In 1973 Daniel Ludwig sold out to Japan's Mitsubishi Corporation. Mexican authorities agreed to permit this, provided a step-by-step plan was followed to bring ESSA under the government's majority ownership. This was effected by 1976, with Mitsubishi controlling 49 percent of the partnership. Japan's commercial interest in Baja actually dated back almost to the turn of the twentieth century. In *The Zimmermann Telegram*, the historian Barbara Tuchman describes a secret treaty made around 1908 between the Mexican and Japanese governments. It concerned Magdalena Bay, San Ignacio Lagoon, and Guerrero Negro Lagoon being leased to Japan for naval, fishing, and fleet-coaling rights. When the United States eventually got wind of this in 1911, President Taft dispatched twenty thousand troops to the Mexican border and sent the naval fleet down to Baja. War did not materialize, nor did any Japanese naval base at Magdalena Bay. But Japanese influence—primarily under the Mitsubishi banner—remained stronger than that of any other foreign nation along this stretch of coast. In 1990, for example, Mitsubishi Heavy Industries built an electric power plant at Magdalena Bay that services most of southern Baja.

Mitsubishi had been founded as a shipping company in 1870, and it was the old Mitsubishi Trust that constructed and owned the majority of

Japan's whaling fleet. In the years leading up to World War II, the Japanese whaling industry stepped up production to support its war effort in Asia, selling whale oil in Europe to acquire foreign currency. As late as 1965 a fisheries company majority-owned by Mitsubishi was making overtures to the Mexican government about establishing an offshore whaling station along the Baja coast. To its credit, Mexico did not grant the permit. Gray whales had not been legally taken in Mexican waters since the Norwegian-Mexican factory ship *Esperanza* departed Baja in 1935. All commercial taking of gray whales had been formally banned by the International Whaling Commission in 1946; Japan, five years later, became the last of the forty member nations to sign the agreement.

After World War II the occupying American forces had disbanded the Japanese *zaibatsus*, those gigantic military-industrial combines that had powered the imperial war machine. Mitsubishi, the second most powerful prewar holding company, fragmented—only to be replaced by today's *keiretsu*, a more loosely structured family of companies tied together by interlocking shareholdings. These operate in all fields of endeavor, from production of raw materials to manufacture and retail sales. The Mitsubishi Group remains the largest of Japan's eight such industrial amalgamations. The bosses of its 28 chief corporations (out of 160 altogether)—Mitsubishi Motors, Mitsubishi Oil, Mitsubishi Heavy Industries, Mitsubishi Electric, Bank of Tokyo–Mitsubishi, and more—meet on the second Friday of every month. Even the world's largest beer company, Kirin, falls under the Mitsubishi banner. The corporation's overall revenues—$129 billion in 1999—are larger than the national budgets of many countries, including Mexico's ($74 billion). And in Japan, as the independent TV journalist Teddy Jimbo puts it, "Mitsubishi is not simply a huge company but a part of daily life. You cannot escape their influence."

Mitsubishi's 49 percent interest in the world's largest solar saltworks, ESSA, falls under the jurisdiction of the Chemicals Division of Mitsubishi Corporation, also known in Japan as Mitsubishi Trading Company. Approximately half of the salt exported from the Guerrero Negro operation is used in Japan's chemical and chloralkaline industries (in production of commodities including PVC plastics, bleach, chlorine gas, road salt, and glass). The story goes that by 1989 ESSA foresaw that the 135 square miles of evaporating salt ponds at Scammon's Lagoon were soon going to reach their maximum potential. So, when then-Mexican President Carlos Salinas de Gortari paid a state visit to Japan, he met privately with the head of Mitsubishi and they discussed potential expansion of the saltworks. Why not the San Ignacio Lagoon? The geographical and geological conditions were similar. Opening a new industrial operation there would more than double production, to over 13 million metric tons of salt a year. An envisioned pier and narrow-gauge rail connection between Scammon's and San Ignacio Lagoons would enable transshipment directly to export markets, rather than the existing means of having first to barge the salt from Scammon's Lagoon to Cedros Island, fifty-four miles out into the Pacific. Mexico would become the world's number-one salt producer, and Japan would be assured of an ample supply.

These plans were proceeding discreetly, and smoothly, until a man named Homero Aridjis entered the picture.

> *The poetry of the Aztecs was brilliant and imaged; it borrowed its comparisons from the flowers, the trees, the brooks, the most pleasing objects of nature.*
>
> —LUCIEN BART, *The Aztecs*, 1886

The first thing you notice about Homero Aridjis is the hands: in constant motion, punctuating the air with intensity, fluttering toward his chest like wings coming in for a landing. As his hands seem to fill the living room of his home in Mexico City's Lomas de Chapultepec section, Aridjis is saying: "Inspiration for poets, painters, and composers has always come from nature. The task of poets, and of holy men, is to tell this planet's stories—and to articulate an ecological cosmology that does not separate nature from humanity."

Aridjis (pronounced "ah-REED-hees") is a prizewinning author of twenty-eight books of poetry and prose. His work has been translated into a dozen languages. Two collections of his poetry *(Blue Spaces* and *Exaltation of Light)*, as well as three novels *(Persephone; 1492: The Life and Times of Juan Cabezón of Castile;* and *The Lord of the Last Days: Visions of the Year 1000)*, have appeared in English. In 1997 Aridjis became only the second non-European and non-American writer ever elected to the presidency of PEN International, an esteemed literary organization founded in 1921 to promote literature and language and to defend freedom of expression. He served previously as Mexico's ambassador to the Netherlands and Switzerland, received two Guggenheim Fellowships, and taught at three American universities—Columbia, New York University, and Indiana University.

Aridjis is also Latin America's leading environmental activist, the founder and president of the Grupo de los Cien, or Group of 100. The name derives from a declaration signed in 1985 by a hundred prominent Mexican artists and intellectuals. They initially sought to raise awareness about Mexico City's dire air pollution. The group's efforts forced the government to limit the circulation of cars, drastically reduce the amount of lead in gasoline, and publish daily reports on air-pollution levels. Far from the confines of the city, the Group of 100 is responsible for saving the country's sea turtles, migrating monarch butterflies, and Lacandón rain forest from complete annihilation.

My first extended visit with Homero Aridjis came in June 1998, two months after my first contact with the gray whales at San Ignacio Lagoon. I'd traveled to Mexico City to do a magazine profile of this poet who had taken on Mitsubishi and his own government to wage all-out war against the proposed saltworks expansion. As my taxi passed through a hilly, wealthy district of Chapultepec—Mexico City's Beverly Hills, the driver said—en route to the suburb where the Aridjises reside, I thought of what the American writer Pete Hamill, Homero's longtime friend, had told me: "Homero has not felt, as Václav Havel has in the Czech Republic, that you either have a commitment to civil society or to art. He's been able to continue to do both. In Mexico in particular, there's a constant conflict between the issues of environment and the realities of the way business is done. Homero brings an amazing decency—and great effectiveness—to a subject that can make people absolutely cynical."

A three-story stucco house, painted predominantly yellow, stood behind a locked iron gate. Lilies, representative in Mexico of the national soul, flourished in the entryway. I was ushered inside by a housekeeper. A friendly German shepherd named Rufus T. Firefly (from an old Groucho Marx movie) was the next to greet me. Homero came running downstairs to say that his wife, Betty, would talk with me while he finished his bimonthly column for the newspaper *Reforma*, because he was late meeting the deadline. Betty Ferber de Aridjis is the international coordinator for the Group of 100 and has translated three of Homero's novels into English. They met, she said, as we sat down on a living room couch beneath an overflowing bookshelf crowned by numerous paintings, in 1963. She was intending to spend a few weeks in Mexico before starting graduate school at Columbia; after she was introduced to Homero, Betty returned to New York only to pack up.

A thunderstorm was stirring outside when Homero entered the room a half hour later. He was wearing a blue-and-white pinstriped shirt and light-colored slacks. He was not an imposing figure, rather short and

Homero and Betty Aridjis in 1964, soon after meeting (courtesy Betty Aridjis)

with dark hair askew above a broad smile. Yet he seemed to pulse with energy. In fluent English he explained that the "emergency" he had been upstairs writing about was the fires that had devastated the Mexican countryside in recent months. Fueled by drought and El Niño, an estimated thirteen thousand fires swept across more than one million acres in numerous regions. In announcing a reforestation plan, the Mexican government had revealed that most of the fires were set intentionally. Loggers, ranchers, farmers, developers, drug traffickers, and soldiers in pursuit of Zapatista guerrillas had all been looking to "clear" the land for their own purposes.

"This is one of the biggest tragedies in this century in Mexico," Homero said, shaking his head and sipping a glass of Jamaica-flower water. "You can't recover an ecosystem like the Chimalapas, the richest in biodiversity, that took thousands of years to form. All over Mexico, lakes and rivers are drying up, and we are going to have a terrible problem with lack of water in the years to come."

Betty added, "Interestingly enough, in Spanish there's really no word for wilderness. The notion here is, nature has to pay its own way or there is no justification for its existence."

As parrots in his backyard garden offered a soundtrack, Homero recalled how his personal awakening to nature—and, simultaneously, his poetry—began. He was born in 1940 in the central highlands region of Michoacán, in a little village called Contepec. His mother was Mexican, his father a Greek who had founded the local Apollo Cinema and Olympus garment factory. Homero was the youngest of five brothers, the last two named after figures in Greek literature and history. He was ten when, early one afternoon, he discovered a shotgun his siblings had left leaning against his bedroom door. He picked it up and climbed onto the roof. Above him flew a small flock of birds, similar to the nearly one hundred mockingbirds, solitaires, and canaries his mother fed daily. Homero aimed the gun. Then his conscience was stricken: "Every morning I awake to hear them singing to the light. Why do I want to kill the birds I love?"

As he lowered the butt of his gun to the ground, the weapon went off. A shot pierced his stomach. His parents rushed him to a nearby village, but the doctor wasn't there. Eight hours after the accident treatment was found at a general hospital many miles away. For nineteen days the boy hovered between life and death. For the rest of his life Homero would retain images of the birds that had soared in and out of his semiconsciousness.

Mexico is a land of many legends, and one came immediately to my mind. In the golden age of the country, according to Aztec myth, there lived a high priest known as Quetzalcoatl, the feathered warrior. When Quetzalcoatl sacrificed himself to the gods, "from his ashes rose all the precious birds, the coringa, the spoonbill, the parrots. . . . Then the heart of Quetzalcoatl rose into heaven and . . . was transformed."

During his long recovery from the shotgun wound, Homero began to write poems. ("Reading Homero's poetry," I recalled Pete Hamill telling me, "I get the sense there's a bird leaning over his shoulder.") No library or bookstore existed in his village and, Homero says, "culture came for me as a personal conquest." Every afternoon he would walk to a hill near his home. During the winters "the sky would be aflame with red, orange, yellow, and black." Thousands of monarch butterflies would alight on the fir trees covering the hillside. In those years it was not yet known that the monarchs of North America migrate many thousands of miles, some from as far away as southern Canada, to reach their seasonal nesting place in Mexico's *oyamel* forests. Homero awaited them every year "as if their arrival marked the beginning of a prolonged fiesta."

Later, during the fifteen years in the 1960s and 1970s that he lived abroad as a diplomat and teacher, Homero would make an annual winter pilgrimage back to his hill. As the trees were cut for firewood, the presence of the butterflies diminished. "I felt that my own childhood was being killed, my memory of a natural beauty that had once overwhelmed me. The possibility of my village becoming a wasteland, a silent country without wind in the trees or animal sounds or birdsongs, made me feel desperate. Butterflies became for me a symbol of life's fragility." So it is not surprising that one of the first crusades of the Group of 100 resulted in the establishment of five protected sanctuaries for the monarchs in 1986.

TO A MONARCH BUTTERFLY	A UNA MARIPOSA MONARCA

(from *Construir la muerte*, 1982, tr. George McWhirter)

You who go through the day	Tú que vas por el día
like a winged tiger	como un tigre alado
burning as you fly	quemándote en tu vuelo
tell me what supernatural life	dime qué vida sobrenatural
is painted on your wings	está pintada en tus alas
so that after this life	para que después de esta vida
I may see you in my darkness	pueda verte en mi noche

There is a breath of the mystic Nahuatl songs and present-day initiation hymns of the Huichol Indians in his verses. Homero does not consider himself an environmentalist but rather "a lover of nature." The difference, he believes, is in terms of the immediacy. Environmentalists in the industrialized United States and Europe "more or less move from policies, agreements, and so on. But here, in defending the butterflies or the trees, you have to move almost like a doctor in a hospital with emergencies.

"In Mexico"—he shook his head again—"everything is political."

For Homero, political awareness came long before the Group of 100. He had risen to fame at age twenty-four when his first novel, *Mirándola dormir (Watching Her Sleep)*, won the Xavier Villaurrutia Prize for best book of the year. In 1966, at the invitation of Arthur Miller and Lewis Galantiere, he helped organize a PEN conference in New York. It made literary history by presenting Latin American writers such as Pablo

Neruda, Ernesto Sábato, and Mario Vargas Llosa to a U.S. audience. Homero spent most of the 1970s in Mexico's diplomatic service. Then in 1980 he was appointed director of a new cultural institute in his home state of Michoacán. There he organized a remarkable poetry festival, including Allen Ginsberg, Jorge Luis Borges, Günter Grass, and Seamus Heaney.

"Other people in the cabinet were very jealous of Homero because he was doing so much," his wife remembers. "There was a lot of intrigue." When the state government abruptly canceled his second scheduled festival, Aridjis went ahead with it anyway. He changed the venue to Mexico City and enlisted artist friends to sell some paintings to help cover expenses. Thousands of people attended. "After that, Homero was more or less on the blacklist for several years," according to Betty. "It was actually providential, because his time was his own, and that's when he wrote *1492: The Life and Times of Juan Cabezón of Castile.*" This novel, *The New York Times* said, "succeeds in magisterially re-creating that woeful and bizarre period of Spanish history that prefigured the discovery and conquest of America."

Homero generally writes in the morning ("When I don't write, I become very neurotic") and devotes his afternoons to environmental work. He finds that his activism has only added to his creative life. "It's given me a different reading of the world. All kinds of people, good and bad. Idealism and betrayal. Confrontation. Ecology, like poetry, should be practiced by everybody."

Asked to name the writer he most admires, Homero responds without hesitation: "One of my models is Dante, because he's a moral poet who took an active role in his time." His wife smiles and adds, "I often think Homero would also like to put certain people in various circles of hell."

It was in 1985 that the Aridjises read a letter to the editor written by a philosopher friend, Ramón Xirau, calling for action against pollution. The air in Mexico City had gotten so bad, birds were dropping out of the sky dead. Homero and Betty conceived a plan: a declaration signed by the cultural elite. Homero discussed drafts over the phone with the Nobel Prize–winning poet Octavio Paz while other prominent friends gathered signatures. "We, who live beneath this viscous mushroom that covers us day and night, have the right to life," stated signatories including the Nobel Prize–winning writer Gabriel García Márquez and the artist Rufino Tamayo. Homero recalls: "We were attacking pollution as the effect of political corruption, going to the forces behind what people were suffering. The government was shocked."

Early in 1987 Homero traveled for the first time to Scammon's Lagoon. Besides spending time among the gray whales, he learned from marine biologists during a weeklong visit that there had been a spill three years earlier from ESSA's salt operation. An estimated 4 million gallons of diesel fuel had leaked into the lagoon. "The accident was never reported," Homero recalls. "The company tried just to recover the fuel to reuse it. There was never any evaluation of the environmental damage. I understood then that there was a problem with the company's policy of secrecy and covering up." When he tried to get Mexican newspapers to publish something, "there was almost automatic censorship"—the Ministry of Commerce was calling the shots.

But the power of the Group of 100 increased that year when Homero's arousal of international press interest thwarted a plan to build four hydroelectric dams along the Mexico-Guatemala border that would have destroyed two of the most important Mayan archaeological sites and inundated twenty-two villages. Early in 1988, having heard rumors about alterations to the coastal habitat as the salt company increased production, Homero returned to Scammon's Lagoon in the company of Manuel Camacho Solís, a friend who happened to be Mexico's latest minister of urban development and ecology. In Baja Sur gray whales were already a national symbol. A huge statue of a gray whale tail marked the entrance to the capital city of La Paz, where there was also the Community Gray Whale Museum. Back in 1971 Mexico had established a gray whale reserve at Scammon's Lagoon, the first of four pieces of legislation enacted to protect the whale's lagoon habitats over the next seventeen years. In 1977 the government had sponsored the First International Symposium on the gray whale in Guerrero Negro. Now, just over a decade later, Homero was petitioning for additional measures to "stop any hostile modifications of the coast and to have more vigilance in the area."

Much to Homero's surprise, Camacho went even further. The minister announced that he would seek legislation to create the Vizcaíno Desert Biosphere Reserve, a nearly 7-million-acre area incorporating Scammon's and San Ignacio Lagoons, a bighorn sheep habitat, and the numerous pre-Columbian cave paintings within the Sierra de San Francisco. The goal was to conserve endangered plants and animals as well as allow only "compatible human activities" within its boundaries. The reserve would be the largest such area in all of Latin America. Homero was ecstatic and set about intensively lobbying Mexican President Miguel de la Madrid to follow through. On the final day of his administration, that

November 1988, the president did. Theoretically, the gray whale's winter homeland would see its biodiversity preserved without new obstructions by humans. "Environmentalists are optimistic," as one newspaper account stated, "that Mexico has found the means to protect the gray whale for generations to come."

Still, there had been ominous portents during Aridjis's and Camacho's visit earlier that year. Even the Mexican minister had been forced by ESSA to obtain a visitor's permit because, Homero would recall, "they said this is the property of the Japanese government." Whale watching on the lagoon, they were allowed only to reach a certain point out of range of the saltworks before being sent back to shore. The then-coordinator of Mexico's marine mammal program, Luis Fleischer, was part of the tour, as were several foreign journalists invited along by Homero. When Fleischer began lecturing the journalists about "how much the gray whale population had recovered and so now we can open the area to commercial hunting," as Homero remembers it, the poet was outraged and said so. He suspected some kind of deal between Fleischer and Japanese whaling interests. And, despite the appearance of success in preserving the lagoons, Homero found himself writing this poem:

GRAY WHALE

(From *Imágenes para el fin del milenio & Nueva expulsión del paraíso*, 1990,
Tr. Betty Ferber and George McWhirter)

Gray whale,
once there is no more left of you
    than an image
of the dark shape that moved
    on the waters
in animal paradise,
once there is no memory,
no legend to log your life and
    its passage
because there is no sea where
    your death will fit
I want to set these few words
on your watery grave:
"Gray whale,
show us the way to another fate."

BALLENA GRIS

Ballena gris,
cuando no quede de ti más
    que la imagen
de un cuerpo oscuro que iba
    por las aguas
del paraíso de los animales;
cuando no haya memoria de tu paso
ni leyenda que registre tu
    vida,
porque no hay mar donde quepa
    tu muerte,
quiero poner sobre tu tumba de agua
estas cuantas palabras:
"Ballena gris,
danos la dirección de otro destino."

For Homero, the ascent to the Mexican presidency of Carlos Salinas in 1988 was a troubling one. Today, on a wall of the Aridjises' home alongside a collection of shamanistic masks, there remains a figurine of Salinas holding armloads of money bags. Homero did not know what to expect when he began writing a series of articles titled "The Marine Turtles to Extinction." He had received documentation about the killing of more than a hundred thousand sea turtles on the Mexican coast, particularly in the nesting area of Escobilla. Photographs of the massacre showed the flippers being sliced off and the turtles dropped alive back into the ocean. The skin was being made into purses and shoes and exported to Japan, along with Caribbean tortoiseshell. Homero named names in his front-page articles for the Mexico City daily newspaper *La Jornada*: marines and inspectors, businessmen dealing in leather products. The minister of fisheries demanded to know his sources. Homero refused. Public outcry forced the closure of the turtle slaughterhouse in Mazunte, in the state of Oaxaca.

In 1990 President Salinas issued a decree that imposed a permanent ban on the capture and commercialization of all species of sea turtles that nest on Mexican beaches or swim in its coastal waters. The ceremony took place at Maximilian and Carlota's nineteenth-century castle in Chapultepec. When the Aridjises arrived, no chairs had been reserved for them. It was an obvious intentional slight. As the other dignitaries assembled, a livid Homero was finally approached by a government functionary offering them seats. He refused.

Homero remembers: "The man came back in five minutes and said, 'I have my orders that you must sit—or else I am going to be fired!' I said, 'I don't sit, I don't care if you are fired.' " All through Salinas's speech Homero and Betty were the only guests standing. Afterward, as the president walked in his direction, Homero was stunned to find himself surrounded by several dozen Mexican journalists holding notebooks, microphones, and TV cameras and firing questions. "The next day, there was nothing in the news anywhere. It was a phony thing, a trick."

The next year Homero decided to take his wife and one of their two daughters to visit the turtle nesting area at Escobilla. As he often did he invited a journalist, this time the correspondent for *The New York Times*. They arrived at an isolated, unlighted beach, where a group of marine biologists had set up camp. A cadre of marines were on patrol. "I knew that I was in enemy territory," Homero says, "because I had mentioned many of these people by name. But I arrived unexpectedly. We stayed there until sunrise, watching the sea turtles coming out from the ocean to lay their eggs."

It was one of the most beautiful spectacles of nature Homero had ever seen. Here was one of the oldest animals on the planet, toothless yet having survived both natural and human predators for aeons. As the high waves retreated in white curtains against the night sky, a black shadow was left behind on the beach. "It is so quiet," the poet says, "and then suddenly the turtle begins to move very slowly. She walks about eighty meters and begins to dig with her flippers. It takes twenty minutes to half an hour for her to make a hole. You can approach her and touch the head. And when she lays the eggs, it is so amazing: you can see tears fall from her eyes. She covers the nest. She returns to the shore and stays just a few moments. As soon as the waves come again, she is gone."

When Homero returned the next day, the biologists were gone too. There was nobody around but the marines. "I had a bad feeling. I always try not to expose my back to them. I told Betty and my daughter, we must stay together. I invited two mechanics repairing our car to come as protection also. I don't know what was the plan, but something was to happen. I told the military commander, 'This is a reporter from one of the most important newspapers in the world and, if something happens to him, this is universal news.' Later on we met one of the poachers, a big man. He came over and pretended he didn't know he was addressing me. He said, 'I would like to meet the Group of 100 son of a bitch Homero Aridjis.' I said, 'Hello, how are you, I am that person.' He said, 'Don't believe that we are going to take care of these animals, because they are *billete*, 'bucks.' "

Homero paused, then said, "Always there has been, in these crusades, an element of danger." In other efforts to blunt the sword of the Group of 100, he has been offered bribes, business deals, his own television program, and further ambassadorial posts. He soon found that officials took his criticisms personally, deluging him with telephone calls (after which the phone often went mysteriously out of order, sometimes for weeks). There are parts of Mexico where Homero can no longer go, because, he says, "I am public enemy number one. The poachers and loggers recognize me." Late in 1997, while visiting their daughter Eva in New York, the Aridjises had received death threats on their answering machine in Mexico City. Since that time when in Mexico they had accepted the presence of two government-appointed bodyguards. "We don't know the origins of the threats," Homero said. "In Mexico environment, politics, organized crime, everything goes together."

It was not hard to conceive that his current campaign—to stop the proposed $120 million saltworks expansion to San Ignacio Lagoon—had something to do with it. To understand the genesis of Homero's latest round of battling for the whales, we must go back a number of years. During World War II whales had been given a temporary reprieve from hunting, especially by the Japanese when naval action destroyed almost the country's entire whaling fleet. In the aftermath, however, General Douglas MacArthur, in charge of the American occupation, encouraged Japan to gear up its whaling factory ships again to help meet a food shortage. Whaling was an honored practice in Japan going back centuries. For a time after the war, whale meat came to constitute a considerable amount of the defeated nation's protein food supply and a staple in school lunch programs.

Gray whales, among other species, had only begun to come back in numbers and would probably be extinct by now had the U.S. government not reacted to the revival of whaling by bringing together numerous nations in Washington, D.C., during the autumn of 1946. After two weeks of negotiations, they drew up a multilateral treaty and formed the International Whaling Commission (IWC) to manage "harvesting." The U.S. secretary of state, Dean Acheson, declared: "The world's whale stocks are a truly international resource in that they belong to no single nation, nor to a group of nations, but rather they are the wards of the entire world."

For its first several decades the IWC remained largely ineffectual. Member nations refused to heed the advice of their own Scientific Committee in curtailing commercial catches, while Japan, Norway, Iceland, and the Soviet Union chose not to adhere to whatever restrictions were enacted. In 1961 the largest single-year catch of whales in history occurred—66,090 slaughtered worldwide. By 1970 the United States had listed eight whale species as endangered, including the California gray. Finally, pressure from the United States found the forty-one IWC member nations voting for a moratorium in 1982 that was to take effect worldwide by 1986: "Catch limits for the killing for commercial purposes of whales from all stocks . . . shall be zero."

The IWC, however, did continue to permit limited whaling for the purposes of scientific research. They granted Japan's request to take 825 minke whales from the Antarctic, supposedly as part of a twelve-year study on population dynamics. In fact, over the ensuing decade, Japanese whalers brought in about 3,000 minkes from the Antarctic and North

Pacific—all headed for their upscale restaurant market, where even small portions of such whale meat were now a hundred-dollar-a-plate delicacy.

In 1993 the IWC held its annual meeting in Kyoto, Japan. *The Christian Science Monitor* reported that Japan had spent $80 million in the year leading up to the gathering, in an effort to influence Japanese public opinion in favor of renewing commercial whaling. The *Tokyo Shimbun* newspaper reported that Japan had offered $234 million in economic aid to various countries to secure IWC votes toward just such a resumption.

At the time Mexico's delegate to the IWC was Luis Fleischer, the same man who had espoused allowing gray whaling on that trip five years earlier to Scammon's Lagoon. Realizing that the IWC's whaling ban was in jeopardy, Homero Aridjis wrote a letter to President Salinas, warning about the Mexican emissary "who was always on the side of the Japanese." The Group of 100 then issued a press release to similar effect, which was published in Mexican and Japanese newspapers. Japan's attempt in Kyoto failed by an eighteen-to-six vote of IWC member nations. When Fleischer returned to Mexico, he was replaced as its IWC delegate.

In 1994 the IWC was about to convene in Mexico's coastal resort town of Puerto Vallarta. Homero was lobbying his government to vote for a new circumpolar whale sanctuary in Antarctica—one that Japan was looking to have reduced in size. Homero and Betty had been in Bellagio, Italy, for a month on a writing grant when they received the text of the Mexican Fisheries Ministry's planned statement. It was in complete support of the Japanese. Homero dispatched it back to Mexico with crossed-out pages—and flew home immediately. When he went to a meeting at the Fisheries Ministry, the second in command was waiting for him—with a check for fifty thousand dollars. "This man said, 'It's for you, do whatever you want. But our position is going to be with the Japanese against the sanctuary.' "

Homero refused the money and walked out of the office—only to receive legitimate help from an unexpected quarter. Emilio Azcárraga, a billionaire who was close to Salinas and whom *Forbes* magazine ranked as the wealthiest businessman in Latin America, wanted to meet with him. Azcárraga was the owner of Televisa, which held a 95 percent share of the Mexican TV audience. He was commonly known as El Tigre and, while a friend of Octavio Paz, was considered by Homero as "a really fearsome person." Azcárraga, it turned out, loved to sail, especially offshore among the whales. He told Homero that he was a government loyalist who didn't agree with him or even give a damn about the environment. Then Homero remembered El Tigre's saying something like, "But I respect you, because you are honest and we can talk. And I am going to support you with the whales." The TV mogul privately agreed to finance a full-page ad that ran in *The New York Times* on May 10, 1994—"Group of 100: In Defense of whales"—pointing out that Japan had "openly bought the votes of small Caribbean nations" and other developing countries, calling upon Mexico and other IWC nations to stand unequivocally behind the French proposal for an Antarctic sanctuary.

Arriving in Puerto Vallarta, Homero noted a third proposal on the table, a divide-and-conquer strategy to avert the sanctuary. He called a press conference and berated the Mexican government's position. Not only did Televisa give Homero full coverage but it published the address and phone number of Mexico's Japanese embassy on the national news. Thousands of calls of protest poured in from the Mexican citizenry.

The assistant fisheries minister who had offered Homero the fifty-thousand-dollar bribe received a phone call from the presidential palace. Carlos Salinas instructed the fellow to go to Puerto Vallarta, where *Homero Aridjis* would inform him of the new Mexican position. At the IWC meeting the Antarctic Whale Sanctuary was adopted by a three-quarters majority.

The Group of 100—and the whales—had won again. Still, Homero realized that Salinas remained very close to the Japanese, even sending his own children to a Japanese school in Mexico City. And the poet wondered where lightning might strike next.

In the autumn of 1993 a tall surfer named Serge Dedina had arrived in Baja with his wife, Emily Young. The couple planned to spend a year doing field research here for their dissertations. Serge's subject was gray whale conservation. To fund their work they'd received a contract from the U.S. Marine Mammal Commission to prepare a technical report on potential threats to the gray whale's lagoon habitat. One of the first things the couple learned about was a plan by Exportadora de Sal to build the planet's biggest salt production facility around San Ignacio Lagoon.

"We assumed everyone must know about this," Dedina remembers, "not just a couple of starving graduate students running around in an '88 Ford F-150. I mean, it was no secret in Baja that one of the world's biggest corporations was going to build one of the world's largest industrial projects in the middle of Mexico's largest protected area. Well, we had to send monthly reports back to the Marine Mammal Commission, so they'd know I wasn't just out surfing and drinking beer. This was the first they'd heard about it, and they were immediately concerned."

Somehow a lagoon fisherman living in a little shack had a copy of the blueprints for the project and, in 1994, allowed Dedina to hand-copy them. Shortly thereafter Dedina and Young traveled to Guerrero Negro to interview ESSA's director, Juan Bremer, as part of their research. He treated the expansion as a fait accompli. Under Mexican law an environmental impact assessment (EIA) was required and had been completed. When Dedina encountered the EIA team, he learned that they'd spent a mere two days in the vicinity of San Ignacio Lagoon. In fact, they'd camped outside the little two-house village of La Laguna, never talked to any residents, never even made it into the habitat of turtles, birds, and whales.

Dedina and Young traveled to Mexico City to meet with representatives of the National Institute of Ecology, which was charged with the ultimate authority over the Vizcaíno Desert Biosphere Reserve. The couple was told that the saltworks expansion was imminent but that certain "mitigation measures" would be taken to ensure no adverse impact.

Informed of the latest developments, the U.S. Marine Mammal Commission in early January 1995 dispatched Dr. Bruce Mate to meet up with Dedina in Guerrero Negro. Mate worked out of the Hatfield Marine Science Center at Oregon State University and had been conducting gray whale research periodically at San Ignacio Lagoon for almost twenty years. Dedina arranged a meeting for Mate with local staff of the biosphere reserve. While they, too, were concerned about ESSA's plans, their hands were tied. The EIA was being reviewed by a Baja official in the capital of La Paz, and that would be that.

Dedina managed to obtain a copy of the 465-page document. He had written similar reports for a living and knew what to look for. It was as flimsy as feared: twenty-one pages of actual impact analysis, five pages devoted to potential mitigation, and a mere twenty-four lines to the gray whales' utilization of the lagoon.

Mate and Dedina went to La Paz to visit with a Mexican functionary who sat at a little corner desk, tasked with reviewing the EIA for a $120 million project. When the marine scientist and the graduate student started making some probing inquiries, the official became furious. "He said, 'You guys have no right to ask these questions!' It was a very, very tense meeting," Dedina recalls. Mate says he left the room realizing that "this was almost a done deal. They were going through the motions and rubber-stamping an incredibly lax environmental assessment. The people reviewing the document did not understand the incredibly important role they had as stewards for Mexico's resources. It was just a paperwork process. They were, at the same time, doing an EIA on a two-year-old power plant."

Dedina and Mate knew there wasn't much time. While Mate returned to the United States to report to the Marine Mammal Commission, Dedina made a fateful decision. He knew about the Group of 100, the role Homero Aridjis had played in creating the biosphere reserve, and Homero's published vow that he would forever defend the homeland of the gray whale. On January 13, 1995, he placed a call to the Aridjises in Mexico City. "What do you mean?" Betty responded to his news. "You're joking, right?"

Homero, too, was stunned. Beyond Baja the project had been kept completely under wraps. Eight days later his column appeared in *Reforma.* Its headline translated: "Mexico Sells Whale Sanctuary." The Group of 100 simultaneously issued a press release to foreign media based in Mexico City. Government officials retorted that the saltworks would not harm the gray whale and would provide development opportunities for a "desert area with limited economic possibilities."

Homero set about trying to obtain a copy of the EIA and received a bureaucratic runaround. Supposedly, the document wasn't in Mexico City but in Baja. It wasn't with this environmental agency, it was with that one. "Nobody seemed to know where it was, because they didn't want to give it to us. I would wait for several hours in different offices, while they played cat and mouse with me. Finally we were able to get it from someone in the United States, who wanted to remain anonymous—but a very high-level person."

The Aridjises immediately went to work studying and summarizing the EIA. On February 15, Homero published another column in *Reforma*, "The Silence of the Whales." The EIA's scant attention to the gray whales, he pointed out, ignored the fact that nonstop pumping of seawater out of the lagoon at the rate of 6,600 gallons per second would lower its temperature and salinity. Each month eight oceangoing ships would dock at a mile-long pier to be built fifteen miles from the mouth of the lagoon, right in the middle of the whales' migratory path. Risks associated with shipping operations included oil spills, frequent winter storms, and summer hurricanes, which could make use of the pier impossible and lead to dredging of the lagoon to accommodate ship traffic.

Mangroves, pronghorn antelope, brant geese, sea turtles, and dozens of other marine and terrestrial species of plants and animals would be adversely affected. The saltworks and its network of roads would have a permanent physical impact on almost 525,000 acres, with an indirect impact on 3.7 million acres, for a total of 67 percent of the biosphere reserve. In 1993 the reserve had also been declared a United Nations World Heritage Site. The saltworks project made a mockery of such designations. The impact assessment described the lagoon's surrounding area as "terrestrial wastelands, with little diversity and no known productive use."

This was strong stuff—especially when *The New York Times* and *Los Angeles Times* picked up the ball. How would the new Mexican government respond? Salinas, who had initiated the project at the beginning of his six-year term, was gone. The newly elected president, Ernesto Zedillo, had served in the Salinas administration and was handpicked to succeed him after another candidate was assassinated. Zedillo had, however, appointed a woman with considerable expertise as his minister of natural resources; Julia Carabias had even taught a university course on environmental impact assessments.

Twelve days after Homero's latest column appeared, Mexico's National Institute of Ecology (INE) denied ESSA permission to develop the project, on the grounds that it was "incompatible with the goals of conservation" of the Vizcaíno Reserve. Other reasons cited in Carabias's ruling included possible negative impacts on fourteen plant species and seventy-four animal species, including gray whales. Nothing, the agency stated, could justify the permanent transformation of landscape and loss of natural environment in so large an area.

"Without the intervention of Homero and Betty, I don't believe for a minute that Carabias would have had the political capital to be able to postpone the project," Dedina says. "The international press reacted not because it was a gray whale issue but because Homero *told* them it was an issue. And this gave Carabias the leverage she needed to put a hold on."

The hold was short-lived. ESSA filed an appeal contesting the decision on March 17, 1995. This was kept quiet, however, for thirteen days—until Homero chanced to notice a copy on a desk while waiting to see the INE president, Gabriel Quadri. Once the Group of 100 had written proof that the appeal had been filed, Mexico's minister of commerce, Herminio Blanco, came forward claiming that the new saltworks was crucial for Mexico to compete with Australia in the world salt market. (Blanco also happened to chair the ESSA board of directors.)

That May, with financial help from several organizations (the International Fund for Animal Welfare, the Animal Welfare Institute, and Greenpeace International), the Group of 100 fought back with a full-page ad in *The New York Times* denouncing the saltworks project. It began: "There's more than one way to kill a whale. Gray whales rock their newborns to sleep in this warm Mexican lagoon. Their only enemy? Mitsubishi, a giant Japanese conglomerate with plans to suck it dry." The signatories included such literary luminaries as Günter Grass, Octavio Paz, Allen Ginsberg, Margaret Atwood, and Peter Matthiessen, along with the scientists Lester Brown and Roger Payne.

Not to be outdone, ESSA and Mitsubishi came back in June with full-page ads of their own in the *Times* on consecutive days. The company announced it was embarking upon a new environmental impact assessment. Mitsubishi, already the target of boycotts over its logging practices in the world's rain forests, trumpeted its commitment to "environmental stewardship."

Battle lines were being drawn for a protracted struggle. Betty Aridjis sent out lengthy dossiers to a number of American environmental groups. The Group of 100 had worked previously with the Natural Resources Defense Council on seeking to strengthen NAFTA's environmental standards. Early in 1996 the NRDC and the Grupo sent a joint letter to Julia Carabias requesting an "open and participatory" environmental review process.

When ESSA hired the Autonomous University of Baja California Sur in La Paz to conduct the new EIA, there was considerable doubt among project watchers whether it could be an objective study. The governor of the state had continued to speak out in favor of the expansion. The only ground for optimism was that Carabias planned to appoint an advisory committee of international marine experts in natural resources and oceanography to oversee the new EIA. She asked the International Whaling Commission for some recommendations, and after the IWC's 1995 meeting in Dublin, turned to the Aridjises for advice. Four of Homero and Betty's suggested choices—including the American gray whale experts Steven Swartz, Steve Reilly, and Bruce Mate—ended up on the seven-member advisory panel.

Mate remembers the scientists' initial meeting with Carabias: "None of us wanted to be part of a process where we might offer an opinion, the government might do something different, and it would appear that we'd endorsed what happened. So we made an agreement. We would serve *only* if she was willing to release our comments intact in English and Spanish, and that any subsequent review process would be publicly available. She not only agreed, she blessed us and said we were providing her with an opportunity to bring impact assessment into scientific and public purview—which, in reality, Mexico did not have.

"We provided her with fourteen pages of concerns, which were called terms of reference, a list of things we felt needed to be addressed when the company came forward with its new proposal. This was not limited to whales; we discussed fish and shellfish and larval forms, freshwater utilization for a community that would have to grow, even coyotes in the desert and garbage disposal." Swartz, who was at the same meeting, recalls: "Up until Julia Carabias, Mexico's environmental authorities would rubber-stamp anything that made money. She put herself in front of the freight train of economics and politics."

Originally, the corporate blueprint was to see the San Ignacio saltworks completed by 1997. Now the deadline simply for the new EIA was the summer of 1999. But as the NRDC initiated a direct-mail campaign of protest letters to Mitsubishi and the Mexican government, and as other environmental groups pitched in, the debate only intensified. The NRDC and the International Fund for Animal Welfare cosponsored media-and-celebrity tours for visitations with the gray whales at San Ignacio Lagoon. Pierce Brosnan, Glenn Close, and others came. So did the Aridjises, and Robert Kennedy, Jr., a lawyer on the NRDC staff.

There had been some positive developments. Early on, the Group of 100 had filed two lawsuits in the Mexican courts. One accused ESSA of environmental negligence at its Guerrero Negro saltworks and of recurrent violations of Mexican laws. The other requested that ESSA be required to carry out an environmental impact assessment for its *existing* saltworks. This pressure had helped force the Mexican House of Representatives to form a special commission to investigate ESSA's performance at Guerrero Negro.

However, according to Betty, Mexico's environmental law pertaining to activities allowed in a biosphere reserve had been revised. "Initially, the law said the only activities permitted were those traditionally carried out by the local residents. They changed this to read that activities will be allowed in a biosphere reserve if they *benefit* the local residents. That is specifically aimed at the San Ignacio Lagoon."

"Another thing you hear from certain people in government," Homero added, "is that the Japanese want to take control of the lagoon area because it is rich in resources other than salt. There are many minerals in the desert. Uranium. Bismuth. Also copper. It is very difficult, because the policy of the Mexican government is to develop. We are not against development. We simply oppose development where it affects natural areas."

So, for the moment, the project was in limbo. But no one could say for how long. Or what it might finally take to stop it, once and for all.

Gray whale,
show us the way to another fate.

As the years have gone by, the Aridjises have increasingly *become* the Group of 100. The nonprofit organization operates out of their home, without budget or salaries, with the government still refusing their requests for tax-deductible status. The other founders have largely gone their separate ways. The liberals wanted fewer conservatives and vice versa; the well-known looked down upon the lesser-known; some, like Gabriel García Márquez and the late Octavio Paz, had become ideological enemies who wouldn't be caught in the same room together. For Homero, what needed to be done transcended all such differences. "I am very pessimistic about the planetary environment," he continued. "But at the same time, I believe that no matter how difficult the problems are, human beings have to fight for life. Always."

The Group of 100's greatest contribution, he believes, has been "changing social awareness in Mexico about the environment." Politicians now integrate ecology into their programs and speeches. Media coverage has increased dramatically. On the street, from taxi drivers and worried families, there is recognition and thanks. For Homero personally the greatest satisfactions come in his continuing private visitations with the butterflies, sea turtles, and gray whales: "Perhaps you are alive because we did a little in your defense."

It is evening in Mexico City. Homero stands and points with pride at the drawings of a monarch and a turtle given to him by his late friend the acclaimed Mexican artist Rufino Tamayo. He walks past the ex voto paintings framed in the entryway, up a set of narrow stairs, beyond a small room cluttered with filing cabinets and papers, and into his study, with its panoramic view of Mexico City. He sits down at the spacious desk where he writes, and continues:

"William Butler Yeats said, 'In dreams begin responsibilities.' Here I am, a man born in the mountains. I don't even know how to swim. Yet I love the gray whales. It is very strange. With these whales, I know there is no communication, at least in the normal ways with human beings.

Maybe, it can be supersensory or metaphysical, I don't know. But for me, to defend these whales is a dream. And a big responsibility.

"You know, there are no more tyrannical dreams than your own. You can't betray your dreams. You are not paid to have them, and nobody tells you to defend them. When people attack us, they say that Homero Aridjis wanted to be the minister of environment, so I am bitter against the government. Or that I am getting $10 million from the Australian marine salt companies and becoming rich. They don't understand that this is a moral commitment, that it is reason enough to be fighting for these creatures, that when you are involved you are a part of the butterfly and the gray whale. You are a part of the universe of life."

The more I was in the presence of Homero Aridjis, the more I felt that this birdlike man with the tousled hair might well be the living conscience of Mexico. He opens a door onto a balcony where a tree extends its branches toward him. "It's full of fruits for the birds. And it's one of my privileges. Every morning I can walk outside—and touch the treetop."

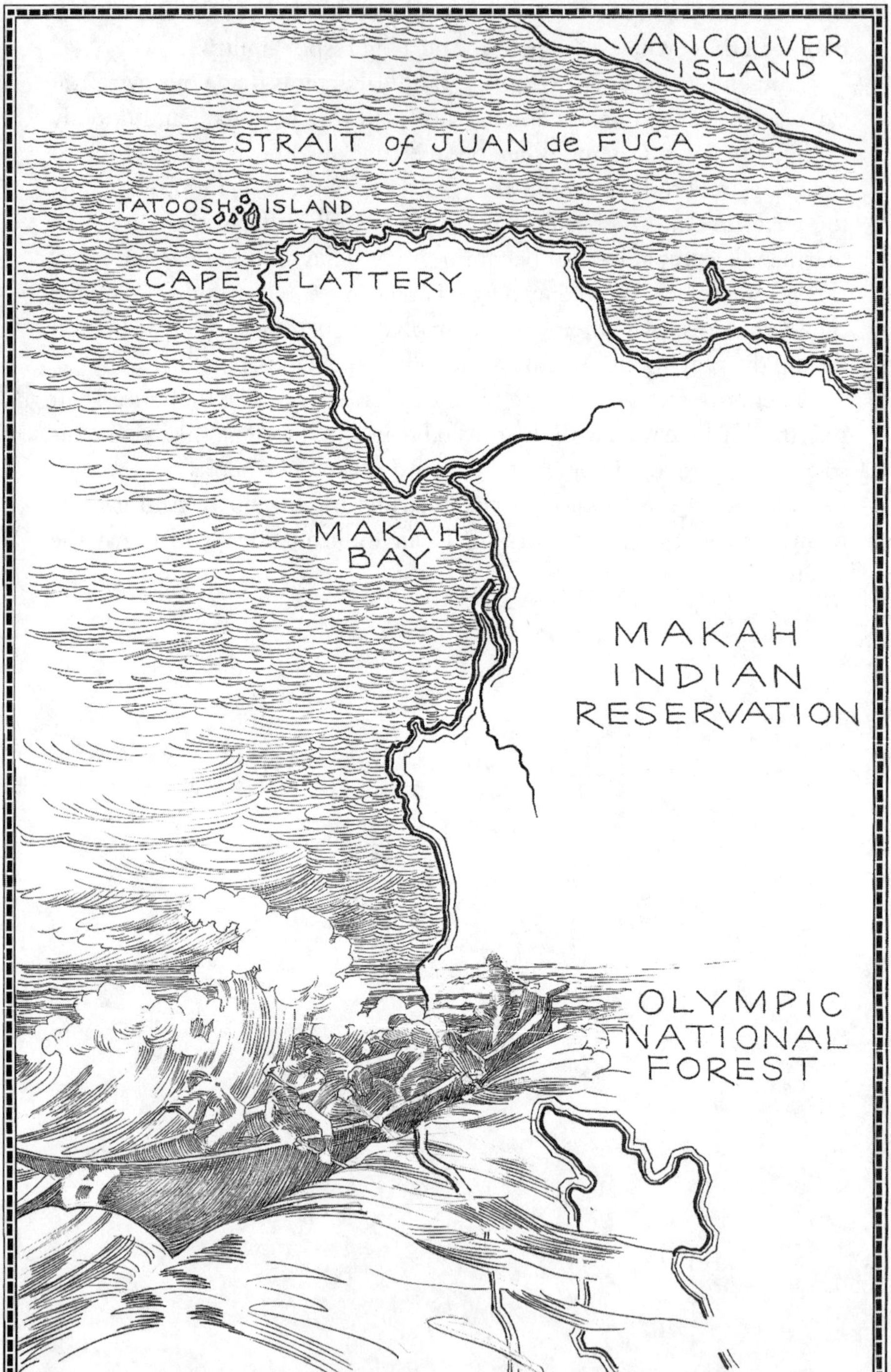
VANCOUVER ISLAND
STRAIT of JUAN de FUCA
TATOOSH ISLAND
CAPE FLATTERY
MAKAH BAY
MAKAH INDIAN RESERVATION
OLYMPIC NATIONAL FOREST

Chapter 4

The Makah Tribe: Hunting the Gray Whale

After evading the civilized whaler and his instruments of destruction, and perhaps while they are suffering from wounds received in their southern haunts, these migratory animals begin their northern journey. The mother, with her young grown to half the size of maturity, but wanting in strength, makes the best of her way along the shores, avoiding the rough sea by passing between or near the rocks and islets that stud the points and capes. But scarcely have the poor creatures quitted their southern homes before they are surprised by the Indians about the Strait of Juan de Fuca, Vancouver and Queen Charlotte's Islands. Like enemies in ambush, these glide in canoes from island, bluff, or bay, rushing upon their prey with whoop and yell, launching their instruments of torture, and like hounds worrying the last life-blood from their vitals. The capture having been effected, trains of canoes tow the prize to shore in triumph.

—Charles Melville Scammon, 1874

The proposed saltworks at the San Ignacio Lagoon was not the only controversy surrounding the gray whale as the millennium approached. With the sanction of the U.S. government, based upon a treaty written in the mid-nineteenth century, the Native American Makah Tribe of Washington's Cape Flattery region announced preparations to resume hunting of gray whales—for the first time in at least seventy years. "Many of our Tribal members feel that our health problems result from the loss of our traditional sea food and sea mammal diet," the Makah explained in a statement. "We also believe that the problems which are troubling our young people stem from lack of discipline and pride and we hope that the restoration of whaling will help to restore that discipline and pride. But we also want to fulfill the legacy of our forefathers and restore a part of our culture which was taken from us."

This outraged many animal protection groups. The Humane Society and several allies sought to stop the hunt through the courts. The Sea Shepherd Conservation Society vowed to place its vessels between the Makah canoe and the whales. In the autumn of 1998, amid news accounts that the hunt was imminent, I journeyed to the remote forty-seven-square-mile reservation where about 1,750 Makah reside in the village of Neah Bay.

From the Seattle airport it's about a five-hour drive to the northwestern tip of the contiguous United States. You first take a ferry across Puget Sound, soon cross a long bridge onto the Olympic Peninsula, then continue north through farmlands to the mill town of Port Angeles. Along the winding, two-lane State Highway 112, you pass through once-spectacular forest now scarred by acres of clear-cuts. It was almost dusk when I reached the coastal sportfishing town of Sekiu, still eighteen miles from Neah Bay, and checked into the Bay Motel.

The next morning I headed on toward the reservation, crossing where the Hoko and Sekiu Rivers run into the Pacific. Canada's Vancouver Island was visible through the mist, ten miles across the Strait of Juan de Fuca. I hugged the curves that wound along the shoreline, past a series of inlets and small bays: Shipwreck Point, Sail Rock, Seal Rock. Steep hillsides brought tall cedars, hemlock, spruce, and Douglas fir down almost to the crest of the road, where ferns wrapped around their roots. The vistas were spectacular. Now I knew why this region was often called America's rain forest. I passed a sign pointing to SNOW CREEK RESORT: THE FUN PLACE. It looked to be abandoned.

Five miles farther a white sign beckoning entry to NEAH BAY—THE MAKAH NATION is crowned by an image of the tribe's mythical Thunderbird, with its formidable beak and broad wingspan, holding a whale in its talons. The reservation's 27,000 acres lie in a valley enclosed by sea and mountains. Twin peaks called Bahokus and Archawat, both of them half clear-cut, enshroud Neah Bay. Until this road was built in 1931, the only access to the village was from the ocean. Scammon wrote in 1871: "It is a snug haven, where ships find shelter behind a pretty island called Wa-dah. The showy, white buildings of the Indian Reservation join the aboriginal village, which is comprised of low and old structures, covering a large ground space, with nothing in their exterior appearance in the least inviting."

I drove by the modernistic Makah Cultural and Research Center, a large concrete-and-wooden building with a pitched roof. It houses a tribal museum hosting a collection of some fifty thousand artifacts retrieved since 1970 from a nearby archaeological site called Ozette, which was once among five Makah villages. I continued into town past Washburn's General Store, a VFW hall, the high school and senior center, the Apostolic Faith Church, the Cedar Shack Espresso and Makah Maiden cafés, the Big Salmon and Tyee Motels. Only the main street was paved. The others were lined with rows of dilapidated houses and trailers, a few displaying satellite TV dishes above overgrown front yards cluttered with old furniture and rusty appliances.

It's three hours from here to the closest shopping mall, and ninety minutes to the nearest movie theater. Neah Bay remains predominantly a fishing village, but a new tribal marina containing slips for about two hundred boats was far from filled. I'd read that the prices of tuna, salmon, and halibut had fallen so low that many of the tribe's fishermen couldn't afford the gas and had sold their vessels. Unemployment on the reservation averages between 55 and 75 percent. The Community Food Bank serves five hundred people, almost one-third of the population.

Today there was little activity in the harbor. Several TV media satellite trucks sat idly in motel and café parking lots. The tribal whaling canoe tied up to the dock was a decoy, I'd heard; four other canoes were rumored to be hidden in coves somewhere along the coastline. I could make out one of the Sea Shepherd protest vessels anchored well offshore, around Waadah Island, near the entrance to the bay. The whale hunt had been permissible as of October 1, but almost a full month had gone by without a move from the tribe. State authorities had been anticipating trouble for a while. Worried about demonstrators trying to bust up the annual Makah Days celebration late in August, Washington's governor, Gary Locke, had stationed eight hundred National Guardsmen for three days—but nothing had happened. The day before my arrival had seen the most confrontational situation so far. About thirty protesters tried to enter the reservation but were stopped by tribal police. As word spread about half the whaling crew drove out to exchange words. Meantime, here at the Neah Bay marina, anti-whalers in ten inflatable craft traded diatribes with the Makah. It seemed only a matter of time before things escalated further.

Captain Scammon had probably been the first outsider to bring the Makah to the attention of a wider audience. In an article for the *Overland*

Monthly in 1871, he wrote: "They are a hardy band, inhabiting a wild, broken peninsula. . . . The Makah men esteem themselves far superior to those of the tribes of the interior, who follow the tame life of fishing in the shaded estuaries and babbling brooklets, or shooting game within the coverts of the forest; 'for,' say they, 'we go far out on the ocean to capture the huge monsters of the deep in our great canoes, while the "salmon-eaters" catch their fish with the *kloochmen*, and, what is more, they dare not fight us.' "

Although Scammon probably became aware of the Makah during his whaling days, it was while serving with the U.S. Revenue Marine during the Civil War that he came into direct contact with the tribe. For six months in 1864 the *Shubrick*, which Scammon commanded, was stationed off Puget Sound. It was the only revenue cutter on duty in the Northwest Territories, charged with checking ships' cargoes, apprehending revenue violators, and rescuing vessels in distress. More than once Scammon was called to Makah territory. A newspaper account from 1864 described him placing into custody an Indian charged with threat-

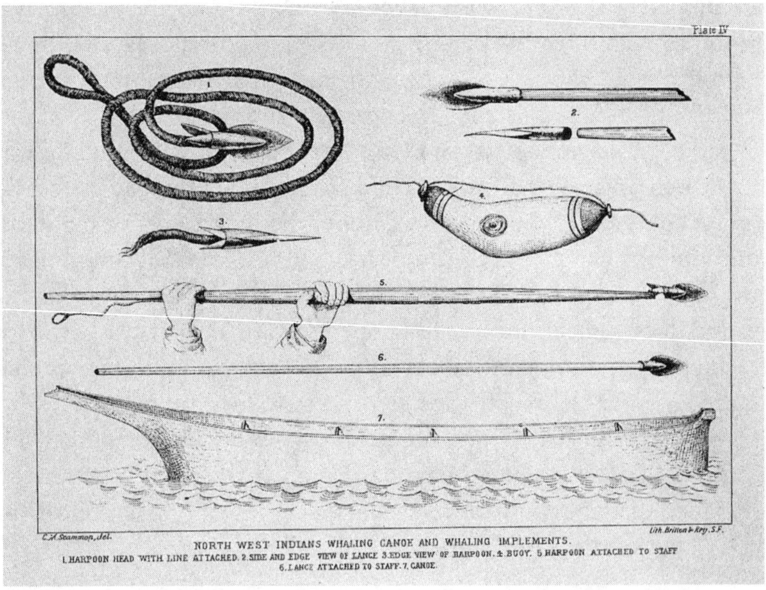

Scammon drawing of nineteenth-century Makah whaling gear

ening the life of a lighthouse keeper on Tatoosh Island. Later reports told of Scammon's rescue in 1870 of a bark that had struck a rock at the mouth of the strait and run aground in Neah Bay. The vessel would have been lost without "the untiring efforts" of Scammon's revenue steamer *Lincoln.* "Capt. Scammon," one newspaper said, "is eulogized very highly."

Scammon's fascination with the Makah is also evinced in other articles he pasted into his scrapbooks: a race between Makah and Clallam canoes in 1868, a lengthy piece about the tribe by a *San Francisco Chronicle* correspondent in 1873. Scammon personally visited the reservation and, judging by the account he provided in his chapter about the California gray whale in *Marine Mammals*, bore witness to Makah whaling on more than one occasion: "The whalemen among the Indians of the Northwest Coast are those who delight in the height of adventure, and who are ambitious of acquiring the greatest reputation among their fellows. Those among them who could boast of killing a whale, formerly had the most exalted mark of honor conferred upon them by a cut across the nose; but this custom is no longer observed."

The Cape People, as they became known, had originally split off from Nootka relatives on Vancouver Island. For at least fifteen hundred years, whaling had been a centerpiece of Makah life and supplied a considerable portion of the food. By early in the nineteenth century, the Makah were kingpins of a vast trading network in whale oil and blubber, which extended from Cape Flattery north to British Columbia's Nootka Sound and south to Oregon's Columbia River. Starting in 1844 and continuing for some years, Yankee whaling ships made annual voyages to Neah Bay. The Makah were producing as much as thirty thousand gallons of whale oil per year. A number of Makah were hired on as harpooners by the industrial fleet.

Then, in 1852, smallpox carried by their visitors resulted in an epidemic that decimated the tribe. "It was truly shocking," a white nineteenth-century trader named Samuel Hancock wrote in his journal. "In a few weeks from the introduction of the disease, hundreds of natives became victims to it, the beach for a distance of eight miles was literally strewn with the dead bodies of these people." (A precontact Makah population estimated as high as 4,000 would fall to 360 by 1910.)

Three years after the epidemic Washington's territorial governor, Isaac Stevens, came to negotiate the Treaty of Neah Bay—which would be cited nearly a century and a half later as justification for the tribe's resumption of whaling. The head chief in 1855 was Tsekauwtl (pronounced "Sakowit"), the son of a slave. For centuries the Makah had held slaves taken either as tribute from vanquished tribes or later by purchase from their neighbors on Vancouver Island. Still, Tsekauwtl's prowess in whaling had resulted in his ascension to leadership. His name, in fact, meant "gray whale" in the Makah language.

Aboard Governor Stevens's schooner the chief proclaimed that "he wanted the sea—that was his country." After twice having departed without reaching an agreement, finally Stevens told the tribe that the Great Father "knows what whalers you are, how you go far to sea, to take whales. He will send you barrels in which to put your oil, kettles to fry it out, lines and implements to fish with." The Makah were guaranteed the right to continue whaling. They gave the governor a white flag of peace and signed a treaty which ceded away thousands of acres of forestland to the U.S. government. The tribe also agreed to cease their practice of slavery but didn't until Lincoln's Emancipation Proclamation eight years later.

The U.S. Congress ratified the treaty, but the "Great Father's" promised maritime supplies never arrived. Instead, government agents continued to press agriculture as the most desirable means of subsistence in Neah Bay. In 1863 an Indian Agency teacher named James G. Swan noted that the Makah "from the first opposed all improvement, and have thrown every obstacle in the way to prevent the young from working the land, jeering and ridiculing them for digging the earth like squaws, instead of going in their canoes and attacking whales, or catching halibut and salmon." The Makah leadership, looking to hold the government to its agreement, wanted to trade their cedar canoes and yew-wood spears for harpoon cannons and whaling steamships. This never happened either. But the Makah soon found a lucrative substitute: pursuit of the fur seal.

"It is but a few years since the Indians have turned their attention to taking seals solely to procure their skins and oil for barter," Scammon wrote in his chapter on fur seals. "And what may seem surprising, it is but a few years since the animals have been known to resort to the vicinity of the strait in such large numbers. We have it from the most reliable source, that there were but a few dozens of Fur Seal skins taken annually by the Indians, from 1843 to 1864; after which period, the number of skins sold by them at Victoria, Vancouver Island, Nee-ah Bay, and points on Puget Sound, has steadily increased, up to 1869, when the number in the aggregate amounted to fully five thousand skins."

White schooner owners in 1869 began hiring Makah tribal members to bring their canoes along on fur seal hunts to the Bering Sea. Wages depended on the quantity and quality of what they caught, but the Makah hunters received two-thirds of the catch and often earned excellent salaries. By 1880 one Makah had invested his earnings in a small schooner. Within thirteen years the tribe owned ten twenty-five- to fifty-ton craft, capable of carrying as many as twenty canoes on their decks and following the seal herds from California to the Pribilof Islands in Alaska. One head chief, Peter Brown, came to own three schooners and was worth a reputed hundred thousand dollars. On paydays, one Makah would recall years later, "seal hunters had to pack their money home in satchels because their pockets couldn't hold it all."

At the same time the fur seal population was declining rapidly, just as the gray whales had a couple of decades earlier. In peak years more than one hundred thousand sealskins were being taken. Then, in one of the federal government's earliest conservation actions, it imposed severe limitations on the sealing trade. In the summer of 1889, a Revenue Marine cutter spotted a schooner taking seals off the Pribilofs, violating a statute that forbade this activity in Alaskan waters. Authorities seized the vessel, which turned out to be owned and operated by a twenty-one-year-old Makah. The tribe decided to file suit. Their attorney argued that the Treaty of 1855 had been violated. He cited Article 4, which contained this clause: "The right of taking fish and of whaling or sealing at usual and accustomed grounds and stations is further secured to said Indians in common with all citizens of the United States."

Did the Makah's treaty supersede agreements made between the United States and foreign governments? That was the heart of the matter, as it would be again a century later. In 1892 Judge Cornelius H. Hanford ruled that the treaty assured the Makah "rights in common with all citizens of the United States, [but] certainly such treaty stipulations give no support to a claim for peculiar or superior rights or privileges denied to citizens of the country in general." The Makah were ordered to forfeit ownership of the schooner. When the tribe persisted in hunting seals, customs officials seized two more of their schooners in 1894 and 1896.

By the turn of the century the Makah whaling traditions were practically nonexistent. Today tribal spokespeople maintain that their ancestors kept whaling until 1926, when they stopped voluntarily as a conservation gesture because of the scarcity of gray whales. That, I soon discovered, was not the way one tribal elder remembers it.

•

At the edge of town, Charles (Pug) Claplanhoo lives in a little wooden house looking out on the sea about fifty feet from his front door. Underneath the floorboards somewhere, he says, is an old Spanish fort. "The Makah drove 'em off, late 1700s," according to Claplanhoo. "Really some tough Indians here then, I guess."

The Makah elder with close-cropped white hair is wearing a T-shirt and overalls. He's retired on a veteran's disability pension—"Korean War got me." In a corner of his living room near an old wood-burning stove is a carving bench, with wood shavings scattered all around the beginnings of a hollowed-out canoe. Along one wall is a tapestry of an elk. A couple of printed plaques are on another wall. One of these says, "Committee: a group of the unfit appointed by the unwitting to do the unnecessary." Another reads: "3 kinds of people; those who watch things happen, those who make things happen, and those who wonder what happened."

Judging by the family heirlooms around the room, Claplanhoo doesn't have to wonder what happened. He reaches out a weathered hand and picks up a wooden harpoon over six feet in length. "That's only

Charles Claplanhoo in Neah Bay (George Peper)

part of it," he says, "there's two feet missing. It's made out of yew wood, that's why it's so heavy. This is the business end"—he touches the metal tip—"it done the job. This harpoon's two hundred years old or more, passed down through my family. Look at all the notches in it. I counted 142, one for every whale."

Claplanhoo holds up a photograph of his grandfather, Jongi, the man he says harpooned the last whale in Neah Bay in 1909, "a big gray, for this Pacific Exposition in Seattle." There are other pictures of towing the whale onto the beach in front of this same house. "My grandfather said it took four or five harpoons to kill that whale. Sometimes they went way out in the ocean, ten to twenty miles out, and then came all the way back again. His father was Maquinnah, last of the real chiefs here and last of the schooner skippers too."

With great care and pride Claplanhoo brings forth a cedar-bark basket that held the seal-gut-fashioned harpoon line, and a long, pointed stick used to stab into the whale's blowhole.

"That over there is a wolf club," he points and continues. "Look at all *those* notches."

"They used to kill wolves with these?" I ask, examining some kind of totem above the many circular carved markings.

"People," he replies with a laugh. "They fought quite a few different tribes down here."

I explain I'd been hearing that the Makah had given up gray whaling during the late 1920s, to help protect the few that remained. Claplanhoo shakes his head. "No, I know when it was. The government shut 'em down, like they did with the sea otters and the fur seals. Grampa told me about it. They were given permission to get that last one, to put on display at the exposition."

The elder adds that both his ancestral harpoon and "stick" had disappeared from his house a couple of years back. "I don't usually lock my door. Somebody from the tribal council came in and took it, but nobody told me. I didn't get it back for a long time. Guess they went all the way overseas with it to one of those International Whaling Commission meetings, to try to claim that their family done the whale hunting. They didn't, though."

Does Claplanhoo believe that reviving the tradition is the right thing to do?

"No, I don't. Because it would be just a waste of good meat really. Nobody knows how to cook it anymore. A gray whale got caught in a fishing net here a few years ago and washed up onshore. Everybody came down and hoo-hahed about eating it, but it went to waste. It all ended up at the dump. No, the whaling died off a long time ago."

"Then why is this all happening?" I ask.

"Just a few people, they're in it for the money. And the crew gets a pretty good chunk from the tribe for fooling around out there, making like whalers. They claim to be descendants of whaling families, but I know otherwise. I say, let's move on, really take care of the tribe. If they fought for our fishing rights like they're fighting for this whale, maybe there would be more jobs. Leave the whales alone. They're not bothering nobody. Lotta people around here feel this way, especially the older ones of us. But most of them are scared to talk about it."

Placing his ancestor's harpoon back in a corner, Charles (Pug) Claplanhoo walks over to his bench, picks up a knife, and resumes carving his canoe.

Was this simply a matter of an isolated Pacific Northwest tribe seeking to revive an ancient tradition? Or might the Makah hunt have wider ramifications, perhaps opening the door to nations like Japan, which have long been pushing to resume commercial whaling? This was the fear of many opponents to the tribe's plans.

Even after the commercial moratorium took effect in 1986, the International Whaling Commission (IWC) continued to allow a small annual quota of whales to be set aside for aboriginal communities that could demonstrate a "traditional dependence" upon whale meat for cultural, nutritional, and subsistence purposes. For the most part this meant the isolated native peoples along the Siberian and Alaskan coastlines who'd long hunted gray and bowhead whales. And this would be the proviso the Makah would eventually seek to exploit.

Gary Ray was the first to raise the possibility of a hunt in 1988, while a member of the Makah Tribal Council. When I went to see him at his ranch-style home on the Sooes River about five miles outside Neah Bay, Ray, now middle-aged, described himself as "a young man born and raised out here, a woodsman, did construction, concerned about our people and where we're going." He explained that not until the late 1970s had the tribe begun to "get some fingers" into management of their own resources. Before that, according to Ray, the Bureau of Indian Affairs "pretty much dictated everything." That included selling Crown Zellerbach (now Crown Pacific) logging rights to half the reservation. Then, under a new "self-governance grant," the Makah began peddling their own timber. According to a 1995 report by the Pacific Crest Biodiversity Project, their last clear-cut of 1.7 million board feet of spruce and hemlock forests had netted the Makah $1.9 million; the money had been intended for schools and community programs, but somehow the funds never materialized. Ray maintains that the tribe was forced to do extensive logging in order to pay the legal fees that finally won back their fishing rights.

At any rate Ray, then a council member, was working at the fish dock his sister managed in 1988 and pondering the tribal treaty. "Here we were in court, fighting against six Columbia River tribes over the chinook salmon that are right out our front door. But whaling was different. We were the only ones with that right guaranteed by the government. So what would happen if we launched a canoe tomorrow and got a whale? Would they throw us in jail?"

Ray began doing some research. The excavations recently uncovered at the Ozette archaeological site showed that the Makah had once harvested five different species of whales, including the gray. Then a huge gray whale washed ashore on the tribal beach in the spring of 1988. Ray salvaged the skull and some of the bones. "I took this as a sign from the creator that it's time for the Makahs to go whaling again. It wasn't really my idea, it was a teaching passed on by the elders. The Makahs at one time sold enough whale oil to light the city of Seattle. That's how industrious this tribe used to be."

So Ray became part of a team that "started knocking on the doors of politicians. It's custom as we go places to offer gifts as a people. So we'd travel to Washington, D.C., and give key chains and necklaces—but these had an emblem of a whale with a harpoon going through it, and the word *Makah*. 'Senator, do you remember that the Makahs are whalers?' A lot of it was education."

When I asked Ray if he was aware of any impetus from the Japanese, he replied that he'd left the tribal council in 1990, "and so I can't answer that one, I was outside being privy to tribal government where an offer may have come up. I know the Japanese have always been very interested because of our geographical location and our fishing resource; nowhere in the world can they get quality fish like ours."

The Makah did indeed have a fisheries contract with a Seattle-based company, Supreme Alaska Seafoods, which owned a fleet of factory trawlers. The co-owner of Supreme Alaska was Maruha, a Japanese multinational also known as the Taiyo Fisheries Company. The company owned one-third of the Japanese whaling fleet and handled distribution of the whale meat sold to subsidize Japan's "research" whale hunts. In 1997 nearly two thousand tons of whale meat from Japan's "research" catch wholesaled through Tokyo for 3.5 billion yen, $26.5 million.

Another source had told me that, in 1991, the Makah's fisheries director was first approached by a Japanese representative of Supreme Alaska. Japan had already trained thirty tribal members in scuba diving and the basics of harvesting sea urchins, whose roe is prized by sushi chefs (at peak prices, one Makah was said to be making as much as two thousand dollars a day). The fisheries director was informed that certain Japanese were aware of the tribe's treaty right and were most interested in the Makah's rekindling their whaling tradition. Indeed, these interests could help with some funding toward such an end. But the first step was to get the California gray whale removed from its designation under the U.S. Endangered Species Act. A meeting was arranged between some Japanese and the Makah Tribal Council to discuss this matter further.

Ray said it was indeed the Makah who started pushing to have the gray whale taken off the Endangered Species List. "A lot of people laughed when we first asked about it, even in-house among some of the Northwest Indian Fisheries commissioners." Late in 1991, however, that commission—of which the Makah are members—filed a formal petition with the federal government. The next year the gray whale became the first marine mammal to be down-listed to threatened status. By 1995 the Department of the Interior deemed the gray whale population to have recovered fully enough to be removed from that category as well.

That's when the Makah made their move. In the spring of 1995 a tribal attorney named Terry Wright brought up a plan at a scientific review committee meeting of the National Marine Fisheries Service (NMFS). According to an NMFS memorandum obtained under the Freedom of Information Act, this is what Wright outlined: "The Makah intend to harvest gray whales (starting in 1995), harbor seals (five already taken), California sea lions, minke whales, small cetaceans such as harbor porpoise and Dall's porpoise, and, potentially in the future, sea otters. The Makah are planning to operate a processing plant so as to sell to markets outside the U.S. The Makah have started discussions with Japan and Norway about selling their whale products to both countries. The plant could be used to process the catches of other tribes as well."

Meanwhile Japan was continuing to push the IWC at every annual meeting to loosen the global whaling moratorium. Its claim, in looking to resume commercial taking of minke whales, was that this was an "integral part of the culture" of several of its small coastal communities. Now the Makah were similarly claiming that the 1855 Treaty of Neah Bay guaranteed them a "cultural need," as well as the right to sell whale products.

Ray recalled officials from the NMFS coming to Neah Bay for a private meeting. "They explained they were in a tight spot," in that the United States provided the teeth behind the IWC's efforts to keep a lid on commercial whaling. Any sale of whale products was simply not going to wash. "Okay, we realize we're in a global world now. By right we didn't *have* to, but we'd go ahead and walk those miles with our government, because of the sensitivity of the whole issue."

On May 8, 1995, Hubert Markishtum of the Makah Tribal Council wrote in a letter to the NMFS: "It should be emphasized, however, that we continue to strongly believe that we have a right under the Treaty of Neah Bay to harvest whales not only for ceremonial and subsistence but also for commercial purposes."

How Markishtum and others got this idea is unclear, since in Article 13 of the same treaty, the Makah "agree[d] not to trade at Vancouver's Island or elsewhere outside the dominions of the United States." Clearly, there is no market for whale products *inside* the United States. At any rate, the NMFS undertook an internal study to evaluate whether the tribe's resumption of whaling might be valid under the aboriginal subsistence clause of the IWC. The NMFS evaluation was finished in August 1995 and identified several potential impacts, notably including sending a signal to "several tribes along the coast of Washington and British Columbia that have indicated a similar interest in obtaining a ceremonial and subsistence quota, and [who] look to the Makah to set the precedent on an international level."

Nevertheless, in March 1996 the NMFS approved an agreement with the Makah Tribal Council to "make a formal proposal to the IWC for a quota of gray whales for subsistence and ceremonial use." The National Oceanic and Atmospheric Administration, NMFS's parent agency within the Commerce Department, gave the tribe $200,000 in start-up funding to establish the Makah Whaling Commission and to begin making preparations.

That autumn of 1996, at the annual IWC meeting in Aberdeen, Scotland, the U.S. delegation presented the proposal, with the assurance that the tribe "does not desire nor has it requested a quota for commercial use." Marcy Parker, an elected official of the Makah Tribal Council, pleaded to the delegates that "even though we haven't hunted the whales on our ocean in seventy years, we have hunted the whales in our hearts and in our minds." According to the IWC's report on the meeting, Japan "commended the USA's presentation and expressed understanding of the welfare of the Makah."

But when an IWC subcommittee considered the matter, several nations questioned whether the Makah situation met the definition of subsistence. Had they really maintained "a continuing traditional dependence on whaling and on the use of whales," as the IWC's guidelines set forth? Back in the United States, two congressmen—Washington's Jack Metcalf and California's George Miller—prodded the House Committee on Resources to pass a unanimous resolution opposing the Makah whaling initiative and the federal agencies' decision to support it. The next day the United States' IWC delegation withdrew the proposal, since it was clear the required three-quarters vote of approval would not be forthcoming.

At the same IWC meeting several Makah elders presented their own petition *against* the resumption of whaling. "[We] think the word *subsistence* is the wrong thing to say when our people haven't used or had whale meat/blubber since the early 1900s," they wrote. "We believe the hunt is only for money." An elder stated that one tribal executive had told the Makah leadership that a single gray whale could fetch half a million dollars from the Japanese. Responding to the backlash, Ben Johnson, then president of the Makah Whaling Commission, was quoted in *The Seattle Times*: "Japan wanted to give us money, to help us buy boats, to show us how to kill the whales, everything. We said no because we knew it would be very controversial, and we want to do everything by the book."

By the book then, the tribe and the federal agencies started developing an alternative strategy. Why not combine the Makah proposal with the annual request by Russian native peoples for their gray whale quota? After all, the IWC had already recognized the Russians' legitimate subsistence needs. And this maneuver would preclude a specific IWC determination on the Makah hunt.

As the 1997 IWC meeting approached, several animal protection organizations began moving ahead with a federal lawsuit aimed at heading off the hunt. If allowed, their lawyers contended, this would be the first whale hunting ever to take place within a designated national marine sanctuary. Indeed, the government was "run[ning] roughshod over federal statutes designed to protect marine resources." The NMFS hadn't even bothered to develop an environmental assessment (EA), as the law required, to determine whether a more thorough environmental impact study needed to happen. So the NMFS finally drafted an EA for other agencies to analyze. The U.S. Marine Mammal Commission commented that the document "suffers from superficial analysis on key issues," although the EA did concede possible negative impacts on the National Marine Sanctuary Program and on local whale-watching operations.

The day before releasing the EA, an NMFS official reported: "The White House confirmed today that the U.S. will pursue a quota for the Makah at the IWC this year." As Elliott Diringer of the President's Council on Environmental Quality later told a reporter, Vice President Gore was being "kept apprised" of all developments. And the issue was deemed sensitive enough for James Baker, a commerce undersecretary and the U.S. whaling commissioner to the IWC, to obtain White House approval before proceeding. One source maintains that "one hundred percent of the U.S. decision to back these guys [the Makah] was based on the U.S. not wanting to be in court" over the treaty rights issue. The Clinton administration feared the Makah claim might open up a situation whereby other Native American treaties could be interpreted as superseding domestic or international agreements.

In the months leading up to the 1997 IWC meeting, the Makah used their government funding—which would eventually reach $335,000—to hire a public relations consultant and a nutritionist to study the importance of whale blubber to the Makah diet. Gary Ray and others traveled to Germany to lobby its IWC commissioner on the proposal. The Makah invited IWC delegates from Mexico and New Zealand, which had opposed the plan, to visit their reservation. "We wined and dined them, showed them who we are," according to Gary Ray.

One day before the 1997 IWC meeting, the NMFS issued a final environmental assessment, which admitted that a harvest of gray whales by the tribe "would be an extremely contentious event" that could have "far-reaching consequences." These included possible strikes of a large number of gray whales without their recovery and possible taking of resident rather than migratory whales. However, the document continued, even "if all the whales landed by the Makah were summer residents, new whales are likely to appear to take their place." The EA also noted "it is entirely possible that additional native groups may start whaling" and that federal agency support "would send the signal that it is possible to carry on traditional whaling." Nonetheless, the effects of the proposed hunt were not seen as "significant," and thus no environmental impact study was required. "The only way to determine positively whether

Makah whaling does qualify [as subsistence whaling] is to obtain a definitive ruling from the IWC itself," the Fisheries Service concluded.

The IWC session convened that October in Monaco. A U.S. spokesman went so far as to say, "More than a third of their people live in poverty and half are unemployed. Their per capita income is less than half of the U.S. national average. The gray whale will help meet the very real nutritional needs on the Makah reservation." The United States then entered a joint proposal with the Russian Federation. This was to allow the killing of 124 gray whales per year by native groups in the North Pacific. That actually represented a decrease—the former annual quota was 140—but now the Makah would be added to the gray whale hunting roster. Over the next five years the Native American tribe could take twenty whales, no more than four in any given year.

The merger plan still didn't sit well among a number of IWC members. Delegates from thirteen countries rose to express doubts about the Makah's having a legitimate subsistence need to take whales while recognizing that the Russian natives' need had been clear for many years. The Australian representative strongly maintained that the IWC shouldn't be asked "to pick up the debris for the demolition of the United States welfare system or to compensate for poor timber practices, or indeed a statement which attempts to persuade us that the Makah have traditionally consumed whale meat . . . when our understanding is that this certainly has not been the case since the last hunt which took place some seventy odd years ago." The delegate added that he wasn't convinced that "purely non-commercial activities . . . would flow as a result of this."

For his part, Gary Ray believes that the vocal opposition from Australia and New Zealand emanated from the fact that "they don't want to deal with the rights of First Peoples in their own lands, the Aborigines and the Maoris. So I see the Makah pricking the conscience of those governments. Some people's hearts are closed, but we've seen a lot open."

At the IWC gathering, however, only five of the member nations openly voiced sentiments in favor of the U.S. request on the Makah's behalf. These included Japan, whose whaling commissioner hastened to "stress that the Japanese coastal small-type operations and activities are far more historically cultured and economically important in that traditional sense compared to that in Makah."

As the gathering wound toward a conclusion, Australia insisted on inserting a clause into the proposed amendment: that the gray whale quota be limited to those native groups "whose traditional aboriginal subsistence needs have been recognized" by the IWC. That wouldn't *seem* to include the Makah, but the IWC member countries faced a dilemma: they didn't want to cut off the Russian subsistence hunting, and it *was* a joint proposal. So in the end no formal vote was taken. As the Chairman's Report of the Forty-ninth Annual Meeting put it, "A broad consensus was reached to accept the amendment."

Immediately the NMFS issued a press release announcing the Makah hunt's authorization, in that the IWC had "indicat[ed] its acceptance of the United States' position that the Makah Tribe's cultural and subsistence needs are consistent with those historically recognized by the IWC." Well, had they? The press release failed to mention the additional clause about previously recognized subsistence needs. The United States simply unilaterally concluded that the IWC had given its tacit approval. Australia responded by saying it rejected this claim as false.

In response to a query about the imbroglio, Dr. Ray Gambell, secretary to the IWC at its Cambridge, England, headquarters, wrote on October 5, 1998: "The IWC has specifically not passed a judgement on recognizing or otherwise the claim by the Makah Tribe, since the member nations were clearly unable to agree." Today the official IWC Web site continues to list the home regions of aboriginal groups it recognizes as having claims to take whales. The Makah are omitted from that list.

The 1997 events had coincided with the establishment, that February, of a new World Council of Whalers. This organization arose after several years of conferences on community-based whaling, the first of which had taken place in Kyoto, in 1993. The last such conference, held in Berkeley, California, had included several Makah among its participants. And it was the Nuu-Chah-Nulth Tribal Council, located on Vancouver Island just across the strait from the Makah reservation, which agreed to provide a home for the secretariat of a global whalers' organization. In a news release the council had declared that aboriginal groups around the world—along with coastal communities in Japan and Norway—have the right to hunt whales as well as to trade and sell the meat and blubber. After the council's March 1998 meeting in Victoria, British Columbia, a large contingent—including whalers from Japan and Norway—traveled across the water to congratulate the Makah on having taken the critical first step.

John McCarty had been the first chairman of the Makah Whaling Commission, and his son Micah had been expected to throw the first harpoon. But I'd heard rumors that the father had been forced out of his position, and the son had abandoned the whaling crew and headed off to college. I asked at the dock where I might find John and was directed back down the road to the Makah housing complex near the entrance to the reservation. He and his wife lived there in a double-wide trailer home.

John McCarty, a mild-mannered gentleman in his mid-sixties wearing red suspenders over a plaid shirt, was fixing a door on his front porch when I arrived. I decided to get right to the point. "I heard your family pulled away from the whale hunt because you don't feel it's being done in the traditional way."

"That's right. Absolutely," McCarty said, and invited me inside.

We sat down at a coffee table. "I'll go back to the beginning," he said. The beginning, for McCarty, was a family whaling tradition similar to Charles Claplanhoo's. Both of McCarty's grandfathers had been whale hunters, and his first and middle names (John and Arthur) derived from the two of them. Before the Bureau of Indian Affairs came and Americanized them, Grandfather John's Makah name was Hishka. The Makah name, passed down through many generations, meant "he who makes the whale blow on the beach." And he was considered among the last of the great ones.

Now the grandson was saying, "One time I heard this story, about a whale that got away. They were out in the middle of the strait somewhere and had three or four harpoons in this big gray. I don't know how long Hishka fought him, but it must have been a long time, because they really took care of their whaling gear, and I know he wouldn't want to lose it. But that whale was too big and tough, I guess. So finally he got away, and they couldn't find him and they came home.

"A day or two later, he heard about this other tribe who'd found a whale with some harpoons in him down the strait. So Hishka got in the canoe with a couple of guys and they paddled down there. He told the tribe—these were Clallams—you guys can keep the whale, I just want my harpoons back. The Clallams said no, we're gonna keep the whale and the harpoons too.

"So Grampa, he didn't say anything, he just turned around and went home. I don't know how many canoe loads of guys he got to come along this time. But he went back there and slaughtered 'em all. He got all his harpoons, and he brought the Clallam chief's head back on a stick, on the bow of the canoe. Yeah, that was Hishka. One fierce fellow."

McCarty continued, "Hishka happened to be there in the heydays, when whale oil was really precious. So he used to sell barrels and barrels of it. He became extremely wealthy. When he'd have what we call a potlatch, it'd last for seven days. And he'd be feeding hundreds for that entire time!"

His other grandfather, Arthur, had been "the one that used to dive down and sew the mouth of the whale shut, so he wouldn't get full of water and sink before they started towing him home." Then McCarty's father, Jerry, had maintained the family heritage and "seen the last whales hit the beach. He told me a story of how one time they harpooned this great big bull. They only secured one harpoon, and he towed 'em straight out to the ocean, clean into the darkness. They kept going and going, until they didn't know *how* far out they were. Anyway they had to cut the whale loose. Daylight come, and they could hardly see land, this high peak was just a little tip off the horizon. They had sails along with the paddles, but it took them all day to get back in."

So when John McCarty's cousin Dan Greene, the tribe's fisheries director, asked him in late 1995, "John, how would you like to go whaling?" McCarty remembered feeling, "Boy, that'd make a complete circle in my life." As a direct descendant of one of the last whaling families, he'd agreed to chair the tribe's Whaling Commission. On the fateful decision day McCarty was in attendance at the IWC meeting in Monaco.

"We knew if they took a vote, we'd lose," he recalled. "What happened was, we made a phone call to Washington, D.C., in the middle of the night and woke 'em up and told 'em they'd better get on the ball because they weren't helping us like they should. Next day, I was sitting in what they call the pigeonhole room, in front of these great big TV screens, and this old [IWC] chairman was reading the proposal out loud. Afterwards, he asked if any nations cared to comment further. Nobody said anything. So, bang, we got it by consensus. I looked at one of the tribal people next to me and said, 'What happened?' He said, 'They didn't vote against us. So we got it. We got the whale.' I said, 'You mean that's all it took?' "

Then, back on the reservation, things started to disintegrate. McCarty found himself being eased out of the picture, allegedly because he wasn't moving forward quickly enough. Another family began exerting control. Which would have been all right, in McCarty's view, except that the crew itself simply wasn't preparing properly. "That's why my son Micah quit, because he got so discouraged. They haven't practiced paddling in over a month. At one meeting the elders got up and said, 'You guys have not talked to us about how you should conduct yourselves as whalers, about your spiritual feelings.' Don't get me wrong, I'm still not against the whale hunt. It's just they haven't trained right."

•

Inside the Makah Cultural and Research Center, I sat in the library studying what had been set down about the whalers' preparations of old. On the road to acquiring *tumanos*, or power, first came the ritual bathing. Each morning at dawn the hunter went alone to a freshwater lake or pond. Entering the water, he sat down until his skin was well-soaked. Then he rubbed himself with bunches of hemlock twigs, first on the left side until the needles wore off and the bare twigs were stained with blood, then on the right side. Having used up four bunches, the candidate dove and stayed underwater as long as possible. Each of the four times he emerged, he would blow a mouthful of water toward the center of the lake, seeking to make a sound resembling the blow of a whale. Always his movements were quiet and slow. All through the winter he bathed only when the moon was waxing.

There was a curious practice of using skeletons, skulls, or corpses during the ceremonial bathing. Some say these were the bones of dead whale hunters, believed to bring power to the living. A whaler would take a skull from a burial place, tie it on a rope around his waist, and swim about "sounding" and "blowing" like a whale. In some cases skeletons from grave boxes were used. They would be carried on the whaler's back when he bathed. Occasionally fresh corpses were used in this way. The body had to be that of a male, dead not more than four days; sometimes a slave or even a small boy was sacrificed for the purpose. A pad of rosebushes and nettles was put on the whaler's back, and the corpse was placed back-to-back. "One time a man put a corpse on his back in preparation for bathing, face forward, and it took a death grip on his throat, and killed him," wrote whaling historian T. T. Waterman.

At night the whalers prayed in whispers for success, to the "Four Chiefs," to the sun, directly to the whales. Any contact with women was forbidden. In the weeks leading up to the hunt, a man and his wife slept separately. Some chiefs demanded even more arduous rites. A prospective whale hunter had to swim out to certain barnacle-covered rocks and reefs. There he would drag his naked body back and forth across the small, jagged shells.

The expedition would begin with the first new moon in May. The whaler in charge of the canoe wore hemlock twigs on his forehead and a bearskin robe as he put to sea.

From T. T. Waterman, *The Whaling Equipment of the Makah Indians*,

A Makah set to harpoon a gray whale, circa 1900 (courtesy Washington State Historical Society, Tacoma; by Curtis, Negative No. 56519)

1920: "On the high seas at night, the Makah steer by the Pole-star. They never get lost, even in a fog, for they are helped by the swells and the wind. In this part of the Pacific the heave runs pretty consistently west and east. They know by experience also that thick weather comes on when the wind is from the southeast. So if the weather is thick, they conclude that the wind is from that quarter. Clear weather, on the other hand, is usually accompanied by winds from the west. By considering all of these matters, the whalers always manage to get back 'somehow.' "

When a whale was harpooned, this prayer was uttered: "Whale, turn toward the fine beach . . . and you will be proud to see the young men come down . . . to see you; and the young men will say to one another: 'What a great whale he is! What a fat whale he is! What a strong whale he is!' And you, whale, will be proud of all that you will hear them say of your greatness. . . . When you come ashore there, young men will cover

A Makah practice-harpooning a float, December 1998 (AP/Wide World Photos)

your great body with bluebill duck feathers, and with the down of the great eagle, the chief of all birds; for this is what you are wishing, and this is what you are trying to find from one end of the world to the other, every day you are traveling and spouting."

From Waterman again: "An old Indian, nicknamed Santa Ana, once volunteered to ride a badly wounded whale. He clung fast to the lines, and dispatched the cetacean with a butcher-knife. . . . When the whale 'sounded' he clung fast and 'went under' with him, stabbing him meanwhile as best he could."

When the village came to receive a whale, the whalers' wives danced and sang. Eagle down was placed upon the dead animal's hump and blowhole. An old, experienced man measured the whale and made the first cut. This was taken to the captain's house, set upon a rack, and decorated with feathers. The old whalers would sit facing the "saddle-piece," and each would sing his own whaling song.

On one of the Makah museum's walls, there also appeared these words: "Makah stories tell of the great beginnings of this world when animals and humans conversed freely, interacted as equals." And I remembered an overture made by one whale protection group to bring the

Makah face-to-face with the creature they were so intent upon hunting. If desired, their way could be paid to the San Ignacio Lagoon. With the exception of a lone tribal elder, the Makah weren't interested.

That elder, a seventy-four-year-old woman named Alberta Thompson, was about to take center stage in the realm of the Cape People.

Chapter 5

A Tribal Elder and the Gray Whales

The gray whale and the Makah have always belonged together. My dream is that I wake up one morning and the Tribal Council has called a conference to make a statement: "We now realize that this whale gave up its life for us a hundred years ago so that we could eat. Now we want to honor and protect the whale until the end of time."

—Alberta Thompson, Makah Elder, 1998

There is no street sign on the rut-laden road where Alberta Thompson resides in a fourteen-foot-wide white trailer home. A totem of a woman with a broad, warm smile, she has black hair with a streak of silver across her forehead. She sat on a couch dressed in a white sweater and blue plaid skirt. On one wall was a framed photograph of Jane Goodall, inscribed "For Binki, together we make a difference." Alberta wore a single earring: a whale's tail. Underneath some family pictures on a nearby shelf, a wooden sculpture of a gray whale rested near a framed picture of tail flukes arcing gracefully into the water.

Approaching the age of seventy-five, Alberta was almost blind in one eye. Plagued with arthritis in her joints, she used a wheelchair when she traveled. Despite this handicap, over the past several years she had journeyed to meetings of the International Whaling Commission in Australia, Scotland, and Monaco—speaking out *against* her people's intent to hunt.

When the Makah Tribal Council initially resolved to resume whaling, Alberta explained, she opposed the decision on principle, because it hadn't been put to the entire tribe for a vote. "My dad felt that, as a female, I didn't belong on the ocean. I'd have to sneak away to the cliffs or the beach. Honestly I'd never seen a whale, let alone a gray whale."

That changed in 1997, when Alberta traveled to Baja's San Ignacio Lagoon. A guest of the oceanographer Jean-Michel Cousteau, she had her trip financed by the International Fund for Animal Welfare. She met Homero Aridjis and Robert Kennedy, Jr., and learned about the salt-works fight. She also met the gray whale.

"I met what I was fighting for, face-to-face," Alberta remembered. "First a mother whale rose up out of those warm waters right under my hand. She looked me straight in the eye, mother to mother. Then I saw a harpoon scar on her side, probably from up north in Siberia, where the native people still hunt the whales for sustenance."

Alberta paused, her eyes brimming with tears. "The mother brought her baby over to our little boat. I talked to them and I petted them. I felt their spirit of trust was somehow being conveyed to me. I laughed and I cried all the way back to shore, and all that night. I've never been the same since. When times get hard, like they are now, I think of those great big wonderful beings."

In many ways this grandmother of seven and great-grandmother of three seemed the least likely of activists. Everyone in Neah Bay still calls her Binki, a childhood name her baby sister invented. She is extremely soft-spoken. Indeed, Alberta recalled being so painfully shy as a teenager that she had to repeat the tenth grade for refusing to speak in class. Now she found herself giving dozens of TV and radio interviews—and being looked upon by many of her own people as a traitor.

Since reaching middle age Alberta had spoken out on a number of issues in Makah political life. "Binki," as one tribal member put it, "has been our resident dissident for the last twenty-five years." After the retirement of her late husband, a logger, she began objecting to clear-cutting on tribal lands. Now most other Makah elders who'd stood alongside Alberta on such matters, including opposition to the whale hunt, had been frightened into silence. Alberta's stance had caused her considerable problems. In August 1998 the tribal council terminated her job as a clerk at the Makah Senior Center. They alleged that while on the job she had spoken on the telephone to the Sea Shepherd protest group. "This clearly shows your participation in non-work-related activities during scheduled work hours," the council's letter read. Alberta now survived solely on a $350-a-month Social Security check.

She recalled meeting the Sea Shepherd's founder, Paul Watson, at

The tribal elder Alberta Thompson (George Peper)

the IWC gathering in Scotland in 1996. Watson had congratulated her for making a courageous stand and said he'd like to come to Neah Bay to meet with the elders. Alberta responded that she'd be pleased to help arrange this. But when Watson showed up, she remembered, "All hell broke loose. The council was sure he was going to bomb our parade! I told the council, 'This fellow you call a terrorist has done more for this world in conservation than all of you put together.' "

It was now late in the day, and we agreed to continue our conversation another time. I needed to get back to Sekiu and turn in early, because I was hoping to locate Paul Watson the next morning. The Sea Shepherds had been patrolling the Neah Bay area for a month in two large vessels, a 173-foot-long flagship and a 90-foot retrofitted Coast Guard cutter. They'd been joined by a number of protesters in inflatable Zodiacs, many of whom had motored over from Vancouver Island.

As I drove past the marina a canoe carrying what I presumed to be the Makah hunting team came paddling in. I thought of Scammon's description of one of its predecessors:

> The Indian whaling-canoe is thirty-five feet in length. Eight men make the crew, each wielding a paddle five and a half feet long. The whaling-gear consists of harpoons, lines, lances, and seal-skin buoys, all of their own workmanship. The cutting material of both lance and spear was formerly the thick part of a mussel-shell, or of the "abelone"; the line made from cedar withes, twisted into a three-strand rope. The buoys are fancifully painted, but those belonging to each boat have a distinguishing mark. The lance-pole, or harpoon-staff, made of the heavy wood of the yew-tree, is eighteen feet long, weighing as many pounds, and with the lance attached is truly a formidable weapon.

In my mind I ran through a quick comparison of the matériel planned for the late-twentieth-century-style hunt. Yes, the eight-man canoe would still be used. But this time around the canoe would be followed by a high-performance "chase boat"—to help maneuver a whale into place and, once it was harpooned, ensure that it did not escape—and by a smaller support boat to help in case of accident or injury. Yes, a harpoon would still be used. But this time around it would have a high-tech tempered steel tip, and, once secured into a whale, the harpoon would be supplanted by a more modern weapon. The Makah, in order to meet an IWC requirement that the kill be as humane as possible, had hired a University of Maryland veterinarian and ballistics expert who came up with a retrofitted World War II .50-caliber antitank rifle. At first glance the gun looked like a bazooka. It was twice as powerful as an elephant gun. The first time a Makah fired it, he got pushed backward six feet by the kick and landed on his backside.

Scammon had recorded: "Their whaling-grounds are limited, as the Indians rarely venture seaward far out of sight of the smoke from their cabins by day, or beyond view of their bonfires at night. The number of canoes engaged in one of these expeditions is from two to five, the crews being taken from among the chosen men of the tribe, who, with silent stroke, can paddle the symmetrical canim close to the rippling water beside the animal; the bowman then, with sure aim, thrusts the harpoon into it, and heaves the line and buoys clear of the canoe."

What would the captain make of what's going on today? I wondered. This time around the Makah hunting crew would venture out to sea chaperoned by U.S. Coast Guard ships. The Coast Guard had promulgated a rule requiring all craft to stay at least five hundred yards away from the Makah's hunting teams. The only other vessels allowed within this "exclusion zone" would be National Marine Fisheries Service boats, whose crews would monitor the hunt. The protesters, not surprisingly, said they had no intention of following such orders.

In 1874 Scammon wrote of another gray whale predator: "In whatever quarter of the world the orcas are found, they seem always intent upon seeking something to destroy or devour." Most native peoples have traditionally considered the taking of orcas, also called killer whales, taboo, lest some terrible misfortune befall their people. Not the Makah. According to Scammon, the Makah "pursue and take them [orcas] about Cape Flattery, in Washington Territory, as they consider their flesh and fat more luxurious food than the larger balaenas, or rorquals."

It was ironic, but perhaps fitting, that the Sea Shepherds intended to use underwater hydrophones to blast recordings of killer whale sounds in the direction of the migrating gray whales, in the hope of scaring them away from the Makah's harpoons.

I awoke early the next morning and drove over to Olsen's marina in Sekiu, where the anti-whaling boats customarily moored for the night. A Canadian with a fast Zodiac—an inflatable model developed by Jacques Cousteau in the early 1970s—said he was about to try to find Paul Watson. The Sea Shepherd commander's ship, *Sirenian*, was supposedly anchored out by Seal Rock. I told the Canadian that I knew Watson and had already been in touch about my showing up here. Handed a life jacket and told to hop in, I hunkered down against a strong south wind.

It had been almost ten years since I'd seen Watson, when I'd spent time with him in writing a magazine article about the Sea Shepherds and other environmentalist "Monkey Wrenchers." He'd been born in Toronto and raised in the hinterlands of Canadian New Brunswick. I remembered his saying that his activism began at age six, when a trapper killed a beaver he was fond of. The next year Paul and five siblings destroyed all the traplines in the area. He'd left home at fifteen, become a Merchant Marine seaman in Europe, joined the Canadian Coast Guard, and in 1973 found himself drawn to Wounded Knee, South Dakota. A new American Indian Movement was then occupying the Lakota Sioux reservation in protest of government mistreatment of Indians, and two AIM members had been killed by U.S. authorities. Watson volunteered to help as a medic. It was there, he said, that he "learned to conquer the fear of dying."

When the siege at Wounded Knee ended after seventy-one days, Watson received honorary initiation into the tribe. Guided by two medicine men, including the grandson of the legendary Black Elk, he experienced a vision that was to frame his future: "I suddenly saw myself in a grassy, rolling field, gazing into the eyes of a wolf," Watson had told me. "The wolf looked at me, then into a pond, and walked away. When I told the Sioux what had happened, they gave me my Indian name: Gray Wolf/Clear Water. Then I went back into the vision, and saw a buffalo standing on a ridge. It began to speak to me. And as it told me that I must protect the buffalo of the sea, an arrow came and struck it in the back. Attached to the arrow was a cord, symbolic of a harpoon."

Not knowing what it all meant, Watson went home to Vancouver. Six months later a fellow he'd known when they organized protests against nuclear testing in 1969 came to see him. Robert Hunter said that a few friends wanted to turn their old "Don't Make a Wave Committee" into a new organization whose mission would be to save whales. Soon Watson was among the leaders of Greenpeace, up against Soviet harpoons from a little Zodiac that would capture the imagination of the world when Walter Cronkite aired the footage. In his first memoir, *Sea Shepherd: My Fight for Whales and Seals*, Watson wrote of his watching a dying sperm whale whose "eye fell on . . . two tiny men in a little rubber raft, and he looked at us. It was a gaze, a gentle, knowing, forgiving gaze."

He'd left Greenpeace in 1977 after colleagues objected to his grabbing a seal hunter's club and throwing it into the sea, an act they regarded as violating their pacifist principles. With financial help from the animal advocate Cleveland Amory, Watson then formed Sea Shepherd to conduct "aggressive nonviolence against pirate whalers" and other ocean marauders. Over the years since he and his growing navy had sunk whaling boats in ports in Norway, Iceland, and Spain; rammed illegal fishing vessels at sea; and cut Japanese drift nets.

There were a number of remarkable tales about Watson. In 1981 a Soviet destroyer was pursuing the *Sea Shepherd II* through Siberian waters, still three hours from U.S. territory. Watson's crew had been documenting the Soviet killing of California gray whales for use as feed in fox farms, and the ship was racing home with the film. The destroyer, more than twice as fast, kept cutting across the *Sea Shepherd*'s bow and soon was joined by a helicopter that began dropping flares. As the Soviets started the engine on a launch suspended halfway up the destroyer, Watson ordered his men to grease the gunwales under the barbed wire that ringed his ship's deck, hoping further to dissuade a boarding party.

Ben White, who was part of the crew that day, continued the story: "The captain of their ship radioed over and said, '*Sea Shepherd II*, stop your engines and prepare for boarding by the USSR.' Paul radioed them back, 'Stop killing whales!' The Russians sent their message again, and again Paul told them, 'Stop killing whales!' Suddenly all hell broke loose. I was standing on the foredeck. I looked over at our crew on the bridge, which was listening to Paul over the radio, and they were cheering like crazy. Then I noticed the crew on the stern were cheering wildly, too. Except they couldn't have known anything about Paul's exchange with the Russians. They had seen something else. Out of the blue, a California gray whale had come up to the surface, right between our ship and the Russians. Everyone took this as an omen that we would be okay. And within a few minutes the destroyer backed off and just followed us out of Soviet waters."

In his 1994 book, *Ocean Warrior*, Watson recalled literally hundreds of gray whales coming to surround his ship in Siberian waters:

> The whales surfaced. Their backs poured—it is the only word—upward, like animated asphalt roadways. Turning. Turning, like great wheels, like mandalas, until it seemed that they had completed a circle and were, after all, mythological serpents devouring themselves. But then, the slow-motion crash, hundreds of minor waves, the tail emerging like a catapult, sunlight glistening in the water on their flanks. . . . They came up out of the sea, their clear Siamese cat's eyes the size of saucers looking at us unblinkingly. We were witnessing the gathering of the whales at the terminus of their long northward migration. And we were awed.

In recent years Watson had been jailed several times and been almost killed by Norwegian whalers' gunfire and by Canadian sealers who hog-tied him, dragged him across the ice, and threatened to throw him into the sea. His navy claimed responsibility for having rammed or sunk a dozen or more vessels. Today Sea Shepherd was a forty-thousand-member organization with an annual budget of close to a million dollars and a small fleet of vessels.

Now, just ahead, I could make out the black hull of the *Sirenian* jutting out from behind a massive boulder. Our Zodiac pulled up alongside, and I climbed a ship's ladder onto the deck. A black Sea Shepherd flag flew on the bow, next to an American flag. A crew, including several women, were fastening down another inflatable with *The Prince of Whales* emblazoned on its side. One fellow was peering through some binoculars at the horizon. I walked up another set of steps into the wheelhouse. Captain Watson was standing at the helm. At forty-seven he still possessed the same round baby face and wavy mane of hair, now completely white. He was wearing black jeans and a navy blue Sea Shepherd Crew jacket.

Watson said he found it ironic that, given his own history, his latest skirmish had pitted him against Native Americans. Even the colorful flag on the *Sirenian*'s stern was a gift from the Gitsanwatsuitu Nation, in honor of the Sea Shepherd's interception in 1991 of a Columbus reenactment voyage; the group had "reclaimed" San Salvador Island on behalf of indigenous people and even coerced an apology out of Spain. "Here we're caught in the dilemma of political correctness," Watson added. "Greenpeace won't get involved in this, because they're afraid of being perceived as taking on the Indians. I'm being called racist by a lot of people. My view is, I'd be a racist if I *ignored* the Makah. If I respect

The Sea Shepherds' Paul Watson with the Makah elder Jesse Ides, at IWC meeting in Monaco, 1997 (AP/Wide World Photos)

them as a nation, I have to treat them the same way I would any other pirate whaling nation. We gave them a challenge two weeks ago: produce any documentation that demonstrates they have the recognition of the International Whaling Commission, and we'll leave."

He spoke of the United States' having backed the Makah's claim of treaty rights and "cultural necessity" for its whale hunt. "Once you change the definition of aboriginal whaling for subsistence purposes, that's a very dangerous precedent. Not just for whales, but whether you're cutting down forests or exploiting any endangered species or fragile habitat. Say somebody wants to do uranium mining in the desert. They just find some local tribe that feels they have the treaty rights to sell that land for mineral exploration and says it's part of our 'cultural necessity' to trade our resources."

We were approaching the precipitous cliffs that loom above Cape Flattery, so named because it "flattered" Captain James Cook's hope of finding a safe harbor on his 1778 voyage here. Up ahead an island was becoming visible through the morning haze. In 1871 Scammon had passed along this same midpoint of the gray whales' journey from the Baja to the Arctic. He'd been

> off the Strait of Juan de Fuca, enveloped in one of those heavy fogs that prevail during the summer months, hiding every thing from view between ship and shore. The wind had lulled during the night, and the morning brought no breeze, except now and then a breath from the land, wafting with it the dismal rumbling of the waves dashing among the caverns of Cape Classet. Once we heard the splash of paddles and the whoop of Indians, as they were passing seaward to their fishing grounds. At length the fog began to break. First, we saw a few tree-tops; then, the deep green forest; next, a bold headland, blended with the lofty sails of several vessels, as they peered above the cloud; and soon Tatoosh island, with its guiding light-house, broke through the mist, when, in an instant almost, the vapor disappeared.

Tatoosh Island, Watson was saying, is named after a Makah chief. "Nobody lives there, even the lighthouse is unmanned now. There are two incredible caves running all the way through the island, you can go right in with an inflatable and out the other side. One of the old lighthouse keepers said that when the ocean surges roared through those caves in the winter, it sounded like a combination of all the thunderclaps in the world and made you feel as if the whole island was being forced up into the air."

Now, beneath the vapor, I glimpsed an astounding sight: a series of rock formations that girdled the half-mile-wide passage between Tatoosh Island and the mainland. Sculpted by storms over the centuries, they were called sea stacks. The tide rushed through the rocks at white-water velocity, and I couldn't imagine how this ninety-foot-long ship was going to make it through without running aground. But Watson showed no sign of changing course.

"When we first got here," he shouted over the pounding surf, "some of the Makah came out in their canoes and tried to run us onto those rocks. They didn't know I'd already charted the way through here, what's called the 'hole-in-the-wall,' a couple of years ago. It's only four or five fathoms deep, but we'll be all right. Might be a little rough, though."

As we listed wildly through the narrow channel, the most spectacular rock of all emerged, a descending column detached from a cliff at the very extremity of the cape. This rock rose almost perpendicularly for more than a hundred feet above the ocean, leaning slightly to the north-west. De Fuca's Pillar, Scammon had called it, after the Greek ship's pilot who'd first sighted it in 1592.

Makah legend has it that a youth once managed to scale the pillar. At the top he carved a two-headed bear, the symbol of his house. But when he tried to descend, it proved impossible. Friends attempted to reach him with rope fastened to arrows or tied to the wings of captured gulls. Finally, convinced it was the desire of the spirits that the youth be left as a sacrifice, the others returned to the village. It's said the bones still remain at the summit and provide warning, by means known only to the wisest elders, of approaching storms. Nobody has ever tried to scale De Fuca's Pillar again.

Not far beyond is another pillarlike rock. Lodged in a deep crevice across the top is a long, bleached log, one splintered end pointing out to sea. It was apparently tossed there by some stupendous wave (or by supernatural agency, as the Makah believed). On Tatoosh there were said to be ancient petroglyphs. When the Makah ceremonies were outlawed, they would paddle their canoes into the mouth of the strait and to the island's hidden northern side, where they held the rituals secretly until the federal law was repealed in the 1950s. Opposite the island along Cape Flattery is a large cave where the pounding breakers make an unusual whistling sound. The Indians think a demon lives there, one who comes forth during storms to seize canoes and bring the crews inside the cave, from which they issue forth again as birds or animals.

From Tatoosh Island more than 230 species of birds, many of them migratory, await favorable weather to cross the strait to their next resting place on Vancouver Island. According to Scammon, "When beholding it [the strait] from seaward, one can not wonder that [the explorer] Vancouver entered this artery of the Pacific with the hope of finding it reach through the continent to the Atlantic side."

"Look!" Watson shouted. "Sea lions!" He abandoned the helm momentarily, slid open the wheelhouse door, and excitedly relayed the news to the rest of his crew. "Sea lions!" He was grinning like a kid in a candy shop, watching as a host of Steller sea lions splashed in the waves ahead. Then, slowly, Watson turned the *Sirenian* around on the far side of Tatoosh and began maneuvering his way back again through the rocks.

"I think," he said, "let's head on into Neah Bay and check things out."

Along the way Watson talked of various ways and means by which the federal government was trying to get him out of here. The Sea Shepherds' two ships were registered in Belize and Canada. "When I went in for supplies to Port Angeles about ten days ago, the port director called me in and said we hadn't cleared customs properly in Seattle. 'You're supposed to clear as commercial vessels,' he said. I said, 'We're not commercial vessels, we're a research vessel and yacht.' He said, 'We've determined you're protest vessels, that's a commercial activity.' My lawyer said, 'Well, since when did First Amendment rights become a commercial activity?' I had to pay the customs guy $850."

Watson shuffled through some papers and handed me a letter from the National Marine Fisheries Service, dated September 23, 1998. It read:

> You raised a question recently concerning the Marine Mammal Protection Act and its application against a person making orca noises in the water to scare whales. Be advised we believe willful acts designed to harass whales are violations of the Marine Mammal Protection Act and enforcement action would be taken to stop the behavior. Unauthorized behavior that harasses or attempts to harass a marine mammal is considered a "take" under the Act, and by attempting to scare whales away from an area you

would be unlawfully harassing (taking) them. Sincerely, Thomas Shettler, Special Agent in Charge.

I laughed at the incongruity of this message, and Watson smiled. "Well, we'll have to see what happens," he said. Five weeks ago a federal judge in Tacoma had ruled against the various groups seeking to enjoin the hunt, stating that the Commerce Department had acted properly in evaluating the potential environmental effects. "So," Watson added, "I guess we're not going anywhere."

Only yesterday a pair of gray whales had swum right into the harbor—"I started to revise my view that whales are intelligent," Watson said—and the Sea Shepherds had gone in to chase them out. They weren't allowed onto the reservation itself; they'd be arrested by the tribal police. But they could enter the harbor, where the Coast Guard also kept constant vigil.

The Neah Bay marina is on the lee side of Cape Flattery, sheltered from Pacific storms. Only a few of the anti-whaling Zodiacs were present when the *Sirenian* sailed in late on the morning of November 1. The Sea Shepherd crew launched *The Prince of Whales*, and Watson put out word of their presence over the ship's radio. Before long a small armada of about a dozen Zodiacs—the *Orca Spirit*, the *Loyal-2*, and more—and a Jet Skier started buzzing the shoreline. STOP THE GRAY WHALE SLAUGHTER, one of their banners said. Slowly but inexorably a crowd of Makah began building along the bluff that overlooks the water. ECO-COLONIALISTS GO HOME, one of their banners said. The media, including all three Seattle TV stations, were gathering as well.

Inside the *Sirenian*'s wheelhouse, one of Watson's compatriots began "broadcasting" messages from "Radio Free Neah Bay" over a microphone: "We didn't see any of you bad old whalers out there this morning." A Makah with a loudspeaker fired back: "You white men killed more whales than we did!" Watson went over to a speaker system and began playing high-pitched recordings of orca sounds at a loud decibel level. A large rock splashed into the water below, narrowly missing one of the Zodiacs. Whoops and applause erupted along the dock.

"Are you going to hang around awhile, Paul?" I asked.

"Until everybody gets bored, I guess," Watson replied.

High noon. The larger of Watson's ships, *Sea Shepherd III*, hove into view. It looked like the leviathans' own ocean liner, with painted whales, including a very long navy blue orca, adorning all four sides and another painting of Neptune with his trident reaching out to touch a whale. Two banners were visible across the hull: STOP JAPAN'S ILLEGAL WAR ON WHALES! and PROTECT THE GRAY WHALE. Anchoring just outside the harbor, the ship lowered a Zodiac. The USCG *Cuttyhunk*, the Coast Guard's vessel moored at the marina, launched its own red inflatable with several crewmen.

Now on the tribal dock a Makah matriarchal elder wearing a long black overcoat walked down toward the shoreline and raised a hand to wave. "Lisa! Do you see Alberta?" Watson radioed excitedly to Lisa DiStefano, his second in command, in the approaching Zodiac. Her boat pulled up alongside the *Sirenian*, and Watson leaned over the side to confer.

DiStefano's Zodiac headed in to greet Alberta Thompson. I stood on the *Sirenian*'s bow about a hundred yards offshore, watching as perhaps fifty Makah closed in at the end of the sloping concrete launching ramp where Alberta was. I could hear chanting and drumming. DiStefano left her two male companions, jumped out of the inflatable, and started to wade ashore. Someone pushed her down into the cold, shallow water. As she was being handcuffed by a tribal policeman, another tribal member grabbed the Zodiac's line. Dozens more joined him in dragging the Sea Shepherds' boat onto the ramp. I saw Alberta cover her mouth with her hand, an anguished expression on her face.

"Oh no, they're taking the boat!" shouted a woman next to me from Minneapolis, whose husband had joined the Sea Shepherd crew a few weeks ago. "What's the Coast Guard doing just standing there?"

"Honey, duck! They're firing rocks. Get back!" her husband exclaimed.

A missile whizzed past my head and caromed against the hull of the ship. On a nearby pier in front of a row of warehouses, within firing range of the *Sirenian*, about thirty Makah were throwing chunks of concrete in our direction. Most looked to be teenagers, and some had slingshots. One kid, I realized, was using a tribal policeman's shoulder to provide him leverage.

We all sought refuge. I rejoined Watson in the wheelhouse. He was talking over a radio to the Clallam County Sheriff's Department. "Two individuals have been struck by stones, the tribal police and the Coast Guard are doing absolutely nothing to stop this," he was saying. "Roger, understand that," crackled the reply.

Back at dockside another Sea Shepherd had gone ashore to try to rescue the seized Zodiac. I watched as a tribal cop grabbed and shoved him onto the cement. He came up with a bloody forehead, and was the fourth

A battered Sea Shepherd protester under arrest by tribal police
(Errol Povah, SEA SHEPHERD—Friday Harbor, Wash.)

Sea Shepherd to be arrested today. Triumphantly, the Makah surrounded the Zodiac and lifted it onto a trailer. Two of the tribal members hoisted the Makah Nation flag to raucous cheering. The trailer pulled away.

"You guys spent so much time figuring out how to protect those whalers," Watson was saying to the Coast Guard over the radio. "You haven't given any thought to protecting us, have you? We're not backing down until we get our property back."

A large rock cracked the tempered glass wheelhouse window. Several helicopters circled overhead.

The Coast Guard's inflatable pulled up alongside and asked for Watson. "We just want you to be aware it's a dangerous situation," one of the Coast Guard said. "Remove your vessel and stand out of the way, that's all I can advise you." Watson said nothing, but one of his crew hollered, "Bullshit!"

Now another inflatable came alongside the *Sirenian*, carrying more than a dozen media. As they clambered aboard and Watson's press conference ensued, the rocks kept flying and reporters were ducking. "We're not retaliating against anybody, we didn't break any laws," Watson said.

Had he planned for this to happen? “No, absolutely not. Alberta displayed such courage in coming down to show her support for us, that Lisa felt she had to go in and greet her personally. Lisa was invited onto tribal land by an elder of the Makah Nation.”

Whack! Another flurry of stones. “Watch out, watch out!” Watson commanded and moved the press contingent to the other side of the vessel, behind the metal hull. “The Coast Guard just told me they have reason to believe there are explosive devices, fireworks, and things more powerful that might be thrown at us,” he continued. Ping! One of the reporters, a thin, bespectacled fellow huddled up against the side of the ship, kept scribbling notes as his eyes darted in all directions, looking every inch the war correspondent. Ping! “Jesus, it’s like taking bullets around here,” Watson said, and flinched.

Not long after the media departed, tribal police made a halfhearted attempt to disperse the youths who were pummeling the ship from the pier. A canoe moved into the harbor. When its paddlers arrived at the dock, a sizable tribal contingent formed a circle around them and began singing and beating deer- and elk-skin drums. I glimpsed a man standing there, looking right at me, pointing a harpoon. I put on one of the combat helmets that Watson had issued to his crew.

Twilight closed in. The Coast Guard pulled alongside again and took Watson away to an emergency meeting. There was some dissension among the crew in his absence over whether they should’ve sent the Zodiac in to greet Alberta Thompson. After all, they had only so many boats to try to get between the Makah and a gray whale.

I talked awhile with an attractive woman in her mid-thirties who’d come to the Sea Shepherds a month ago. For nine years she’d been a camerawoman for a Toronto TV station. She’d journeyed to Neah Bay for a week to do a story on the hunt. She said she’d looked into the eye of a resident gray whale everyone called Buddy, and something happened to her. Things weren’t the same when she went back to Toronto. She’d packed her bags and returned to join the Sea Shepherd crew.

About an hour after nightfall the Coast Guard brought Watson back to the ship. The other Sea Shepherds had been released from custody; the tribe had no jurisdiction to hold them. But the Makah refused to return the Zodiac, saying it would now be used to help guard their whalers during the upcoming hunt. Watson decided to head back to Sekiu for the night. The twelve-mile run along the coast felt interminable. Watson said he’d heard from the undersheriff of Clallam County that some Makah had been purchasing rifles with laser scopes. He had his crew walk around the ship picking up all the rocks that had been thrown at them. They counted more than two hundred and saved them as "evidence."

Watson managed to reach Alberta Thompson on his cell phone. "Hi, Binki, how are you? They're going to be arresting you? On what charge?" A pause. "They say they would banish you from the reservation. All you're doing is standing up for your constitutional rights. . . . *You're* the one that's responsible for this? Well, if you are, that's quite a compliment."

The Sea Shepherd commander laughed. But it had been an ugly day in Neah Bay, and nobody knew where things might go.

The morning after the confrontation dawned in rain and fog. In the coffee shop of Sekiu's Breakwater Restaurant, the waitress took my order and said, "Do you feel like you're at the end of the world?"

I returned to Neah Bay. Just beyond the town an unmarked road circles behind Cape Flattery and follows the Waatch River downstream. Three miles down is the tribal center. Until 1988 it was an Air Force radar station and barracks, now converted into Quonset-like administration offices. I was hoping to talk to Keith Johnson, chairman of the Makah Whaling Commission. As it happened, he and Tribal Chairman Ben Johnson were about to hold a press conference.

About a dozen media gathered around a large conference table. Gray-haired Ben Johnson, wearing a pink pin-striped suit and a handsome turquoise bracelet watch, sat at the head. He was flanked by Keith Johnson and Marcy Parker, a tribal council member and mother of the whaling captain Eric Johnson, a husky Makah in his early thirties who also joined the group.

Ben began by describing a meeting the previous evening in the community hall, where about 160 tribal members had shown up on a half-hour's notice. "A lot of people are very concerned about what's happening, how it's affecting our families and children." One of the big topics was what ought to be done about Alberta Thompson. "She never was arrested, that's just rumor," the chairman said. "As far as banishment, yes, we've been discussing that. There's a vast amount of penalties in our code, civil and criminal."

Keith continued in his high-pitched voice, "Our officers believe Alberta willfully, knowingly incited a riot and also obstructed governmental officials trying to do their duties. Our people saw firsthand yesterday how she supports Sea Shepherds breaching our shores. Here's an individual who's been here all her life, now doing this to our people, our land, our waters. There was a lot of concern about her being used by this group of terrorists, sadness and fear—for her. Alberta is the Sea Shepherds' big-ticket item right now—for media coverage, for more donations and sympathy."

Wasn't the tribe worried about perceptions of intolerance? "Yes, we're concerned about the perception we're trying to shut people up or not allowing them to exercise their freedom of speech. That's untrue, we want them to do that. But when it gets into acts that violate our local laws, it's another story. Lines have been crossed."

One reporter suggested he thought he'd heard loud voices during the tribal council's meeting this morning. "Nooooo," Keith replied, "that's our ancestors, disguising what we're saying." He laughed.

Eric Johnson was asked whether, given the increasingly bad weather and the events of yesterday, the whalers were still intent on going ahead this year. "Well, that can only be answered through our prayers and through speaking with our elders," he responded quietly.

Did the tribal leadership have any regrets about yesterday's incident in the marina? "I don't," Keith said. "It was bound to happen, and I think our people handled themselves most appropriately." Ben quickly jumped in, saying: "Well, we did get a pretty clear message from our elders last night. They said we fouled up, we should've stayed away. If this ever happens again, you won't see any Makahs down there, just enforcement officers and the Coast Guard. . . . It's lucky it didn't get out of hand. I mean, what if all them boats had tried to come ashore? Then you'd really have had a big fight."

"A free-for-all," Keith said.

"And I think at that point it would have been the end of Sea Shepherd." Ben couldn't seem to resist. "I don't think you'd see 'em anymore."

A secretary brought in a tray of smoked fish to pass around. The two men went on to talk about making some sort of legal agreement, whereby Paul Watson could have his Zodiac back if he'd agree to stay out of the marina. "We want them out of the bay, at any and all costs," Ben said.

A little later, outside the Makah Maiden café, Keith informed reporters that the tribe had issued a ten-day permit to the whaling crew at their captain's request. "The Fisheries Service told us they think the gray whale migration has begun. Everything is in place, and the weather is fair today. So let's go get a whale!"

Inside her trailer home the phone rang, and Alberta Thompson reached for it. "Yes, I'll be here," she told the caller, "unless they come to arrest me."

Why had she gone down to the dock yesterday? I asked. "I wanted to see what was going on, because I'd heard about what happened the day before and it made me feel so sad. The Makahs didn't throw rocks then, but they were hollering obscenities. So I wanted to go let the Sea Shepherds and the other environmental people know I backed them." Alberta paused, parted the curtain behind her, and took a quick, nervous glance out her trailer window. "Now I don't know what's going to happen."

The phone kept ringing, and at one point Alberta broke down in tears as she hung up. "Sorry," she said. "I'm finding friends I didn't know I had." A white teacher from the Makah high school dropped in, saying he'd been threatened for his own anti-whaling stance. The teacher said one of his students had ridiculed Alberta today in class, and he'd responded, "If we are to treat the elders in our village this way, we've lost our soul."

"Thank you," Alberta said in a whisper.

It was getting dark, and she was clearly exhausted from the past two days. We agreed to meet again tomorrow. I wanted to learn more about her sense of Makah tradition.

Back in Sekiu I ran into Paul Watson coming into the dock. He said he'd heard eggs and a firecracker had been thrown at Alberta's car. Someone had also scraped paint off the length of the car with a sharp object. She might be moving tonight, Watson added, for her safety.

The following morning I picked up the Seattle paper and learned that Watson was right. OPPONENT OF MAKAH HUNT FLEES RESERVATION, the headline read. Sometime during the night Alberta had departed her trailer home, and nobody knew her whereabouts. Her attorney, Helga Kahr of Seattle, "said she asked Thompson to take shelter with friends out of state for several days." Meantime, the tribal police chief, Lionel Adhunko, "said he planned to file papers today criminally charging Thompson with inciting a riot." The newspaper added that "an investigation is under way by the FBI, Clallam County Sheriff's Office, and U.S. Attorney's Office to determine if the detention of protesters by tribal police was legal, if their civil rights were violated, and whether the boat was legally confiscated or stolen."

My time with Alberta felt very incomplete, but there seemed nothing I could do. As I was about to head for the reservation again, a freelance photographer suggested that I stop off at the Snow Creek Resort on the way in. "Snow Creek? It's off-season. I didn't think anybody was there," I said.

"No, a couple is staying there you ought to talk to. They've got an organization called In the Path of Giants. They have a grant to document the entire migration and life cycle of the gray whale."

East of Neah Bay I pulled off the road into a gravel driveway, past a blue house trailer with drawn shades, and knocked at the door of a white wood-frame beach house. Inside a woman who introduced herself as Heidi Tiura was alone at a computer, dashing off e-mails about the latest events in Neah Bay. She eyed me suspiciously at first. As I finished describing my day on the Sea Shepherd boat and my initial interview with Alberta, she seemed to have decided something. "Well, Alberta's here," she said. "We're hiding her until someone can drive her to Seattle tonight, and then she's flying out to California. Wait here, and I'll go ask if she wants to talk to you."

Alberta agreed. As I entered the trailer she smiled warmly and clicked off the TV. "I'm just taking it easy," she said. "I need to get away for a while; it won't be long."

We started talking about her personal history. She was the descendant of three Makah who'd signed the 1855 Treaty of Neah Bay, the granddaughter of a man who was "in the whaling canoe," and her father had been a seal hunter. He spoke only English to her, but as a young married woman Alberta did considerable driving for the elders and learned the Makah language from them. "I'm forgetting some, because I don't have anybody to talk to." She'd lived here all her life, except for working in Seattle as a welder building destroyers during World War II, when she met her husband. "He's the one who wanted to stay in Neah Bay, and I'm glad I listened to him, because when my children were growing up they had free range and there was no danger here. Not like it is now. They rode horses all day long and swam in the rivers, and the rivers were clean then."

Over the years Alberta had been a teacher's aide and secretary at the tribal school, then worked at the Indian Health Clinic before her last employment at the senior center. She'd served on the Makah's law-and-order committee and election board. On the wall of her home was a Clallam County Community Service Award she'd received in 1997. She was raised in a time when tribal tradition was paramount—when "we still had teaching pumped into us, how we were supposed to live."

Alberta closed her eyes and continued, as if in a reverie: "In the old days, we had a lot of medicines that took care of us. The tradition was, you didn't even share these with your friends. This was your family medicine. One that's well known is wild crab apple bark. That cleanses your blood. I remember a man who was immobile in bed with arthritis, and within months of taking that, he was healed. You've seen Marcy Parker, one of the council members behind the whale hunt. Her grandmother's boy was in a TB ward, and there seemed like nothing more to do. Then the grandmother made this special tea from roots and cured him. And there was devil's club, for skin diseases."

Alberta grew up eating traditional foods—seal meat, dried fish with salmon eggs, the sprouts of salmonberry bushes gathered from the hillside. "We had our own potatoes, a long wide potato with shoots coming out that we called Ozette. We didn't have to buy bread because we made what we called buckskin bread, like a biscuit that you flatten out and put in the oven. We've survived just fine without whale meat for as long as I've been alive. The salmon, not the whale, is our real subsistence food. Why doesn't the council put their energy into fighting for our fish? Our kids don't have problems with violence, drugs, or alcohol because we stopped our whaling tradition. We simply became more like the outside world."

She shook her head and continued, "All this beautiful scenery we have here—if we started a whale-watching business, we could tell people about our culture. And we wouldn't have to kill anything to do it." The previous year Alberta and four of her grandchildren had gone whale watching here on the Olympic Peninsula. This trip was sponsored by the Humane Society, which was exploring the possibility of such trips being run by the Makah. They'd sailed out toward Tatoosh Island, where she remembered going in the spring as a child to gather seagull eggs. They'd seen six gray whales. One even surfaced alongside the boat, a rarity in the rough North Pacific waters. One of Alberta's descendants cried: "Look, Grandmother! Is this the same gray whale you touched in Baja? He looked right at me!"

When I left Alberta, I walked out past the boat ramp, onto the beach behind Snow Creek Resort, and sat down on the sand. I don't know how long I was there, maybe fifteen minutes or maybe half an hour, when a gray whale surfaced alongside a small boat moored about fifty yards offshore. It looked almost white and had a large mark on its side, in the shape of a wing. I watched Angel Face—as I'd later learn the Snow Creek resident had been named—play in the kelp and roll around in the surf. And I thought of Alberta Thompson saying, poignantly, gratefully, "The whales have taken me many places."

Later that week the seventy-one-year-old Washington congressman Jack Metcalf showed up in Sekiu and stood beside Paul Watson at a press conference. Metcalf had just won reelection. He said he was "gravely concerned about the recent civil rights violations at Neah Bay," decried "the denial of Alberta Thompson's freedom of association," and called for "an immediate, aggressive investigation by the U.S. Justice Department." Metcalf was a conservative Republican who, on every issue except hunting whales, was scarcely known as an environmentalist. Still, when I spoke with him about why he felt so strongly about this, his eyes took on that "look" as he remembered. His father was a commercial fisherman, and Metcalf recalled being awestruck at the age of eight while surrounded by a pod of orcas.

About a dozen years later, "I was in Alaska after World War II, running a little boat across a bay. I was about half asleep when a huge gray whale came up out of the water right beside me. I thought, Oh boy, he sure knows I'm here! Startled me and scared the heck out of me at first. I was sitting pretty low in the boat, and I actually looked *up* at its eye. Somehow I knew then everything was gonna be all right. The whale went down, came up again, and followed me for a long time. As if it was trying to tell me something. I never forgot it."

Metcalf was not known as an advocate of Native American rights. The congressman's appearance, Tribal Chairman Ben Johnson told his own press conference the next morning, "turns this into an anti-Indian thing now." Johnson raised the possibility that AIM—the still-existing American Indian Movement Paul Watson had joined forces with at Wounded Knee—might come help protect the reservation from "terrorists" like the Sea Shepherds.

Jean-Michael Cousteau, eldest son of the late, legendary Jacques Cousteau, came to the reservation as well. He had a two-hour meeting with the tribal council. Cousteau sought to persuade them to create a ritualized hunt that would sustain their culture and heritage but wouldn't harm any gray whales. Afterward Ben Johnson said, "We will still go hunting."

By the end of the first week of November, they hadn't. The weather remained decent enough, but the whalers stayed dry-docked. At the daily media briefings various reasons were given: a dead radio battery, equipment failures on the support boat. Eventually Keith Johnson told the increasingly impatient press contingent: "Our whalers feel we've lost our focus, that we need to back up a minute, see where this whole thing is headed. I don't blame them."

I left before Alberta Thompson returned to the reservation. Nobody came to arrest her. The Sea Shepherds sat down with the Makah leadership for a macaroni-and-cheese lunch dialogue aboard a Coast Guard ship, but not surprisingly they found no common ground. As gale-force winds started to buffet the Cape Flattery region, the Sea Shepherds left Neah Bay. Eventually the tribe would return their seized Zodiac.

From a distance I read about a leading aide to the communications billionaire Craig McCaw helicoptering onto the reservation. McCaw had spent $12 million on helping life imitate art—returning Keiko, the famous killer whale of the *Free Willy* movies, from captivity to the wild off the coast of Iceland. Now his emissary came to offer the Makah a considerable amount of unspecified aid if they'd agree to drop the hunt. At the end of November the tribe called a general assembly to discuss various offers they had received, including a wind-generation system as a power source, a program to replant deforested Makah land, and the purchase of land once owned by the tribe (perhaps as a gift from McCaw to the Makah). Keith Johnson emerged from the meeting to say: "We're not selling our treaty right. There will not be any deals that will stop us from whaling."

In 1998 anyway, the gray whales apparently had other ideas. One morning shortly before Christmas, I went to the *Seattle Times* Web page and saw a front-page story that began with this paragraph: "Scientists don't know how they did it, but thousands of gray whales apparently skirted the coast of the Pacific Northwest unnoticed, despite the media, federal authorities, and Makah Tribe whaling crews and tribal fishermen looking on."

Already gray whales were being seen as far south as San Diego—safely en route to the lagoons of Baja.

Chapter 6

Return to La Laguna

. . . and in the tropics tremble they with love
and roll with massive, strong desire, like gods.

—from D. H. Lawrence, "Whales Weep Not"

THE more I pursued the path of the gray whales, the more I would be struck by the diversity of extraordinary people who seemed to follow in their wake. Another of these was Robert Francis Kennedy, Jr., second-eldest son of the late senator. He is a senior attorney for the Natural Resources Defense Council (NRDC) and had traveled for the first time to the San Ignacio Lagoon in February 1997. The Aridjises had met him then, and again in 1998, when all stayed at a campground near Rocky Point, as part of the tour arranged by NRDC and the International Fund for Animal Welfare. "He's so articulate," Betty Aridjis told me, "a fabulous weapon in the fight. This year, he even went diving with the local clam diggers. They were really astonished, blown away—because Kennedy is probably fifteen years older than they are, and he took in more clams than they did!"

On a day-to-day basis Kennedy was codirector of the Environmental Litigation Center at Pace University, located about an hour north of New York City. He and the law school students had successfully sued more than 150 companies for polluting the Hudson and Croton Rivers. Kennedy had also helped to establish some twenty-five Riverkeeper operations around the country, private groups modeled on the original Hudson Riverkeepers that patrol local waters looking for "environmental criminals." So how had he come to be so involved in the battle to protect the gray whale's Baja habitat?

I arranged to hook up with him in New York City and find out. He said we could talk in his car as he drove back to his home in Mount Kisco for a barbecue with his students. Bobby, as he likes to be called, was instantly recognizable as I approached him at the wheel of his Plymouth Voyager in midtown Manhattan. Now in his mid-forties, he looks eerily like his father: the sandy brown, wispy hair, the toothy smile, and, above all, the deep-set, fiery light blue eyes. He was dressed in a T-shirt and slacks. I began by asking him about going clamming with the lagoon fishermen. He grinned and said, "Yeah, I spent a day on one of their boats catching and shucking hatchet clams. They had an air compressor with a sixty-five-foot hose, and they actually let me go down and walk around on the bottom and harpoon those big clams."

As he made the turn onto the Hudson Parkway, Bobby continued, "My uncle who died recently, Jimmy Skakel, was the last American to harpoon a whale. There's a December 1962 *Life* magazine article about that. He harpooned a sperm whale in the traditional way from a little dory in the Azores, and went on the Nantucket sleigh ride. It pulled him something like twenty-one kilometers, and then he rowed it back to shore."

There was another Kennedy family connection to whales, in particular gray whales. *Time* magazine's obituary for Bobby's cousin John Kennedy, Jr., after the fatal plane crash, began with a story of John, his girlfriend, and two friends vacationing during the mid-1980s and "communing with the extraordinary gray whales in Magdalena Bay." As they all brought their ocean kayaks onto a deserted beach, John had jumped back into the water and begun to swim farther and farther out into the Pacific, "well beyond the big lines of breakers rolling toward shore—until the friends suddenly realized they couldn't see him at all. They stood onshore, panicking and scanning the horizon, wondering whether and where and how to go after him." Suddenly, John simply reappeared, "emerging from the heavy surf—dripping, exhilarated." Had he, I could not help wondering, gone back out there to "commune" some more with the gray whales?

Bobby recounted how, when he had his own first visitation with the gray whales at San Ignacio Lagoon, he "recognized this guy there, but at first I didn't know from where. His name is Tim Means. He owns the Baja Expeditions campground, where we all stayed. He's an expatriate American, with a Mexican wife and kids, and a real character. And it turns out that, when I was a kid and my father was whitewater rafting all the western rivers, it was Tim who used to be our guide."

The cellular car phone rang incessantly, and we could talk about whales and salt only intermittently. At one point Bobby was almost shouting into the receiver: "You think I'm after your job! Well, I'm not after your job, you're just supporting a bad bill!" When he hung up he informed me, "That was my congresswoman."

I asked if he'd describe his experience going out in a *panga* among the gray whales. He nodded and continued: "You know, I'm not a very sentimental person. I don't think we ought to save animals because they're cuddly or pettable. But it's simply an amazing experience having those whales roll over and look you eye to eye. There really is an interspecies contact there. There's an intelligence. And it's undeniable. It's different from any experience I've ever had, and I've been around animals all my life. You sometimes see that kind of meaning and significance in the eye of a dog, occasionally maybe a chimp or something like that. But seeing it coming from an animal that huge and so different than us, coming from a whole different universe—I mean, it's like the theme from *E.T.*, you know? Where they reach out and touch fingers. It's like two universes touching and finding a commonality. That's about as far as I want to go with that, but—it's truly an extraordinary experience."

I told him what I'd felt about the eye-to-eye moment, that it was as if my entire being was somehow being read by the gray whale. Bobby smiled. "You know, the whales can't look to the sides, and they can't look above. That's why they have to turn their heads, because their vision is directed down. They can only look below. That's where the world is for them."

He turned off the highway into the Hudson Valley, only about forty miles north of the city but rural enough to host coyotes, turkeys, and white-tailed deer. As we entered an area of tall trees shielding stately homes, Bobby described his idea of bringing some high-level Japanese, perhaps Mitsubishi executives, to the lagoon to meet the whales. He'd been in Tokyo recently giving some talks on the environment and had visited with Japan's foreign minister about the saltworks. "The Japanese people actually love whales and have tremendous antipathy against whaling. At the same time, they're very accepting about their political structure. It's assumed that government really works for the people; they're not cynical. And I talked to a lot of people over there who said it's the big whaling guys who are running the show."

He pulled into a circular driveway. He and his wife, Mary, have lived at this estate since 1984. Bobby's study is just to the left of the entryway, his desk surrounded by walls of handsomely bound old books. The walls of one room down the hall are covered with Kennedy family campaign posters. Yet the spacious grounds seem to dwarf the house. A garden leads down to a large swimming pool. Way out back, tucked among the pines, is a thirty-acre lake, where Bobby stocks fingerling striped bass that he gathers by seine net from the nearby Hudson River. All this is visible from a picture window in what Bobby calls his junk room. It's filled with everything from a huge snakeskin across the ceiling to a collection of bones. He asks, "Would you like to see the alligators I keep in that aquarium there?"

After his "wild kingdom" tour, we sat down and talked some more about the saltworks. "They might be able to engineer themselves into a situation where the whales can be fine at San Ignacio," Bobby said. "But nobody's gonna go there to see the whales if it looks like Bayonne, New Jersey. This is a place where people come for a unique and special interaction. If you destroy this place, you're destroying something that's been unchanged for thousands of years. This would diminish all of humanity. We preserve wilderness because, ultimately, it enriches us historically, culturally, and spiritually. The fact that San Ignacio Lagoon is remote and everybody's not gonna get a chance to go there isn't a reason not to preserve it. The fact that *anybody* can go enriches *everybody*. Well, nobody will if it's a salt factory. And once they can be sure nobody's gonna visit anymore, they can do anything to the whales they want—and who's gonna notice? The whales' constituency are the human beings who go there to visit them."

There was silence for a moment. When Bobby spoke again, it was with the passion and intensity I'd witnessed thirty years earlier, as a student at the University of Kansas watching his father initiate his 1968 presidential campaign. His father was speaking about ending the Vietnam War. Bobby Jr. was speaking about stopping the world's largest corporation at a Mexican lagoon. "Well, if they want to destroy it, they're gonna have a war on their hands. If they keep escalating, we'll keep escalating. I think we can stop this. We'll do whatever we have to do."

Bobby Kennedy intended to be back at the lagoon early in March 1999. So did Homero and Betty Aridjis. And, I found out by way of a phone call to Idaho, so did my rescuer of the previous year, John Spencer. The gray whales seemed to be playing quite a role in my life. I knew I had to join them all once more at the San Ignacio Lagoon. Maybe this time I could arrange to go out with Pachico Mayoral.

Much had transpired since my first trip to the lagoon. In November 1998 the annual meeting of the UNESCO World Heritage Committee was held in Kyoto. A coalition of Mexican and international groups had formally petitioned the committee to place San Ignacio Lagoon in its "in danger" category, a designation that could result in scuttling the saltworks plan. As officials and diplomats from fifty nations gathered, a column by Bobby Kennedy appeared in *The Japan Times.* He pointed out that the 552 World Heritage Sites around the world included the Grand Canyon, the Egyptian pyramids, the Galápagos Islands, the San Ignacio Lagoon, and a number of historic places in Japan. Twenty-five of these places, such as Yellowstone National Park, had been given "in danger" status. "I cannot conceive that Mitsubishi Corporation would ever propose to put an industrial plant amid the shrines and temples of Kyoto," Kennedy wrote. For the first time the Japanese media provided a plethora of coverage of the issue. The World Heritage Committee decided to dispatch a delegation to look at the lagoon. By now more than six hundred thousand people had also written Mitsubishi urging the corporation to call a halt to the proposed expansion.

The letter-writing campaign had been fueled the previous summer when Mexico's environment secretariat, PROFEPA, announced the conclusion of a study into the mysterious deaths of ninety-four endangered black sea turtles in December 1997 at Scammon's Lagoon. The cause wasn't red tide or poaching, the investigators determined after studying the turtles' skeletons and detecting cellular damage at the microscopic level. It was saline stress. The turtles had died as a result of salt brine and other contaminants apparently dumped into the lagoon by the Exportadora de Sal company. Fish and plankton had been killed as well. What's more, PROFEPA discovered another fish kill caused by a brine spill of over 4 million gallons from the ESSA plant in May 1998.

The Aridjises' organization, the Group of 100, issued a press release appealing to the Mexican government to fine the company and call an immediate halt to any expansion plans for the San Ignacio Lagoon. Bobby Kennedy and the NRDC's Joel Reynolds coauthored an op-ed piece about the turtle kill that appeared in the *Los Angeles Times,* concluding that "Mitsubishi must abandon, once and for all, its reckless plan to turn this glittering natural jewel into salt."

Shortly thereafter someone's reckless plan for Homero had forced the poet into a public airing. PEN LEADER IN MEXICO TELLS OF THREATS, *The New York Times* headlined after Homero called a press conference in Mexico City late in August 1998. Messages on the Aridjises' answering machine clearly showed that whoever was calling had access to information about their movements and personal lives, information Homero maintained could only have come from a government tap on his phone.

In one case a threat was received just hours after a telephone interview Homero had given a *Washington Post* reporter, and it repeated several phrases from that conversation. Recent references to the safety of the Aridjises' two daughters—along with a pointed "you are going to die very soon" message for Homero—had prompted the press conference, at which the recordings were played.

Twelve prominent writers—including Arthur Miller, Nadine Gordimer, Edward Albee, Jamaica Kincaid, Susan Sontag, and Mario Vargas Llosa—responded by sending a letter to Mexican President Zedillo, calling on him to investigate and assure the Aridjises' safety. The *Los Angeles Times* followed with a lengthy article: PROTECTIVE POET FINDS HIMSELF IN NEED OF BODYGUARDS IN MEXICAN MYSTERY. Homero was quoted: "I used to have the freedom to go out alone, to take public transport, the subway. I loved to walk. . . . It's part of my life as a writer. But now I can't do it. My mind isn't free. If I go into the street, it's in a car with two armed men, turning every which way." And Homero was quoted in a *New York Times* story about bodyguards: "We live in a culture of fear. And we don't really know if the men watching out for us are good or bad, honest or criminal."

I feared for the safety of Homero. He feared for the safety of the gray whales. At another press conference, in January 1999, he called attention to the fact that, since a lone gray whale had shown up in Baja as early as November, hardly any had been seen. "There has been an undeniable change in the global climate, and it's possible this has had an impact on the migratory patterns of the whales," Homero said. The whale scientist Bruce Mate concurred that "we have never seen the migration begin so late."

Toward the end of February I flew into San José del Cabo and rented a Mexican Volkswagen at the airport. I drove north along a dusty road that winds through mountains and desert past small herds of cattle to the seaside home my friends and I had made ours ten years earlier. In two cars, four of us set out the following morning. We continued north through Cabo Pulmo, a divers' paradise hosting the only living coral reef in this part of the world; through La Paz, where the large statue of a gray whale's tail beckons all visitors; on to the old mission town of Loreto. After stopping there for the night, before noon the next morning we'd reached the heart of the Vizcaíno Central Desert. Approaching the oasis town of San Ignacio from this direction, a trio of spectacular volcanic peaks known as Las Tres Vírgenes (The Three Virgins) crowns the horizon at more than six thousand feet. These are cloud-capped and startling, rising ponderously over a sea of cactus. The volcano last erupted 27,000 years ago. Yet deep within its ancient being, it continues to hold seething molten rock, seeming to promise a future return to power.

I was again in Scammon country. In the *Overland Monthly* for March 1870, the captain had written: "The west coast of Lower California is high, broken, and extremely barren: no green hill-side or verdant valley cheers the tired mariner, when closing in with its shores; not even the solitary habitation of a ranchero appears, to give evidence of human or animal life. Many lofty peaks of the Sierras have a sombre, reddish hue, when seen through the hazy atmosphere which often hangs about the land, in this temperate latitude; and the variegated, precipitous cliffs, which, at many points, look as if overhanging the ocean, far surpass in picturesque beauty the 'Gay Head' of our New England coast."

This year I'd made no advance plans for whale watching. The Kuyima office told me that their camp was booked solid. John Spencer had given me the names of some of his friends, fishermen like Pachico Mayoral who had their own operations. We'd take our chances. The road to the lagoon seemed much improved, though this probably was owed to the fact that I was riding in a VW Bug rather than a whale-sized vehicle. We made it without incident in less than three hours. I pulled over at the outer limits of the little village of La Laguna, the same village I'd wandered into so desperately in the dark a year ago. It was empty in a sweltering sun. We headed on about a half mile to a sign reading, MALDO'S WHALE-WATCHING TOURS.

Romualdo (Maldo) Fischer and his wife, Catalina, John Spencer had told me, were two of the finest people he knew at the lagoon, besides which, Catalina was the best cook around. She answered our knock at their little wooden house. Two of my companions spoke near-fluent Spanish. Catalina, who looked to be in her mid-thirties, walked outside and pointed to a palmetto-roofed *palapa* about a hundred yards away and right above the beach. This was her restaurant whenever the tour groups called upon her. Otherwise, nobody was there, and we were welcome to camp inside. If we didn't mind sleeping on a shell-covered floor, we'd probably be more comfortable than in a tent, especially given the strong winds at night. Just stack the tables and chairs in a corner. Feel free to use the stove. The outhouse *baño* is that plywood shack nearby.

We unloaded our gear at our new residence and drove past the small airstrip and on down the road. I stopped at the Kuyima camp. Carlos and Teresa were in the kitchen and asked right away about my young nephew. Tonight, down the road at the American-owned Baja Expeditions camp, they said, there was to be a big meeting between local people and a group of politicians arriving from Mexico City. They gave us directions—basically follow the salt flats for another half hour, as far as we could go before running into the lagoon.

When we pulled up at Baja Expeditions, with its campground of large domed tents and solar-heated showers, Homero and Betty Aridjis were just about to take a *panga* out for the second time that day. They'd flown in yesterday from San Diego, part of the first of two VIP contingents to be hosted this year by the NRDC and the International Fund for Animal Welfare. Their group included a number of members of parliament with the European Union, here to visit Mexican counterparts—and gray whales—for the first time. Already this morning, Betty said, they'd had several friendly encounters—no, not the politicians! The whales, of course!—and one mother had even come to scratch her back against the boat. They'd see us at the meeting tonight. Off they went.

By 7:00 P.M., Baja Expeditions' big army mess tent was standing room only. Kerosene lamps and candles cast shadows over the faces of close to fifty lagoon residents, primarily fishermen, who blended in among the politicians and journalists (even *The New York Times* was here). Andrés Rozental, formerly Mexico's ambassador to Great Britain—and Mexico's official delegate to the yearly International Whaling Commission meetings—acted as host and translator. It was an eclectic gathering of humanity in one of the planet's more remote places. Through the flaps of the tent, the wind carried the resounding *ppfff-whoosh* of the gray whales.

The dialogue was about the future of the lagoon, which necessarily meant how the local people felt about ESSA's expansion plans. The first speakers talked about a lack of communication between the company and the populace. They simply had no clear idea what the plans *were*. An older man allowed that one thing he *didn't* want was any change in the topography and ecology of the region. Homero, seated in a folding chair next to Betty, recalled a letter sent by one of the local cooperatives (known as *ejidos)* in 1995 to President Zedillo, in which they suggested a number of sustainable projects they'd like to see implemented here. Had they ever received a response? he wondered. No, they hadn't. Still, another man said, if jobs are offered here by ESSA or anyone else, the people are faced with the unfortunate dilemma that they need the work. They have families to feed.

I heard what seemed a familiar voice coming from a far corner of the tent. It was a man in a red cap and red sweatshirt. Yes, unmistakably, Pachico Mayoral. "Everything that happens on Laguna San Ignacio," he was saying, in a tone quietly plaintive, "from the point of view of fishing and development, has been for the people *outside.* Not much of it has ever stayed to benefit those who live here."

I soon found out more specifically what he meant. The next speaker looked very much like Pachico, and I'd later learn this was his son Ranulfo. In an equally plaintive way, he asked: "Did you see all those mountains of shells along the road, between the airstrip and the camp? These are our *callo de hacha,* and our *almeja.*" Scallops and clams, the lagoon's most valuable seafood commodities, had been completely depleted in only four years. Fishermen had come from considerable distances, dozens at a time, and illegally dredged around the clock for Pacific calico scallops and fan scallops along the northern shoreline. Most of them were employed by urban-based entrepreneurs from Ensenada and La Paz, who bribed the Baja fisheries authorities to issue permits for harvesting the restricted shellfish. Later, when lagoon fishermen tried to start an aquaculture project to raise the same shellfish, they were unable to obtain permits from the government. All that remained, all across the salt flats, were the mounds of empty shells.

Then there was the turtle poaching. An old man with rheumy eyes began speaking. Between May and August, after the gray whales and the tourists had departed, the poachers came from San Ignacio town and other places. They set their nets in the lagoon, captured all the green turtles they could, and trucked them alive to the northern resorts, where they became soup. I saw Homero's face turn ashen. Because of him all capture of sea turtles had supposedly been banned since 1990 along the Mexican coast.

Maldo Fischer, my lagoon host, identified himself to the crowd as president of the local tourist association. He was a handsome man with bright eyes and a dark mustache. He spoke of approximately 70 families who lived along the lagoon, with another 100 to 150 involved in the fishing cooperatives. Members of these eked out a living catching corbina, snook, halibut, sierra, sea bass, and grouper. If a saltworks came, he believed that the fisheries would "no longer be feasible." Industry would drain off their way of life.

Perhaps the most disturbing revelation was this: a management plan for the Vizcaíno Biosphere Reserve had long since been prepared, but it was being held up by the Ministry of Trade—the Mexican "owners" of Exportadora de Sal. So resources were unavailable that would otherwise aid communities like this one in establishing sustainable projects such as aquaculture.

Adolfo Aguilar Zinser, a Mexican senator who'd been elected on a Green Party ticket, was outraged by everything he heard. "We seem to operate in two completely schizophrenic dimensions," he told me afterward in English. "We come here attracted by the whales, and by a foreign company planning a project with the Mexican government that's going to attack the whales. Now we realize we've come very late. This is already an area where the people are devastated by neglect and abuses that have a lot to do with wider corruption. What these fishermen are asking us is, How can we take care of the lagoon? Because somebody is always going to come and rip it off!"

You could never tell about gatherings like this. The politicians who came were those predisposed to be sympathetic toward the concerns of the local people. And you didn't hear many voices of the people speaking in favor of the advent of big industry. Whether they represented majority opinion about the saltworks or not, it was hard to say. Nonetheless, as one of my friends pointed out, these were the most articulate campesinos he'd ever seen. He'd been to many similar meetings during his decade living in Baja, and generally such people were extremely reticent.

When the group adjourned after nearly three hours, I walked over to Pachico. I didn't know if he'd remember me. To my surprise he smiled warmly and saying, "Amigo," greeted me with a hug. Yes, he said, I could come see him tomorrow in the late afternoon, and we could talk about going out together among the *ballenas.*

Outside the tent Venus was conjunct Jupiter. A bright corona surrounded the moon. A number of the politicians walked down onto the beach to set up a telescope. I followed with the Aridjises. Homero, listening to their palaver from an increasing distance, finally muttered in exasperation, "Can't they *ever* be quiet?"

The gray whales were out there, splashing, breathing, singing, basking under a radiant night sky. We stopped, far from the madding crowd, stopped and listened. In silence, alone together, we stood for at least ten minutes under the halo of the moon. Sometimes, strangely, the trumpeting sounds of the gray whales reminded me of elephants. Other times they were a kind of unearthly music. "It is magic, no?" Homero said.

The modest home where Pachico Mayoral and his wife, Carmen, raised their three sons is the first one on the perimeter of the settlement called

La Laguna, just before Domingo Mesa's place, where I'd stumbled into John Spencer the year before. A corrugated tin roof is held to the house's plywood frame by several tires and assorted other car parts. A rooftop solar panel provides power for an interior radio and battery-activated lights. Old boats and barrels line the outer yard, and off to one side are a number of motors under an open *palapa*.

Entry is through a gated vegetable garden, where whale vertebrae serve as newel posts. The garden is graced by hanging plants and hummingbird feeders, beautifully tended by Carmen. Pachico's combination office and dining room is just inside a screen door. Maps of the lagoon cover the wall behind him. He rests his arms on a plaid tablecloth and holds a cigarette.

With my friend Richie to translate for us, Pachico is much more comfortable speaking in Spanish. His wife's family, the Aguilars, had been the first settlers of the entire lagoon region. They'd emigrated, first to Guerrero Negro and then here, from drought-parched ranches in the interior of the peninsula, hoping to find a better life harvesting from the sea. Pachico came to San Ignacio Lagoon in 1961 for a fishing trip, by

Francisco (Pachico) Mayoral (George Peper)

ship all the way from his hometown of Santa Rosalía, on the opposite coast, along the Gulf of California. "It was a very long vacation," he says. "The ship, I never saw it again. There are two reasons that I have in my heart. One is, I fell in love with my wife. And I fell in love with the lagoon. The whales I met later."

In those years, Pachico continues, there was something that passed for a road, but often months would go by between vehicle arrivals. Until the 1950s water was carried by pack mule to the ten families living along the lagoon. Even today, in the hot summer months, the population was comparatively sparse. "Most of the people who are here now will not be here then. They only come when the cow is fat."

I ask if he could elaborate on the first time a gray whale came up to him, in February 1972. Pachico points to Luis Pérez, a neighbor who has just joined us. "He was my partner. We were fishing at Punta Piedra for grouper and cabrilla. Suddenly, we were among many whales. Hundreds of gray whales swimming in this three-mile-long, one-mile-wide inlet. They would not leave us alone. They would spy-hop on one side of the boat, raising their heads to look at us. Then they would go under the boat. At first, we were very scared."

Pachico had good reason. In the dunes less than a mile across from Punta Piedra (Rocky Point), the narrowest sector of the lagoon, were a series of crosses. They marked the spot where, only a few years earlier, several fishermen steaming along in their *panga* had accidentally struck a whale and been killed. Since that time fishermen passing through this area would slow down and pound on the hull to let the whales know they were coming and, they hoped, keep them at a distance.

Pachico continues: "One of the whales was kind of scratching itself on the front of our little *panga*, and most of her body was underneath, so we could not move. This continued for almost an hour. I couldn't fish because we were using handlines, and I had to lift the line so I would not hurt the whale. Or else anything could happen."

Pachico stops momentarily, as if to catch his breath.

"I don't know what finally compelled me to reach out my hand. The moment I touched the whale for the first time, I felt something incredible. I lost my fear. I was amazed. It was like breaking through some kind of invisible wall. And I kept touching.

"That moment I compare with when my first child was born. It leaves a deep impression in my heart."

Pachico places a hand over his heart, just as he had done when we met a year ago. The gesture is equally genuine. He remembers returning home that night and saying to Carmen, "No fish today." He laughs. "This was a day that the whales owe me, because I could not work!"

Word had spread through the cluster of small wooden shacks edging the lagoon. It felt as though a miracle of sorts had occurred. One of their fishermen had reached out to touch a whale and had returned unharmed! How could this be? In the nights that ensued, by flickering kerosene lamps, Pachico and Luis repeated the story many times. They and the other fishermen struggled to understand. What did the whale want? What was it trying to tell them?

"I feel the whales came looking for us," Pachico continues, "the people from the lagoon. After this, we are better prepared for service. We analyze that the whale—which was named the fish of the devil because it attacked the whalers—now forgives us for hunting it. They are demonstrating the friendship that we never showed them, and that the animals are compatible with us. I think they are smart and that they are showing us how to live. And I believe this is a great lesson for us."

I could not resist asking directly: "How much time does it take you to train the whales to come to the boats?"

Pachico smiles. "Everyone has his own way, and I don't think it would be wise to tell of mine. That is the professional secret."

A little later he elaborates: "First there was one friendly whale. The next year there were more. They teach their young to be friendly, too. It is like the first friendly whale told the others what it was like. How they do this, we don't know. They speak and communicate under the ocean."

Had he also visited the whales in Scammon's Lagoon? "I went on one occasion. But I didn't like the way they ran it, the whale-watch guides. I believe that the sound of the motor is important. If the motor is finely tuned, so that it does not have rough noises, I believe this can be something pleasant for the whales. It is like music, you see. If the music is out of tune, it is not enjoyable. I believe the whales can tell things by the way you use the motor. If you approach them too fast, they will run. When they feel comfortable, they come to the boat. That is a decision of the animals, not of us, the people."

Pachico became the first Mexican to offer whale-watching tours of the lagoon, and since the late 1970s he had worked with several of the leading gray whale scientists: Steven Swartz, Bruce Mate, Bruce Reitherman. Later Swartz would recount how Pachico had come to his rescue at the beginning of his first research season. "When our original commercial outfitter failed to support us like he'd promised, Pachico showed up out of nowhere. 'Put everything in my *panga*, let me show you around,' he said. He told us to stay away from Parmiter Island, just west of Rocky Point, the only island in the area with rattlesnakes on it. He always made sure we had enough gas. When he'd come cruising to the entrance shortly after dawn to pull his lobster traps, he'd stop in for coffee and bring us some of Carmen's tortillas—the best in the entire peninsula."

Pachico recalls of his time with the scientists: "They had learned from books, and I had everything at my sight. We united their thoughts with mine, the theoretical and the practical. I could not ask for more. Now I have three children who are naturalists. They know the English language and the history of the region. That is the most important work, because I will leave behind people who are prepared."

What is it, above all, that the gray whales have taught Pachico?

"The lesson, I apply to the family. To love and appreciate my family, like the whales take care of their young."

> *A gentle joyousness—a mighty mildness of repose in swiftness, invested the gliding whale.*
>
> —Herman Melville, *Moby-Dick*

The next morning, around ten, four of us met Pachico at his house. It was low tide and cloudy. A cool breeze blew from the northwest. At the shoreline along the upper lagoon, Pachico's boat, the *Suzy Q*, waited with its anchor sunk into the mud. He pulled on a pair of hip waders and motioned us to climb aboard. He pushed us out through the shallows by plunging one oar continually into the muck. As the water deepened Pachico lifted himself over the bow and rowed for a while. One time, he told us, he'd broken down and realized he didn't have any oars. So he took his coat and shirt and made a sail out of them.

He started the motor. It was a long haul from here to the north inlet, where whale watching was allowed. I noticed that Pachico steered around the shorebirds as we headed out. We went by the two other small settlements along the eastern shoreline, La Base and La Fridera. After about forty minutes, as the surf near the mouth of the lagoon became visible, I began to see the gray whales ahead. They were breaching and spy-hopping.

Pachico slowed down. He rolled up his sleeves and leaned over the side. He began rapping his knuckles on the metal hull. I wondered what this sounded like underwater. "When I am near the whales," he said, "I have a habit to make a noise that is low and strong, but also gentle and enjoyable. Trying to do a variety of music with my fingers. The sound should be always in the same place, this strong part of the *panga*, but it should change."

About ten other boats were in the vicinity. There didn't seem to be a lot of action. A sunny day is better than a gray one, Pachico said. But some days the whales just don't feel like coming up to boats anyway. You never knew. We waited. Pachico maneuvered slowly around the lagoon, unsmiling, his square jaw set, his obsidianlike eyes riveted on the water. More breaches near us. A spy-hop that seemed to last forever, maybe twenty feet from our stern. Pachico had an uncanny sense of where the whales are. Was I surprised? Yet they would come only so close.

We'd been out about two hours when the lone whale came calling. I watched it approaching for well over a minute, from a considerable distance. Pachico's rapping of knuckles seemed to take on a more distinctive rhythm. I tried it, too. About to bypass our stern, the whale turned to face us. It was a fully grown adult, at least thirty-five feet long. Its flipper momentarily touched the boat, like a caress. It skirted within inches of my outstretched hand, and the hands of my friends, and went right to Pachico. He rubbed its head lightly and smiled. I could actually see the whale's eye close, like that of a cat in the blissful state of being stroked. Then the whale bowed its head and dove.

It was like two old friends who could meet only occasionally exchanging a greeting whose profundity was in the silence that followed.

Pachico's eyes were moist. Now I understood fully why he was called Guardian of the Lagoon. In that moment I didn't care if I touched a whale this day. And I wouldn't. I could only bear witness.

The sun emerged from behind us. Through a vaporous spout that lingered, backlit by the morning light, came the arc of a rainbow.

When we returned that afternoon, I raced to my Volkswagen and drove like mad toward the Baja Expeditions camp. I was supposed to meet Homero and Betty to go on another whale watch at two, and I was late. Somehow I managed to cut ten minutes off the usual travel time. The Aridjises and several others were just walking toward the *panga* when I arrived. As we fastened our life jackets and waded out to the boat, I heard one of the men say: "Well, on our trip this morning, we didn't have anybody come hang out with us. But we saw the Pink Floyd!"

The fellow who mentioned the Pink Floyd was Chris Peterson, the Baja Expeditions naturalist. A protégé of two of the leading gray whale scientists, Theodore J. Walker and Steven Swartz, Chris was formerly the education director at San Diego's SeaWorld and about to become head of the Norfolk, Virginia, zoo. He'd been coming to the San Ignacio Lagoon for twenty-five years and had seen it all. "It was pretty wild this morning," he explained, sitting next to me near the stern, "classic courtship routines. I'd anticipate more later this afternoon. During the slack tide, when the current isn't going in or out, a lot of times the whales will take that opportunity to start courting and breeding."

Scammon was the first to describe aggregations of whales at lagoon entrances, later to be called staging areas for mating. He was also the first to note that, among grays, females are somewhat larger than males. However, as a gentleman of the Victorian era, Scammon never commented specifically on any mating habits he observed, except to mention briefly that the gray whales' polygamous ways were "of the Turkish nature." Even this rather casual statement—which appeared in his article about gray whales for *Overland Monthly*—was omitted from *Marine Mammals* a few years later. He was the first to point out, however, that the female's period of gestation is a little over a year. Scammon wrote: "This statement is maintained upon the following observations: We have known of five embryos being taken from females between the latitudes of 31 degrees and 37 degrees north, on the California coast, when the animals were returning from their warm winter haunts to their cool summer resorts, and in every instance they were exceedingly fat, which is quite opposite to the cows which have produced and nurtured a calf while in the lagoons; hence we conclude that the animals propagate only once in two years."

John Dean Caton, in an article titled "California Gray Whale" that appeared in *The American Naturalist* in 1888, made this behavioral observation: "Although the fiercest fighters of all known whales it has not been known that they are quarrelsome among themselves. So far as [is] known, peace and quiet prevails among all the members crowded together in the upper ends of these water enclosures or lying-in hospitals [the lagoons]. But few males have been observed to intrude themselves into the privacy of these retreats."

Roy Chapman Andrews, of the American Museum of Natural History, expanded upon male-female gray whale relations in 1914, writing specifically about the Korean or Western grays. "The male . . . at all times shows strong affection for the female and Captain Melsom tells me that during the migration, when a school of males led by one or two females is found, if one of the latter is wounded, often the former will refuse to leave until she is dead. One day when hunting a pair he wounded the cow and the bull would not leave, keeping close alongside and pushing his head over her body. Later he struck the male with a harpoon but did not get fast and even then it returned and was finally killed. Captain Melsom assures me, however, that if the male is killed the female will seldom remain."

In 1927 Laurence M. Huey made his way to San Ignacio Lagoon late in the winter season and recorded witnessing a "frustrated" and "desperate" male attempting to mate with each of two females with calves. Almost thirty years later, in 1955, during the springtime journey north from the lagoons, the Associated Press reported from San Diego that "the sex life of Pacific gray whales is being spied on as never before during the largest migration of the barnacled mammals ever officially observed." In some areas Californians watched from their front porches as "at least two bulls [are] interested in each romantically inclined cow." According to Dr. Raymond Gilmore of the U.S. Fish and Wildlife Service, one male generally "seemed to have priority," but "she accepts another admirer or two without any trouble resulting." Gilmore observed that the French would call such a relationship a ménage à trois. Each female appeared to receive about equal attention: "There are no belles and no wallflowers."

This was the first indication that mating took place beyond the lagoons. Today scientists have come to realize that *most* females become pregnant either during the latter phases of the southbound migration or outside the lagoons themselves, in slightly offshore waters. This is not to say, though, that reproductive *behavior* doesn't continue in the lagoons. For some time one popular theory among scientists was that a "helper" male in the vicinity often served as "a stabilizing agent keeping the copulating pair together," as William Samaras wrote in 1974.

This notion was dispelled by Jones and Swartz, who observed courting and mating on many occasions near the entrance to San Ignacio Lagoon between 1978 and 1984. They wrote: "From our eighteen-foot observation tower we watched as many as five hundred whales a day entering and leaving. Here amorous giants lunge through the water, trailing plumes of vapor from their blowholes. . . . We learned that gray whale mating is a far more complex and uninhibited affair, with males and females copulating with an assortment of partners. This promiscuous activity sometimes blossoms into giant free-for-alls involving as many as eighteen to twenty individuals at a time." Preceded by boisterous foreplay, these mating bouts "of truly herculean proportions" lasted for hours. Males did not appear to be either helping one another or competing for a certain female but simply taking turns. Sometimes the males would gently nudge each other, but their interactions never seemed aggressive.

It's the females, in fact, who are more likely to exhibit aggression. Bruce Mate, who's also spent many years studying gray whale behavior at San Ignacio Lagoon, has observed: "The females are in control of mating—and they don't mate when they aren't ready." Both Mate and Swartz have seen females with calves being chased by groups of males. Swartz has observed the pairs attempting to avoid this by "hugging the shore, 'tiptoeing' by, not to attract attention. When one of these freight train chases does occur, the problem is that the calf can get left behind or stressed. I've seen mothers turn around and confront a group of males with tail slashes and other means of fending them off."

Of gray whales, the majority of which reach sexual maturity around their eighth year, Mate adds:

> I've seen mating happen with a whole gang at one time, but not *necessarily.* It's not uncommon that a female has more than one male trying to gain her attention. I've seen females lay at the surface on their backs, so their genitals were not exposed in the water, and males parked on either side of her with their penises out of the water and probing around trying to gain some advantage. Of course, they know she's got to roll over eventually to breathe, and then one of them's going to have an opportunity. However, I've also watched one female get into mating relationships with three different males in a forty-five-minute period. I mean, *white water!* Very much appearing to be a consenting relationship, going from one to the next.

Which leads us to a theory advanced in 1986 by Dr. Robert L. Brownell, Jr., then of the U.S. Fish and Wildlife Service, and Dr. Katherine Ralls, of the Smithsonian Institution's Department of Zoological Research, in a paper titled "Potential for Sperm Competition in Baleen Whales." They noted that, "in species where a female usually copulates with more than one male," males tended to have relatively large testes. Males did not compete directly but rather through "attempted displacement or dilution of the sperm of rival males." Animal behaviorists had previously looked at this phenomenon in chimpanzees, which possess large testes and are known to be lascivious; by contrast, gorillas have much smaller testes and each male has his own harem. In the whale world it turned out to be a similar story. Male humpbacks, with rather small testes, compete fiercely for females, going so far as violent physical contact, which can wound their rivals. Blue whales and bowhead whales, also with smaller testes, are far more monogamous than right whales and gray whales. (For the record, male gray whales are quite well-endowed. The testes of an adult weigh over forty-five pounds apiece, and their fibroelastic penises are more than six feet long. It is also noteworthy that, according to Brownell and Ralls, "the average testes weight in gray whales on the migration to the breeding grounds is about 70 percent greater than on the return journey.")

However, none of this can explain the astonishing sight witnessed on more than one occasion by the marine scientist Dr. Marilyn Dahlheim at the San Ignacio Lagoon. While conducting acoustical studies there in the 1980s, she saw bottlenose dolphins "mixing right in among gray whale courting groups. I mean, it's six o'clock in the morning and you're just getting up, and, oh man, this is better than Greenwich Village! One time, there was a little bottlenose wedged right between the whale's Pink Floyd and its belly. I could have sworn the dolphin had a smile on its face."

Steven Swartz had a rather dumbfounding encounter of an even more intimate variety. "We were out floating around one day watching an amorous group of gray whales," he remembers, "when we drifted into the middle of them. A male suddenly came belly-up, rolled over, and flopped his penis right into the boat. Fortunately, he left. As far as I know, we are the only people ever to have a Pink Floyd enter our boat when we were *in* it. It was an inspiring moment."

Now Chris Peterson suddenly pointed to a thrashing around a hundred yards off our bow. About ten whales were speeding together into the heavy currents toward the mouth of the channel. "Look at that one animal leading out front!" Chris shouted. "He's always got his ventral side to the group. Sometimes they'll do races around the lagoon. Oh, look! They're starting to spin around. Check out the sparring and spy-hopping!"

Sure enough, the whales had slowed down, several with their heads craned out of the water. Some scientists have speculated that spy-hopping may serve as a means of sex identification in such moments. I watched awestruck as several large whales, apparently males, surrounded a female. I heard the splashing of tail flukes. One of the males looked to be placing a flipper over the female's back. She slowed. The whales

The "Pink Floyd" (© Bob Cranston/Innerspace Visions)

bobbed up and down and rolled over each other and then quieted and seemed to be touching tenderly with their bodies and flippers.

Then, could it be . . . ?

"Thar she blows!" Chris shouted. "Pink Floyd to starboard!"

It was only a momentary glimpse. These whales weren't in the mood for an audience, and who could blame them? They swam on toward the channel opening.

We motored over to an island with a huge sand dune, where a forty-foot-long gray whale was said to have washed ashore the previous week. Some of those from our boat went off with Chris to try to find it. I had no desire to do so. Nor did Homero. We walked along the beach together, looking for shells. He gathered up a handful of sand dollars and handed them to Betty for safekeeping. Homero was subdued, contemplative. I didn't dare intrude on his thoughts.

Later he would send me the poem that had come to him on this day. It was called "The Eye of the Whale" (Genesis 1:21) and dedicated "To Betty."

Outside the Baja Expeditions tent that night, several members of the

European Union (EU) Parliament were sitting with Homero on canvas folding chairs and sipping drinks. Homero was reminiscing about his years as Mexican ambassador to the Netherlands (1977–79), when he became acquainted with Prince Bernhard. "He spoke Spanish, more or less, and he loved Mexico. He said it was one of his favorite countries." They'd been out of touch for years, though, when Homero's phone rang one evening. "The voice said, 'El Príncipe is calling you.' I thought it was a joke. But yes, it was Prince Bernhard! To my surprise, he said to me that he had been a few times to the San Ignacio Lagoon. He'd heard that I was leading opposition to the saltworks and he wanted to support me. 'What can I do?' he asked. I told him to write a letter to Zedillo."

Prince Bernhard is said to have reached out to England's Prince Philip, the titular head of the World Wildlife Fund (WWF). Prince Philip then enlisted the Mexico City–based branch of WWF to prepare a series of studies on the economic contributions of the lagoon's fisheries and management of marine resources in the area. As it happens, his own interest may originally have been sparked by his wife.

One morning, the whale biologist Marilyn Dahlheim told me, her research team was awakened at Rocky Point by "a very loud black helicopter. It landed, whipping the flaps of at least eight tents. Out jump these Brits in white coveralls. 'Bloody good morning to you. The Queen wants to go whale watching.' We could see three or four big ships at the mouth of the lagoon. One of them was the *Britannia*—the emblem was visible on the side—with these two Mexican warships as escorts. The handsome prince was onboard. And pretty soon, there she was, Queen Elizabeth, out there in a *panga* wearing her little white gloves—and touching gray whales!"

Now, some years later, a Dutch parliamentarian named Hemmo Muntingh began talking excitedly to Homero, in English, about his afternoon's experience. "We were in the middle of the water, nowhere, when suddenly this huge female gray whale came, together with her young one. They stayed with us for half an hour. The young one was on this side of the boat, then in front, then on the other side. We could caress it—and I *kissed* it! I really *kissed* it! This was a fabulous event."

Homero sat there, nodding his head over and over, a smile across his face. Muntingh went on, "The mother was just lying there as if she was watching the young one, and sometimes she came up and rocked the front of the boat. I must say it was sometimes a little bit frightening. But then when she came and looked at us, you were not scared at all, just happy. I can't explain it. In fact I don't want to, because at that moment it has nothing to do with rationality, it's just emotion."

Homero gazed into Muntingh's eyes and nodded again. The group was joined by Luis H. Alvarez, chairman of the Commission on Environment and Natural Resources in the Mexican Senate. "Once you get the chance to see these whales," Alvarez said, "I think it is a natural reaction to fall in love with them. And to want to do the utmost so this continues to be a place where they can come and feel safe and secure."

It was astounding, and heartwarming, to hear these politicians describing the whales so paternally. The EU's Parliament had 626 members from fifteen countries, speaking eleven languages. Most of its representatives here were part of the fifty-member Committee on the Environment. Before long they would be voting on a new European free trade agreement, which the Mexican government desperately wanted. "During the negotiations," Muntingh said, "we will try to bring such attention to this problem of Exportadora de Sal that they will stop the salt project." Great Britain's David Bowe said he was part of the EU's delegation to the Japanese Parliament, where he intended to raise the issue as well.

The parliamentarians, many of whom are lawyers, believed that it ultimately came down to a legal issue: could the Mexicans really allow such an industrial development in a protected biosphere reserve? They'd read the legislation, and they didn't think so. Homero wasn't so sanguine. "The problem in Mexico is that the politicians don't respect democracy," he replied. "You have a country where the legal processes are slow and the media is manipulated. We have to win this fight through world public opinion, through the people who love the whales."

A rose-colored sun was setting above the tabletop mesas. Homero turned to me and said quietly, "Always, as my daughter said to me, it is very important to come back here. To touch the whales, you get energy for your cause ahead. Fresh energy. Then you know you have to fight very strongly again."

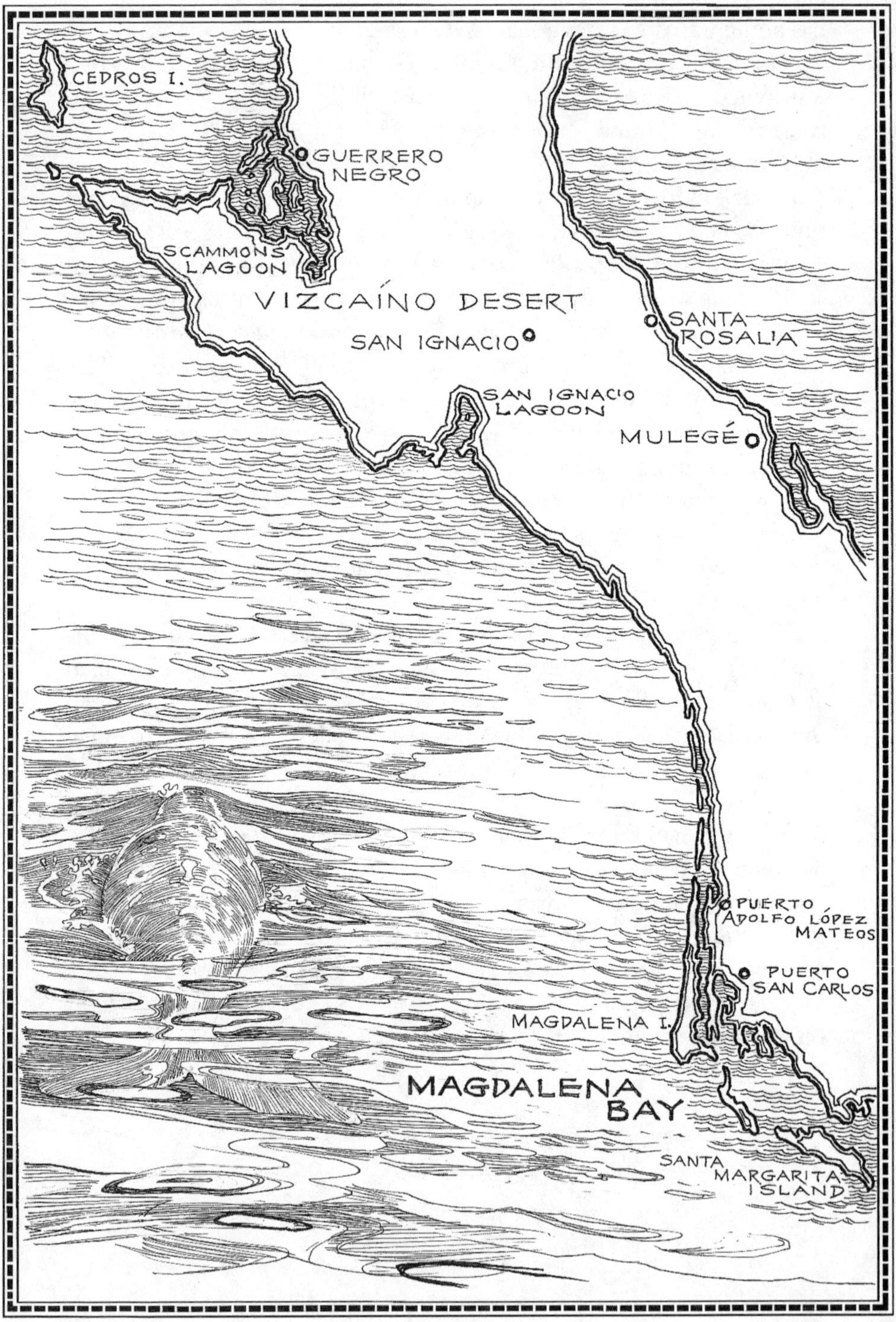
CEDROS I.
GUERRERO NEGRO
SCAMMONS LAGOON
VIZCAÍNO DESERT
SAN IGNACIO
SANTA ROSALIA
SAN IGNACIO LAGOON
MULEGÉ
PUERTO ADOLFO LÓPEZ MATEOS
PUERTO SAN CARLOS
MAGDALENA I.
MAGDALENA BAY
SANTA MARGARITA ISLAND

Chapter 7

Journey to the Pillars of Salt

There go the ships; there is that Leviathan whom thou hast made to play therein.

Psalm 104:26

Then the Lord rained on Sodom and Gomorrah brimstone and fire. . . . But Lot's wife behind him looked back, and became a pillar of salt.

—Genesis 19:24, 26

The first group of distinguished visitors, including the Aridjises, would leave San Ignacio Lagoon the next morning. It would be several days before the next contingent, including Bobby Kennedy and several Japanese dignitaries, showed up. I'd been told to expect John Spencer's arrival by then, too. In the interim I decided it was time to take a firsthand look at the existing salt fields by Scammon's Lagoon and to glean the perspective of Exportadora de Sal. Betty Aridjis suggested that, on the way, I stop at the fishing village of Punta Abreojos, where ESSA planned to build its new pier and loading facility. "None of us has had a chance to go there," Betty said, "and you could find out how the local people feel about the expansion."

Scammon, in his 1869 report on the resources of Lower California, had provided a geographical description of the then-uninhabited region of Punta Abreojos. After many miles "bold, rough coast" gave way to a "more inviting appearance." A ship's traveler came

> near to Point Abreojos, which is in latitude 26°4', longitude 113°42'W; then again occurs a sandy coast, behind which are found small lagoons, with passages into them from the sea, that

> will only admit the smallest craft, or ordinary open boats, in very smooth weather at high tide; the shore line at this point makes a sharp turn, running to the northeast, a distance of twenty-eight miles, then turns abruptly again to the southeast, forming the open bay of Ballenas [the entry to San Ignacio Lagoon]. . . . Off Point Abreojos, an outlying reef of rocks extends six miles. Abreojos, or "Open-Your-Eyes Point," seems a fitting name for it; there is a passage between the reef and the main, that may be used in case of necessity.

From San Ignacio Lagoon with a veteran guide, it's possible to reach the settlement relatively quickly over a series of desert roads. Otherwise you're likely to get hopelessly lost. Far better for a lone traveler to take the roundabout, five-hour route: return to the town of San Ignacio, head west down Highway 1 for sixteen miles, then turn left onto a decent enough graded road known as Sierra de las Mulas for another forty-five miles.

No matter the dusty streets, Punta Abreojos is a pretty little place. In 1998 the town had celebrated its fiftieth anniversary. Quaint homes of wood and occasional cinder block are colorfully painted and often have backyard gardens. Where the village meets the sea, adjoining the fish-packing plant are the offices of the fishermen's cooperative. Its president, Antonio Zúñiga, is a lean, mustached man in his thirties. For our conversation he was joined by Jorge Salorio, head of a vigilance committee, and the corporation's secretary, Isidro Arce.

It quickly became clear that all three men were vehemently opposed to ESSA's proposed expansion to their vicinity. Indeed, they maintained, the only supporters were people living in a nearby *ejido* who planned to "rent" space for new housing. Yes, the company had sent representatives to explain the alleged merits of the project to the people of Punta Abreojos. The road back to Highway 1 would be paved. Drinking water, electricity, and sewerage would vastly improve. Indeed, a whole new adjacent town would be built—subsidized housing for over fourteen hundred temporary workers and as many as two hundred permanent employees and their families. "They say it will help our community evolve," Antonio said.

In his view evolution was already proceeding quite well, thank you. The town's population had grown to more than a thousand. Fishing rights were passed on from one generation to the next; the cooperative had an exclusive concession from the Mexican government to run a fishery that generated a gross profit equivalent to $2 million annually. The cooperative had told Exportadora since the beginning that they feared the salt project would have a detrimental impact on their abalone, clam, and lobsters, as well as on oyster beds and the potential for cultivating scallops. They didn't need any more problems—the poaching situation was bad enough.

Jorge explained his role as supervisor of the Cuardrine Watch, teams of local men who took turns keeping an open eye on the Abreojos shoreline. It wasn't only the fish "pirates" they had to worry about. It was the heavily armed, high-speed boats of the narco-traffickers that cruised the shoreline. Out back of the cooperative the vigilance committee had installed a radar tower in 1997. This enabled them to distinguish between the drug vessels moving north, which the cooperative had no choice but to ignore, and the nighttime shellfish raiders from places like El Cardón, an "outlaw fish camp" on the southern fringe of San Ignacio Lagoon. Over the past several decades El Cardón had been settled largely by fugitives from the Mexican mainland.

One of Punta Abreojos's best-producing shellfish locations was 6.5 miles south of town—precisely where ESSA wanted to build the pier for the saltworks. "We understand, of course, why it is more cost-effective for Exportadora to load the salt here," Antonio said. "Because at Guerrero Negro, they must first send it on barges more than fifty miles out to Cedros Island, then store the salt for transferring onto the big container ships. They say there is no problem at Cedros with contamination. But Cedros is an island with very deep water and a strong current. It's not the same here. This is an area, too, of powerful hurricanes, almost every year."

I remembered the whale scientist Bruce Mate's having raised the same issue when the project became public knowledge early in 1995. "This is one of the major concerns of the entire plan," he wrote in a letter to colleagues. "The dock is exposed to the southward winter storms and summer hurricane-generated waves. Should it fail, the backup process of shipping salt would probably have to be a tug and barge operation located near the mouth of the lagoon. With the dredging of the mouth entrance and a turning basin for the tugs and barges, this could be a major impediment to gray whales continuing to use San Ignacio Lagoon."

Now Isidro added, "Where they plan to put the dock here, the whales also pass. Would you like to see?"

I followed Jorge and Isidro in their pickup, past a red-and-white beacon lighthouse and along a sandy coast road. We passed a group of schoolchildren all in uniform, the boys in khaki and the girls in white. They were carrying antidrug signs in some kind of school demonstration. Eventually we reached Punta del Pateki, whose name means "a place where there's a spring" or, as Jorge charmingly put it, "an area where the water comes out of the land." Here was the blueprinted spot for a 1.2-mile-long pier jutting out from a rocky shoreline into the Pacific. I tried to picture it: long conveyor belts moving crystallized salt from a massive land stockpile to deepwater cargo ships that would arrive every three days and take a full three days to load.

Suddenly Isidro shouted, "There's a whale!" Less than a mile out a gray whale could be seen moving rapidly in the direction of the lagoon entrance, some fifteen miles away. Then came another. And another. Last week, they said, 250 gray whales came by here. It was obvious the pier's pilings would be directly in the path of their migration. Not to mention the potential risk of the whales colliding with commercial ships.

We drove back into town, where I toured the fish-packing plant. About forty people work there. I was impressed with the cleanliness and efficiency. A half ton of abalone gets processed in a day. Women shell the abalone at one table and put it in cans at another. Much of it goes to Ensenada, where it's shipped on to Europe. Out back the rainbow-colored shells constitute a small mountain.

Bonifacio Arce lives on the main street of Punta Abreojos. When he arrived fifty-five years ago from the town of San Ignacio, there were two houses on the shore, and even these were soon lost to a storm. He's spent his adult life manning the oars in a three-man fishing *panga*, and his massive arms show it. He's still working at the age of seventy-three. Over the years Bonifacio had seen literally thousands of gray whales pass close to his boat. "If the salt factory comes," he says, "it's going to scare them."

Jorge Salorio invited me to spend the night with his family. After a delicious dinner of broiled lobster, he brought out a series of aerial photographs. Viewed from above, the various landscapes took on animistic forms. Scammon's Lagoon looked like some kind of dog or hyena, its head turned as if to see what was coming from behind. The mountainous region of the pre-Columbian cave paintings appeared to contain the wizened face of an Indian shaman. As strange as it seemed, San Ignacio Lagoon actually took on the shape of a whale.

Jorge pointed out where ESSA planned to put its industrial salt operation. "This is the Estero de Coyote," he said, "with the richest mangrove wetlands in this region. And this"—he pointed to an area just east of the *estero*—"is where the concentrated salt brine will be pumped from the northwest side of San Ignacio Lagoon to further evaporation areas." The final area for salt crystallization was to be north of the *estero*. From there the crystalline salt would be harvested by loaders onto conveyor belts carrying it to a cleaning facility and a one-million-metric-ton stockpile by the Abreojos pier.

Jorge next pointed to the nearby arroyos. "During the fury of a *chubasco* storm," he said, "all that salt and brine will wash right down into the Estero de Coyote." Now I understood what Francisco Flores had been talking about. Flores is a biologist specializing in mangroves, who had spoken up at the meeting of lagoon fishermen and politicians at the Baja Expeditions camp. "I'm very worried about the discharge of brine to the Estero de Coyote," Flores said, "because mangroves will not support that kind of salinity. The ocean is 35 parts per thousand. Mangroves die after 70 parts per thousand. The brine is around 240. The *estero* is too small to have the capacity of diluting all that. And in these mangroves you find many juvenile fishes, and crustaceans in the larval state, that are vitally important to the fisheries of the region."

Jorge insisted I see the mangroves in the morning. With his *panga* hitched to the back of the pickup, we drove out across the dunes. When we launched and set off along the edge of the estuary, I realized these must be Scammon's "small lagoons, with passages into them from the sea, that will only admit the smallest craft." Dozens of little islands were ringed with red-and-white mangroves, growing at their northernmost extent in the Pacific. They were tall and thick and healthy, looking not unlike three-leafed clovers, their tangles of branches extending deep into the low marshland. At their "feet" played hundreds of tiny silver lisa fish. On some of the mangrove stems, Jorge said, lived a small but delicious "tree oyster."

The abundance of bird life was breathtaking. Brown pelicans dove for lisas and clams. Long-billed white ibis strutted on the sandbars. Where the local people were seeding Japanese oysters, dozens more shorebirds perched beside the cultivation beds. As we passed over the ubiquitous eelgrass into more open waters, a bottlenose dolphin suddenly surfaced beside our boat. We gave chase as the dolphin stayed sportively out of reach.

The last thing I wanted to do was leave my meanderings among the "mangles." But I had an appointment 125 miles away, at the corporate offices of Exportadora de Sal.

•

The journey to Guerrero Negro still looks much as Scammon once described it: "The surrounding country for miles . . . is a sandy desert, of decaying trap formation, with occasional clusters of dwarf shrubbery, and the universal cactus and prickly pear, struggling, between an arid climate and sterile soil, to maintain existence. . . . Immediately at the shore line, low, drifting sand-hills predominate, behind which lie three lagoons, bearing names given by whalemen, as follows: Upper Lagoon, Black Warrior Lagoon, and Scammon's Lagoon."

The captain never revealed how he learned about this region. He could have heard stories from Mexicans or overland travelers he met at Magdalena Bay. The future Scammon's Lagoon, some 230 miles north of Magdalena, is known to have been visited from land by Spaniards, Mexicans, guano collectors, and foreign fur traders in search of sea otter pelts. It was then known as Laguna Ojo de Liebre or Jackrabbit Spring and was the sole source of freshwater for hundreds of square miles along the Vizcaíno Desert. In centuries past the lagoon had been an important fishing ground for the Chumash Tribe. There were legends of Indian women wearing skirts fashioned from whale tendons, and of young tribesmen stringing their bows with similar material.

It was after a dismal season at sea that Scammon decided to try to find this lagoon. In November 1857 his latest brig, the 135-ton *Boston*, returned to San Francisco virtually empty, having spent more than six months seeking whales, seals, and sea elephants along the California shoreline. Scammon was then thirty-two, and he had a reputation to uphold. He convinced all but three of his thirty-five men to remain aboard for the winter. The whale-oil market price was as high as forty dollars a barrel. Any gray whale taken could be expected to yield between twenty-five and forty-five barrels. If they could locate a yet untapped whaling ground, a bonanza might be reaped.

Scammon reckoned that the lagoon must be about midway down the Pacific side of the peninsula, below a whaling anchorage at Cedros Island and somewhere in the low country he knew existed on the eastern shore of Sebastián Vizcaíno Bay. This lagoon, he was told, would be very dangerous for a whaler both to enter and to leave. Passage was accessible only through a narrow entry channel and over a shallow sandbar, between breaking waves that pounded against a long, white dune-capped shore.

Three wooden whaleboats were stowed aboard the *Boston*. It was a three-masted brig, with square sails rigged to the upper portions of the mast nearest the stern as well as a fore-and-aft sail called a spanker. A whale ship carried less sail than a merchant vessel, thus requiring fewer men aloft and allowing more hands for the whaleboats.

The *Boston* sailed south for California's Catalina Island, where a small schooner tender called the *Marin* made rendezvous. Scammon reasoned that the tender, which drew only four or five feet of water, would be particularly useful in lagoon whaling. It could follow the whaleboats into the channel and assist in towing captured whales back to the "mother ship" at anchor in the lagoon.

Probably sometime around Christmas the *Boston* paused for the night at Lagoon Head, where the lee side provided shelter from the elements. The whalers were not far below the twenty-eighth parallel, which today delineates the border between northern and southern Baja, and a time-zone change from Pacific to Mountain. The following morning Scammon sent the *Marin* and three whaleboats south, to "sound out" a channel he believed might lead into the lagoon.

Two days passed. A messenger returned to Lagoon Head with the news. Yes, there were three inlets at the far end of Sebastián Vizcaíno Bay. Each was filled with gray whales. Two of the inlets appeared impassable for the *Boston*. The tender, however, had passed into one estuary without difficulty. There appeared to be enough water on the sandbar just outside the mouth of the lagoon for the larger ship to follow, if there was a brisk breeze.

That afternoon the brig and the tender met at the heaviest-shoaled sector of the lagoon's entry point. The winds died, and both were forced to anchor. A dark, misty night came, accompanied by increasingly heavy swells. "Nothing could relieve us from our perilous situation but a strong land-breeze, to take the vessels back to the first anchorage or to sea," Scammon would recall. "Not a soul on board slept during that night."

The next day at noon a north wind carried Scammon's ship along the edge of the barrier bar and southeast along the tricky tidal currents. He was inside the outer reaches of the lagoon. "The waters were alive with whales, porpoises, and fish of many varieties," Scammon recorded. "Turtle and seal basked upon the shores of low islands studding the lagoon; and game of many species was so abundant that shoals of acres in extent, left bare by the receding tide, would be closely covered with geese, duck, snipe, and other species of sea-fowl."

It was too early to commence whaling. So, while the *Marin* headed off on another scouting mission, the *Boston* crew cast anchor behind a sheltered point and went ashore. They would need firewood. A bit of a hike up the beach, they found a shipwreck. This, Scammon realized, must be the *Tower Castle*, a British whaler that had run aground thirty years before. He knew the story. A date cut into a rock indicated that the *Tower Castle* had been lost in 1827. The ship was bound from the Pacific to Europe when it foundered. The crew managed to salvage enough matériel to build a comfortable shanty house. An officer then headed off with several men in one of the whaleboats, looking to obtain another vessel and come back for their companions.

"But before their return," Scammon reported being told, "the supply of fresh water became exhausted, none could be found by digging, and a fruitless search of the back country for springs or standing pools in the ravines only hastened their end." All that remained when the rescue party arrived was a journal kept by the stranded officer in charge. His record grew shorter and less intelligible each day. It noted that the others had all died of starvation and thirst. The last line read: "Feeling the same symptoms as did my dead shipmates, it is but reasonable to expect that my time will come soon."

Now, strewn along the beach, were the remains of casks, broken crockery, and some of the ship's spars. But if Scammon saw stumbling upon the *Tower Castle* as any kind of omen, he kept it to himself.

Scammon's three whaleboats were moored to one another near the shore. It was impractical to haul them up at high tide, and a lone boat keeper had stayed behind. At some point while the others were onshore, the boat keeper decided to take a bath. By pulling out the bottom plug, he converted one of the whaleboats into what seemed a handy tub. The surf came up again. Quickly the boat grew waterlogged and, when the frightened keeper moved too far to one side, capsized. Thrown into the lagoon, he started swimming toward the beach. A swift current uprooted the anchor. Suddenly, all three whaleboats began drifting through the breakers on the ebbing tide, out of the passage toward the open sea.

Several Polynesian crew members, called *Kanakas* by whalemen, witnessed the trouble from the shoreline. They gathered pieces of plank as surfboards and headed off through the rolling waves to try to salvage the small boats. The ship's carpenter, an expert swimmer, shed his clothes and plunged into the surf. The men had almost reached the three floating boats when the anchor, dragging along the bottom until then, came loose altogether. The currents swept both the *Kanakas* and the carpenter out of reach. The *Kanakas* managed to swim back to shore, exhausted. The carpenter was nowhere to be seen.

A young lookout climbed to the top of a sandhill and waited. Finally he glimpsed one of the boats being carried back on the flood tide. A dozen men dove into the sea and swam out again. Risking their lives as they grappled with the boat in the surf, they managed to pull it well beyond the high-water mark. Now the second boat appeared in the flood tide, and the third. While the rest of the crew cheered, all three vessels were secured onshore. The voyage could proceed after all.

Still, almost another week would pass before whaling began. First came a vain search for the lost carpenter's body. It took three days to transport the boats along the beach to a spot where they could be safely launched into the lagoon. At this point a northern gale struck, not abating for another three days. But Scammon remained determined to move forward. The gray whales were out there in abundance, surfacing and spouting, moving back and forth between the outer basin and the inner reaches of the lagoon.

Soon the slaughter commenced. When the *Boston* finally arrived back in San Francisco in the summer of 1858, it was carrying seven hundred barrels of oil, "so deeply laden that her scuppers were washed by the rippling tide." Scammon had taken only eight months to fill his ship, a task that often took as long as four years. The voyage earned the profit he had been hoping for. About fifteen thousand dollars from the cargo was divided among the *Boston*'s owner, Tubbs & Company, and the captain and his crew.

Scammon did not name this lagoon after himself; rather he dubbed it Boston Lagoon, after the ship that brought him there. The Scammon's Lagoon designation soon came from other whalers, for his secret did not take long to leak out. One of his mates wrote to a brother, commander of a North Pacific whaler, about the discovery. The news spread quickly among whalemen making an autumn rendezvous in the Sandwich Islands. When Scammon returned to the lagoon the following winter, he was far from alone. Of eighteen other vessels that showed up, half crossed the passage. Scammon operated this time from a much larger ship, the *Ocean Bird*, with a small squadron of accompanying vessels. His wife, who sometimes accompanied him on his voyages, looked on as the crew killed forty-seven gray whales in three months, yielding a whopping 1,700 barrels of oil.

After four years of intensive whaling, Scammon's Lagoon produced, by his estimation, some 22,250 barrels of oil, worth a total of $333,750. By then it would be virtually devoid of gray whales.

Another source of potential wealth remained to be exploited. Scammon wrote that "a good channel is found along the south shore, reaching to near the head, where is found an extensive salt-field, called Ojo Liebre." One of the first nonwhaling vessels to arrive here, he added, was "the brig *Advance*, from San Francisco, California, bound to Scammon's Lagoon, for salt, [and] wrecked between Black Warrior and Upper Lagoon."

As recently as 1955, when Dr. Paul Dudley White came to Scammon's Lagoon to try (unsuccessfully) to record the heartbeat of a gray whale, he wrote that "official charts show only an uncertain outline for much of it. Neither its twisting channels nor its wide shoals and tidal sand flats have been accurately sounded or marked with buoys. . . . Only an occasional turtle hunter visits Scammon's barren shores. No one lives there. We saw a few deserted weather-beaten shacks and the blackened remains of campfires surrounded by charred mounds of turtle shells and bones. How such camps exist for more than a few days is hard to imagine. The nearest fresh water is a rabbit-haunted spring four miles from the head of the lagoon."

Dr. White noted that he followed the same course into the lagoon as Scammon had almost a century earlier. When the gray whales failed to cooperate with the doctor's experimental plan, the voice of White's partner, Donald Douglas, boomed over the radio:

Old Captain Scammon,
He sure was right—
These mama whales
Can stand up and fight!

The team called themselves the Captain Scammon Club and intended to reconvene the following year for another attempt. They never did.

"Years ago," White added of his trek, "salt was dug from extensive flats near the lagoon's head, and large ships negotiated these tricky channels. Now an abandoned pier and rusty remnants of a narrow-gauge railway are all that remains."

All that had changed abruptly, however, in the mid-1950s. Soon after my arrival in Guerrero Negro I heard the story from Jorge Cachuma, the portly and loquacious owner of the Ballenas Motel, where I checked in for sixteen dollars a night. He has resided in Guerrero Negro for forty-two years and remembered one old fisherman, Don Miguelito Aguilar, who lived over near where the airport is now. (I realized this must be the Aguilar family related to Pachico's wife.) Don Miguelito caught fish and sea turtles every day. He preserved them in the naturally occurring salt around his little house. He'd haul his catch to barter at the old mining town of El Arco, twenty-eight miles away, bringing along his black dog, Pedro Viejo, and leaving his wife alone for a week. There wasn't another human being living for miles in any direction.

When ESSA came to Guerrero Negro, Jorge continued, "They started out in tents. There was no port for the machinery coming in by barge to level out the land. They made a dock. They built the facility at Cedros Island to keep the salt. They needed to dig a canal, but they couldn't make it straight because of the tides. Someone said to the engineers, follow the coyote's path. Because on low tide, the coyotes went out to the island to feed on the birds."

Jorge had gone to work at ESSA the year it started salt shipments, in 1957, working his way up in the accounting department before retiring in 1993 and going into the motel business. Who could have imagined how it all would grow? he marveled. Today ESSA had 1,000 employees, 750 here in Guerrero Negro and 250 more on Cedros Island. The company built and maintained the workers' houses. They'd give you a water heater and they'd fix your bathroom. At the company store you'd get up to a 50 percent discount on groceries and anything else you wanted. Sort of like the American Army PX, Jorge figured.

I set out to have a look. Guerrero Negro now has a population of almost thirteen thousand. Emiliano Zapata Boulevard, the dusty main drag, is unpaved and lined with taco stands, motels, pharmacies, restaurants, and bars. The street ends at a bank, a big grocery store—Tienda ESSA—and a small, palm-filled park on the right. The brick-and-concrete complex of the ESSA headquarters is on the left. The company's logo, emblazoned across the front gate, is a gray whale.

Behind ESSA's gates tidal flats begin their irregular, crescent-shaped wanderings to the north through the salt ponds and, ultimately, to Scammon's Lagoon's gray whale domain. After thirty-some miles a series of channels and mudflats connects to Guerrero Negro Lagoon, site of the original ESSA saltworks. This is all in one of the Vizcaíno Biosphere Reserve "buffer zones," where by law activities that are "compatible" with the local environment are permitted.

The reserve's office is in Guerrero Negro, in a nondescript two-story building on a dirt road a few blocks off the main thoroughfare. Aaron Esliman, its twenty-six-year-old subdirector, met me for lunch. A graduate in fisheries engineering at the Autonomous University of Mexico in La Paz, he's been with the reserve for three years. Esliman described an effort to bring back the endangered peninsular pronghorn. Last year the staff had captured five babies and placed them in a large, watered nursery, to prevent their being eaten by coyotes. A similar conservation program was under way for the bighorn sheep. The office conducted courses for the whale-watching guides as well.

"The main problems of the area are the diverse activities—mining, agriculture, livestock, fishing, and tourism," Esliman said. "The idea is to organize all these productive activities under schemes that are sustainable for this reserve." How did the saltworks here and proposed for expansion fit into that equation? "What most worries me," he continued, "is the misinformation being disseminated about the new project. Like it's going to dry up the lagoon, or that by making the dock the whales won't be able to pass by, or that the waste can affect the lagoon. None of these things is true. The way the company does solar extraction of salt is one of the cleanest industries in the world."

As it turned out ESSA is one of the leading financial backers of the biosphere reserve. The company assisted with monitoring of the various wildlife populations, including supplying personnel, boats, and fuel. Indeed, the vast salt fields themselves abounded with flora and fauna. "This is probably one of the largest bird refuges in the world," Esliman went on. "Thousands of birds come to winter here every year, because there is so much to eat. And the salt evaporation ponds have an algae that puts out as much oxygen as a fifty-thousand-acre forest."

In this company town, clearly the Vizcaíno Biosphere Reserve's administration was very much a part of the company.

I wondered if the views of the local fishermen here might be different from those in Punta Abreojos. At Pescados y Mariscos de Guerrero Negro—one of three fish-packing plants in town—I sat down with a half dozen men. They explained that, as a result of uncontrolled exploitation by outsiders, you had to have lived here five years before you could start fishing. Everyone had a relative who worked for Exportadora. Pascual Martínez, one of the oldest original fishermen, remembered: "In the beginning, the company wouldn't even let us enter their property on the roads. We had no access to the beaches, so three of us went to see the governor of Baja Sur. Everything we gained was because of fights and fights." Manuel Antonio Flores, a younger fellow the size of an NFL tackle, expanded: "Now things have changed. Exportadora has put a lot of attention into the problems."

I asked if they'd observed any environmental problems that had detrimentally affected their fisheries. "When there are spills from the ponds, yes, there is death," Manuel continued. "We can't say how much the salt affects, but we believe it affects a lot. But we don't have the proof. Today, the company is using better preventative methods, trying to avoid spills in the lagoon."

The company had also begun financially supporting the fishing sector, including subsidizing aquaculture programs. "Look, the town has been growing because of the economy that comes out of Exportadora," said a man named Espinoza. "The people who work there must buy fish and meat and clothes. In that way, we all live interrelated."

"If one day Exportadora de Sal should disappear," said Manuel with finality, "I think the town would end."

The next morning I was ushered into an ESSA conference room, surrounded by a large map of the world, smaller colored maps of the operation, and a portrait of President Zedillo. The head of ESSA's Environmental Affairs and Industrial Safety Department, a biologist named Julio Peralta, took up a position at a blackboard holding a piece of chalk. His father had worked for the salt factory, and he'd been born here thirty-nine years ago. "First of all," Peralta began, in English, "I like to explain to you what exactly is a solar salt field like Exportadora, what we do here and how. After seeing our current operation, you will more easily understand what is San Ignacio. Feel free to ask any questions. I will magnify the doubt."

I wasn't sure if Peralta actually meant that, but he laughed, and so I did, too. He ran through the forty-year-plus history of the plant and the various means of salt extraction. Fortunately, he said, a new experimental method tried by ESSA's main customer, Japan, had proven too expensive. Thus, ESSA's—and Mother Nature's—way proved best. Peralta asked me to move up to a front seat for my lesson in "production of salt by evaporation of seawater." There followed an explanation filled with percentages and grams and a map delineating the concentration ponds in blue and the crystallization ponds in pink. Basically, it takes two years from the initial stages until salt "harvesttime." Ocean water enters the lagoon with a salt content of 3.5 percent; at the far end of the lagoon, where the water gets pumped out, it's already raised naturally to 4 percent. The water is then moved across the tidal flats, gravity-fed over the next eighteen months through a series of concentration ponds, using solar and wind evaporation to increase the salt content to 27 percent. This salt water is then moved into shallow crystallization ponds, where it sits for another six months until, fully crystallized, it can be harvested, cleaned, and loaded onto barges for transshipment to Cedros Island. On the island the salt is unloaded, sorted into grades, stockpiled, then loaded again onto oceangoing freighters. Besides Japan these touch port in the United States, Canada, Korea, Taiwan, and New Zealand.

"We are [a] completely profitable company," Peralta said proudly. In 1998 ESSA showed a profit of $10.5 million on $80 million in sales. When his annual end-of-the-year bonus comes around, Peralta added, "I will get from fourteen to twenty months of pay average." Then, of course, there was the company store, with half off all merchandise and the three-hundred-some company houses in a paved minisuburbia complete with children's playground behind the local park across the street. Here ESSA had planted rows of trees to serve as windbreaks to reduce the dust levels. "Rent? Almost nothing. I have a nice house, three bedrooms and my studio. I pay monthly about eighty pesos. That's it. Eight dollars."

Peralta laughed. I remembered one observer's comment that this was more like a traditional Mexican hacienda or even a Soviet factory town, where dependency is created by subsidizing everything. Meantime, the ESSA operation remained quite separate from the rest of Guerrero Negro, providing minimal benefits to the community. After all, ESSA even had its own church and school.

As I raised the concerns of environmentalists and residents about the expansion to San Ignacio Lagoon, Peralta had a ready answer for everything. The way he explained it, ESSA had made some major changes in its original plans. To reduce noise and potential pollution, the pumps would be moved farther inland and be electric-powered rather than diesel. A double dike would be built to avoid any potential break in the ponds that could send an excess of salt back into the lagoon. The gaps between the pilings on the concrete pier would be widened to eliminate any potential interference with migrating gray whales. There had never been intent to do any dredging in the lagoon itself, or any development along its shores. The revised environmental impact assessment would look carefully at potential problems in the abalone and lobster areas. The pier near Punta Abreojos was being designed to withstand even the waves of a hundred-year storm. "If it's shown we are going to affect any of the resources of the area—fisheries, gray whales, sea turtles—we won't proceed with the project," Peralta said.

Even here at Guerrero Negro, he continued, all the controversy had forced certain improvements. For example, consider the leftover brine. This is a reddish liquid, high in magnesium chloride, a residue of the salt crystallization process. It used to be simply discharged into the sea. But, yes, there had been occasional fish kills. Now the company was storing and then recirculating the brine to produce table salt, after which a new dilution system reduced any toxicity to almost nil.

Still, I noted, many marine biologists had expressed concerns that the nonstop pumping of saltwater out of San Ignacio Lagoon—at 6,600 gallons per second—was likely to lower the lagoon's overall water level, salt concentration, and temperature. Aquatic life, of course, depends upon all three of those elements being relatively stable. Warm water serves as insulation for newborn gray whales, and higher salinity increases their ability to float. "No problem," Peralta insisted. "We will take out only a small fraction, about 0.5 percent, of the daily seawater exchange between the lagoon and the ocean. The tidal exchange of seawater into the lagoon is about two hundred times the amount we remove each day for evaporation." I wasn't convinced. I recalled the Mexican biologist Francisco Flores's saying at the recent meeting at the lagoon: "The project calls for such powerful pumps that in the course of one year 130 percent of the water contained in the lagoon will be pumped through the system. The effects of this upon the San Ignacio ecosystem are very difficult to measure."

Peralta and I argued such points back and forth for a while. When I asked whether the world's demand for salt justified such a move, the biologist nodded his head vigorously. "If San Ignacio is approved, it will take us about seven years to produce the first ton of salt. Gradually the demand is growing for salt, to produce PVC, glass, aluminum, not only in Japan but in countries like China and Malaysia. Somebody will have to supply it." Expand to San Ignacio Lagoon, and Exportadora would become the world's number-one salt producer, surpassing Cargill of the United States.

Peralta suggested a lunch break, after which I would be taken on an automobile tour of ESSA's operation here, where "you will see for yourself how we work in harmony with nature." Fair enough.

Remember Lot's wife.

—Luke 17:32

Out behind the corporate offices, off-limits to the general public, the latticework of ponds begins. My driver-guide, Pedro Peña, informed me that we'd actually be starting at the end of the line, where the final solution gets collected for dispatching to the washing plant. If not for the desert sun, I could've sworn I'd been transported directly into the Antarctic. Or was it a moonscape? It was surreal. I got out of the car disoriented, not so sure I wouldn't immediately be shivering, and marveled at what looked like a vast field of ice crystals. Some kind of huge yellow machine was loosening up the stuff and making mounds out of it, after which an even bigger combinelike harvester scooped up great heaps and spit them into a 120-ton trailer truck with eight-foot-diameter tires. Everything felt disproportionate out here: the machinery, the ponds—almost 75,000 acres of evaporation ponds—and the pumps that sucked lagoon water out at a rate of 528.6 million gallons per day.

Pedro drove up to seventy miles an hour across the dirt roads on the wide levees between the ponds. There's a kind of eerie beauty about it all, one that Bruce Berger described like this in the book *Sierra, Sea and Desert*:

> To penetrate the vast network of salt ponds is to enter a phantasmagoria of light, billowing color, and Euclidean geometry. Rectangular ponds suggesting snowfields stretch on each side of causeways of packed sand. Some basins are so mistily blue that only a white line of salt divides water from sky; others are pink with algae or tinged a corrosive green. Some evaporation ponds are so wide and their horizons so low that they seem extensions of the sea, and their spume blows onto the causeways like meringue. Because each pond has a different salinity, as well as distinct natural and introduced microorganisms, each supports different fish and crustaceans that attract different species of birds.

The bird life was indeed astonishing. The ponds seemed like magnets for clouds of white pelicans. Waterfowl sometimes speckled the sky like confetti and came to light in the man-made wetlands. My van flushed a flock of sandpipers. A snowy egret fluttered its downy feathers. A peregrine falcon glided overhead. We passed by navigation towers that guide the company's transport barges, now used by ospreys as nesting sites. Over the last twenty years the region's osprey population has increased fourfold. Altogether more than 110 species of birds, including such endangered species as the black-necked stilt, the peregrine falcon, and the great blue heron, have been cataloged on the site. The birds feed on *Artima salina*, a tiny crustacean sometimes called a sea monkey. Millions of these thrive here, in turn feeding upon a layer of microalgae that shows up orange or brown at the borders of the concentration ponds and returns oxygen to the air. Peralta had called all this "the most adequate example I know of nature's linkage to an organic stage to produce salt."

We drove for what seemed an endless time. Pedro pointed out the trestles of an old rail line set up by English and Dutch "salt pirates" in the nineteenth century. Just beyond is where fifteen diesel engines requiring nearly 4 million gallons of fuel each year pump water out of the lagoon into the evaporation ponds. Not far from the lagoon's shoreline, I watched two dolphins skirting by. Over nearly three hours we traversed the fringes of all forty-nine crystallization ponds and all eighteen concentration ponds before making a big circle to Chaparrito, the port back near the corporate headquarters.

There the crystallized salt from the trailer trucks was being unloaded into the washing area, a churning Maytag of brine (99.7 percent pure upon emergence, so they say), then moved by conveyor belt to add to a literal mountain of salt. Yes, a six-story white dune of sodium chloride. "You can eat it," Pedro said. I walked over, poked my finger into a section, and licked. It tasted like Morton's all right. I was told I had just watched 360 tons being deposited. Each waiting barge could carry up to 6,500 tons across to Cedros Island. "The mountain of salt is five times bigger on Cedros," Pedro said—a million tons awaiting oceanic transport.

I tried to envision an operation of this scale transplanted to the region around the San Ignacio Lagoon. It would cover more than 62,000 additional acres of the Vizcaíno Biosphere Reserve. The expanse of evaporation ponds would be larger than the San Ignacio Lagoon itself. In fact, the ponds would cover an area the size of Washington, D.C. This would be accompanied by a dramatic surge in population and "human services," amid the desert foxes and leatherback turtles and burrowing owls, threatened species all.

Exportadora would respond to such a statement by pointing to what it's preserved—and even enhanced—at the Guerrero Negro saltworks. Yes, it was undeniable that considerable flora and fauna flourish along those ponds. Yet I never could shake an awareness of something terribly false about it. Pigeons thrive in New York City, after all, but at what cost? Whether ospreys nesting on barge navigation towers are "organic," perhaps only philosophers of a Road Warrior future could say. This wasn't

The saltworks at Scammon's Lagoon (Steven Swartz)

really the issue. Heritage, natural heritage, was. There isn't much left on the planet that remains relatively unspoiled by humankind. San Ignacio Lagoon is one of those places. It was as simple as that.

I thought of Homero Aridjis standing on the lagoon shoreline, talking about gray whales to a reporter for *The New York Times*: "What we are defending is the right of this species to exist in a poetic world, to live in paradise." In his latest column for Mexico City's *Reforma*, Homero had elaborated, "There are values in the natural world which cannot be transferred or translated into commercial values. . . . For a nature lover to discover San Ignacio Lagoon . . . is comparable to the emotions felt by an art lover who visits the Sistine Chapel for the first time, or for a music lover who listens to a Mozart sonata. What I want to say here is that natural values can be translated into cultural values, but never into the exploitative greed of a corporation whose primary goal is to make a profit."

From Scammon's journal, kept aboard the *Ocean Bird*, 1858: "Monday Dec. 20th . . . It is agreed between the masters of the *Sarah Warren* and the *Ocean Bird* not to whale in main lagoon till Thursday next so as to give the whales a chance to come in without being disturbed."

Today it's about twenty miles from the town of Guerrero Negro to the whales in Scammon's "main lagoon." It's like night and day compared with trying to get to San Ignacio Lagoon. Soon after leaving the main highway onto a well-graded road, you sign in at the salt company's gated entry point and follow the salt flats toward the water. At the end of the road you pay three dollars to park. There are camping facilities along the edge of the lagoon, where you buy your whale-watching tickets at a kiosk. The twenty-dollar price for a ninety-minute *panga* cruise includes a box lunch and beverage afterward at the little *palapa* restaurant near the lagoon entrance.

Given the proximity to town and all the amenities, it's not surprising that far more visitors come here: 9,130 in 1996, compared with 2,969 that year at San Ignacio Lagoon. Given the considerably larger size of Scammon's Lagoon—covering a roughly boot-shaped area of 180 miles—it's also not surprising that gray whales concentrate in the highest numbers here. More than half the population gives birth in Scammon's. The average frequency of the overall gray whale population that is in San Ignacio Lagoon is around 25 percent, with just over 14 percent at Magdalena Bay and a little over 7 percent at Black Warrior (Guerrero Negro) Lagoon, north of Scammon's.

The differences extend to the behavior of the whales. The observation area at this lagoon is four times larger than at San Ignacio's, giving the whales ample room to roam. Yet the only disastrous collision between a gray whale and a boatload of whale watchers took place in Scammon's Lagoon in 1983. There are different accounts of what made it happen. According to one version, the *panga* was pursuing a leaping calf, got too close, and its mother turned on the craft. Another version is that no chase took place, the boat was drifting with its engine off and apparently spooked a whale. Whichever way things transpired, a large gray definitely came up under the *panga* and sent it flying into the air, then thrashed the boat with its tail when the *panga* hurtled back down. Two of the ten people aboard died. Both were men from Los Angeles. The first suffered heart failure, the second lay in a coma for several days before succumbing to head injuries. One of the men was a retired science professor, who'd come to Baja to research a book on whales.

The morning I went out onto the lagoon, there were plenty of gray whales in what appeared to be pairs and little family groups. I observed lots of spy-hops and flapping flukes. But the whales kept their distance, as they generally do here. In Scammon's Lagoon people seemed to be spectators of, but not participants in, their lives.

I thought of Scammon's describing the scene in the winter of 1858–59, the boats "gliding over the molten-looking surface of the water, with a portion of the colossal form of the whale appearing for an instant, like a spectre. . . . Numbers of them [the boats] will be fast to whales at the same time, and the stricken animals, in their efforts to escape, can be seen darting in every direction through the water, or breaching headlong clear of its surface, coming down with a splash that sends columns of foam in every direction, and with a rattling report that can be heard beyond the surrounding shores."

Here, where their first holocaust took place, I imagine the gray whales may not have the capability to, in the words of Macbeth, "pluck from the memory a rooted sorrow." At least not when it comes to what's been wrought upon them by our species.

Chapter 8

WHALE WATCHERS: THE SCIENTIST AND THE ARTIST

Remote and imperturbable, the lives of whales are somehow enough to match any fantasy humanity can create. They are what we have lost, what we yearn for. They are in some ways the last wild voice calling to the consciousness of terminally civilized humanity, our last contact before we submerge forever in our own manufacture and irretrievably lose the last fragments of our wild selves.

—ROGER PAYNE, *Among Whales*

AROUND a little campfire he'd built, John Spencer was talking about his early days visiting Scammon's Lagoon. That was where he'd started out, not here at Laguna San Ignacio. For the first dozen or so years, driving down with his brother-in-law, he'd always gone to Scammon's. In particular he remembered three whales that seemed to come back every winter season.

"They were distinctly marked," John was saying. "One had about a four-foot wound in his side, all pink, with what appeared to be a penetration in the center. Whether that was an old wound from a harpoon I don't know, but it sure looked like something exploded and the cavity just never healed up. One of the other whales looked like it had either gotten into a coral reef or had teeth marks all down one side, just like an Indian pictograph. And still another one we always saw had a great big white spot on the side, looked like a continent on a map."

None of these whales—or any other whales in Scammon's Lagoon—would ever come over to John's boat. He'd be there as long as two weeks, spending at least six hours a day on the water. "They would *not* let you play with 'em. We even used to float around out there in inner tubes hoping to do it. Nada. And I never saw 'em do it with any other boats either."

Did he have a theory on why? I asked.

John sighed and shrugged. "I think the whales told each other that Guerrero Negro was an awful place," he said, in utter seriousness.

John had arrived at San Ignacio Lagoon the day before my return from Punta Abreojos and Guerrero Negro. This year he was accompanied by his family—his wife, Nancy, their two young children, and a grown son from his first marriage who'd brought along a girlfriend. Attached to John's modified pickup was an open trailer jam-packed with supplies and mountain bikes. The license plate read BALLENA. At the moment Nancy was inside a tiny kitchen slicing steak—enough to feed the small village that had already begun to gather near the campfire. Dogs rambled and chickens scurried across the yard at the village of La Laguna.

We had quite a bit to catch up on. After he left San Ignacio Lagoon last year, John had headed down a hundred miles of rough road to the Malarimmo coast near Turtle Bay. There he'd measured a gray whale that had washed ashore. "It was fifty-seven feet long. That was after it'd been laying there on the beach for several weeks, so I sense he was probably bigger than that. When the scientists tell you that their maximum length is probably forty-five or fifty feet, well, it's hard for anybody to estimate the length of a whale in the water."

Last October, John had driven from his home in Idaho to the Makah Indian reservation—"with one purpose in mind: to sink their canoe." He'd even concocted a plan. Once he spotted the whaling canoe in the marina, he was going to run his truck off the dock, into the water, and right over it. He'd simply slump down in the front seat faking like he'd suffered a heart attack—he'd already had bypass surgery—and then have somebody call for the nitro pills, which were in his glove compartment. Once John arrived and had been there a week talking to the tribal people, however, he began to waver. He reasoned that the Makah weren't the ones who drove these whales almost to extinction; a whale was just another fish to them, so maybe it was okay. "Then as I was leaving, I thought, Wait a minute! These are the whales my kids play with, and they're gonna kill one! Fuck *them.*"

It had been only eight years ago that John Spencer made the shift from Scammon's to San Ignacio Lagoon. Only eight years since he'd met Pachico Mayoral. "Seems like we've been brothers forever," John continued. "There was a time—and this time never goes away—when I learned as much from him as from anybody I've known. We used to talk and talk all night long, till three or four o'clock in the morning, just him and I. Of course Pachico, in the beginning and maybe still now, pioneered these situations with the whales in this lagoon. For many years the whales have been his life. And he's probably seen more whales than anybody alive."

So what of the legend about Pachico's teaching the whales to come to us? John hesitated, and I watched the firelight flicker across his face. "Yeah, there could be some magicism involved there. I really believe these animals get used to seeing your face. I mean, they'd *have* to. They do everything else, you think they're not gonna recognize Pachico? Come on."

Pachico wasn't among them, but his friend Luis and other men from the village were gathering closer to the fire now. They didn't understand English, but maybe they understood something else. "Here's some thought for you," John went on. "We probably play with the same whales year after year. I figure it's like with people. Some niños like my son Kanga are a little bit crazy, and others like my daughter Signe are

Pachico Mayoral and John Spencer (Dick Russell)

more standoffish. Some mothers love to go to a party and throw their babies at one of their friends for the rest of the night. Other mothers, you can't get the babies out of their hands. Just like here, some mothers push their babies toward your boat, and others push 'em away.

"The only thing I ultimately know is, we're dealing with a creature of intense intelligence. They've proven that. We just haven't figured out how to talk to them yet. They're waiting for us to do that. Probably laughing every time they look up at us in that boat."

John Spencer laughed. Everybody else laughed, too. One of Pachico's sons, Jesús Mayoral, came around, and at one point John went chasing after him with some kind of spray cologne. Nancy emerged with the first tray of steaks. "Around here," she said, "they call him Juan Loco." It was time to eat.

The next day more of my own friends and their kids showed up from Los Angeles. We went out whale watching on two *pangas* manned by our hosts, Maldo Fischer and his brother Cuco. It was an extraordinary trip. For at least twenty minutes the same mother and her calf hung around us. As the young whale lifted its head to eyeball one of the women, Maldo called out: *"Besa! Besa!"* As he instructed, my friend leaned forward and exchanged a kiss with the whale that sent her reeling backward. Another female friend, who spoke halting Spanish, called out something to Maldo. "You're saying I drive the boat like a whale?" he shouted back, looking quite baffled. "No, I said you dance the boat *with* the whales!" Maldo cracked up laughing and exclaimed, "That's true, that's true, it's a dance! *Bailar con ballenas!*"

After the dance I drove over to the Baja Expeditions camp. Jeff Pantukhoff, who'd been filming on the Sea Shepherds' boat that wild November day in the Makah marina, had shown up. He was a former telecommunications executive. Now forty, he'd left it all behind after a visitation with the whales at this lagoon and formed the Whaleman Foundation in 1995. Jeff had made the underwater film *Gray Magic*, which gave the World Heritage Commission its first look at San Ignacio Lagoon. "If you look into the eye of these creatures, it somehow changes you," he said. "I was hoping to get some of the Makah whalers to come down here and get eye to eye with the whales, and see if they'd feel the same way."

Bobby Kennedy was standing outside the main tent, huddled over several maps with a small group of Japanese. The NRDC had hoped to have more come. But, as the director of the NRDC's Marine Mammal Protection Project, Joel Reynolds, explained: "About two months ago, we received a message that members of Japan's business roundtable wouldn't be able to join us after all, because perhaps it would reflect negatively on Mitsubishi." Still, there were a few news reporters, a distinguished-looking older gentleman whom I understood to be the David Geffen of the Japanese music business, and a prominent artist. The idea was, their presence would help generate pressure within Japan against Mitsubishi's expansion plans.

"See, they'll pump the water through here into this area of evaporation," Bobby was explaining, while an aide translated for the record executive. "When you suck the water out of the lagoon, you're also sucking out a lot of the animals that live here, the larvae. Less food for the whales, less fish for the fishermen." He went on to raise the specter that, if the salt factory should be allowed in such a protected area, every national park in the world would be open to industrial activity. "It's a terrible precedent, maybe next year someone would try to do it at Mount Fuji." The executive raised his eyebrows and said, "Ahhhhhh."

Bobby was sure that, if the San Ignacio expansion happened, the less efficient operation in Guerrero Negro would eventually close down. "It's all about saving on transportation costs. Mitsubishi will be able to increase its profit margin by taking the salt here instead of there. If you're the bean counter in Tokyo, ultimately you curtail operations at Scammon's Lagoon. Because that's the way the free market works. If you look at the maps we've obtained from Mitsubishi, it's clear this area at San Ignacio could be dramatically expanded. In fact, it's been surveyed apparently at great expense and could be doubled in size on the existing salt flats."

Over lunch I sat across from the actress Lauren Hutton. Bobby entertained the table, trading Irish jokes with the whale scientist Roger Payne. The Kennedy accent was, not unexpectedly, flawless—even doing leprechaun imitations. Afterward, during the siesta time preceding the afternoon's whale watch, I spoke with Bobby and a few others about something I'd heard that morning. I'd been talking to Jorge Peón, custodian at the lagoon airstrip. The problem that "doubly worries us," Peón said, was that Mexico's government-owned oil company, PEMEX, had done some test perforations and discovered petroleum less than 650 feet in from the beach. "So it turns out that the Laguna San Ignacio is also an oil reserve," Peón continued. "If Exportadora de Sal creates a precedent, then PEMEX can enter too."

Hearing this story, Joel Reynolds looked at his colleagues and responded, "Some people have said that Mitsubishi's real interest in all this is establishing a connection to get oil out of Mexico."

Later I would do some archival research. Indeed, back in the mid-1970s, PEMEX had drilled about a dozen test wells near the mouth of Scammon's Lagoon. According to a page-one story at the time in the *Los Angeles Times*, "The test drillings have indicated that the top of a sedimentary dome—where oil deposits are usually found—lies directly beneath the lagoon itself. . . . Other figures close to the situation said there also has been intensive prospecting in San Ignacio Lagoon and Magdalena Bay, two other primary breeding sites for the gray whale." All the Mexican government would say was this: "There are good probabilities of extensive deposits south of the twenty-eighth parallel in two areas on the western side of the peninsula." There had followed a private conference in Guerrero Negro arranged by Mexican and American environmental groups, about which a law journal reported: "Mexican officials encouraged complete protection of the whales and environmentalists endorsed the need for development of a profitable tourist industry as a buffer against damaging oil development in the future." That conference was in 1976, the year gray whales started approaching American tour boats in San Ignacio Lagoon. As far as I could tell from the press accounts, that was the end of the petroleum problem. Or was it?

Jorge Peón's even bringing up the possibility made Roger Payne more desperate to find a means to stop the saltworks. Now in his sixties, Dr. Payne is among the world's most knowledgeable whale experts. He'd become prominent during the late 1960s, when he discovered that the vocalizations of humpback whales could be classified as songs. Today he is president of the Ocean Alliance in Lincoln, Massachusetts, and a scientific adviser to the International Whaling Commission; his numerous honors include a MacArthur Fellowship. Over the years Payne had led more than a hundred expeditions into every ocean and studied every species of whale. He knows everybody. Working with NRDC, he'd recently drafted a statement to be signed by many leading international scientists urging Mitsubishi to abandon the proposed saltworks. (Eventually this would appear as a full-page ad in *The New York Times* under the heading "An Unacceptable Risk . . . ," its thirty-four signatories including Nobel Prize winners Philip Anderson, David Baltimore, Murray Gell-Mann, Roger Guillemin, Sir Aaron Klug, Sir Andrew Huxley, Brian Josephson, Mario Molina, and James Watson.)

Here at San Ignacio, Payne was trying to get proof of something long debated by marine biologists—that gray whales do at least some feeding in the lagoons. If that could be shown, it would be another weapon to use against the saltworks. Payne had gone up once in a small plane, but, he explained, "it was too early in the morning, the light was too low, and the angle of the sun wasn't penetrating the water, so we couldn't really see the bottom." He'd have to try again, but maybe not until next year.

Payne had written about the "friendly gray whale phenomenon" in his book *Among Whales* and stated he "would not be particularly surprised if it turns out that this species is not any more friendly than some other whale species." I figured this was a perfect opportunity to explore that subject a bit further. We sat down together inside an old white school bus that the Baja Expeditions camp uses to shuttle folks in from the airport. Much of Payne's work had taken place among the right whales of Patagonia, and I'd heard he'd had a few similar close encounters there. "Well, the right whales are coming along, but they haven't yet become quite as friendly as the gray whales here. A few of them are. If you just sit out there on the water, they'll keep circling closer and closer until finally they bump the boat. The trouble is, most people try to pet them on their heads. Which is the favorite place to pet *gray* whales, but right whales have these thickened patches of skin there, called callosities, where they're very sensitive. They wince when you do that."

So among the whale community the gray remained unique. I asked Payne if he'd conjecture on why they'll visit you up close and personal in San Ignacio but rarely in Scammon's Lagoon. "It would have to be the purest speculation," he replied, "because I don't know what is the pattern of occupancy of the lagoons each year by individual whales. But *if* the whales are largely faithful to one lagoon, I would suppose it means that this cultural change has not yet spread to the rest of the population. But it is now starting to. There's a girl here in camp who told me she knows two other places where gray whales are initiating friendly behavior.

"I know a wonderful story that occurred off Vancouver. A policeman called up Jim Darling, a marine biologist there who's a former student of mine, and said, 'There's a gray whale here we're going to have to destroy.' Jim asked why. The fellow said, 'Well, it's crashing into boats, ramming boats.' Jim said, 'Does it quickly roll over on its side after it's rammed a boat?' The policeman said, 'Well, as a matter of fact, it does.' Jim said, 'Pet it. That's what it wants.' So this is apparently one of these friendlies from down here, which, on its way north, was stopping and soliciting petting."

How then did Payne look at what was going on? Were the whales, as some imagined, trying to communicate with us? "I don't believe so, necessarily. I think you need a language in order to have the sort of communication that implies. But they may be trying to communicate love, affection, and so forth. My feeling is, more than anything it's just pleasing to them to be scratched. And we, of course, enjoy rubbing them."

Much of what these whales are doing eludes us. This became evident when I asked Payne whether we know why they spy-hop and breach. "We absolutely do not know about spy-hopping. It has not been studied in a systematic enough way. It should be. What I've seen of it suggests that it's not spying *above* the water but below the water. A gray whale has eyes which point downward and slightly ahead. So if you're swimming along on a horizontal course, you're basically surveying a little in front and mostly below you. Now you're approaching something on your same level, for example, an outboard motor making noise right on the surface. It's very loud, and my God it's getting closer—you wonder, how close *is* this thing? In fact, you would have to tilt your body upward in order to see well in that direction, to examine things. You could do it first with one eye and then with the other. So that would make the sorts of spy-hops you see that tend to be forty-five degrees to the plane of the stern of the boat. Or you can look directly from beneath your body and get binocular vision. I don't know about gray whales, but right whales absolutely do have binocular vision. Well, if your head comes up in the air, it's just happenstance that you're close to the surface when you did this looking-forward maneuver. That's what I think spy-hopping is, but again it's pure speculation."

Payne continued: "I do have some evidence about breaching, from observing the right whales in Argentina. If you sit on a cliff and look out across the bay, here is what you see. For two hours, no breaching. Then somebody suddenly starts to breach and, in the next five minutes, three miles away another one breaches, and six miles away in another direction you see it, too. Quite a bit of breaching goes on for five or ten minutes. Then it all dies down. Another three hours go by and you see nothing. Finally somebody breaches in a totally different area, and *that* spreads around. So breaching is catching. Breaching begets breaching.

"That kind of thing is very similar to the sort of behavior you get, for example, in howler monkeys, up there in the trees making a lot of noise. Basically what I think they're doing is announcing their position to other howler monkeys. My feeling is, breaching could have a very important function. The entire herd of right whales in Argentina has a center of distribution and concentration which moves up and down the coast, slowly, during days, weeks. Breaching may be a way of checking how far behind the whole herd you are. It'd be like walking along a jungle path where you couldn't see anything, and shouting every now and then to make sure you're still with the group. You don't have to know it's Joe who's shouting, you just hear his voice call back to you, well there's somebody that close at least, and then a couple of others check in—you start to get the shape of the herd."

We talked whale politics for a while. It turned out that Payne knew Paul Watson well and possessed a strong admiration for him: "There are things Paul does that are so extraordinarily brave, you wonder how anybody does them. A person like Paul is absolutely essential at the early stages of getting the world's attention focused." Payne was also convinced that Japanese whaling interests were using the Makah and the potential resumption of whaling by other native peoples to open a door for themselves. "It's their only hope, and they know it. Japan has been fighting the moratorium on commercial whaling since day one. They didn't tell us what they were going to do, but they told us they had a plan that was going to carry them into the following century. That's how patient they are. I have no doubt they are adhering to this plan, and it includes going around and establishing a precedent among these native groups for breaking down the moratorium worldwide."

As I walked back to the main tent, everything was quiet. I went down to the beach and watched the whales spout and gazed into the near distance, where Mexican and Japanese businessmen wanted to invade. And I thought of something else Roger Payne had written, in reference to *Moby-Dick:* "I have wondered whether the underlying genius of Melville's book isn't that he knew just how and by what steps whales would enter our minds, and how once inside they would metastasize and diffuse throughout the whole engine of human ingenuity, mastering and predisposing it to their purpose. And I think he knew that having penetrated the last defenses of the system, that whales would reconstitute themselves, reintegrate at the point of origin of all the meridians of the imagination, its very pole, and there tie themselves forever into human consciousness by a kind of zenith knot."

Out in the middle of the desert, a couple of miles inland from San Ignacio Lagoon, where boxthorns are the only vegetation, there lives a little man in a little shack with a little dog. His name is Francisco Hernández Zamora. People in the region know him as Gerardo. I'd run into him showing some of his artwork at the Baja Expeditions camp, where he invited me to visit him. The off-road twists and turns to his humble abode make him almost impossible to find—a bit like a leprechaun, which in some ways Gerardo resembles. He specializes in *arte de la tierra,* or land art. One of his works graces a mountainside along the beautiful stretch of coast road between Mulegé and Loreto, above the Bahía Concepción. It's a vast network of stones laid out in the form of a whale and called *Geoglifo de la Unidad (Geoglyph of Unity).*

Now Gerardo is walking me through his new work in progress. It's a gray whale, from head to tail. It's being fashioned out of the thousands of scallop shells left along the lagoon by pirate fishermen. The design calls for eventual dimensions of 2,000 by 1,200 feet. You'd only be able to see its full outline from the air. Gerardo did his aerial and terrestrial surveys to find this location in 1996. Then he built his one-room home out of woven straw mats and began "planting" his whale in 1998. So far it measures 1,000 feet long by 16 feet wide in one direction. Jojoba and other desert greenery have been placed at intervals along its path.

It's called *Geoglyph of the Whale Kuyima.*

In his early forties, Gerardo is originally from Mexico City. Both his parents are artists. His father is a controversial muralist who depicted social upheaval: "the army, the peasants, the gunmen contracted by the regional caciques. The majority of my father's paintings," Gerardo says, "were destroyed by the government." His mother worked in conceptual art: "minimalist, primitive, abstract, until she arrived at the maximum reduction. She has more than twenty years working with mineral carbon, painting black." Gerardo continues, "For me, it was always the question where *is* the art? In the social fight, or in the more internal and more isolated introspection? I understood that simply what's happening is that each individual is limited. These two extremes of expression are both important. But we have to choose *something.*"

Gerardo made his choice. In 1989 he said good-bye to his family and announced he was going to Baja California. Asked to elaborate on his plans, he responded he was going to the desert—to capture the song of the whale. Whatever that meant. Gerardo admits he had no idea.

"The curious thing was, I came here in October. The fishermen saw me as a little strange, someone coming to see the whales in a time when there are no whales. They were suspicious. But I wondered, even though this was not whale season, what did it have of the whale? What it had was the space that she likes—for giving birth and for being born. My first connection with the whale was to feel and to understand this space. For this, I did some exercises. First of emptiness. I told the fishermen I was very occupied in not doing anything. Then of walking, to be in silence."

Every day for a month Gerardo walked from La Fridera to El Cardón, a distance of almost twenty miles between the two lagoon fish camps. Nobody knew what to make of him. Some looked upon him with pity, thought he was a lost soul who didn't know what he was doing. Others suspected he was a fugitive. "Each time they greeted me, people projected a different emotion." Gerardo simply took it in stride. He was teaching himself the meaning of acceptance.

When Gerardo returned late in the day, he would pass the dump at El Cardón and pause to pick up some of the trash. He took the pieces home and began creating a collage of the trash on mats of *carrizo*, a kind of cane. By the end of a month the collage was in sections, each about thirty-two feet long. He called his work *Códices of La Fridera* and dedicated it to the gray whale and the fishermen. "It was important for me to communicate that they were giving me something very valuable," Gerardo recalls. "A different way of thinking than what I brought from the city."

When he finished his collage, Gerardo asked a friend to help carry it to a schoolroom in El Cardón. He was standing on a bench, hanging the last of it in the morning light, when he heard the children approaching. As he describes it, when they entered the room, "what I hear is a great silence. They stand there looking, contemplating." The teacher granted Gerardo two hours each day to do exercises with the children—building vocabulary, inventing stories—focusing on the composition.

For a time Gerardo worked alongside the lagoon fishermen digging clams. Then he traveled to Ensenada in the north of Baja. "There a friend said to me, 'If you really want to find the whale, look at this.' She showed me a photograph of a cave painting of a whale. This was a great shock—what is a whale doing in the mountains? I said, 'Yes, then that is where I must go.' "

Gerardo took a bus back to the town of San Ignacio. From there he hitched a ride on a pickup truck carrying fruits and vegetables. It dropped him off at the foothills of the Sierra de San Francisco. A local man guided him, by mule and on foot, into the region with more than two hundred painted caves. Radiocarbon dating had found differences of more than three thousand years within the overlapping images of a single wall. Some of the murals were 160 feet long and up to 33 feet high. Once said to have been created by an ancient race of giants, the cave paintings had been unknown to the modern world until 1962. That's when the mystery writer Erle Stanley Gardner—author of several adventure-travel books about Baja in its prehighway days—took a helicopter expedition into the most remote mountain recesses and presented his rediscovery as a cover story for *Life* magazine. The pictographic art resembles nothing else ever seen in the Americas. Today it is thought that the paintings were conceived in the context of shamanic rituals, among a prehistoric hunter-gatherer people called the Comondú.

Near the head of the Arroyo del Infierno, Gerardo climbed to a high, steep bluff with a cave visible near the apex. It turned out to be a complex of caves. Along one wall was a massive red form, outlined in white and lying in a horizontal position. Gerardo had no doubt it was a gray whale. The name Kuyima came to him. He stayed with the whale until almost sunset. He walked on to the nearby Arroyo de Santa Marta. On a wall of one of its caves appeared a serpent with the head of a deer and the tail of a whale. "These two beings accompanied me on my walks, the mythic whale and the serpent, as I meditated upon the primitive men of Baja California. They were seminomads. Half the year they lived on the coast, and half the year they lived in the mountains. They were great walkers."

And huge people? I asked. "No. The legend that they were giants, I took in another sense—that the soul of a community is giant. The act of walking put me a little closer to the situation of the ancient people. To live in the desert, besides food and water, perhaps the most valuable thing to carry was the color to paint. Art is just as important for human beings in a restricted environment. In my meditations I asked, Why are the animals painted so much in the caves? Because they were much better adapted to the desert conditions, and so could be teachers of human beings. The animals portrayed represent not only food but the beings by which people are nourished internally."

Eventually Gerardo walked all the way back to the lagoon from the oasis town of San Ignacio. Once again it was not whale season. He had yet to see a live whale. "Still, one day, a good friend told me that the other fishermen were saying I talked with the whales. This made me laugh. I said to him, 'Look, they don't know me, they have just seen me walk.' "

Gerardo returned once more to Ensenada. He sealed himself in a small room and went to work on "a mythical synthesis of the exercise I had done." He called it *Maijuanui Codex (The Big Book of the Gray Whale)*. It was a painting much like a cave painting, with a whale in the middle and a serpent beneath, and scroll-like lettering which read, in part: "Then and because of this, the great monster of the sea took form: the whale Kuyima, that which dances in the clouds. In her fits all of man and his destiny. Men learn in this way the magic, the great power of love."

The next time Gerardo came to San Ignacio Lagoon, in 1993, it *was* the time of the whales. "I returned at almost the end of the season, but they were still in the lagoon. A group of fishermen had begun to have a relationship with the whales, and one whale particularly had caused a great attraction among them. It was easy to recognize her. She had so many spots, from all her barnacles, that she looked not gray, but white. She was one of the biggest whales. A fisherman friend said, 'I especially want you to go and get to know this whale.' He took me out in his *panga*, and we found her easily. We could hug her and kiss her. In that moment, I baptized her as Kuyima—the name that had come to me in the cave."

Kuyima, I remembered being told during my first visit to San Ignacio, means "light in the darkness."

"No," Gerardo continued, "that is another story. What happened was, I left some souvenirs for friends. They were crafts made from the baleen of a whale. On the baleen all I put was 'Kuyima,' in phosphorescent paint. On the other side, I wrote, 'Light in the Darkness.' Because in the esoteric literature this was said to be the year of the light in the darkness. So my friends thought that was the literal meaning of *kuyima*. One of them, Raúl López, came to me in Ensenada later that year and told me he was starting a camp for tourists and asked if he could use the name Kuyima. I said, of course.

"So in 1994 they began to receive tourists, who would ask what *kuyima* means. They were told, 'light in the darkness.' They would ask why. Well, the camp had to tell the tourists *something*, so they began to make up stories. Fishermen are very good at inventing stories. They would tell the tourists that a friend had come, a painter who enjoyed walking. One night at Punta Piedra, in the darkness when there was phosphorescence in the sea from the plankton, I had seen how a whale jumping made the waters shine. That's why I had called this whale Kuyima, light in the darkness."

Had Gerardo really seen this?

"No. Another story was that this same painter had seen the whale, but it was a moonlit night. Upon jumping, the whale with the reflection of the moonlight looked white. And *that* was the light in the darkness."

Gerardo went on: "I didn't come back again until 1996, and that's when I found out about these stories. I said, 'Fine, but *kuyima really* means "the one that dances in the clouds." ' And they said, 'Now what do we do with the name?' I said, 'There is no problem. Now *kuyima* simply means two things.' "

The whale could light up the darkness. Or it could dance in the clouds. Gerardo nodded.

What did he think the gray whale was trying to tell us, in coming up to our boats? "I think that it is only reflecting its being. It is a very powerful being, and it does not have natural fear. When there is a dialogue, of course it takes two to communicate. So what happens when dialogue is established between the great whale, which has no fear, and the human being, which by its history is afraid? There is a great emotion in the human being when we touch this other being. Gentleness is the whale's way of being. Much with which we could enrich this dialogue will be in how much we may learn not to be so afraid. At the same time, to learn more love and respect for all other beings. In this sense, the whale is a symbol of this great power that exists in everything else. The whale is interlocutor and translator—to be able to communicate in this larger way. The whale has always been there. Waiting for us to begin to talk, with happiness."

Gerardo continued: "Now one thing that worries me in this new relationship between man and whale is that here we are fomenting contact in the moment of its birth. Yet in the next five years, traditional communities of the north are going to kill more than six hundred gray whales. In Washington, Alaska, and Russia, this is authorized. Beyond that, there is the illegal hunting. As a species we have a schizophrenic relationship with this whale. On the one hand, we embrace it when it's born. After that, we kill it up north. The same human beings. It's not possible that this other being can understand what's happening. The moment can arrive when the whale won't want contact here in the lagoon. Because this relationship is not correct."

So Gerardo believed it was his mission to try to rectify the imbalance the only way he could. Here. At San Ignacio Lagoon. With the *Geoglyph of the Whale Kuyima* spread across the landscape. For almost two hours we had been standing in the desert among the scallop shells gathered from all around the lagoon. Gerardo pointed to where the shells formed the outline of a whale's fin and said: "I take the children to collect the shells in a pickup truck. Then we place the shells here with hoes and shovels. I tell the children we are planting the shells like seeds. So that afterward in the lagoon, there can be scallops. These are the children of the fishermen. What we are trying to seed in their hearts with this symbol is the necessity that the new fishermen have to be aquaculturists. Not only to take from the sea. We will soon collect whale bones to go on top of the shells.

"Once you break the barriers of 'This is so big, this is too much work, this is impossible, this is insane,' then the possibilities of the dream begin to flower. The size of the whale represents the soul of this community, and this project is dedicated to develop values in the community. Love of the earth. Identity of the community. Cohesion of the community. Community dignity. These four elements fortify the soul of a community. If these elements do not exist or are weak, the community also is weak and divided. Because of that, I consider this kind of work a way to find new possibilities for art in the society. The thesis is that art should be known as an axis of social development, as it was for those who made the cave paintings long ago.

"At the same time that the community participates, also the elements of the environment participate. The wind and the sun, for example, work in this project. The wind cleans the shells, the sun bleaches the shells. This is a symbol of a different way for us to relate with the environment. We are human beings, and the others are organic or inorganic beings that are equally as alive as us. Right now the most developed concept about the environment is called sustainable development. It is a good concept, but it is limited. Because it still conceives of nature as resources, not as other beings. And it does not contemplate the spiritual aspects of *human* beings."

We walked against a strong north wind along the edge of the scalloped whale, back toward Gerardo's little home in the desert. He recalled how, during his exercises and meditations, he had begun to "work with the harmonic song" of the gray whale. "One doesn't listen *directly*," he said. "The gray whale does not have a richness of songs like the humpbacks. What is inspired in the gray whales, I hear in Tibetan music. And also sometimes in the songs of the cicadas."

Gerardo went inside his straw-matted dwelling and returned holding a small black dog. This was Gaia, named after the Greek goddess of mother earth. Gerardo said, "In the beginning there was Chaos, Gaia, and Eros. From Gaia and Eros, earth and love, was born Uranus, the sky. So now I am dedicated to making my geoglyph and to taking care of Gaia. Because she is little."

It was a memorable meeting with an old soul, an engaging mythmaker who might have emerged out of one of Carlos Casteneda's books. Still, I could not escape the feeling that something less elevating had changed at the lagoon in 1999. Not that there weren't other moments to savor—the whale's communion with Pachico, a friend exchanging *besos*, a glimpse of the "Pink Floyd." But this year, on six voyages, no more whales had directly approached my *panga* than on that first *day* the year before. There didn't appear to be as many mother-calf pairs. There seemed to be considerably more males. On one trip our guide Maldo even became frightened at a sudden thrashing of tail flukes very close to our bow. He thought this might be a male, scolding us for getting too close to a female and her calf. Several other times I observed what was apparently a male head off an approach toward us by a mother and her recently born. On another occasion, after a pair had been playing around us for quite a while, a third protective whale came rapid-fire under the boat.

Maldo also told us he'd seen two dead gray whales in the lagoon the previous week. Both were fully grown females, and they had floated out to the entrance with the high tide. Then, going out one morning with Pachico, I saw a couple dozen seagulls perched on something floating motionless in the middle of the lagoon. Could it be a big log? No, as we came closer we could see that it was definitely a large adult whale. Pachico muttered something about how it wasn't unusual to see them die of old age, but he motored past as quickly as he could and never looked back.

John Spencer said that, on his way here, he'd noticed a number of mothers with calves off the secluded Malarimmo coast, an unusual place to see *any*, which made him suspect gray whales might be going elsewhere to give birth. "Pachico told me there are fewer whales here this year," John added. "It's an off year with the weather. The water temperatures are different. It's this El Niño thing maybe."

The day before I was to leave the lagoon, a marine biologist showed up at the Baja Expeditions camp. This was Jorge Urbán, professor of research in charge of marine mammals at the University of Baja California Sur in La Paz. Nobody in Mexico had studied gray whales more intensively than Dr. Urbán in recent years. He had some findings to announce. There was indeed a distinct change in 1999, both in the timing of the gray whales' migration and in their distribution pattern. They'd arrived late. Even so, Urbán said, "the water temperature in the lagoons in January was two or three degrees colder than the year before. Our guess is, the mothers needed to look for a better place to have their young."

Thus, many of the whales had continued south into warmer waters. Not only had twice as many as usual stopped off at Magdalena Bay but others had been seen around Cabo San Lucas, at the very tip of Baja, and they even rounded the bend north again into the Sea of Cortés, going up as far as the Bahía de Los Angeles. Altogether, that's nearly a thousand miles farther than San Ignacio Lagoon. Some were also seen off the Mexican mainland near Puerto Vallarta and Acapulco. "With the increase in the numbers of gray whales," Urbán said, "maybe we could expect they are looking for historical areas where they might have been distributed before the time of whaling, such as to lagoons on the mainland of Mexico." (During the 1950s, the whale biologist Raymond Gilmore had discovered gray whales calving in small numbers in two entirely new locations on the eastern side of the Sea of Cortés, along the coasts of Sonora and Sinaloa, which Gilmore felt "explains survival of the gray whale during the critical years of heavy whaling from 1850 to 1880 and again from 1924 to 1938.")

Yes, Urbán's conjecture was one possibility. But in response to prodding from Roger Payne about what *else* might be going on, Urbán continued: "We suppose that with El Niño bringing warmer waters to the gray whale feeding areas of the northern Bering Sea, the productivity was slower there. So the mothers didn't eat enough. The ones that did arrive here at San Ignacio, some of them gave birth to smaller animals." In 1998 biologists had counted a peak point of 230 whales at one time in the San Ignacio Lagoon. (Far more than that, as many as 2,500, are estimated to visit the lagoon during the course of the season.) At the end of February 1999, the biologists counted 161 adults in the lagoon, but only seventeen pairs of females with babies. "Normally by this time, almost 80 percent of the whales are mothers and calves. Instead, this year we saw many single adult whales and mating behavior continuing. So, since mothers and calves are generally the friendly ones, there are very few friendly whales this year."

This explained a good deal about what I'd been witnessing at San Ignacio Lagoon. It also served as a warning of how fragile is the gray whale's world, in both its feeding habitat and its breeding areas, where life is so utterly linked to weather patterns and water temperatures. It was too warm up there, too cool down here. Many whales had to forage longer and then travel farther in search of the right conditions to give birth. The timing was off. Was nature—or, perhaps, humankind's impact *upon* nature in the form of global climate change—putting the gray whales in potential jeopardy?

I listened as Jorge Urbán talked more about Magdalena Bay. By the beginning of March nineteen dead gray whales had washed ashore there. All except two of these were calves, which were noticeably smaller than usual. This was about three times the "normal" mortality observed in the bay. Then again, twice the customary number of mothers with calves had shown up at Magdalena, so one couldn't necessarily jump to conclusions.

The next morning, as I returned from one last whale-watching excursion on the lagoon, I saw two black Sikorsky helicopters coming in for a landing at the Kuyima camp. Soon the road heading that way from Maldo's place was completely blocked off, guarded by soldiers. We joked it must be Mexico's President Zedillo arriving to see for himself what these whales (and all the controversy surrounding them) were about. Later I'd find out that's precisely who it was—Zedillo bringing his wife and children for a weekend outing. As far as the lagoon's future was concerned, that seemed like a hopeful sign.

I said my good-byes, resecured the loose rear bumper onto my Volkswagen, and headed off to see what I might find at Magdalena Bay.

Chapter 9

SOUND CHECK: ECHOES FROM MAGDALENA BAY

Magdalena Bay is probably more generally known than any other on the Lower California coast, and by many regarded not only as a spacious and safe harbor that might generally shelter the navies of the world, but the adjacent country toward the gulf is generally capable of producing abundantly, if properly cultivated, and there are other tracts valuable for grazing. The following is based upon information obtained from the most reliable sources and personal observations: The bay is forty miles long, greatest breadth fifteen miles; points making from Margarita Island and the mainland divide this grand sheet of water into two bays, named by the whalemen Weather and Lee Bays.

—CHARLES MELVILLE SCAMMON, 1869

THE first known explorer of the Baja coast, Spain's Sebastián Vizcaíno, sailed into Magdalena Bay in 1602. According to the eighteenth-century historian Miguel Venegas, the explorer "gave it the name of Bahia de Bal[l]enas or Whale Bay, on account of the multitudes of that large fish they saw there." When Scammon first came here in 1855, as captain of the 370-ton *Leonore*, the scene he observed was an exotic one. There were "oysters that grow on the trees," hanging from the trunks of mangroves, which themselves offered an unlimited supply of wood. The bay swarmed with fish dubbed mangrove-groupers; clams and mussels fanned across the flats.

The one scarcity was freshwater. According to Scammon, it could be found, curiously enough, only by digging in the sand. "The usual process of obtaining water is to take both heads out of a cask, then place it on the beach where the water is found; work the cask down through the loose sand, and removing that on the inside of the casks, till sufficient depth is reached for the water to ooze in, and convenient for bailing. The water, when first brought on board ship, had a white or milky appearance, but after settling for a few days and pumped off, seemed quite clear and drinkable."

Scammon noted that a few whalers had pursued gray whales around the bay between 1846 and 1848, killing thirty-two. The discovery of Arctic bowhead whales in 1848, however, had taken many vessels far to the north. Eventually frustrated by winter storms, American whalers decided to make permanent "between seasons" cruises to Baja. They hailed from Honololu and San Francisco, as well as New Bedford, Martha's Vineyard, and other eastern ports. In the winter of 1855–56 they honed in on the gray whales at Magdalena Bay.

The whale hunter, Scammon wrote, "now finds the object of pursuit not in the fathomless blue water, but huddled together in narrow estuaries, the banks on either hand lined with the evergreen mangrove. Frequently the hollow sound of the spouting whale is heard through the trees, and the vapor ascending is seen above them."

Through three passages the gray whales moved in and out of the bay's forty-mile-long reaches. Ships anchored everywhere about various points, capes, and islands. The wooden whaleboats were swift, but the grays were considerably faster swimmers than either the sperm whales or the Arctic bowheads. They also seemed to possess more intelligence. Chased through the beds of kelp just offshore, the grays soon learned to shun these. Driven to the bay's outer shores, they proved a ferocious prey.

Scammon would recount a tale, quoting "the king of skippers in Devil-fish lore," a fellow he referred to as Captain L:

> I'll tell you what happened to me in my own boat, up in the "mudhole" [Magdalena Bay], season afore last. We was chasing a cow and calf, and I charged my boat-steerer to be careful and not touch the young sucker, for if he did, the old whale would knock us into chopsticks; but no sooner said than done—slam went two irons into the critter, chock to the hitches, and that calf was "powmucky" in less than no time, and the boat-steerer sung out: "Cap'n, I've killed the calf, and the old cow is after us." Well, just about this time, I sung out to the men to pull for the shore as they loved their lives; and when that boat struck the beach, we scat-

> tered. I'll admit I never stopped to look round; but the boat-steerer yelled out: "Cap'n, the old whale is after us still," when I told all hands to climb trees!

Most of Captain L's crew became so discouraged that they went off on a fruitless expedition into the mountains in search of Baja gold. But if the tree-climbing tale was perhaps apocryphal, what Scammon witnessed in Magdalena Bay in the winter of 1856 was very real indeed. Sixteen gray whales were set upon one morning, the men rowing into harpoon range in wooden boats no more than thirty feet long and six feet wide. The whales destroyed two boats entirely and staved the others fifteen times. Six of eighteen crew members were hurt: one came away with two broken legs, another with three fractured ribs. All before a single whale was captured.

In a report for the U.S. Fish Commission, Charles H. Townsend described cruising along this coast in search of sea elephants in 1884: "I heard many stories told by the natives of the ferocity of the female gray whales when attacked in their breeding places—stories amply attested by the number of graves of ill-fated whalers one meets with all along these desolate shores. . . . That fatalities were of frequent occurrence may be emphasized by the statement that in the vicinity of the now deserted lagoons a leading feature in the landscape is the solitary grave with its conspicuous fence of weather-worn whale-ribs."

Long afterward a trading vessel docked at one of the bay's two large islands. In 1914 the *Mary Dodge* was primarily looking to buy guano for use as fertilizer, but the captain told the local people he'd also pay ten pesos a ton for whale bones. They went to work collecting. The *Mary Dodge* left with 125 tons, promising to return for more in a few months. A massive additional pile, 200 tons of gray whale bones, was gathered on the beach. The ship never came back. The bones remained, a kind of monument to the carnage that had occurred there.

Between 1845 and 1874 American and European whalers had killed an estimated 3,290 gray whales in the lagoons and bays of Baja. More than half of these—some 2,100—were taken in Magdalena Bay. Scammon reported that the period between 1856 and 1861 was the most lucrative for whalers in the bay, with 34,425 barrels of oil at fifteen dollars a barrel bringing in $516,375. Fifty whaling ships anchored there in the winter of 1858. Nine years later, when Scammon's friend J. Ross Browne traveled to Magdalena, there were only two. Gray whales, Browne wrote, were "becoming scarce, so much, indeed, as to render their pursuit no longer profitable." Scammon added, "Where thousands of barrels of oil were taken annually, now only a few hundred are obtained."

It took a full day for me to drive from San Ignacio Lagoon south to the deepwater fishing port of San Carlos on Magdalena Bay. Along the last stretch, through the Santo Domingo Valley, irrigated fields of corn and alfalfa blend with orange trees, amid jumping choya and cardon cactuses, where Mexican eagles perch. On the outskirts of the town, I found myself passing a sprawling industrial complex. This is Termoeléctrica C.F.E., a new power plant designed to balance more than half the electrical grid of the Baja peninsula, from Guerrero Negro down to Cabo San Lucas. It was built, in a joint venture with the Mexican government, by none other than Mitsubishi Heavy Industries. While the plant is advertised as thermoelectric, it's really diesel-fueled. A restaurant in San Carlos was built out of the heavy-duty timbers used as crating for shipping it here. The engine is three stories high. It's a 45,000-horsepower, twelve-cylinder in-line motor, and it churns a generator that's thirty feet in diameter.

The Japanese are rumored to be biding time on erecting a resort complex south of Magdalena Bay—a couple of hotels, a golf course, a marina, aquaculture farms for shrimp and oysters. They're said to have bought the land, but run into that basic problem Scammon wrote about: a dearth of water. Today, water for the four thousand–some residents of San Carlos is pumped from a large well about twenty miles inland on the Magdalena Plain.

At the edge of town, where pavement turns abruptly to sand, vendors at little roadside stands waved their arms and did their best to flag me down. They were peddling whale-watch tours. Each February since 1994, San Carlos and Puerto Adolfo López Mateos—another bayside town eighty miles to the north—have held festivals celebrating the birth of their *ballenas*—complete with concerts, fireworks, and Miss Gray Whale beauty pageants. With an estimated twelve thousand foreign tourists showing up each winter season, whale watching has become big business on Magdalena Bay.

Instinctively, I was uncomfortable. My first encounter with a local guide came in a restaurant, where the fellow was hustling several Americans at the next table, bragging about how he'd twice actually hopped on a whale's back and been taken for a short ride. Later I'd hear that this same boat driver had recently rammed into a whale. A local whale watchers' union of between forty and fifty skiffs had been formed. On weekends twenty-four whale-watching *pangas* at a time were allowed out on the waters. An American hotel owner told me he'd seen as many as fifteen people crowded into a single boat. Oftentimes whales got chased, and he said it was amazing there weren't more accidents.

It was suggested I seek out Francisco Ollervides, whom everyone calls Paco. He was Mexican and working on a doctoral thesis for Texas A&M University. For the last three years he'd been coming to San Carlos trying to get a better handle on how whale watching affects the gray whales. Specifically, Paco was analyzing how the whales move and behave around boats. Also, do they respond differently to the engine noise emanating from tourist-laden *pangas* than they would to a shrimp trawler or one of the oil tankers bringing fuel for the power plant?

I found Paco Ollervides at his wintertime base, the School for Field Studies, a complex of palm-tree-shaded buildings right on the bay. Headquartered in Beverly, Massachusetts, the School for Field Studies can be found in half a dozen locations around the globe, including Kenya, Australia, Costa Rica, and here in San Carlos. The students, primarily Americans, spend six months in small communities working on environmental problem solving. In San Carlos they were helping the fish cannery do something about its effluent discharge and studying the impacts of ecotourism on the gray whale population.

"This bay is different than Guerrero Negro and San Ignacio," Paco explained, "because it's the only area the whales come to which is not part of the biosphere reserve. That was why I was interested to observe in Magdalena Bay, since it does not have the same protected status. It has the same whale-watching regulations but different pressures. There's a bigger surface area and less enforcement. Boat captains do a lot of violations of the regulations, and this goes unchecked. With the Mexican authorities, we're trying to design some core areas where whale watching should not occur."

At nine the next morning I meet Paco and his assistant, a graduate student from the University of Kansas named Jennifer Pettis, at the gated entry to the Port of San Carlos. This is Baja's only deepwater port besides Ensenada, and vessels of all shapes and sizes are moored at the pier. A fishing trawler has just tossed a huge shark onto the dock. Paco, who's in his early twenties, is dressed in shorts and wearing an Australian bush hat. He beckons me to follow them up about a hundred feet of metal stairs to an observation tower. Nineteen ninety-nine is an unusual year; not only are more whales—some three hundred compared with a maximum count of sixty-six in 1998—occupying this southernmost sector of the bay but also many are opting to stay in the vicinity of the port. "Generally they prefer to be at the mouth," Paco says, "which is about an hour away from here. So this year has been incredible for the whale-watching companies—and for us." As we reach the enclosed tower, Paco points out several large vessels winding down the bay's shipping channel through dunes and mangroves, past one of the longest islands in Mexico. "The whales have been seen to avoid areas like San Francisco or L.A., which could be a result of ship traffic," he adds. "It could happen here, too—if it goes unregulated and the number of boats just increases every year."

The view from up here is spectacular. I gaze into the near distance at a complex of inland tidal channels protected from the Pacific's rollers by misty volcanic peaks and the dunes of Magdalena and Margarita Islands. The spouting of gray whales just below looks like smoke signals wafting up. "Right now it's good sea conditions because the wind is not blowing," Paco says. "Otherwise the blows get taken away really fast."

He commences adjusting the angles on a surveyor's theodolite. This method of monitoring whale movements was pioneered by Roger Payne. A theodolite is basically a transit—the same kind you see obtaining level measurements along a road—which measures the azimuth and downward angle. The bubbles inside it, Paco says, tell you how balanced the theodolite is. Jennifer records all the data onto a computer for later analysis.

Paco zeros in on something and motions me to take a look. It's a jarring close-up: a dead gray whale, floating belly-up, its body adorned with seagulls. "We think it's the same one we've been seeing for a couple of days," he says. "It just moves up and down with the tide. We've even seen one right down here in front of the pier." Six of the seven dead gray whales documented here this year have been females. Paco has been plotting the exact locations on a map. On the one that washed up by the dock, he conducted a necropsy to try to determine the cause of death. The inner-ear bones were extracted and sent to an expert with the Woods Hole Oceanographic Institution and the Harvard Medical School laboratory, where, Paco continues, "they will do CT scans and MRI's to see if there's been sound damage that might have caused the whale to die."

We would do well at this point to consider the remarkable auditory sensibilities of the gray whale and its brethren. Scammon observed: "The ear, which appears externally like a mere slit in the skin, two and one-half inches in length, is about eighteen inches behind the eye, and a little above it." In the human brain the centers for sight and hearing are approximately the same size. In the brains of whales, however, the sensory nerves responsible for hearing are much larger than those for seeing. It's a matter of environment. Beneath the ocean's surface darkness and turbidity impose extreme limitations upon sight. Sound, however, not only travels farther underwater than in air, sometimes for thousands of miles, but also travels four and one half times *faster* than in air, at up to 5,000 feet per second. Within the inner ears of gray whales as well as other cetaceans, features have evolved that are capable of picking up a spectrum of sounds inaudible to people. At the same time, as Aristotle put it long ago, "Even a small noise . . . sounds very heavy and enormous to anything which can hear underwater."

In the spring of 1856, in what may have been the earliest consideration of the gray whale's sensitivity to noise in its environment, the *Monterey Sentinel* wrote: "The whales are every year getting more shy from the use of the bomb lance [whose load of powder explodes on impact]. It is said that they hear the bomb explode in the water, even though ten or twenty miles off."

In his description of whaling on Magdalena Bay, Scammon offered additional insights: "Every ship's cooper and his gang were busily at work with their heavy hammers, driving the hoops on the casks, and the whole combined produced a deafening noise upon the water, which echoed from cliff to crag along the mountain island of Margarita. This, with the chase and capture of the animals, the staving of boats, and the smoke and blaze from try-works by night, soon drove the whales to the outside shores." The ship's captains talked over the situation, and many decided to go look for whales elsewhere. "After suspending whaling for a few days, and a number of ships leaving meanwhile, the [gray] whales again returned to their favorite haunt."

There is evidence that ship traffic may have directed the gray whales away from another nursery area, the warm waters of San Diego Bay. A 1922 *History of California Shore Whaling* records: "Smythe's *History of San Diego* states that in the early forties [1840s] San Diego Bay was a favorite place for the female whales in the calving season, and at such times on any bright day, scores of them could be seen spouting."

Marilyn Dahlheim, who today works out of the National Marine Mammal Laboratory in Seattle, was one of the first to scientifically point out that sounds—both natural and man-made—could have a dramatic impact upon gray whales. She spent five winter seasons, 1981–85, at San Ignacio Lagoon, researching a dissertation for the University of British Columbia titled "Bio-Acoustics of the Gray Whale." Off the narrowest sector of the lagoon, at Rocky Point, she submerged a cage housing a little transducer that could broadcast different types of sounds underwater in the direction of the whales. When outboard engine noise was played, Dahlheim was fascinated to find the whales attracted to the transducer. When she went up to the Middle Lagoon area, where sound transmission was reduced because of the extensive sandbars, she discovered something equally intriguing. "I'd heard about whales hiding behind islands or icebergs to avoid increased levels of sound in their environment, but I never knew whether to believe it," Dahlheim says. "But when I played oil-drilling sounds, not even at a very high decibel level, I found clusters of whales in the Middle Lagoon behind those sandbars, where there are what I call 'sound shadows' and the noise wasn't as intense."

After two years profiling the whales' acoustical habitat and characterizing their calls, in 1983 Dahlheim started conducting short-term playbacks. The next year she increased the duration of the playbacks. Her goal was to determine how gray whales might alter their own sound structure when faced with longer-term and increased levels of man-made noise. She played 120 hours of prerecorded sounds, ranging from industrial noise to the vocalizations of the gray whale's primary predator, the killer whale. What happened was telling. To escape increased levels of noise, the mothers and calves took their leave early. By the beginning of March, their count in the lagoon was 81 percent lower than the mean of 223 for the period between 1977 and 1982.

Dahlheim says: "At first I didn't quite believe what was happening, that my little transducer was causing this. I remember some of my team members sitting around the camp at night discussing it, saying, 'Oh my God, what's going on?' As a group we thought, Well, we've got to continue, because if what we're seeing *is* a response to these types and levels of noise, we need to know—and we don't want to do it again."

Because of this striking effect upon the whales, the Marine Mammal Commission funded a follow-up study in 1985. This time no artificial sound projection took place. Dahlheim was strictly looking at the numbers and distribution of whales in the lagoon. The grays' numbers were back up, though not as high as they'd been during the previous several years.

Now Paco Ollervides has been able to establish some very "significant correlations" between numbers of boats and behavioral reactions of whales in a given area. In the presence of larger vessels in Magdalena Bay, gray whales were seen to change their swimming speed and direction, as well as their breathing intervals. In general, groups of half a dozen would congregate in the mornings, when there was no ship traffic. With the arrival of boats the whales would disperse into pairs or singles, coming together again after the boats departed.

What Paco is observing is, of course, taking place within a relatively confined space. In a broader ocean context, Dr. Christopher Clark, of Cornell University's Bioacoustics Research Program, has described what marine mammals face today as an "acoustics traffic jam." The noisiest offenders are supertankers and cargo ships, whose propellers emit a pervasive low-frequency hiss, which penetrates more deeply in warmer water. The shipping lanes pretty much parallel the gray whale's migratory route. Along the central California coast alone, more than four thousand large vessels transit annually. Baleen whales such as the gray are presumed to hear—and to vocalize—in the same ranges occupied by ship noise. Might the din be causing hearing loss? Or drowning out vital communication, say for a humpback whale calling to a prospective mate several hundred miles away?

"In terms of behavioral disruption, shipping is what I'm most concerned about," says Dr. Peter Tyack, a senior scientist with the Woods Hole Oceanographic Institution and an expert in whale acoustics. "In terms of potential for ear injury, it's the most intense sound sources, such as explosions, sonars, and the air guns used for seismic surveys by the oil companies. I don't know whether there's been an explicit cozy agreement between the oil and commercial shipping industries with the regulators, but it's certainly been a policy of 'don't tell us, we won't look or ask questions.' It's very important there be a strong push to recognize the real risks to marine mammals and come up with a reasonable policy that covers and regulates all these sources of noise."

During the 1990s several new sources of oceanic "noise pollution" came on-line and under fire. One was acoustic deterrence devices deployed by commercial fishermen looking to drive seals away from their catch, with only minimal guidance from wildlife agencies. Another was an experiment in acoustic thermometry, in which the Scripps Institution of Oceanography embarked on a $40 million effort using underwater loudspeakers to boom high-intensity (195 decibels), low-frequency sound waves across expanses of ocean. The aim was to get a better handle on long-term climate change by measuring the speed of sound to "take the temperature" of the sea (the warmer the water, the faster sound travels). Several studies of potential impacts on whales and other sea denizens were carried out, to inconclusive results, and the project has foundered.

Gray whales became the centerpiece of research focusing on the third—and most controversial—recent deployment: the U.S. Navy's Low Frequency Active (LFA) acoustic sonar system. The Navy would like to deploy this across 80 percent of the world's oceans. Designed to detect enemy submarines by scanning the seas with sound waves, LFA would flood thousands of square miles of ocean at a time with intense sound. After a dozen Cuvier beaked whales beached themselves in 1996 during NATO antisubmarine exercises in the Ionian Sea, and faced with a potential lawsuit from the NRDC, the Navy set about undertaking an environmental review of how LFA sonar might affect marine mammals.

That's where Peter Tyack came into the picture. As part of a three-phase study commissioned by the Navy, in January 1998 he and a colleague conducted playback experiments in low-frequency sound to probe the behavioral responses of gray whales migrating off the central California coast. I'd met with him some months later to discuss his findings. Tyack is a slightly rotund, blond-haired, white-bearded fellow with soft blue eyes and a pleasant demeanor. He did his thesis on the songs of humpback whales and has been studying how whales and dolphins "use communication signals in the context of their social behavior" since coming to Woods Hole in 1982. He'd been part of a team in the 1980s that performed similar playbacks of oil industry sounds at migrating gray whales. These included air guns used in seismic explorations. The biologists had concluded from this study that the whales were more sensitive to continuous noise—they'd tend to avoid exposure at levels of around 120 decibels—than they were to the short pulses of the air guns. Dissenting scientists claimed the whales didn't really care about loudness.

So Tyack hoped that his tests for the Navy would provide some more definitive answers. "If gray whales showed an avoidance response," he said, "my main concern was that this Navy sonar could affect other animals over huge areas in the open ocean if the ship were operating in the middle of the North Pacific or North Atlantic. That's because low-frequency sound can propagate over a very big range."

Migrating gray whales made an ideal case study, because hundreds a day were sojourning close to the coastline, "and you know it's always new animals, since nobody's circling back around north." They could also be watched from shore, instead of from a boat or airplane that might al-

Peter Tyack (rear) studying gray whale behavior off the California coast (courtesy Peter Tyack)

ready be disturbing their behavior, by using a theodolite. As they moved past, the whales' path was plotted by computer. Meantime, a sound source was placed on a ship moored in the center of the migration corridor. Its underwater speaker blared sound at the whales at varying decibel levels, of forty-two seconds' duration and repeating every six minutes. The question was, would the whales noticeably change their course and at what range of noise?

The procedure is more complex than I'm outlining, replete with scientific controls and assorted attenuated simulations of stimuli. Suffice to say that the experiments as designed were highly sensitive and were conducted over 150 hours during eighteen days that yielded tracks on about fourteen hundred whales. I sat next to Tyack at his computer as he ran through his data accompanied by a few low-frequency imitations of the sonar sound—a higher-pitched "ooo ooooo ooooo" and "whoooooooo," followed by a deeper, slower "whoooo whooo whooo."

The hypothesis, Tyack said, was "that the behavior of the whales would be disrupted and they'd avoid exposure at received levels in the range from 120 to 155 decibels." Whales had avoided these sound levels in previous studies, even though these levels are not very loud by the standards of human noise in the ocean. The source level of a small outboard, for example, is 155 decibels at about one yard; the boat would have to be a hundred yards away to reach the lower end of this exposure range. In this latest test the whales did move away from the sonar sound imitations—but not significantly. They skirted the sounds by only about 300 hundred feet. Even with a source level of 170 decibels, according to Tyack, "they were a little less sensitive to the LFA sonar than we'd expected."

He clicked his computer mouse to another diagram. "But *here*," Tyack continued, "comes the big difference." When the sound source level was increased by another fifteen decibels, the whales suddenly avoided the noise by six-tenths of a mile. They swerved upstream away from the sound, an avoidance reaction so dramatic that "blind" onshore observers (who didn't know whether a playback or a control was happening) could instantly ascertain what was transpiring. "See that giant gap on the screen?" Tyack asked excitedly. "That's a huge difference in the scale of response. That's saying the whales sure as hell *do* care about how loud the sound is, and they're choosing to move big-time to get to a lower exposure. All the critics of our earlier work would have claimed there won't be any discernible difference. So this was totally satisfying. Almost never in whale science do you have a couple of days when you can resolve a question like that."

Then the scientists decided on one more test. They kept the same 185-decibel sound level, but they had their ship move twice as far offshore, "away from the center of the migration corridor, but a place where there are still some whales." What happened when the sound was turned on again was a complete surprise. Tyack recalled: "Many whales passed near the source, but they paid no attention to the transmission. We even raised it to a 200-decibel playback, and still the whales had no problem with it whatsoever."

Tyack sighed deeply. "Well, every experiment that works usually opens up some whole new can of worms. I don't know what exactly is happening here. My hunch is, it relates to the behavioral ecology of what the animals are doing. For certain reasons, gray whales stick right next to the coast as they're migrating. So I wonder if they're listening to surf, or using some acoustic cue in the near-shore areas for migration. And maybe putting a loud noise right near *that* interferes with the process of orientation in migration—in a way that an offshore sound source doesn't do. We'll have to learn more about the sensory basis of migrations to really get at that."

The scientists' conclusion? The LFA sonar should be kept away from inshore migratory corridors and near-shore areas in general. The Navy's response was to limit the operation of the system to more than twelve nautical miles from shore. "That's a bit of a safety zone from my perspective," Tyack said, but he cautions that much more needs to be learned.

Other marine biologists are not as sanguine. Another researcher of whale acoustics, Dr. Lindy Weilgart of Dalhousie University in Halifax, Nova Scotia, points out that Tyack's tests used only a fraction of the full operational power level of the LFA system. "Tyack may well be right," Weilgart says, "but it would be safer to assume that if inshore whales are clearly shown to avoid LFAs, then the problem may not be with using LFA just in that particular environment but everywhere. Perhaps the offshore migrating whales—those that reacted less—were already more damaged or marginal individuals. Anything that has the potential to change, even slightly, a whole population of migrating whales should be viewed with great caution. If something serious befalls these migrating animals, it means that the whole population is doomed."

For his part, Tyack is most concerned about deep-ocean, deep-diving toothed whales, such as the sperm and beaked whales, in areas where sound refracts downward and animals could face jeopardy when foraging in the depths, where the LFA energy concentrates. He told me: "There continue to be anecdotes of military sonar activities associated with beaked whale strandings. The facts that they dive deep, are globally distributed, and we know so little about their behavior make me at least want to find out—what really happens when a sperm or beaked whale hears a sound of 140 decibels? Does it stop its normal feeding? Does it show avoidance even if the source is a long way away?"

In mid-March 2000 more answers came about Navy acoustics—and the results were tragic. Shortly after the Navy began conducting mid-frequency range sonar tests in the Bahamas, seventeen whales from four species beached themselves over a four-day period. Seven died, and two had bleeding eyes, which suggested acute shock trauma. Despite the Navy's initial statement that its testing and the strandings were coincidental, a study by the National Marine Fisheries Service found six of the whales had suffered hemorrhages of varying degree in or around the ears, possibly caused by "a distant explosion or an intense acoustic event." Late in May the Navy canceled another test scheduled off the

East Coast of the United States and announced "a priority need" to examine the LFA issue.

The question of what drives whales either toward us or away from us, of what they hear as we beam or approach them with our mechanical toys, leads to another question: the means by which whales might "talk" to one another about the various situations they face. We know that all whales "speak" acoustically, but how they do so is still largely a mystery. They produce sound by squeezing air through either their larynx or their blowholes, or by bursts of air from their lungs. Humpback whales, as Paco Ollervides puts it, "sound like violins," and are known for inventing complicated songs that sometimes continue for more than an hour. Killer whales, or orcas, make high-pitched sounds at frequencies as high as 24,000 hertz. By contrast, humans generally speak in a range between 2,000 and 4,000 hertz. Gray whales have a wider range of communication than people and also span the same frequencies we talk in.

In the 1850s whalers were the first to report gray whales making sounds that could be heard above water. H. L. Aldrich wrote in 1889, "It has been known for a long time that humpback whales, blackfish, devilfish [gray whales], and other species of whales sing." Yet when the marine scientist Carl Hubbs tried listening through a hydrophone to gray whales at the San Ignacio Lagoon in 1950, he couldn't detect any signals at all. "All we picked up was the chatter of shrimps," Hubbs reported in one newspaper article headlined SEA SLEUTHS, UNDERWATER WIRE TAP FAILS IN EFFORTS TO MAKE WHALES TALK. Hubbs wondered if gray whales might, like giraffes, be voiceless.

So at that point the gray whale was labeled "the quiet whale." Then, in 1955, what appeared to be pulses from a gray whale were recorded off Point Loma near San Diego. A small landing craft was maneuvered purposefully into an oncoming whale's path, then its engines shut down. To avoid collision the whale changed course slightly and emitted a series of sounds. Two years later Soviet whalers in the Bering and Chukchi Seas reported hearing "low-pitched roars" coming from gray whales.

The 1960s saw the first breakthroughs in cracking the gray whales' sound barrier. Some of the most extensive early research was conducted by the Anti-Submarine Warfare and Ocean Systems division of the Lockheed-California Company. They wanted to gather data on gray whales "because of their obvious antisubmarine warfare importance as potential 'false targets' for passive and active sonar." An expedition was mounted to Scammon's Lagoon. While it established that individual gray whale pulses last about one-tenth of a second and usually occur in groups of four to six, as faint whistling sounds or "croaker-like grunts" and "low-frequency 'rumbles,'" excessive ambient background noise made by snapping shrimp in the shallow lagoon waters continued to make underwater listening difficult. Follow-up studies detected intense sounds that scientists likened to hammering against the hull of a wooden ship; crunching and scratching noises; metallic "blip blips" like coffee percolating, and defined cries from a juvenile gray whale in the course of being captured and transported from Scammon's Lagoon to a research tank at SeaWorld.

Then, in 1967, came a report by the Office of Naval Research and the University of Rhode Island titled "The Controversial Production of Sound by the California Gray Whale." Hydrophones lowered into the water off Point Loma and La Jolla in Southern California discovered the migrating whales to be continuously vocal over thirteen days and nights. The researchers had decided to try to force a highly controversial issue. There had been many recordings of toothed whales using echolocating sonar but very few reported instances of baleen whales doing so. So the scientists met the southbound whales head-on from a seventy-foot-long Navy transport vessel. The wake of the ship's propeller, along with an adjacent kelp bed, created underwater visual and acoustic interference around and through which the whales had to maneuver. The closer the whales came, the louder a series of "notes" could be heard, "like those from a comb as one runs a thumbnail down the fine teeth." The elicited sounds were under 2,000 hertz, and thus would normally go undetected or be passed off as other unidentified noise. But the researchers were now convinced that the gray whale "does produce trains of pulse sounds similar to the echolocation bursts of toothed cetaceans," on an as-needed basis. One Navy Electronics Laboratory oceanographer described the whales' making a deep moaning, "like the left hand sounds of a piano," and suggested these low noises might cause echoes enabling the whales to detect large objects—not only ships but landmasses or other whales.

By the late 1960s scientists analyzing some 60,000 feet of recording tape concluded that gray whales are not quiet at all but in fact are among the most vocal of whales. Researchers at Scammon's Lagoon noted that it wasn't uncommon for gray whales to make a series of clicking noises when a plane or helicopter passed close overhead. Gigi, a young gray whale held in captivity for two years during the early 1970s, was also heard to make rapid clicks when she was released back into the wild—a sound she'd never uttered while being studied at SeaWorld. Were these clicks indications of anger? Excitement? Fear? Attempts to communicate with other migrating whales?

The marine biologist Steven Swartz later had this to say about his six-year-long (1978–84) observations at San Ignacio Lagoon: "If a female wants to recall her calf from a group of calves, or if she's sleeping and decides to leave, she utters some sort of sound. The calf responds immediately, and off they go. So there's definitely communication going on amongst them. Exactly how it's carried out and what's the significance under various circumstances I don't know. No one's done the definitive work on the gray whale's song, if you will, or on gray whale dialects."

Swartz noticed the whales would habitually approach his boat from the rear—the first indication, he says, "that something about the engine sound interested them or was providing a beacon for them to orient on. Then they'd flip over upside-down and look at us. For a while, that was puzzling. We put on our masks and stuck our heads in the water and watched them." Similar to what Roger Payne conjectured about the grays' spy-hopping in the proximity of boats, Swartz soon realized that the whales were checking them out in a rare moment of having to look *up*.

Swartz and his co-researcher Mary Lou Jones elaborated in a jointly written study: "They appear to show an interest in the submerged portion of the outboard engine running in neutral, such as the nonrevolving propeller and exhaust ports from which sound emanates. Some whales repeatedly bumped the still propeller of the engine with their rostrum and even took it into their open mouth. The whales avoided Mexican fishing boats with forty-horsepower engines running at high speed. However, they approached and even followed twenty-horsepower engines running at moderate or low speed. Curious whales usually left the vicinity of idling outboard engines when they were shut down, and they also avoided nonmotorized vessels including kayaks, canoes, and small sail boats."

In one "experiment," Swartz said, "we could actually get them to trumpet blow, like a humpback, by revving the outboard engine. They expel bubbles underwater and there's quite a bit of sound associated with those big bubble bursts. They'd blast, we'd rev the engine, and they'd blast again. We'd get this kind of rapport going back and forth."

Shari Bondy, another longtime gray whale observer who resides in Guerrero Negro, made some related observations in a paper titled "Behavioral Changes in Gray Whale Population in Ojo de Liebre [Scammon's Lagoon]." Gray whales, she wrote, "can react evasively to small changes, like alterations in speed or direction of a boat or a switch of captains or motors of a particular *panga.* For example in 1997, the Ejido Benito Juárez replaced the old motors they had used for years with new Yamaha engines on all three *pangas.* Almost all whales were very evasive and even dove deeply, throwing their tail (which is very unusual there). It took a full week for the whales to become accustomed to the new sounds of the motors and gain enough confidence to allow the boats to approach again."

At San Ignacio Lagoon in 1999, John Spencer had also talked with me about the whales and the boat motors. "It's amazed me for twenty years," he said one night around the campfire, "how can animals this big not bump the boat more? Not that they'd be aggressive, but just their size. You stop to think about it, they're water acrobats. When a forty-five-foot animal can rub his nose on one side and his tail on the other side without bumping you, you've gotta figure they know where they're at in the water. What I think the motor does, along with telling the whales where we're at, is to give them a sense of direction about the size of the boat. And how they can play with it. And with us."

It was Marilyn Dahlheim who first found it "noteworthy that these 'curious' whales respond to underwater engine noise that occupies the same frequency ranges as their own signals." However, Peter Tyack shares John Spencer's view that's not *all* the "friendly" gray whales are responding to. "I wouldn't think that just an overlap of frequency range would necessarily make something attractive," he told me. "It certainly would make it easier for the whales to detect the boats. But what's amazing is simply how social and interactive these animals are. The fact that they're willing to reach out toward our species is a remarkable thing."

Tyack also takes issue with scientists who assert that, because what we hear of gray whale "communication" consists primarily of grunts and pops and clicks, these whales are necessarily more "primitive" than, say, the humpbacks. "I think it's very dangerous to assume that if you hear what sounds like a simple acoustic repertoire, that means the communication system is primitive or very simple. Humpbacks produce their song in an ecological setting where sexual selection has caused there to be a strong pressure for that song to be very complex."

Then there are sperm whales, which, Tyack notes, "basically just make simple clicks, but in different rhythmic patterns. Some appear to be individually distinctive, others group distinctive, and still others are shared over a broad geographical area. These simple clicks solve very complex problems in amazing ways, so it would be a grave error to say, 'Oh well, they *just* make clicks.' " Jim Nollman, an author and founder of Interspecies Communication, Inc., told me he believes the sperm whales' clicking to one another constitutes "some kind of a symbolic language, based on echolocation. At the least it's sending holograms to one another. I believe one sperm whale can turn to another and show a 3-D movie of what it just did in a deep dive, actually display the giant squid it just ate. This kind of communication is based not in grammar but in physics."

Gray whales' communication—and what's been misinterpreted as their muteness—may in part be a means of avoiding detection by their inshore enemies. In terms of man, Charles H. Townsend described in his 1886 report for the U.S. Fish Commission how so many gray whales had been wounded by inexperienced hunters using the bomb-lance gun that the whales "became wary and in general more quiet in their movements, leading some of the whalers to a suspicion even of their 'blowing' more cautiously." Another report, published by the naturalist Roy Chapman Andrews in 1914, recounts the story of a whaling captain "hunting a gray whale in a perfectly smooth sea. The whale had been down for fifteen minutes when suddenly a slight sound was heard near the ship and a thin cloud of vapor was seen floating upward from a patch of ripples which might have been made by a duck leaving the surface. The whale had exposed only the blowholes, spouted, refilled the lungs and again sunk, doing it almost noiselessly. The gunners assert that this is quite a usual occurrence when a single gray whale is being hunted." (Steven Swartz would later term this "snorkeling behavior," a phenomenon exhibited by no other species of whale.)

A no less compelling example of the grays' reaction to a predator focuses on the orca, or killer whale. Two researchers for the Naval Undersea Research and Development Center conducted an experiment in the late 1960s, subjecting gray whales migrating south to three sound sources: killer whale "screams," a pure tone of the same frequency, and loud random noise. The grays didn't react to the latter two, but fully 95 percent of the time "blowing whales, or those running at the surface, immediately swirled around and headed directly away from the killer whale sound source." Hiding close to inshore kelp beds, "in many instances their blows were invisible and even blows at close range were scarcely audible."

Peter Tyack finds it striking just how reactive gray whales are to the playbacks of killer whales. "I'd expect that, when they're migrating and very concerned about being detected by an acoustic predator, that may put a lot of constraints on the gray whales' communication signals. That doesn't mean the signals aren't important. Within a small group, they could very likely be making a set of faint contact calls. When we scientists hear the sounds, these are very much like ambient noise. A lot harder for us to study, and less flashy. But flashy doesn't always mean inherently complex. It's not like the gray whales are being silent. My hunch is, they're just being cryptic. I'd call gray whales exquisitely acoustic animals."

Marilyn Dahlheim's experiment at San Ignacio Lagoon found gray whales would cease vocalizing altogether and do snorkeling behavior in the presence of killer whale sounds. She taped gray whales for more than two hundred hours at the lagoon, using a hydrophone off a small Zodiac. Dahlheim's overall analysis found the grays "more soniferous than expected from published accounts." One of the sounds she detected—akin to the ringing of a Chinese gong—has never been heard outside the lagoon. Altogether Dahlheim was able to define seven distinct sounds and identify numerous situations in which gray whales are the *most* vocal. These included when they were in a small area or interacting with bottlenose dolphins; when they were on a collision course with a boat or another whale; when single whales were chasing mother-calf pairs, and when boat noise was prevalent.

Dahlheim noted that gray whale signals were consistently produced below biological ambient noise, beneath the constant clatter of the lagoon's snapping shrimp, and where it's relatively quiet. The whistling calls of bottlenose dolphins were just the opposite. They hit the "ceiling" of the snapping shrimp and bounced back up. "These data imply," Dahlheim wrote, "that these cetaceans have different acoustical 'niches' possibly dictated by the constant high levels of biological ambient noise in Laguna San Ignacio."

The acoustical niche idea represents an adaptable strategy that would ensure an animal minimum interference with its own signals. The peak of gray whale sound production occurs in the Baja lagoons. Studies haven't shown the whales to be very vocal along the migration. Off Alaska's St. Lawrence Island, Dahlheim found the grays make almost no sound whatsoever. Unlike the dynamic acoustical habitat within the lagoons, she points out, there aren't a lot of fish making noise in the whale's northern territory. The natural ambient sound level is very low, too, consisting of little more than sea noise and rain.

Dahlheim believes the types of calls whales make have evolved over time to reduce competition with the natural acoustic levels in their environments. They can be exposed to such sounds at high levels—rain, for example, can be heard to depths of two thousand feet. Dahlheim compares this with terrestrial situations, for instance where monkeys in the Madagascar forests utilize whistle calls to cut through the dense vegetation.

All of these findings imply, of course, that anything that reduces whales' receiving-and-transmitting capabilities could adversely affect their reproductive potential or even their survival.

In certain ways Paco Ollervides is picking up where Marilyn Dahlheim's work left off with gray whales (she's now into her second decade studying killer whales in southeastern Alaska). Besides Paco's theodolite tracking to observe the interactions between whales and boats, he's looking to discern whether the whales change their vocalizations to suit different situations. To do this several days a week he goes out into Magdalena Bay in a *panga* with a portable hydrophone. The engine is turned off, and the hydrophone is placed underwater at predetermined stations. Paco records what it picks up for five minutes at each site.

He explains: "From Marilyn Dahlheim's work, we have the *known* signals of gray whales and a basis for how gray whale sounds change in the presence of increased noise. We're trying to see if, when there's boat noise, they do any changes in the initial frequency or the duration of the call or eliminate some calls. It's a little hard, because the sound frequency of the boats is very similar to the sounds of the whales. There's this phenomena called masking, where the frequencies overlap in both amplitude and intensity. You're not really sure if the whale is actually quiet, or if it's vocalizing but the boat noise is overriding or masking it. So we have to do a lot of recordings with boats and without boats. Then, after our field season ends in April, we go back to Texas A&M and I do my computer analysis to try to pick out the whale signals and filter out other noises."

On the eight tapes made so far this season, Paco says he's been lucky in recording certain specific "events." He's come upon a group of gray whales feeding. "It used to be thought their feeding was nonexistent in Baja, but now more and more evidence supports the fact that they feed opportunistically," he says. There were no boats in the area at the time, and Paco was able to pick up some "feeding signals" through the hydrophone. He's also recorded several instances where the whales are socializing together or exhibiting sexual behavior.

He'd begun by successfully matching the seven types of gray whale signals that Dahlheim had detected at San Ignacio Lagoon. Since then Paco has been able to classify three entirely new signals in Magdalena Bay. "The last one was my most interesting," he says, his excitement building. "Because I could *only* record it when there were boats around. I did this just three times on two different occasions, so admittedly it's a very small sample. But I call it the 'annoyance signal,' because it's a kind of negative response. And this is the same signal that was recorded at SeaWorld a year ago, when they were taking blood samples from the juvenile gray whale, J.J., in captivity there. So it was not even the same source, but it produced the same effect!"

This was more than intriguing. On the one hand, gray whales that come right up to small boats obviously aren't "annoyed." Indeed, precisely the opposite would seem to be the case. On the other hand, here at San Carlos, besides the industrial ship traffic, the whales often found themselves *surrounded* by whale-watching vessels —probably many more than at the San Ignacio Lagoon. Thinking back over my time at San Ignacio, I couldn't help but be impressed with how conscientiously those skippers maneuvered around the whales. While the number of *pangas* on the lagoon at any given time had been allowed to increase from twelve to sixteen in 1999, overcrowding was never a factor. Guides like Pachico and Maldo certainly took people into the proximity of the whales, but nobody ever chased them. In Magdalena Bay that awareness didn't exist. That is what Paco's studies were aiming to improve.

Paco was saying: "I get criticism from some of my fellow marine mammalogists, who say, 'Why do you spend so much time with gray whales? They're so ugly, they look like giant cucumbers or slugs!' " Who says that? I asked. "People that are jealous of our research," Paco responded with a laugh, then continued: "Well, it's true that gray whales are not streamlined like humpbacks or blue whales. Or cute like belugas. But they are so graceful. Filled with expression."

Some of them were beginning to become "friendly" in Magdalena Bay. Paco said he'd noticed that, early in the season, "mothers actually put their bodies in between the boat and the calf. Then as the calf grows older and more curious, the mother allows more contact." After he had been working around gray whales for eight years, 1998 marked the first time Paco had touched one. "It was a calf, and I could see its eye looking into my eyes. I knew we were talking. I didn't say anything. This is an experience a lot of people have had in different ways, but I know it's there. They're trying to save us from our human side, I guess."

I asked what he meant by that. "Well, I firmly believe that we have detached ourselves from nature, with automobiles and satellite TV and microwave ovens and so on. That to me is one of the reasons why we don't protect nature. We have to feel a part of it first. The gray whales serve to connect us again."

As the sun faded over Magdalena Bay, I sat on a promontory not far from the School for Field Studies. I watched for whale spouts and mused upon their echoes. Roger Payne, in describing the commonalities between human songs and humpback whale songs, has written: "This commonality of aesthetic suggests to me that the traditions of singing may date back so far they were already present in some ancestor common to whales and us. If this is true it says that the selective advantage of singing and the laws upon which we humans base our musical compositions (laws we fancy to be of our own invention) are so ancient they predate our species by tens of millions of years."

That was a wild and wondrous thought for a scientist, not dissimilar from Paco's feeling that he and the baby gray whale were "talking." I reflected on two of my friends having been suddenly inspired to sing as the whales approached them. For one, it was a chant she had not thought of in years. Jim Nollman, among the pioneers in interspecies communication, tells a story in his latest book *(The Charged Border)* about donning a wet suit and taking a "huge, floating drum" out among the gray whales off the California coast in the mid-1970s. When I met with Nollman, he told me the same story but with a twist:

"I was going into the water ten miles off the coast of Mendocino in January, in fifteen-foot swells and a boat about the size of this table. It's also the white shark breeding ground for the West Coast. I can't believe I was doing that, I was nuts! Anyway, there were times when I'd be floating around out there with this big wooden drum that had tongues cut out of the top, like you see in craft fairs today. I'd made outriggers for it, and a seat, so I could actually sit inside the drum." Nollman called it a "whale singer" and, he wrote, was hoping to evoke "whale songs heard on record, in the ocean, and especially in the human imagination." However, "if the drum tipped too much to either side, water flowed in through the tongue slits."

He said he wasn't having a whole lot of luck, and "the swells are getting up to twenty feet. One moment you're at the crest and the next moment in the valley, it's very disorienting. My boat is gone, I don't know where my crew members went. For some reason, I wasn't freaked. All of a sudden, I had this jolt of being examined. It was very strong, and I don't know how to explain it, except that I felt like my brain was a room and somebody was walking around inside of it. A few seconds later this gray whale came right up beside me and made eye contact. Just stared at me. I've never felt so scrutinized, almost invaded, but there wasn't any feeling of danger. No, it was a feeling of trust. Of surrender."

I sent David Gude a transcript of the conversation I'd taped with Nollman. David had been a musician and a sound engineer for a record company and long been fascinated by the acoustics of whales. He'd been the one who called the whales over with his song on my first visit to the San Ignacio Lagoon. In response to the Nollman interview, he wrote me,

> Very mysterious. They may "hear" or visualize by electrical stimuli. Maybe they see auras ("Nollman: 'They like to stare.' ")
>
> [In Baja] four went by as I was swimming (my last morning). Dove under and called to them. (Always wondered what I would say.) Said, "Hello," which came out, "Oh-oh." "Greetings from my world to yours," which came out: "ah ah, ah, ah ah, oh, ur." Very interesting. It was reduced to digital and tones. . . .
>
> Maybe grays learned to project electromagnetic waves. Has anyone tested? . . . Huge mystery here. They rule the waterways. Man is 95% water! Lots to learn from them.

As I finished reading the letter, that last phrase lingered. "Lots to learn from them." I realized it was time for me—and getting time for the gray whales—to leave the Baja. Time to heed the call north. And I knew I'd have to follow, wherever the call took me.

AÑO NUEVO
STATE RESERVE
SANTA CRUZ
MONTEREY CANYON
POINT PINOS
MONTEREY
CARMEL
POINT LOBOS
STATE RESERVE
BIG SUR
↓

Chapter 10

Orcas and Grays Along the Shores of Monterey

We love being out there with the whales, and after our time at Neah Bay, we were desperate to find a way to infuse people with a sense of stewardship for the ocean and its creatures. What better way than by taking them out there? To transport people into this other world, a place where dolphins dance on your bow, luminous jellies pulse by, and great whales go about their slow-motion business is the best way we can think of to enlist them as foot soldiers in the whales' navy.

—Heidi Tiura, Monterey, California

My northbound tracing of the gray whale's path began in San Francisco. I drove through the neighborhood northwest of downtown where Charles Scammon and his family lived between 1856 and 1874. The original house on Ellis Street, I learned, had been destroyed by fire during the 1906 earthquake. I continued down to the harbor and looked out over San Francisco Bay, where Scammon had "cast anchor before the Golden City of the Western Slope" for the first time on February 21, 1850. From here he had set sail on his eight whaling voyages, as commander of seven different ships with the Revenue Marine and on the Western Union Telegraph Expedition. At the wharf a charter-boat captain told me a strange story. While gray whales rarely linger in the bay, he said, for five years in a row at least one had swum in during the annual springtime festival known as the blessing of the fleet.

I decided to follow Route 1 out of the city, back south along the coast past the sites of some of the old shore-whaling stations, and into the onetime whaling center of Monterey. According to Scammon, California's "whaling year begins on the first of April"—I was only a few days late. In

1874 Scammon had listed eleven "whaling parties" scattered along these shorelines. I drove by Gray Whale Cove and into the first of these at Half Moon Bay. Population 10,600. Lots of surfers headed for the beach. Lots of RVs and horse paddocks. Lots of Sunday afternoon traffic. At the entry point to the state beach, a park ranger said he'd never heard of any whaling station still standing around here. I kept on going, past artichoke and strawberry stands and into the verdant rolling hills that lead down into a whistle-stop town called Davenport Landing, population 200.

This had been the last residence of Captain John D. Davenport, the man who conceived the idea of California shore whaling. He was, Scammon tells us, "a whaling-master of much experience and enterprise" who hailed from Tiverton, Rhode Island, where he'd been half owner of a 180-ton schooner that plied between California and Hawaii, trading and whaling from 1845 to 1852. Then, in 1854, Davenport had established himself with a new bride in Monterey and recruited a dozen men to pursue humpbacks and gray whales with harpoons and hand lances from two small boats. It proved from the outset a lucrative business, which Davenport sold to a group of Portuguese before moving in 1868 to an area fourteen miles north of Santa Cruz where he built a large wharf and named it after himself. Primarily, Davenport Landing had handled shipping for several shingle companies and lumber mills.

I turned off the highway onto Davenport Landing Road and followed a big horseshoe curve back to Route 1. The paint was peeling badly on the old Davenport place, a clapboard house and barn, which I learned had recently been sold to a private owner with plans for renovation. I stopped in at the roadside Whale City Bakery Bar & Grill. It was a terracotta building accented with a colorful wall painting—of a smiling gray whale swimming through a life saver ring. In the window was a handwritten notice informing me that I'd arrived "where the blubber meets the road," a spot that "puts Davenport Landing on Highway 1's destination dining and drinking map." I settled for a cup of black coffee as I gazed at the gray whale bumper stickers and whale's tail pins and migration collector's buttons.

Across the highway and some long-obsolete railroad tracks, a small path festooned with wild, sweet-smelling white alyssum leads out toward a spectacular cliffside. At the top of the hill a sign warned: "Keep away, dangerous unstable cliffs, a fall can be fatal." I went to the edge anyway and peered down. Enclosed by leaning pine trees was what remained of Davenport's wharf. A Native American fellow was sitting nearby with his two dogs, looking out to sea. "Yeah, that's the old pier from the whaling days all right," he said.

A shore-whaling company, as Scammon described it, consisted of "one captain, one mate, a cooper, two boat-steerers, and eleven men; from these, two whale-boats are provided with crews of six men each, leaving four hands on shore, who take their turn at the lookout station, to watch for whales, and attend to boiling out the blubber when a whale is caught. . . . The cruising limits of the local whalers extend from near the shore-line to ten miles at sea. At dawn of day, the boats may be seen, careening under a press of sail, or propelled over the undulating ground-swell by the long, measured strokes of oars."

However, Scammon hastened to add, "this peculiar branch of whaling is rapidly dying out, owing to the scarcity of the animals which now visit the coast; and even these have become exceedingly difficult to approach." California's shore whalers, he reckoned, lost about one-fifth of their struck whales. I realized, as I continued on toward Monterey, that Scammon was really the first to undertake what's become a fixture in marine mammal biology today—the annual whale count and estimates of abundance. In his chapter on gray whales, Scammon recorded the calculations of shore-whaling observers that "a thousand whales passed southward daily, from the 15th of December to the 1st of February, for several successive seasons after shore-whaling was established." He estimated this meant an "aggregate" gray whale population, at that time, of about 47,000. From his own observations between 1853 and 1856, he continued, the number actually "did not exceed 40,000—probably not over 30,000." The count of grays since killed by the whalers, Scammon went on to speculate, "would not exceed 10,800, and the number which now periodically visit the coast does not exceed 8,000 or 10,000." That, of course, was still a precipitous drop, and by the 1870s Scammon cited a Captain Packard's guess that "the average number seen from the stations passing daily would not exceed forty."

On rounding Point Pinos, the quiet town of Monterey, with its naturally unique surroundings, is revealed in all its loveliness. The relics of old Jesuitical times can still be seen, but they are rapidly passing away. . . . Monterey still retains the aspect of an antique California settlement. It was at this place—in "Colton Hall"—that the Constitution of the State of California was framed. . . . Near the pier that projects into the bay stands the custom-house, which has withstood all the changes in governmental affairs since the time the pueblo became an entry-port. . . . In rambling through the town you meet many superior adobe structures, admirably adapted to the wants of the first settlers, being cool in summer, warm in winter, and proof against the attacks of Indians.

—From Scammon's "Pacific Sea-Coast Views,"
Overland Monthly, March 1872

The capital of California under Spanish and then Mexican rule, Monterey sprawls loosely around the gently curving arc of its vast, prolific bay. "It is impossible to describe the number of whales," a Spaniard named De La Perouse wrote of Monterey Bay in 1786. "They blowed every half minute within half a pistol shot from our frigate." An issue of the *Monterey Californian* later added: "These marine monsters were so numerous in Monterey Bay that whalers could fill up lying at anchor." Most experts believe that both references are to gray whales.

Downtown, near Fisherman's Wharf, is the First Brick House of California, where it turns out Captain John D. Davenport was the initial occupant. It's down the block from the Old Whaling Station, an adobe dwelling with an adjacent rose garden and arbor. Local legend has it that whalers kept their lookout from the upstairs windows, which had unobstructed panoramic views of the bay. Decorated with antique furnishings, today the Old Whaling Station's house and grounds are rented out for weddings and other fancy functions by the Junior League of Monterey County. Guests tread across a whale-bone sidewalk entryway and pass inside under a whale vertebra. On the second-floor veranda sits a rocking chair constructed from ribs and more vertebrae.

I recalled an article Scammon had pasted into a scrapbook, written by Prentice Mulford, who sailed down the California coast on Scammon's Revenue Marine ship *Wyanda*. In an 1869 edition of the *San Francisco Bulletin*, Mulford described the Monterey beaches "strewn with the bleaching bones of dead whales," looking from a distance like cast-up driftwood. "The great backbone joints are even used for pavements. They are buried in the earth, leaving only the round, upper surface exposed; and very good pavements they make. These blocks of bone, once the leviathan's spinal column, the seat of that strength by which he coursed from the Arctic seas to the equator, come to these base uses at last."

Such reminders remain abundant. Around the corner from the Old Whaling Station is California's First Theater, a historical monument at the intersection of Scott and Pacific. An arch of whale ribs and vertebrae forming newel posts for steps marks the entrance to the low-slung adobe building where nineteenth-century melodramas like *Troopers of the Gold Coast* were once staged. Reportedly the theater had also served as the equipment storage area for Davenport's whaling company.

I proceeded a couple of miles farther to the San Carlos Cathedral, advertised as "the oldest continuously functioning house of worship in California." Its "Self Guided Tour" notes that, when Father Francisco Pacheco renovated and enlarged the chapel in 1858, "whale vertebrae sections were used to pave the outside sidewalks. By the early 1940s this had become dangerous"—slippery when wet, I wondered?—"so it was taken up and stacked at the rear of the church." Nothing was indicated about what next became of the whale bones.

I maneuvered through the commercial maze that was once Steinbeck's Cannery Row and headed for the outskirts of town. Steph Dutton and his wife, Heidi Tiura, live twelve miles outside Monterey. I hadn't seen the couple since Neah Bay five months before, when they'd been among the most vocal opponents of the Makah hunt. They'd provided refuge to Alberta Thompson after the fracas between the tribe and the Sea Shepherds. Now their outspokenness had caused the loss of a substantial research grant to their organization, In the Path of Giants. The funding would have enabled them to track and document the entire migration of the gray whale, using a kayak as their primary research vessel. So they'd come home to central California and taken stock. Their efforts at Neah Bay had exacted quite a toll. Certain that they'd done what they had to do, they still didn't know how they'd pull themselves out of the apparent wreckage of their lives.

I drove up a winding road into the hills overlooking Monterey Bay. The single-story ranch-style house is surrounded by pines, oaks, ferns, and, as Heidi puts it, "more poison oak than should have been allowed by nature." The couple share these surroundings with white-tailed deer, quail, Steller's jays, coyotes, and rabbits. "Our garden has clearly been touted as the deer version of an all-you-can-eat salad bar," Heidi says, giving me a quick tour, "but every year I plant lobelia and petunias anyway."

When they found this place in 1996, Steph Dutton went straight to the master bedroom's bathroom and found an old ball-and-claw foot bathtub. For a leg amputee, he says, such a bathtub is a thing of beauty.

They are an unusual couple, drawn to each other and to the gray whales in ways that seem fated. Both are in their late forties. Heidi is a native Californian, and Steph, born in Austin, Texas, grew up near San

Diego. He is tall and quiet, with deep blue eyes ("the color of storm surf," as Heidi describes them), a long mustache, and wavy graying hair that overlaps his ears. She is freckle-faced, with long strawberry-blond hair and a vivacious smile and manner. Each of them wears a handmade whale's tail necklace, reminiscent of the earrings I'd seen on Alberta Thompson and John Spencer.

"Let me tell you what happened when I was a little girl," Heidi had said to me in Neah Bay. "I grew up in a rural area on the coast of Northern California. At night I used to ride my donkey across the highway and out onto the headlands above the ocean. I'd just stand there for hours. Especially when it was somewhat foggy offshore, it would be black and velvety and calm. I could smell the salt, hear the buoys and sometimes the foghorns up the coast. I was just entranced. It was partly fear, and partly fascination with what was out there. Now here I am, forty years later, and I'm wondering if even back then—because these whales were passing by my whole lifetime—were they pulling me and calling to me? Because I've always felt this pull, always."

For thirteen years Heidi Tiura started and ran a multitude of businesses: custom T-shirts, windsurfing, a deli and bar. Then her affinity for the ocean took control. She sold her businesses and went to sea. She headed north, running her own small boat-towing service over Washington's infamous Columbia River bar, working as a deckhand on charter boats and then on progressively larger tugboats on Puget Sound and in Alaska. Having become a licensed master of offshore vessels, by 1994 she was setting up sea kayak excursions for tourists who came to the southeast Alaskan village of Sitka on cruise ships.

That is where she met Steph Dutton. Divorced with three children, Steph at various times had been a contractor, a paramedic, a firefighter, an undercover cop, and a private investigator. The experience that changed his life forever had taken place in 1978. Steph calls it his "brush with extinction." He was driving home from working half a night shift as a favor for another firefighter when he came upon an accident in the hills near Escondido, California. Beside a wrecked truck, Steph recalls, "a guy was lying on the road, covered with a blanket. I thought he was probably dead. But when I pulled it off, amazingly he was alive, and begging me to help him. That's when I heard a squeal of tires, looked up, and saw headlights coming right for us." He'd responded by pushing the accident victim off the road, then trying to dive out of the way himself. "But the oncoming driver hit me right as I was in midair, and smashed my leg. That threw me in under the wrecked truck. I was lying there in a puddle of gasoline when I saw a fireman walk up holding a lit flare. That's when I knew I'd die if I lost consciousness. I had to stay awake long enough to direct my own rescue."

It took eighteen months of failed operations before Steph demanded amputation just below the knee of his right leg. He wanted to get on with his life. After virtually no rehab, he taught himself to walk with a prosthesis. He did concrete work, a physically demanding and grueling job even for a person with both legs. He became a paramedic. He also resumed skiing, earning accreditation as an instructor. The examiners had no idea he was skiing on a prosthesis. Steph's intent was to teach skiing to amputees and paraplegics, but there was little demand in that area at the time.

Then he discovered sea kayaking. Not only did any trace of his limp vanish the moment he sat down in a kayak hull but his upper body strength could flourish when he was paddling. In 1993 Steph set out on a solo paddle from Victoria, British Columbia, all the way to Ensenada in Baja California—a voyage of 1,600 nautical miles that would take him two full months. Nobody had ever tried this before. And he almost didn't make it. As Steph was passing along the Oregon coast, a raging storm struck. He remembers: "It was a full-on gale, and I was filled with a mix of awe and terror. Mostly, I was worried about keeping myself upright. Suddenly, I was stunned to see the heaving water all around me start to make some strange moves. Then I realized—I was surrounded by a pod of spouting whales! A whale on my right dove, and then surfaced again on my left side. My gosh, I thought, forty feet of whale just swam under my kayak. It was as though they were guiding me, helping me to stay oriented to the coastline. I considered that they might be a pod of grays, but I couldn't be sure. Right afterward, I made it my business to find out a lot more about them."

The following year Heidi Tiura heard a story from her sister about a one-legged ex-firefighter who had solo-kayaked the entire Pacific coast of the contiguous United States. She thought, "How inspirational! I said to myself, 'I want this man in my life.' " But Heidi had no idea when she went to pick up a new kayak instructor at the little airport in Sitka that she was about to meet him. "Even when we went drinking that night down at the Pioneer Bar, and Steph ended up telling stories of his long paddles down the coast, I still didn't put it together that he was the very man I'd dreamed about. I did notice he had a slight limp, and the next day, as we all piled into the truck, I glimpsed a steel bar poking out between the end of his pant leg and the top of his shoe. I touched it and asked, 'Hey, what's this?' Steph just quietly replied, 'Ah, it's nothing,' which suddenly spoke volumes to me. So, we fell in love in a day. And we've been together ever since."

They married and settled first in Astoria, Oregon. Steph's growing reputation as one of the world's premier kayakers led to financial support from the U.S. paddling industry. With a partner he circumnavigated the northern Channel Islands off California. With Heidi as expedition manager, he paddled the Oregon coast in the winter of 1995, which took a full thirty days in treacherous weather.

In January 1996 the couple were in Big Sur when they saw the misty whale spouts offshore, dozens of them. They decided to launch kayaks from Monterey the next morning and try to follow the gray whales' trail. Even a strong storm that blew ashore that night failed to dissuade them. They set off at 7:00 A.M. and paddled for hours. Heidi had never paddled so far, and her hands blistered. The blisters broke and stung from the salt water. They ate nothing and drank only a little water. They were miles offshore, in sixteen-foot seas and twenty-knot headwinds when, Heidi remembers, "suddenly we were surrounded by barnacle-encrusted

Heidi Tiura in her kayak (Steph Dutton)

Steph Dutton in his kayak (Heidi Tiura)

freight trains, just blowing and thundering past us, all around us, underneath us."

"When these whales dived beside us, about ten feet down, we could see an incredible, blue-green luminescence surrounding them," Steph says.

"At one point, one stopped, a full-grown gray whale," Heidi says. "It turned its head ninety degrees and rotated its eye up and looked at me. And it was like a lightning bolt. Had I been standing, it probably would have knocked me off my feet. It just grabs at your soul."

When their kayaks slid onto the beach at 3:30 that afternoon, the couple were exhausted. Before she got out of her kayak, Heidi turned to Steph and said, "I want to go. I want to go with them."

"That solved a huge problem for me," Steph would say later in presentations they've given all over the United States and Canada, "because I'd already decided I wanted to craft a project around these magnificent animals, and I was wondering how to break the news to Heidi!"

Their research organization, In the Path of Giants, was under way. They moved to Monterey, spending the first several months in a small

RV with their huge Chesapeake Bay retriever. For the next two winters they tracked the gray whales migrating along the central California coast. "We got the snot kicked out of us by El Niño," Heidi says, "me in *Sea Dog*, our twenty-six-foot powerboat, and Steph in a twenty-two-foot kayak in thirty-five-foot seas!" Steph paddled literally thousands of miles with the whales as the pair filmed and photographed them, learning their patterns and nuances. He seemed to possess a phenomenal ability to adapt to a single whale or even a pod, often knowing their speed underwater, their course, and when they'd surface. Much of this emanated from tracking them by a muscle-powered craft, where you must pay attention to the finest details—such as the angle of a fluke as the whale sounds, signifying a change in direction—or else you end up half a mile from the animal the next time it surfaces. One goal of these endeavors was to document a standard gray whale route past Monterey Bay, which could lead to establishment of rules for human navigation that would minimize disturbance to the whales and their young. Information keyed into charts supplanted older data.

Heidi says, "We've never witnessed friendly whale behavior, you know. Most people assume we've been to the lagoons and had this wonderful soft experience with these whales. Instead, all we've done is beat ourselves up on the open ocean, where they want very little to do with us. But we're learning their rhythms, their habits."

The couple had taken their boat up the coast to Neah Bay and the Makah reservation in late September 1998, with an offer to help the tribe develop whale-watching and sea-kayaking tourist ventures as an alternative to the hunting plans. Heidi had created a curriculum called the Sea Dog School in Monterey, where middle and high schoolers were taught navigation and seamanship skills and even the rudiments of operating a kayak or powerboat. "We offered all of this to the Makah to be woven through their culture," she says, but they were turned down on all counts. "We were branded as troublemakers, treated as outlaws, and thrown out of a federally funded marina." So they'd stayed on just outside the reservation. They established a base camp, aiding the Sea Shepherds and using their boat to scout for any tribal canoes that might be after the whales.

Until then the National Marine Fisheries Service was ready to grant In the Path of Giants a research permit. Several other groups had been prepared to provide financial support for a planned voyage along the whales' path all the way to the Bering Sea. Heidi would pilot the mother ship while Steph and a team of paddlers would use kayaks as their primary research vessels. They planned to attach to the whales small dart tags with the capacity to log all sorts of data, as well as to transmit it to satellites.

"When our biologist called us and told us to get out of Neah Bay and eliminate the advocacy stuff from our Web site—otherwise, he'd have to withdraw—we almost complied," Steph says. "But finally we knew we couldn't leave, even though we'd face losing all our funders. Somehow, we'd find ways to continue our work."

And so they had, despite the project's forced cancellation. Heidi already had her Coast Guard master's license; in the spring of 1999 Steph obtained his. The pair found employment in Monterey and began running whale-watching cruises, eventually forming their own company. "We bought a great boat and named her *Sanctuary*," Heidi explains. "Steph has completely gone through her engine room and she just purrs. I put life-sized Pacific white-sided dolphins on her bows, because so often we are surrounded by them. Much like a phoenix, Sanctuary Cruises was born out of the ashes of In the Path of Giants."

As we sit on the deck of their home, looking out across tall pines, gnarled oaks, and twisted manzanitas all the way to Monterey Bay, the sun is setting. Steph is unusually talkative. "I remember going to Magdalena Bay several years back, being out there in my kayak and seeing spouts and thinking, Thar she blows! I'd go racing toward the spouts, and then I'd find myself saying, Wait a minute, where does this *come* from? It was a little scary to me. Is this a past-life thing? I'm not a New Age kind of guy, but I was rocked by that. Then, last year, Heidi and I were in Boston at the Union Oyster House, where Daniel Webster drank brandy and ate oysters. We were seated at table two in the Pine Room, drinking Harpoon ale. It was hot, but, at the same time, we looked at each other and we had goose bumps. We both thought, We've been here before, we *know* this."

Steph is silent for a moment. Strangely, I'm thinking of Ahab and recalling lines from *Moby-Dick:* "Here be it said, that this pertinacious pursuit of one particular whale, continued through day into night, and through night into day, is a thing by no means unprecedented. . . . For such is the wonderful skill, prescience of experience, and invincible confidence acquired by some great natural geniuses among the Nantucket commanders, that from the single observation of a whale when last descried, they will, under certain given circumstances, pretty accurately foretell both the direction in which he will continue to swim for a time, while out of sight, as well as his probable rate of progression during that period."

Then Steph continues: "My fascination with the gray whales is not anthropomorphic. The bond I sense with them comes from something deeper. I think I feel a strong connection to the gray whale because of my disability. I wouldn't say they're flawed physically, but they're not everybody's idea of a whale. They aren't pretty. They aren't the high-glamour species like humpbacks, blues, or orcas. Yet they've survived two near extinctions. They've withstood the ravages of greedy humans for more than a century. Somehow, they've found a way to endure. Deep in my mind, I have an image from an old fairy tale called 'The Steadfast Tin Soldier.' He's the one who's all beat up and missing a leg, yet will never shirk his duty. That term, *steadfast*, means so much to me because that's what the gray whales are. They have this incredible will to survive. There's a sense that this whale won't quit. Nor will I."

Along the California coast, there may be no better place to contemplate how the gray whales "withstood the ravages" of Melville's era than at Whaler's Cove in the Point Lobos State Reserve. It's south out of Monterey, across the Carmel River and down a pine-shrouded winding road, nestled in the lee of southwesterly breezes. These craggy headlands are among the most spectacular spots on the entire California coast. When Spanish explorers sailing past four centuries ago heard sea lions barking off the rocks, they called it Punta de los Lobos Marinos, Point of the Sea Wolves.

Inside what may be the oldest wood-frame building in California, a cabin built by Chinese fishermen around 1851, is the West Coast's only on-site whaling exhibit. On the far wall is a large framed sketch of this very spot, depicting a team of shore whalers. It was drawn in 1865 by Charles M. Scammon, who later described the scene:

> The one [station] which most interested us is half-hidden in a little nook, on the southern border of the Bay of Carmel, just south of Point Pinos. Scattered around the foot-hills, which come to the water's edge, are the neatly whitewashed cabins of the whalers, nearly all of whom are Portuguese, from the Azores or Western Islands of the Atlantic. They have their families with them, and keep a pig, sheep, goat, or cow, prowling around the

> premises; these, with a small garden-patch, yielding principally corn and pumpkins, make up the general picture of the hamlet, which is a paradise to the thrifty clan in comparison with the homes of their childhood. It is a pleasant retreat from the rough voyages experienced on board the whale-ship. The surrounding natural scenery is broken into majestic spurs and peaks, like their own native isles, with the valley of the Rio Carmel a little beyond, expanded into landscape loveliness.
>
> Under a precipitous bluff, close to the water's edge, is the station; where, upon a stone-laid quay, is erected the whole establishment for cutting-in and trying-out the blubber of the whales. Instead of rolling them upon the beach, as is usually done, the cutting-tackles are suspended from an elevated beam, whereby the carcass is rolled over in the water—when undergoing the process of flensing—in a manner similar to that alongside a ship. Near by are the tryworks, sending forth volumes of thick, black smoke from the scrap-fire under the steaming cauldrons of boiling oil. A little to one side is the primitive storehouse, covered with cypress boughs. Boats are hanging from davits, some resting on the quay, while others, fully equipped, swing at their moorings in the bay. Seaward, on the crest of a cone-shaped hill, stands the signal-pole of the lookout station. Add to this the cutting at the shapeless and half-putrid mass of a mutilated whale, together with the men shouting and heaving on the capstans, the screaming of gulls and other sea-fowl, mingled with the noise of the surf about the shores, and we have a picture of the general life at a California coast-whaling station.

Sweeping Monterey cypress still tower here over the front yard of the museum. A park ranger tells me that the last two native stands in the world are found here and on nearby Pebble Beach, the last two holes of whose famous golf course are visible from Whaler's Cove. Under the boughs a series of whale bones are anchored into the earth alongside a huge metal whaling try-pot. The floor joists of the museum were fashioned from six whale vertebrae, which, the curator says, "were lying all over the place when the original dirt floor got replaced around 1900. They're strong, and the termites don't bother them, and they were free."

There had been several small villages around this cove at different junctures—first the Chinese fishermen, next the Portuguese whalers. "When kerosene replaced whale oil," Ranger Chuck Bancroft tells me, "the Portuguese moved out by the mid-1880s, tearing the cabins down and taking the lumber with them." Then had come Japanese at the tail end of the nineteenth century. They were abalone fishermen who also temporarily established a joint venture with some Portuguese called the Japanese Whaling Company. "There was a large scaffolding down here, where they'd bring the whales and tie them off to Window Rock," Bancroft adds. After shore whaling ceased for good in 1901, it wasn't long before Hollywood discovered Point Lobos. Almost fifty movies have been filmed at these vistas since 1914.

I'm standing in front of the Scammon sketch, reading an inscription about how, in 1848, Lewis Temple had "invented a new type of harpoon head that quickly became the standard of the industry. Because the harpoon head was able to pivot or toggle within the whale's flesh, it was less likely to pull out and thus resulted in capture of many more whales." I realize that's the same basic type of harpoon head that the Makah Tribe intend to use for their first strike, now, one hundred fifty years later. I walk outside and across the road to an open meadow, where red-hot poker

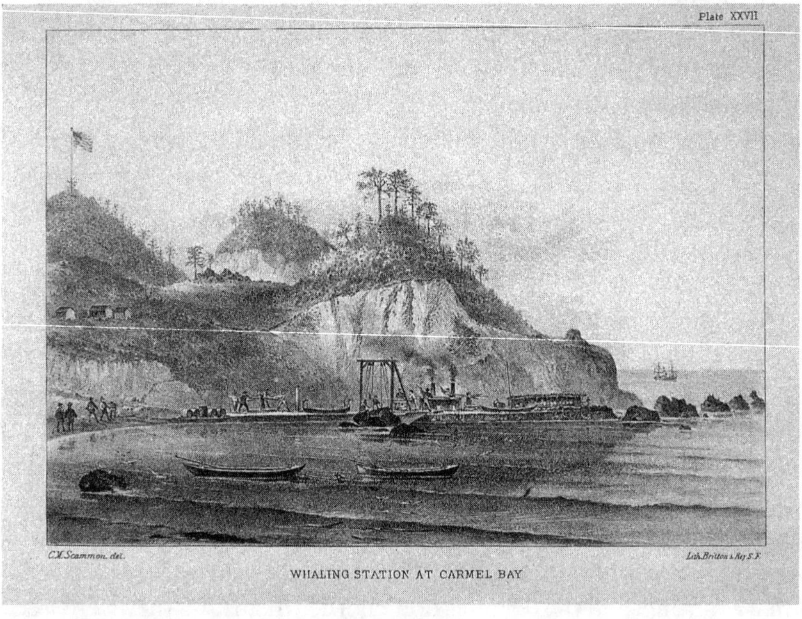

Scammon's Whaling Station at Carmel Bay

plants blossom, and envision two dozen try-pots boiling amid the cacophonous din of Scammon's "general life at a California coast-whaling station."

By 1886, "of the eleven whaling stations mentioned by Scammon as established along the coast ten or twelve years ago, only five remain," according to the Townsend report "Present Condition of the California Gray Whale Fishery." John Dean Caton's article for *American Naturalist* (1888) added that, in Monterey, "the old whale boats may now be seen leaning up against the sheds useless and abandoned." Another writer, Edward Berwick, recalled in a 1900 issue of *Cosmopolitan* magazine how "the shores of Monterey Bay were frequented by that greatest among seabirds, barring the albatross, the giant fulmar *(Ossifraga gigantea)*. This bird then found it to his interest to leave his haunts in the Southern Hemisphere, and feed his fill on dead whale, in abundance and variety, off California's coasts. With the decline of whaling he again withdrew southward, and for many years not one has been seen here."

What gray whale catch remained during the latter part of the nineteenth century—a total of 167 taken along the California coast between 1883 and 1886—consisted chiefly of males. Females traveling north with their young were staying well offshore. For the only known time in its history, the gray whale's will to survive had driven it far from the harpoon guns of its human adversaries. In Monterey Bay, as I was soon to discover, the animals still needed every bit of that will to fend off another voracious species.

As gray whale mothers and their young travel north together for the first time, they are able to remain in proximity to the near-shore kelp beds on most of their early sojourn across California. Upon reaching the vicinity of Point Lobos, however, they are abruptly faced with the coastline opening into a large bay. The sea here turns a deep cobalt. A vast underwater abyss commences. It snakes to a depth of 10,663 feet, as deep from rim to bottom as the Grand Canyon. Monterey Canyon was forged mainly by the erosive power of silt-laden underwater currents, and it is an extremely fecund area. The nutrient-rich cold water wells up along the gorge. The warmer water of the shore shallows mixes with the deep. The resulting explosion of algae and plankton forms the base of a complex food chain, which supports everything from tiny copepod crustaceans to hundred-ton blue whales.

Upon arriving at this juncture the migrating gray whales face a choice. They can try to take a shortcut across the canyon—one of the deepest bodies of water a gray whale will ever traverse. Or they can hug the shoreline, along the sloping beach and the kelp forests that ring the thirty-mile-wide Monterey Bay as it stretches from Pacific Grove to Santa Cruz.

Those who select the shortcut may not make it. For there, above a seascape of eternal darkness, the orcas wait.

Orcas—killer whales—may have been the "*aries marinus*" of the ancients, and possibly the "horrible Sea-satyre" of Edmund Spenser, since the white marks on their heads might be fancifully interpreted as closely appressed horns. *Orcinus orca* was the name bestowed by Linnaeus in 1758, *orca* being Latin for a type of whale and *orcinus* being derived from the Latin *orcus*, which means "of the netherworld." The whalers called them "whale killer," a title later transposed to "killer whale." For they are not whales themselves but the largest species of dolphin. "The length of the adult males may average twenty feet, and the females fifteen feet," Scammon informs. Other than human beings, orcas are the most cosmopolitan mammals on the planet in their distribution, being found in every ocean basin. And, with their distinctive blend of coloration—jet black above and white below—they are one of the most identifiable. Despite our doting upon Keiko—the film star of *Free Willy*—this species possesses a uniquely predatory appetite for its fellow marine mammals.

This isn't true of *all* killer whales. Scammon had pointed out that several species of orcas could be found in every zone and hemisphere. Jim Nollman, who's concentrated on analyzing their vocalizations in recent years, described it like this: "There are two orca cultures. One group eats fish and vocalizes all the time. They have a social structure based on frequency-modulated whistles, in other words a melody. Then there's this other group, which are stealthy and are called transients. And they eat marine mammals. Recent genetic studies have shown that, even though they travel the same waters sometimes, these two groups of orcas have not reproduced together in at least one hundred thousand years."

It's the transient pods, as scientists call them, that pursue gray whales throughout their entire range. Few have ever evoked these killer whales so eloquently as Scammon. He wrote about them at some length, originally for the *Overland Monthly* and then in his book. Reading his anatomical description, I was struck by how much orcas seem the mirror opposite of gray whales. Whereas the grays are covered with barnacle-raising parasites, the orcas are free of such altogether, the "scarf-skin being beautifully smooth and glossy." Whereas the grays possess no dorsal fin whatsoever, the orcas possess a "dagger-shaped" prominent upper limb. Whereas the grays are toothless, the orca mouth "is armed with strong, sharp, conical teeth, which interlock." Whereas the grays "speak" in the lowest of tones, the orcas have shrill, high-pitched voices. Whereas the grays subsist on the smallest of bottom-dwelling organisms, the orcas subsist on the largest of surface-swimming marine mammals.

"The Gray Whales," Roy Chapman Andrews of the American Museum of Natural History wrote in 1914, "seem to be objects of continual persecution by the Killers; much more so than any of the other large whales."

Scammon tells us how these "wolves of the ocean" would grapple with the grays and other baleen whales of far greater size than themselves.

> And it is surprising to see those leviathans of the deep so completely paralyzed by the presence of their natural, although diminutive, enemies. . . . The habits of the Killers exhibit a boldness and cunning peculiar to their carnivorous propensities. At times they are seen in schools, undulating over the waves—two, three, six, or eight abreast—and, with the long pointed fins above their arched backs, together with their varied marks and colors, they present a pleasing and somewhat military aspect. But generally they go in small squads—less than a dozen—alternately showing themselves upon the surface of the water, or gliding just below, when nothing will be visible but their projecting dorsals; or they disport themselves by rolling, tumbling, and leaping nearly out of water, or cutting various antics with their flukes. At such times, they usually move rapidly over the surface of the sea, and soon disappear in the distance.

Their attacks "may be likened, in some respects, to a pack of hounds holding the stricken deer at bay. They cluster about the animal's head, some of their number breaching over it, while others seize it by the lips and haul the bleeding monster under water; and when captured, should the mouth be open, they eat out its tongue."

Scammon witnessed one such moment "in a lagoon on the coast of Lower California, in the spring of 1858." It was an attack by three killer whales on a gray whale and her calf. The battle "lasted for an hour or more. They made alternate assaults upon the old whale and her offspring, finally killing the latter, which sunk to the bottom, where the water was five fathoms deep. During the struggle, the mother became nearly exhausted, having received several deep wounds about the throat and lips. As soon as their prize had settled to the bottom, the three Orcas descended, bringing up large pieces of flesh in their mouths, which they devoured after coming to the surface."

The predator in Scammon made no attempt to disguise his admiration for the orcas, which have rarely been hunted by man. As he put it, "They are seldom captured by civilized whalemen, as their varied and irregular movements make the pursuit difficult, and the product of oil is even less than that of the Blackfish, in proportion to their size." Indeed, Scammon and other commentators allude to a strange kinship—and sometimes a competition—between whalers and killer whales. As early as 1725, it was noted in an English "Essay on the Natural History of Whales," "These Killers are of such invincible Strength, that when several Boats together have been towing a dead Whale, one of them has come and fastened his Teeth in her, and carried her away down to the Bottom in an Instant." They had even been known to wrestle over a whale "till it draws the Boat under Water." Scammon also cited instances on the northwestern coast where a band of orcas would lay siege to whales being towed to the whaling ships "in so determined a manner, that, although they were frequently lanced and cut with boat-spades, they took the dead animals from their human captors, and hauled them under water, out of sight."

An unusual reciprocal relationship is said to have taken place during the late 1800s and early 1900s among an orca and some shoreside whalers in Australia. Old Tom, as the orca was affectionately known, came to realize that both he and the whalers liked to go after humpbacks. Eventually Old Tom started coming into the bay, lobtailing and slapping the water to let the men know when there were whales offshore. The whalers would row out and follow the killer whale to where the prey was, harpoon a humpback, and allow Old Tom to feed upon his favored portion—the tender tongue—before hauling the rest ashore. It even got to the point, the legend goes, where once the whalers put a line in the water, the orca would pick it up and tow the boat out.

Killer whales assisted the human hunters in another way. One captain reported that, shortly after he began to hunt a group of seven gray whales, fifteen killer whales had shown up. The grays became so terrified that the captain had no difficulty in killing three of them: "When the

Orcas gathered about, the whales turned belly up and lay motionless, with fins outspread, apparently paralyzed by fright."

Just as the vastly diminished numbers of gray whales had learned, in Scammon's words, to "shun the fatal shore"—breaking their time-honored migratory patterns and traveling well away from the California coast and their human predators for many years—so it seems did gray whales develop means of avoiding killer whales. Grays were observed to "go into such shallow water as to roll in the wash and even try to hide behind rocks" when sensing orcas nearby.

The first large-scale whale watching by people also focused on an interaction between grays and orcas. It happened early in 1947, when many marine experts still believed the gray whale to be extinct. A San Diego newspaper reported: "Trapping two southward-bound whales inside kelp beds along La Jolla coastline, a vicious pack of six killer whales yesterday awed more than three thousand beach residents in an unusual four-and-a-half-hour aquatic life-or-death struggle. One of the toothless Mysticeti species of whales eluded the desperately flailing killers (conical-toothed Odontoceti type), but the fate of the second was not determined because darkness set in."

Then, in January 1966, humans intervened directly on the whale's behalf. THREE GRAY WHALES FIGHT OFF KILLERS IN OCEAN BATTLE, the *San Diego Union* headlined in a page-one story. Dr. Theodore J. Walker, an associate oceanographer at Scripps, was leading a whale-watching cruise with nearly a hundred people aboard when he received a radio alert from some lifeguards about the attack. While hundreds of Sunday strollers packed the shoreline at La Jolla Cove, Walker instructed his skipper to make for the scene. The water was a maelstrom as they approached. Walker observed that the trio of gray whales didn't panic but were staying in close formation. They lay on their sides and swatted at the killer whales with their tail flukes. As Walker's ship closed in the orcas broke off with a dramatic series of tail slaps and headed north. They'd abandoned the battle, Walker figured, because the engine noise must have made it difficult for them to hear each other. The exhausted gray whales kept coming up to blow every few seconds.

As Walker's vessel turned parallel to shore, he reported, "the grays stayed between us and the land, and I think they were using us for protection. Normally they wouldn't let us get that close." This went on for about two miles, almost to Bird Rock, where the three whales pointed their rostrums toward the kelp beds and the surf. There, according to Walker, the bubbles in the surf and in the plants' anchors acted as an "acoustical screen." The killer whales' echolocating sonar wouldn't be able to find them. The sun was going down. The gray whales were safe. The whale watchers departed.

A decade later Walker would be among the first marine scientists ever to reach out to—and touch—a "friendly" gray whale at the San Ignacio Lagoon. Walker was known as a crusty sort. A companion observed this was the only time he ever saw the man weep.

In Baja killer whales are known to appear in the Pacific waters outside the breeding lagoons but, with the lone exception of Scammon's account, never to enter the nursery domains.

Now, above Cannery Row and just down from the Monterey Bay Aquarium, Alan Baldridge is sitting in a conference room at the Hopkins Marine Station. Recently retired after almost thirty years as the Hopkins science librarian, Baldridge probably knows as much about gray whales as any layman in California. He's sailed to their Baja breeding grounds and their Bering Sea feeding areas. His fascination with the grays had begun in April 1964, when he was standing atop a huge cliff along the Oregon coastline, "an unusually calm, clear day when I could actually see the whales underwater as they traveled so close to shore." Baldridge was working at the Portland Public Library at the time, and he searched the shelves to see what he might learn about gray whales. There he came across an original edition of Scammon's *Marine Mammals of the North-Western Coast of North America.* "Of course, there's a legion of material today," he says. "But in those days, the sixties, Scammon was still the only major account about the gray whales' natural history."

Baldridge was equally intrigued by Scammon's description of the orcas. In the years since, he has witnessed three attacks on gray whales off Monterey and published the most detailed account since Scammon in the *Journal of Mammalogy* in 1972. "The mothers and calves taking those shortcuts across the bay," Baldridge emphasizes, "are the most vulnerable. It may be that these are simply inexperienced animals. The calves make a lot of noise, of course, and there's considerable vocalization between the pair as they swim along and play around. Presumably the Orcas hear this from some considerable distance, and can home in on them."

Baldridge expects it's the more seasoned gray whales who stick to the inside of the breakers along the shoreline. For one thing, killer whales seem to fear getting beached if they probe too far into the surf. For another, as they do scan the near-shore region with their sonar, the round air sacs in the kelp forests tend to dissipate the echo. So for the grays it's a decent enough place to remain undetected.

They've been known to adopt other strategies as well. In the Bering Sea in 1981, scientists watched from a small plane as sixteen killer whales in two distinct groups approached twenty-seven feeding gray whales. The orcas moved in a crescent-shaped formation, synchronizing their breathing patterns, perhaps to give the impression of a smaller pod. The gray whales suddenly ceased feeding, formed compact groupings of between three and six, and began swimming slowly away. It was speculated that their methodical, tight-knit interactions might provide protection similar to that of fish schooling. Meantime, the scientists had dropped a sonobuoy nearby to try to monitor sounds and transmit these to an onboard VHF broadband receiver. Over a ninety-minute period, not a single waterborne sound was picked up from either species. Both the killer whales and the gray whales, it was believed, were remaining silent to avoid calling attention to themselves. No attack was witnessed. The gray whales were presumed to have escaped.

When an attack *has* been recorded, as Richard Ternullo, one of Monterey Bay's first whale-watching captains, described it to me, "It sounds kind of eerie. You can hear the gray whales talking back and forth. "When you put it on a spectogram, you can see the frequency differences in their calls. Superimposed over that are these killer whales making some of the weirdest sounds you've ever heard."

We were sitting in an outdoor café on Fisherman's Wharf. For the past dozen years Ternullo and a naturalist named Nancy Black have also been conducting a photo-identification project on orcas. They look to identify individuals by their markings and work with others along the Pacific coast to try to determine where they go. And they've probably witnessed more attacks on gray whales than anyone, about one a year since the early 1990s, generally in the submarine canyon area of Monterey Bay. In 1998 Ternullo and Black had a contract with the BBC, which managed to film a couple of these events from Ternullo's fifty-three-foot *Pt. Sur Clipper.* In 1999 they had a similar deal with National Geographic TV.

Ternullo is a short, stocky fellow who wears his long white hair tied back. He had on a Sportfishing Sam Monterey cap (the name of his charter boat operation) and blue jeans. I wanted to know what he'd been learning about the gray-orca interactions. Alan Baldridge had told me that "the observations of Richard and Nancy show that some of the same individual orcas are involved in these kills from year to year. We don't know whether the orcas have followed the whales up the coast, or whether they know the gray whales will be heading north and are intercepting them as they migrate."

Ternullo nodded and expanded on the theme. He said there seemed to be a distinct pattern of hierarchy as well as cooperation among the orcas. "Some killer whales will hang around the edges, as if they're positioning to block an escape route, while others try to separate the mother from the calf." Precisely like Scammon and the other lagoon whalers, I thought to myself. "Then one killer whale will kind of take the lead, seem to be putting in more effort than the others during the direct attack. The next day the one doing the hard work won't be the same one as the day before."

The orcas generally attack first from underneath, ambush-style. Ternullo has seen several adult males take out a gray whale calf in a couple of hours. He's also seen a female orca, and what was apparently her own offspring, take as long as six or seven hours to separate a calf from its mother and kill it. Ternullo figures there is "probably some instructional element" transpiring in the latter instance. "At the end," he continued, "they actually physically get on top of the calf to hold it down. I mean, right on its back, like riding a horse. What they're doing, of course, is drowning it."

An examination of gray whales taken for research purposes at a California whaling station in 1971 had shown that 18 percent exhibited evidence of having been attacked by a killer whale. From what he's observed Ternullo thinks the number is most likely higher. He recently photographed a gray whale with track marks from the front of its rostrum all the way down its back along the dorsal ridge. "If you look at the tails of gray whales, which often have chunks taken out, you realize a *lot* of them have had experiences with killer whales."

What percentage escape, nobody knows. Killer whales often aren't seen for months until gray whales begin showing up off California on their southbound migration. Then, like cheetahs that follow the wildebeest migrations across the African Serengeti, the orcas close in when their prey are most vulnerable, either just before or just after a calf has been born.

Ternullo's voice lowered as he began to reflect on the most memorable encounter he's witnessed. It happened during the spring of 1998, in the deep, unprotected waters across the canyon. A pack of five killer whales had zeroed in on a mother and her calf. The mother innovated a desperate strategy. Time and again she would roll onto her back and swim belly-up. The calf would climb on top of her, between its mother's flippers and just out of reach. Simultaneously, they would roll over and breathe, then resume the same position. The orcas did everything they could to separate the pair. The calf did get bitten but was never hamstrung and maintained its vigor.

"I've never seen anything like it," Ternullo said. "I've seen one trying to support a calf on her back before, but never holding the calf to the chest like this. At the same time the mother took advantage of a manmade object in the vicinity. A boat. My boat. The killer whales kept circling around us, but she stayed real close and used the hull for protection. After a couple of hours, the orcas just gave up."

There followed a perfect spring day in Monterey Bay. Florian Graner, an underwater photographer from Germany, had arrived and was also staying with Steph Dutton and Heidi Tiura. He was hoping to get some more footage here for a documentary, *Wonderworld of the Kelp Forest*, that he was working on for German TV. Heidi would take us out on *Sea Dog*, the twenty-six-footer they'd trailered to Neah Bay last year.

The three of us left the harbor around noon. "You can see Cannery Row over there," Heidi pointed out. "Some of the older buildings have been dandied up and made into restaurants and hotels. And those boats on moorings are two of the original sardine fleet; they go back to the thirties and forties." As we approached a Coast Guard vessel, Heidi went on, "That's the newest class. They're made to self-right if they capsize, for the most horrendous conditions. Finally, our tax dollars well spent!" Next we came alongside a colony of sea lions lying in the water with flippers extended up in the air, "doing some thermal regulation to keep themselves warm," Heidi said.

We cruised past the amber glades of lush kelp forests growing not far from shore, home to various rockfish and mollusks and crustaceans, and sanctuary at times for gray whales avoiding orcas. "This is the orcas' main stage," Heidi said. "They know they can come down here and find the grays." Had she ever seen an attack? I wondered. "No, and I don't want to. I'd probably be out there trying to separate them." But we were heading for Monterey Canyon, where anything might happen.

Florian readied his camera gear and prepared to don a wet suit. He was in his late twenties and built like a welterweight fighter, a good thing since his two diving tanks weighed eighty pounds apiece. The tanks came complete with a rebreather, because gray whales don't like bubbles. Attached to a small boat, Florian had done filming in fifty to eighty feet of water at San Ignacio Lagoon, and stayed down for as long as seven hours in a day. He'd swum there amid lobsters and grouper and beautiful soft corals, and once counted sixteen horn sharks under a rock. He'd accidentally awakened a sleeping whale, which lashed its tail fluke and came straight for him to give Florian a little wake-up call of his own. He'd also had a baby gray gently pick him up—and lift him on its nose straight out of the water! We'd watched some of his footage the night before, including some mesmerizingly long takes on the eye.

"My idea is, when they're migrating that's *business*," Heidi said as she made a sharply angled turn. "But when they're down in Baja—'Okay, let's do a little sunbathing.' "

Some Dall porpoises, which look like small orcas, came racing toward us and began cavorting around the boat. Heidi said that a week earlier, while running a whale-watching trip, she'd seen a mysterious interaction between a group of these porpoises and a very large gray whale. "This whale was swimming very slowly, and we were concerned for it," she recalled. "Usually Dalls move like lightning, but they stayed with her, encircling her. Perhaps she was just resting, but the spectacle of her minimal movement, surrounded by the Dalls, was really touching. We had two little boys onboard. One of them turned toward us and declared, 'She is the princess of whales, and they are her knights, escorting the princess.' Steph turned to me and said, 'That's the name of our next boat.' " While they started their company with *Sanctuary*, a catamaran later added to their fleet has indeed been christened *Princess of Whales*.

Now Heidi got on the radio with Richard Ternullo, who reported seeing a couple of humpbacks. "*What I suggest*," Heidi replied, "*is we go north-northwest and see if we can find some gray whales over the canyon.*" I heard Ternullo agree to follow in our wake.

"These guys don't have much of a fishing industry anymore," Heidi said. "Most of them would rather fish, but rockfish have declined tremendously. Catches keep getting smaller, so they run whale-watching trips. Back when we started doing gray whale research, they didn't talk to us. Then slowly, they started. We've been out here at times when they were desperate to find a whale. It makes sense to use all of your resources: if people are out there every day, talk to them. Tinker, on *Check Mate*, was the first. One day, he called me on the radio and said, 'Heidi, I need a whale.' "

She stopped and put *Sea Dog* into reverse, circling back to retrieve somebody's floating plastic bag. She explained that leatherback turtles and other sea creatures ingest plastic bags and balloons, mistaking them for jellyfish; it's a fatal meal for many. Then, as we started out again, Heidi wheeled around and brought the binoculars up. "I don't know what it was, but something *very* large just breached about a half mile off my bow. It looks like it was headed back toward the point; maybe we can go get a better view of it." She turned and charted a course southwest.

Sure enough, within ten minutes a lone gray whale was clearly discernible at the southern edge of the Monterey Canyon. Heidi alerted a party boat of our location. Soon the *Magnum Force* hove into view, its foredeck crammed with people holding cameras. The whale was moving at a fast clip along the upper fringe of the swells. I could see an ethereal blue-green coloration reflecting off its barnacles. Washing along directly ahead of it, leaving a ripple like a snake, was a reddish brown cloud. "This gray whale is surface-feeding on krill!" Heidi exclaimed. "This is something you hardly ever see!"

Indeed, gray whales are considered classic benthic feeders, and they feed almost exclusively in the Arctic. But here, thousands of miles from their feeding grounds, the ocean is hundreds of feet deep and there's no way they could be bottom feeding. The whale watchers aboard *Magnum Force* hadn't made it in time to see what we did as, throwing up its flukes, the gray whale sounded. *"Should stay down about three and a half minutes,"* I heard somebody say over the radio. Twelve minutes went by, however, and we hadn't seen the whale come up again. I watched the large vessel turn toward port. There was another group of prospective viewers to pick up. *"I'll try to keep track of the gray for when you get back,"* Heidi radioed Leon, the captain, over the radio.

"When the whales are on a mission like this," she explained to Florian and me, "they'll pop up here one minute and next time they're a half mile over there. We were tracking one whale with a very distinctive scar on its back yesterday, but there was no telling where it was going to be next."

We kept rolling. Nothing happened for another half hour or so. Then, as we began retracing our path along the edge of the canyon again, a distinctly rank odor wafted up from the sea. I remembered the breath of gray whales at the lagoon's being inoffensive and even almost fragrant, but this was something else again. At least we knew we were on the right track.

"Gray whales still there?" I overheard one captain asking another.

"Roger, Richard, just about same area, bud."

"You know if it's the same whale?"

"Heidi out here's been on it."

"Look look look look!" Heidi exclaimed again and reached for the radio to call one of the other skippers. *"Oh wow, Leon, there's something very interesting going on in front of our bow! This gray whale looks like it's rushing under the water!"*

"It's going almost in circles!" Florian cried out. "Whale going sideways," as if he were talking to some invisible personage in code.

"Whale blew right in front of me! Right on that tide rip!" Heidi radioed again.

Concentrated in the rip, the whale was trailing a ball of krill that appeared at the surface as a pink swarm. Krill is a minute form of crustacean. Each individual is about the size of a human's little fingernail, and such swarms are known to be the major dietary staple of blue whales in the Antarctic. But gray whales? This one's mouth was swinging like a barn door, ingesting all it could in a pell-mell chase that Heidi informed us is "sometimes called Pac-Manning—gulping in vast mouthfuls like in the video game Pac-Man."

"I think I recognize this guy! I think I've filmed him before!" Florian shouted again.

"Holy shit! A 180-degree turn, just like that, right next to us! It looked more like a salmon!" Heidi exulted. *"We're in Whale Central again,"* she radioed an approaching tourist vessel, *Star of Monterey.*

I helped Florian adjust his tanks, and he jumped over the side holding his camera. It didn't take more than a few minutes, however, for him to realize the waters were far too murky to capture anything underwater. The hungry gray whale had disappeared again, but the bay seemed to be teeming with life. Stormy petrels and small active gray birds called Cassin's auklets dove for prey. A humpback whale breached ahead, briefly displaying its long, winglike pectoral fins. Another boat reported seeing a minke whale. We cruised a tide rip, finding ourselves surrounded by more than fifty Pacific whitesided dolphins. Just outside their perimeter more Dall porpoises circled.

I hadn't felt so exhilarated since my first morning on San Ignacio Lagoon. "So they don't eat till they get to the Bering, eh?" Heidi said. "Well, we shot the shit out of that theory!"

The next morning I went to visit with Alan Baldridge again at the Hopkins Marine Station. The librarian asked me quite a few questions about what I'd seen, then shook his head and said: "You know, I've talked to a number of zoologists here who know a lot about krill. But when I brought up the subject of these daytime krill swarms, they said, 'Alan, that's not right.' According to them, when the sun comes up the krill sink from the surface down to the edge of the dark zone about three hundred to four hundred feet, which hides them from predation. Then as the sun goes down, they rise to the surface again to feed on the phytoplankton. So what you saw proves the scientists off base on a couple of points. It's been observed here before, gray whales in association with some of these spring krill swarms, which I believe are reproductive swarms. But it's not been published in the literature. You were very lucky to see that."

For the gray whales, Baldridge thinks, this represents "opportunistic snacking." Their traditional feast was still waiting thousands of miles to the north, beyond the coastlines of California, Oregon, and Washington, on past Vancouver Island and most of the vast Alaskan shoreline, along the shallow bottoms of the Bering and Chukchi Seas. More than ever, that's where I longed to accompany them.

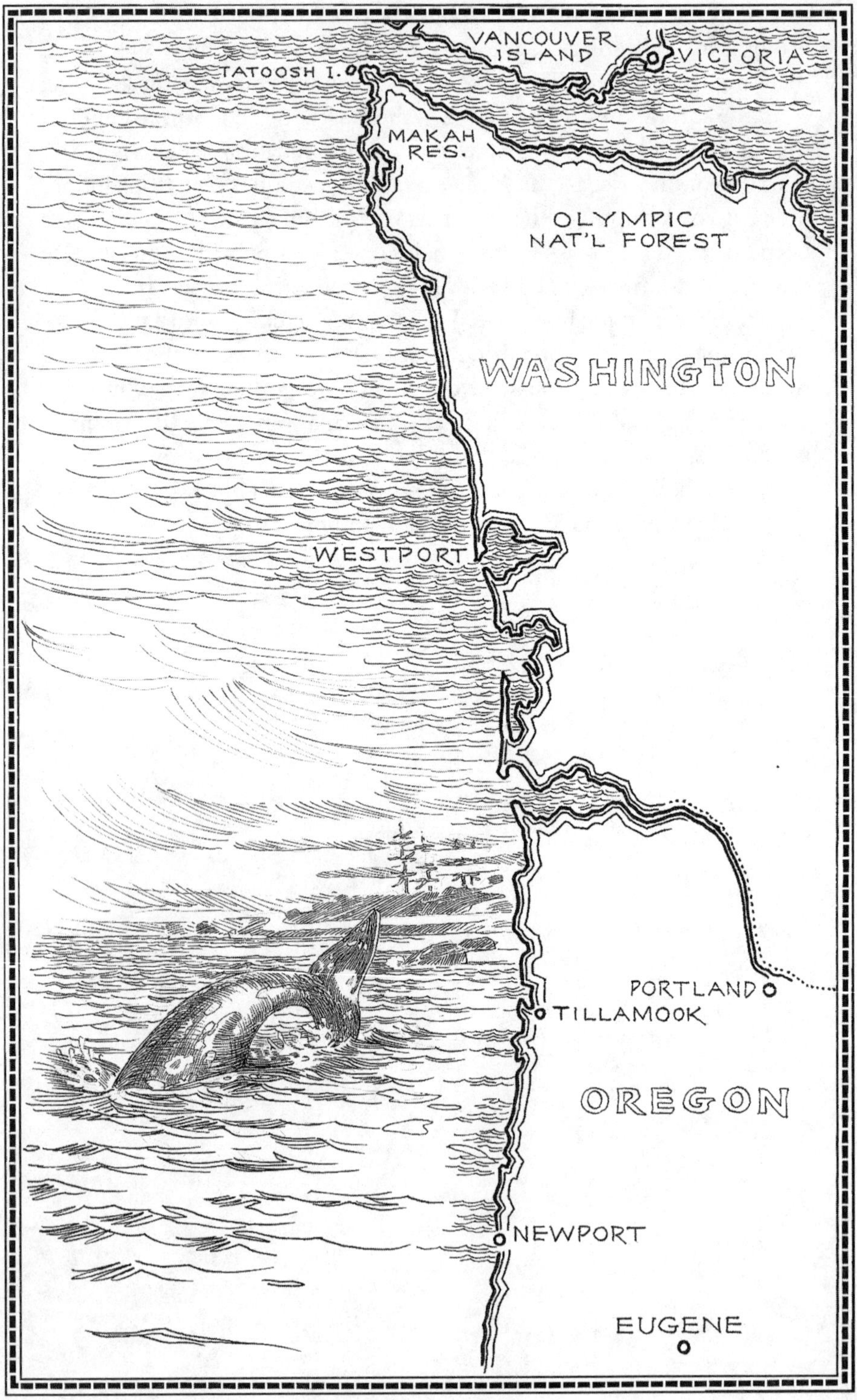
VANCOUVER ISLAND
VICTORIA
TATOOSH I.
MAKAH RES.
OLYMPIC NAT'L FOREST
WASHINGTON
WESTPORT
PORTLAND
TILLAMOOK
OREGON
NEWPORT
EUGENE

Chapter 11

Oregon and Washington: Scholars of the Great Migration

In my own experience I have found that at their best the revelations of science are as inspiring as art. . . . I feel that science has a role to play not just in conservation (in my view the most important human activity) but also in the more spiritual parts of our lives.

—Roger Payne, *Among Whales*

Leaving Monterey and passing San Francisco, the gray whales move north along one of the planet's most stunning coastlines. It's nearly three hundred miles to where Redwood National Park meets the shoreside cliffs above Eureka, California. Here the relentlessly pounding surf of the Pacific echoes in one of the few remaining groves of *Sequoia sempervirens*, majestic, ancient trees rising to more than three hundred feet. Thousand-pound Roosevelt elk also grace the grays' path. The antlered beasts graze in the prairies high above the whales, seals, sea lions, and sea otters.

Onward into Oregon just below Highway 101, the gray whales hug the shorelines of Pistol River and Gold Beach alongside the Siskiyou National Forest, up past Coos Bay and through the Oregon Dunes National Recreation Area, beyond the Suislaw National Forest, and to Newport. It will have taken a gray on the order of eight days to get to the resort center and commercial fishing port on Yaquina Bay, almost five hundred miles north of San Francisco. The region has a long history.

"Notwithstanding the forbidding aspect of the contiguous coast," Scammon wrote in 1874,

> several shipping-ports have sprung into existence between Port Orford and the Columbia [River], among which are Umpqua, Yaquina Bay, and Tillamook. In fact, life along this whole borderland has changed from the old aboriginal times when the sons of these wilds led a life of unrestrained freedom, pursuing with rude weapons their game through the depths of the forest, or fishing in the streams and about the shores with primitive implements of their own make. But at the present day, under the hand of civilization, where once the smoke of the wigwam-fire curled above the tree-tops, now puffs the steam from the mills; and, instead of the war-whoop, the shrill whistle or the clanging horn calls the workmen to and from their toil. The mountain slopes and bluffs, the valleys, clothed in wild luxuriance—where once grazed the elk and the antelope in countless numbers—are at this time covered with lowing herds and bleating flocks. In the more sheltered nooks, or upon the gentler undulations, are the dwellings of the settlers, surrounded by corrals and cultivated inclosures.

A few miles north of Newport's Yaquina Bay, where trails wind their way around a small mountain to Yaquina Head, the gray whales' numbers are counted from the tallest lighthouse on the Oregon coast, 162 feet above sea level. From that vantage point Oregon State University scientists observed that the grays' southbound winter migration in 1998–99 began three weeks later than normal and that the majority were traveling five miles or more offshore; two decades earlier most of the population had been observed within three miles of the coast. This was the first indication that a change might be occurring in the gray whales' time-honored habits.

Depoe Bay, eight miles above Newport, is probably Oregon's premier spot to see gray whales. From what may be the world's smallest harbor, whale-watching cruises maneuver straight out into the ocean through a lava cut in the rock. Skippers are so confident that, if you don't witness grays for your ten dollars, they'll take you out again for free. Within five minutes you're likely to be among the whales, a few of which will stay and forage all summer. Often they seem to show up in particular places at certain times of day.

When the tide is out you're likely to see furrows in the sand. Detritus collects in the furrows, and tiny crustaceanlike creatures known as amphipods eat the detritus. Amphipods, in turn, are the grays' favorite food.

You might see whales come right up into the surf zone, lying on their sides and sucking up the depressions of amphipods like vacuum cleaners. Around rocky headlands the grays also eat swarms of mysids, small shrimp-type plankton that live in the water column amid the kelp beds. Mysids themselves are attracted to the bait in crab pots, so it's not unusual to observe a whale by a float. Marine scientists have received calls about a gray being snared in crabbing gear only to find it's merely moving right down the buoyed line with mysid-seeking mouth agape.

Once in a while, grays do get entangled in fishing nets or lines. Forty-seven such incidents were reported between California and Alaska from 1990 through 1998. Of these, thirteen whales appeared to have survived, while the remaining thirty-four were either mortalities or of unknown status. A week before my April arrival in Newport, one had been seen swimming slowly and attached to portions of a longline vessel's gear. Since longliners don't work the Oregon coast, the whale must have picked it up somewhere in California. The gray was first spotted off Depoe Bay on a Wednesday, and by Friday off Tillamook, sixty miles farther north.

The marine biologist Bruce Mate figured he could catch up. He drove a truck pulling a rigid-hulled inflatable with the gear needed to disentangle the whale, something he had done many times before. For two days he searched the migration corridor between Tillamook and Astoria, even enlisting aerial help from the Klamath County sheriff. The search was to no avail.

"Well, sometimes that's what happens," Mate is saying. "You know, despite the fact we know the gray whales' critical habitats and that they've recovered numerically, I still think they are in some jeopardy. The reason is, the entire population twice a year moves in near-shore proximity to human activity. If you ever did have a big catastrophe, you could expose them all."

We are sitting in Mate's paper-strewn office at the Hatfield Marine Science Center, where he's been based since 1975. This is right around the corner from the Oregon Coast Aquarium in Newport, which housed Keiko the killer whale for a while. Some three hundred researchers work out of the Marine Science Center, which is part of Oregon State University but also hosts a campus of state and federal marine agencies. Mate himself wears several hats. He holds an endowed professor's chair and directs the university's marine mammal program. He's also an extension Sea Grant specialist and member of a collective called the Coastal Oregon Marine Experiment Station. About 40 percent of the time Mate is offshore somewhere, in places like Nova Scotia or Hawaii or Alaska, doing fieldwork with endangered species of large whales.

It's a curious devotion for a man who grew up in Wheaton, Illinois, and never saw the ocean until his last year of college. That's when Mate came west to check out the biological curriculums of graduate schools. He happened to attend a seminar by George Bartholomew, a noted expert on pinnipeds, "who started out by saying nobody knew the migration habits of seals and sea lions." Young Mate was incredulous. "I thought, how could anybody *miss* these big animals? Obviously, the guy doesn't know what he's talking about." Mate had seen a nearby rookery of Steller's sea lions on a field trip, so, he says, "I started poking around in the library, and the man was right—I couldn't find a thing about their migrations." Choosing the University of Oregon's Graduate School of Biology, Mate made this the subject for his doctorate. "Ever since," he says, "most of my true love has been in figuring out what marine mammals like the gray whale do to make a living, where they do that, and why those areas are important."

If anyone goes down in the history of cetology as the pioneer of tagging and tracking whales, it will be Bruce Mate. Hunting implements were the first "tags" to discern whales' movements. As Mate has written: "In the 1850s, Captain Charles Scammon identified Alaskan Eskimo harpoons in the flesh of migrant gray whales, *Eschrichtius robustus*, along the California coast, and concluded that the Alaskan population moved south in the winter to the Mexican calving areas."

Until recent years what scientists knew about where whales went hadn't progressed much beyond Scammon. It remained largely based upon the catch records of whalers and observations on their flensing decks. By the 1930s numbered shafts called Discovery tags were being intentionally fired into whales and recovered during the commercial harvest. Not until the mid-1960s were projectile radio tags developed for scientific purposes. But the low-powered transmitters could be heard only over short distances, they rarely stayed affixed to the whales for long, and it was expensive to follow tagged whales by ship.

By the late 1970s more reliable radio packages for tracking and telemetering data from whales had been developed. Bruce Mate decided to travel to the San Ignacio Lagoon, for the first of many times, in 1979. He'd heard about the "curious" or "friendly" gray whales that had begun

About to tag a gray whale in San Ignacio Lagoon (courtesy Jim Sumich)

to approach boats. It seemed an ideal situation for inserting a tagging device into the outer blubber layer. Otherwise, you needed to shoot it through the air from a crossbow and hope that, *if* you hit your target, the electronics would withstand the impact. For two years Mate had been trying without success to secure funding for a lagoon sojourn, "simply to demonstrate to the scientific community that we could keep these things attached. The agencies would say, Well, if we knew it worked, we might invest in it," he remembers. "Finally, my wife, Mary Lou, suggested we take out a second mortgage on our house and sell our second car. We spent fourteen thousand dollars to do that experiment on our vacation time with one of my students and his girlfriend. And it worked."

Not only did it work but Mate's out-of-pocket experiment would result in the longest tracking distance and duration *ever* in using conventional radio telemetry on a cetacean. He'd gone to San Ignacio with three VHF tags containing transmitters. They could be applied to a nearby whale's back using a handheld sixteen-foot pole. If oriented properly, the transmitter's antenna would then clear the water each time the whale took a breath. That year Mate managed to apply all three tags on gray whales. Their new appendages didn't appear to bother them in the least; the whales remained equally inquisitive around the boat before and after the procedure.

The northbound migration commenced. If the tagging apparatus stayed attached to even a single whale, this would be an opportunity to learn something about its travel time. Mate had his receiver placed with colleagues at the Southwest Fisheries Science Center in La Jolla, just outside San Diego, "and bet that at least one of the three whales would pass during daytime business hours Monday through Friday—which it did." At this point, the gray wasn't moving all that fast. Mate calculated it averaged not much more than thirty-three kilometers (twenty miles) a day from when it was tagged in the lagoon and Southern California. He then had the receiver sent on to his office in Newport, Oregon. "About twenty-one days passed, and I started getting nervous that the whale should be arriving," Mate recalls. "So I went up in a small Cessna airplane and started cruising south with my receiver. I found the whale about eighty miles away, at Coos Bay."

The whale's speed had accelerated; it was covering an average of 130 kilometers (80 miles) a day on this leg of its journey. The gray stroked past Newport the next day and on beyond Tillamook one day later. Mate still gets excited as he describes what happened next: "We sent the receiver on up to Alaska, to folks who were going to do some bear work out at Unimak Pass. They had delays and a plane cancellation. Finally they got out there at night, dog tired, and said, 'Shall we put up Bruce's experiment or shall we wait? Ah, let's go ahead and put it up.' Next morning—beep, beep, beep!"

As the whale transited for twenty-nine days between Oregon and Unimak Pass, en route to its feeding grounds in the Bering Sea, its pace maintained a daily average of 127 kilometers (79 miles). All told, it had been tracked for ninety-four days and over 6,680 kilometers (4,150 miles) before its location became too remote to follow.

"Getting the attachments worked out on gray whales was the first step, the bridge," Mate says. "Being the only coastal whale, we've known their basic migratory pattern since Scammon's day. But with most other whales we've only had a handle on where they are for half the year—either the feeding part or the reproductive part. For quite a few species it's critical to learn their abundance and distribution so we can adequately protect key habitats."

With the advent of new satellite-monitoring technology, this is finally becoming possible. Back in 1970 the Nimbus satellite system was developed for oceanographers and meteorologists to track drifting buoys and high-altitude balloons. Its refined phase is known as Argos. It's used primarily for environmental studies, everything from monitoring river water levels and seismic events in remote locations to tracking wildlife such as whales. Argos provides the only civilian service to locate specialized transmitters from space. There are four Argos receivers onboard several National Oceanic and Atmospheric Administration weather satellites. These orbit high above the earth taking 101 minutes for each circumnavigation. All information received by the satellites is stored onboard and transmitted to earth whenever the satellite passes over one of three ground telemetry stations—in France, Virginia, and Alaska.

While conventional radio telemetry could track only one whale at a time and required on-site observers, hundreds of satellite-monitored whales can be followed simultaneously, day or night, without any field logistics or site personnel. "The messages cannot go through seawater," Mate has written, "so whale transmitters use a saltwater switch to initiate transmissions *only* when the whale is at the surface. To further conserve power, the transmitters are also programmed to transmit only during times when the satellites are overhead."

This means, of course, that the tagging equipment itself has had to come a long way. Mate ushers me into a laboratory down the hall from his office to display "our case of used-to-bes, the historic development. Here was our first success with Argos in whale tracking, on a humpback in 1983. You can see the tag was about the size of a three-pound coffee can, and it lasted for a week. The humpback moved five hundred miles, and we were really excited! Well, today the whole miniaturized device would slip through my ring finger and isn't much longer than a Bic pen. With that tag we're tracking blue whales right now at 205 days and they're still going."

Although he does buy some components, Mate and his Marine Science Center team design the housings, attachments, antennas, software, and all the ways of deploying the tags. Only within the past couple of years, a 75 percent reduction in volume and weight has been achieved. This, Mate says, has made it possible to actually embed a tag in a whale. Only surgical-quality materials are used in preparing to fire it from a crossbow; the latest tag is but a small dart. "Implanting it takes away the hydrodynamic drag of water rushing over it and trying to pull it off. This has meant a great deal in terms of added longevity."

Mate reaches into his briefcase and brings out a state-of-the-art example. "Out of two batteries, each the length of one digit of your finger, we get 48,000 transmissions to a satellite that at its closest is five hundred miles away and at its farthest is about two thousand miles away. It's only half a watt, not much better than a kid's walkie-talkie—but it's *very* sophisticated. With some of these tags, we can also collect sensor data to help us evaluate whale behaviors and environments. We can measure the depth and duration of the dives, the temperature at the surface and at the deepest part of the dive, and how much time a whale spends at the surface. We've even developed sensors with Cornell University to record at what depths blue whales vocalize. We've seen whales attracted to oceanographic features like warm-core rings, and structures that would help concentrate prey. The discoveries coming out of all this are nothing short of remarkable."

The first real success, Mate says, came with endangered right whales in the North Atlantic. "They used to be considered slow-moving, near-shore surface-skim feeders. We quickly found out that they oftentimes moved long distances at high speed, went offshore, and dove deeply for their food."

Bowhead whales tagged in the Canadian Arctic near the end of their feeding season were found not to migrate en masse but to take different routes with different timing. One was tracked moving through waters off Canada, the United States, and Russia in less than a month—not only a surprisingly long distance but as rapidly through very dense ice (covering as much as 90 percent of the sea) as through open water.

Blue whale winter breeding and calving habitat is being identified for the first time. "It looks like blue whales go from California all the way down to the Costa Rican dome, a nutrient-rich region of upwelling near the equator," reports Mate. "Until recently humpbacks were all thought to migrate from Hawaii to southeastern Alaska. Now we believe probably only one-quarter of the humpback population makes that trek." The first one to be satellite-tracked went instead from Hawaii to Russia's Kamchatka peninsula. Since then, others have been located going to British Columbia or Alaska's Aleutian Islands chain. "We're also finding that, once they start, most travel consistently at relatively high speeds, as do gray whales. They get to where the food is and *then* settle down."

Mate continues: "The experience I've had with gray whales gives me some ability to interpret what I'm seeing with other whales. And what I'm learning with other whales, I bring back to gray whale issues. There's a whole lot more to discover about *them*, too."

•

In March 1999, at a scientific meeting to review the status of gray whales five years after their removal from the Endangered Species List, Mate had been the first to hypothesize that the grays might be a growing population with a food supply on the decline. This coincided with what Jorge Urbán was observing in Baja, where gray whales gave birth to fewer calves—and smaller ones—in a trend Urbán supposed might be linked to slower productivity in the northern feeding grounds. Mate believes the latest satellite technology could be crucial in addressing what's going on in the Arctic. "The way I envision it happening," he says, "we would tag gray whales here on the Oregon coast during the northbound migration. If when they get into the Bering Sea the animals go to one spot and stay there, it says they've found a good place to feed. If they keep moving from place to place, that suggests they're not finding enough to eat."

Mate has used the new satellite technology with gray whales only once, at San Ignacio in 1996. He was trying to get at how they utilized the lagoon. "Can anybody really comprehend what it must be like to be a female gray whale?" he marvels. "Once she's sexually mature, she spends 80 percent of her life pregnant or lactating. Think of the energy demands to produce a one-ton baby in a year—that's over 150 pounds a month devoted to baby development! And then nursing with 50 percent fat milk, which continues until the calf is eight or nine months old. There's such a short period between when she weans the calf on the feeding grounds and when she goes back down again to breed. It's truly amazing."

Every year since 1989 Mate has taken small groups on cruise ships out of San Diego to spend time with the gray whales at San Ignacio Lagoon. He estimates probably about three hundred people have made the trip, including a number of repeat visitors. Each night Mate spends about a half hour telling them about the progress being made with various species, "what kinds of dreams and hopes I have for changing the conservation and management of whales." Nobody gets asked to donate, but some do nonetheless. Mate's research is expensive. Each tag costs about $4,000. In a typical field season, by the time he's paid for twelve tags, satellite time, staff, and transportation, the expenses are around $125,000. With government money having dried up considerably during the Reagan era, Mate estimates that from one-half to as much as three-fourths of his funding comes from private individuals and corporations. "And most of the donations come from people who've been to San Ignacio Lagoon," he says. "I view the gray whales as the ambassadors for the whale world in that regard."

Even twenty years later Mate's memories of his first couple of years at the lagoon are vivid. "It was hardly glorious—bathing in salt water, trying to wash dishes, dragging boats across the mudflats dog tired, having to go into town for drinking water along a road which in those days customarily took eight or nine hours. My wife, Mary Lou, has a rodent phobia and wore hip boots during an explosion of kangaroo rats. When word got out that she was a registered nurse, we became the medical center for the ranchers and fishermen. I learned things from people there that were priceless."

Their nearest neighbor toward the lagoon mouth was a fisherman named Ramón Seseña, who lived in a plywood shack with his wife and two small daughters. Much like Pachico Mayoral and John Spencer, Ramón spoke no English and Mate spoke no Spanish—"but we'd sit and somehow talk philosophy into the night. One night Ramón got out some tools and sat there taking apart this little disposable wind-up toy. He cleaned the mainspring, put WD-40 on it, and made it work again. In our society we'd never think about taking a couple of hours to restore a one-dollar toy. The whole time his daughters were at his feet looking up at him like he was God. Of course, when Ramón finished he wound it up—I still remember this little dolphin moving across the wooden floor—and those children were ecstatic. Wow, there's a value lesson—about time investment. It wasn't just recycling, it was doing something for *them.*"

When Ramón noticed Mate's small crew setting up a volleyball net on his first Sunday in the lagoon, Mate recalls his saying, " 'You provide the beer, we'll provide the food.' Well, about sixty people came out of the arroyos and the ranchos. It became a fiesta that involved some serious volleyball playing all day long." (The lagoon fishermen's team was los Tiburones—the Sharks—and the scientists were las Orcas.) Mate continues, "The local folks laid out an array of large clams on the ground, in an area about the size of a sheet of plywood. They covered them with cactus wood, lit it on the downwind side, and the wind took the flames across the clams. And as the embers glowed out, the clams opened. Perfect timing. Talk about utilization of natural resources—they knew exactly what to do! It was wonderful."

Then, of course, there were the gray whales. When I ask what has surprised him the most, Mate responds without hesitation: "These animals have different personalities. It doesn't surprise a biologist to see variability, which is inherent and makes evolution possible: not all animals do the same thing the same way all the time. But I've gotten to know some individual gray whales well enough to see that their habits are pretty consistent between the times I meet them—whether it's one or two years, or even five or ten years apart."

While Mate estimates that probably only 10 percent of the grays in San Ignacio Lagoon are "friendly," those that are seem to exhibit the behavior year to year. "For the females with calves, I think it's pretty clear that sometimes they're using the boats as a baby-sitting service," he says. "There's no rest for a new mother. So they push the babies forward and stay close by. Oftentimes they're quite content to have the calves entertained by us."

Mate recalls one adult female they named Bopper. "She'd sneak up on us. One season she'd suddenly lift our Zodiac with three or four people in it out of the water three feet, and we'd come crashing down. She didn't appear to be malicious at all, just playing, but it was very disconcerting. We didn't learn she was a female until she had a calf later that same season. And then the calf would do the same thing! It couldn't lift our boat more than about six inches, though. We finally put a radio tag on Bopper so we knew where she was and we could avoid her."

Mate never expected to encounter two friendly whales bearing the unmistakable markings of having been harpooned. The first possessed a small entry wound on one side and a craterlike wound on the other, "almost certainly from an explosive harpoon, probably fired off the Siberian coast." The second whale has been named Scarback and is the subject of a large mural on the Newport bayfront. "She was friendly when I first met her in the lagoon and didn't have that kind of mark. A couple of years later you could see the fresh wounds on each side, the one on the right being the size of a bushel basket—and she was *still* friendly. Over the years she's been healing. Scarback continues to be friendly not only in the lagoon but here along the Oregon coast as well during the summer feeding season. She visits us off Newport, Oregon, every year, just offshore of the south jetty. She'll come over to tour boats or fishing boats which pass by. I've met six of her offspring. Before she has weaned them, she'll bring them right up to boats too."

Mate shakes his head at the wonder of it all. "To have been harpooned and still come to us! I wouldn't let my children go up and try to handle a wild chipmunk, much less a raccoon or a rhino. At the lagoon I can easily take people out and we can reach over and make contact with an animal that weighs thirty tons! In a century the gray whale has changed from being the 'devil-fish' to a whale we can approach without any fear."

The first known account of a "friendly" whale—with the possible exception of Jonah's tale—dates back to Scammon's day. In his book Scammon quoted at length from a journal kept by a Dr. J. D. B. Stillman of San Francisco, recounting a voyage made to Realejo, Central America, in 1850. A Sulphurbottom (now blue whale)—so called after its "yellowish cast or sulphur color"—estimated by Stillman at eighty feet long, had followed his ship *Plymouth* for twenty-four consecutive days. Breaking off from a pod of several others, this particular leviathan began "keeping under the ship and only coming out to breathe." The doctor recounted how, fearing the whale might damage the vessel, the crew had tried pumping bilgewater in its direction and followed up with volleys of rifle shots, then pieces of wood and bottles "thrown upon his head with such force as to separate the integument; to all of which he paid not the slightest attention, and he still continued to swim under us, keeping our exact rate of speed, whether in calm or storm, and rising to blow almost into the cabin windows."

The doctor's narrative concluded: "We long since ceased our efforts to annoy him, and had become attached to him as to a dog. We had named him 'Blowhard,' and even fancied, as we called him, that he came closer under our quarter, when I felt like patting his glabrous sides, and saying: 'Good old fellow.' As the water grew shoaler he left us, with regret unfeigned on our part, and apparently so on his."

In the twentieth century, however, the "Sulphurbottom"—once named by Scammon "the swiftest whale afloat"—had been hunted to near extinction. Yet in 1995 Bruce Mate, his wife, and another crew member witnessed off the Southern California coast an example of the blue whale's capacity to seek renewed contact with humans.

The Discovery Channel was filming a two-hour special called "Eyes in the Sky," using blue whales and elephants as their two examples of tracking wildlife by satellite. The camera crew was overhead in a plane, preparing to document Mate's efforts to tag the whales. Blue whales usually feed on krill, brought to the surface via wind-driven upwelling. When the wind had died for several days, Mate observed four whales "freight-training around really fast." There was no way he could think about getting a tag inserted. Instead, he spent half an hour simply trying to keep up with them.

Mate remembers: "Then one of them twice did a little maneuver, like I'd seen gray whales sometimes do when they're going to stop. It's a sort of slight lifting of the caudal peduncle. I said, 'Stop, turn off the motor, let's just wait.' "

About ten minutes passed. Mate figured the speedy whales were probably over the horizon by now. Then, without warning, came a huge splash. "Whoaaaa, about thirty feet away, two of them!" A seventy-five-foot-long whale came over to Mate's eighteen-foot-long boat, rolled on its side, swam underneath, and lay there just ten feet below. Mate looked over and right into its eye. He stood there in "stupefied amazement."

From overhead came a radioed message from the TV crew: not enough action, break off and go do something with the whale. "We said, 'You don't understand, this just doesn't happen every day, we're not going *anywhere.*' They said, 'Well, we're burning up fuel.' We said, 'Then go back to shore, we're staying right here!' "

For the next forty minutes, they did. The blue whale would continually surface nearby, come toward them again, and sink down within a few feet of the boat. "We never had any physical contact, but clearly it was an examination kind of relationship. We never started the motor, and the whale knew right where we were. Mary Lou got some spectacular footage on video. It was one of those days you don't ever forget."

The marine biologist from Illinois is quiet for a moment, then adds: "In fact, there aren't many days out with whales that I forget. I think when it loses that special feeling, I'll go do something else. I feel very blessed. I look at gray whales as the pioneers of something we're getting little glimmers about. I think probably, in our children's generation, we're going to see remarkable changes in our relationships with certain forms of wildlife."

Above Astoria, Oregon, the gray whales pass into the waters of Washington. Just past Grayland, about one-third of the way up its coast, the state's whale watching is based in the town of Westport. It's not a big business, maybe half a dozen small boats with captain-naturalists. But they can often stay right inside Grays Harbor, where about half a dozen gray whales have taken to leaving the migration and spending the summer feeding. Others linger during the migration a few miles offshore, in an area called the Whale Hole. They have been seen milling about over the same areas, sending up mud plumes during bottom-feeding activity, and swimming on their sides in shallow water looking for prey.

While the great majority move on to the Bering Sea, gray whales have been documented staying in Washington for at least 112 days of the year. Some individuals move up and down the coast, feeding in a variety of habitats between Northern California and Southeast Alaska. These regional residents, as well as the migrants, travel past the Quinault Indian reservation and into the Olympic Coast National Marine Sanctuary. This nearly sixty-mile stretch is the longest wilderness coastline in the continental United States—sheer bluffs lined with Sitka spruce and overlooking gray sand beaches dotted with coves and arches. Along the gray whale's path harbor seals haul out on the rocks, sea otters float in the kelp beds, and tide pools teem with as many as four thousand creatures per square foot. Just inland somewhere between 120 and 170 inches of rain fall annually upon the world's largest temperate rain forest, creating moss- and lichen-covered trees as tall as thirty stories.

Beyond Ozette, where the cedar longhouses of the Makah Tribe were buried by a mudslide around A.D. 1500 until their excavation in 1970, the gray whales come at last to Cape Flattery. Here the majority will cross the Strait of Juan de Fuca and continue a northwesterly course along Vancouver Island. A few, however, will make a right turn and keep straddling the Washington shoreline to where the strait widens gradually at Port Angeles.

According to Scammon, this was a "fine harbor, well protected from the prevailing westerly winds but somewhat exposed in the opposite quarter." There he had assumed his first command with the U.S. Revenue Marine in April 1863, aboard the steamer *Shubrick.* Scammon's whaling days had come to an end that March, following an unproductive three months' voyage to San Ignacio Lagoon and then Magdalena Bay. His whaling ship owner, Tubbs & Company, decided to abandon the trade and concentrate on making cordage. Scammon's wife, Susan, was pregnant for a second time, and the family settled into a pleasant residential area of San Francisco.

Almost immediately, though, Scammon found himself bound for the Northwest Territories. The *Shubrick,* a side-paddle-wheel vessel carrying sail, bore a crew of twenty-one men. In 1864 Scammon would be stationed at the entry to Puget Sound, the only revenue cutter on duty in the region during the waning days of the Civil War. When the call came for him to go to Fatauch Island, Scammon first had to find it. Puget

Sound and the Strait of Juan de Fuca were riddled with islands, and the available charts were less than accurate. We know that he stopped a mutiny aboard the *Frigate Bird.* He was also ordered to protect the New Dungeness Lighthouse, a sentinel outpost of the United States, from potential attack by a Confederate rebel ship called the *Shenandoah.*

Later Scammon would write two lengthy articles for the *Overland Monthly*—"Lumbering in Washington Territory" and "About the Shores of Puget Sound"— introducing this little-known sector of North America to his Western readers. The captain's reflections upon the region marked some of his finest writing, as in this excerpt describing the area near Seattle:

> Across the harbor, in full view, were the remaining lodges of an Indian village, tenanted by the few who remain of aboriginal descent; and near the foot of a high, green bank were ranged, in varied structure, the monumental graves of deceased chiefs and warriors, with torn banners waving above them; while others appeared screened with scarlet cloth, and still another stood out like a tomb with glass windows, as if to admit the light of heaven to his otherwise dark, dank resting-place. But our reverie was broken by the harsh steam-whistle of the mills, calling the workmen from their labors to the noonday meal, and we left the hallowed place guarded only by a flock of ravens, perched upon the firs and cedars at some distance.

A century later, every spring in recent years, a small seasonal group of about six gray whales has returned like clockwork to the northern Puget Sound around Whidbey Island. "The whole of Whidby Island is regarded as a fertile spot," Scammon wrote in 1871, "blooming in spontaneous production, in striking contrast with the sombre forest that borders it to the east, and the waste of water that dashes against its sandy and rock-bound shores." Here the grays have discovered dense beds of bottom-dwelling ghost shrimp. Some of the beds are intertidal, and you can see gray whales feeding in seven or eight feet of water at high tide.

Gray whales are known to frequent all the spots Scammon once denoted in this vicinity—Mucleteo, Hat Island, Port Susan, Camano Island. Not long ago, the marine scientist John Calambokidis recalls, "I followed these three gray whales in my boat, across Saratoga Passage and up off an area called Hat Island. I was some distance behind them, and I'm not quite sure what this encounter was all about. A big group of people was on the shore, having some sort of party. Anyway, they spotted the whales and lined up on the beach and started screaming and yelling, all excited! And these three whales just stopped. They came in really close to this beach and started spy-hopping in front of this group of people! This went on for about half an hour before they finally moved on."

Calambokidis says, "The bizarre thing is, this particular pod of gray whales shows up in early March and is usually gone by June. So they're going someplace else, but we've never seen any of them off the outer coast of Washington. The two resident groups do not mix. And we don't know *where* these disappear to."

John Calambokidis was among the founders in 1979 of the non-profit Cascadia Research Collective in Washington's capital, Olympia. The collective's primary focus today is human impacts upon marine mammals. Largely under contract to federal agencies, his organization has authored more than a hundred reports appearing in books, journals, and government publications. Now forty-five, Calambokidis is a tall, gangly fellow with a salt-and-pepper beard who bears some resemblance to—and has a voice very much like that of—the late actor Jimmy Stewart. He's dressed casually in faded jeans and sneakers, looking out his office window at Olympia Harbor on the southern end of Puget Sound. Last year, he says, two gray whales showed up right outside and spent most of a day under a little bridge. Traffic came to a standstill while hundreds of people gathered.

Ordinarily, though, it's on the high seas that Calambokidis encounters his whales. Doing counts of humpbacks and blue whales, he goes as far offshore as forty or fifty miles, generally in a low-running, rigid-hull inflatable raft slightly more than eighteen feet long. "That far out of sight of land," he says, "you have a feeling of aloneness and isolation as well as a respect for the size, power, and grandness of the ocean. But for me there's also a feeling of self-reliance. If you're facing a sea of fog that requires navigation, you really have to trust yourself."

To reduce the element of surprise, Calambokidis equips his craft with duplicate navigation and communication gear. There can be sudden summer storms, or a freighter approaching head-on in a blinding fog. Or sometimes "friendly" whales—especially, in his case, humpbacks. It began in July 1992, when Calambokidis was out alone some twelve miles offshore in California's Santa Barbara Channel. He'd previously observed humpbacks as curious animals, who enjoyed playing with kelp and exploring objects in the water. But never before, in six years of going out

The marine scientist John Calambokidis (courtesy Cascadia Research)

hundreds of times and approaching thousands of humpbacks, had he had a "friendly" encounter.

Suddenly two humpbacks surged up within a few feet of him and exhaled spray into his face. One swam under and held its pectoral fins on each side of his sixteen-foot boat. It continued to surface and dive, swimming around faster and faster, raising its huge tail flukes more and more rapidly. "I was very frightened," Calambokidis would later write,

> and after a half hour I ended up motoring a mile away. I couldn't quite let myself leave, so I went back to the same general area, unsure if I'd find the whales. Two humpbacks surfaced and headed away from me, about a quarter of a mile. I cut the engine and they did a U-turn, charging right up to the boat. They resumed their previous behavior and started pushing me around. Initially, one animal put its pectoral fin against the side of the boat, made contact, then proceeded to push. It was very deliberate and at that point I realized I was in no danger. Then one of them started to lift me—almost steadily—on his back: he arched his back, lifted

> me up in the air, and then put me back down. I don't know how high I was.
>
> The animals were amazing with complete control. They were able to surface, maneuver, and swim within a couple of feet of the boat, and the only contact that was made was clearly of their own control and choosing.

Calambokidis recalls "flipping out" as he was spun in a circle by the humpbacks. The next year a similar experience happened three times with other humpback whales; in 1994, six times; in 1995, a dozen times. Calambokidis says, "That first one remains the most dramatic, but humpback whales approaching or spy-hopping next to boats suddenly became common. And not just our boat. All the whale-watch vessels and other people reported similar experiences, something they'd never seen before, up and down the California coast. You could see it spread through the humpback population. There would be one known 'friendly' in a group of three which would come over, while the other two hung off in the distance; then, slowly, those would move in closer and do this as well."

Were the humpbacks that approached him in 1992 older whales? I ask. "Well, that's the bizarre thing," Calambokidis replies. "We have just under a thousand humpbacks identified off California through our photo-ID work. But the two whales that first did this had never been seen *before*—and they have never been seen *since.* I'm still completely puzzled about whoever initiated this phenomenon."

With gray whales, Calambokidis says, there have been a few times when individuals have circled his boat, and a couple of instances where they've bumped up against it. But in the waters beyond San Ignacio Lagoon, "I wouldn't say the behavior is yet very prevalent." There seems, in fact, to have been a mysterious kind of transfer between whale species—from the grays to their humpback counterparts and even, as Bruce Mate recounted, to a blue whale.

Calambokidis recalls one other strange incident, from 1987, when he was doing a survey in the Norwegian Sea. "It was the first year of the global moratorium on taking whales. I was aboard a Norwegian ship that normally hunted minke whales—when a friendly minke approached and circled the boat." Was the whale, I could not help wondering, flaunting its newfound freedom? Or perhaps saying thanks?

•

John Calambokidis came to his calling by a circuitous route. The son of a Greek sailor and an American mother, he was born in Egypt and attended grade school in Italy. His father died when he was seven, and a few years later his mother brought John to the United States. They lived in Florida, then Washington, D.C. After high school he set out on a long tour of Europe and Africa. Ignoring warnings about hyenas and thieves, Calambokidis traveled across Ethiopia by bicycle. When his pack and shoes were stolen, he wrapped his feet in rags and kept on going. "I didn't realize it at first," he says, "but I came back from that trip with the feeling you can do things people think are too difficult or should not be done."

He headed west to Washington and Olympia's Evergreen State College, "a strange institution that leaves you with a lot of freedom. My undergrad years, I took one year of course work and three years of independent research." He was able to obtain two grants. His initial look at mussels in 1974 became one of the first studies to examine the widespread distribution of PCB pollutants in Puget Sound. "That research got me interested in the marine environment, and I realized from studying the literature that those species most at risk would be the ones eating highest on the food chain." In heading a team of nine students to analyze pollutants in harbor seals, John knew he'd have to immerse himself in their biology: "How could I determine any effect on reproduction if I knew nothing of what normal rates are, or had no population data? A lot of people get drawn to this field by love of a species, but for me it was these other questions and then the large numbers of unknowns regarding marine mammals. As a scientist, I get excited about finding answers nobody's come up with before."

He'd formed the Cascadia Research Collective with several other Evergreen graduates. In the beginning they worked exclusively for the National Marine Mammal Laboratory. But the group soon decided to work independently, "to have more control over what projects we did and how—we called it objective scientific research with an environmentalist's perspective. But we found early on that some of our results didn't always promote the politics we wanted. When faced with such conflicts, we quickly concluded we were scientists first—tightly wed to putting out clear, unbiased information that would be accessible to everyone."

The collective still has only three full-time employees, also utilizing student interns working toward their undergraduate or master's degrees on numerous projects. All funding emanates from either government agencies or other nonprofits (no money is accepted from private industry), with the collective's budget ranging between $200,000 and

$500,000 annually. They work on up to a dozen projects a year, related mostly to whales: population assessments, strandings, the effects of contaminants and underwater noise. Half of the projects now take place outside Washington, especially in California but also in Alaska and as far as the waters off Central America.

Searching for whales in a vast expanse of ocean, Calambokidis says, is following a feeling. He and his colleagues do this well enough to have been among the pioneers of whale photo identification, initiated for gray whales by the scientist Jim Darling off Vancouver Island in the early 1970s. The estimated two thousand blue whales Calambokidis has cataloged off California are more than scientists had previously estimated for the entire North Pacific, and may be the largest remaining blue whale population in the world. In voyaging recently to the Costa Rican dome, Calambokidis's team came up with seven out of fourteen matches on blue whales they'd photographed in California, proving that many of these make at least a 5,800-mile round trip. The collective has compiled long-term photo catalogs on blues, humpbacks, gray whales, and, to a lesser degree, on fin whales, sperm whales, and killer whales.

Grays are harder to identify than the humpbacks with their distinctive and often-visible tail flukes. At the same time, grays often present more pigmentation markings. "The struggle with grays," Calambokidis says, "is that the barnacles and scarring change from year to year, with new layers being added. If I haven't seen a whale for five or more years, it's a difficult challenge to match it with a previous sighting."

Most of the gray whale ID effort takes place within a few hundred yards of shore. Because seasonal residents are relatively few, it's easier even than taking such a count in the Baja lagoons. Calambokidis approaches a whale from the side, staying slightly behind. He takes pictures of the back of the whale around the dorsal hump, usually waiting until the last surfacing of a dive series, when the whale arches its back high. Typically this begins in March and goes into October or even November, with most of the outer coast work taking place in June, after the migration has passed.

Calambokidis brings out a thick catalog of gray whale photos, ongoing since 1984. Here is a plainly visible discolored patch—from a killer whale attack—on the same whale seen in 1989, 1992, 1993, and 1995. Through 1998, 172 gray whales in Washington have been positively identified. For 31 whales seen on more than a day in areas of regular surveys, the average tenure was forty-seven days, and 6 were documented staying for three months or more.

Since 1998 the collective has been collaborating with other researchers and naturalists from California to southeastern Alaska to document the broader movements of these identified whales. Some of these, such as Vancouver's Coastal Ecosystem Research Foundation and Northern California's Humboldt State University, are now maintaining their own catalogs for comparison purposes. Recently the National Marine Mammal Laboratory has begun a research program on gray whales in and around the Makah reservation. The hope of all these efforts is to better understand the range of regional residents.

There have been some surprises—not least of which is the central finding that gray whales seen in the summer and fall off Northern California, Oregon, Washington, and British Columbia appear to number a few hundred animals who return annually to feed. It used to be thought that all the grays moving through the migratory corridor off Westport, Washington, in March would be continuing north. It turned out that fully 25 percent were actually stopping off as residents. The photo surveys have also enabled Calambokidis to conclude that the grays entering the southern reaches of Puget Sound are *not* residents—because they never show up again.

Puget Sound starts a southerly course from Admiralty Inlet at Port Townsend down past the ports of Seattle and Tacoma to Olympia. It is an estuary, a semienclosed glacial fjord where salt water from the ocean mixes with freshwater draining from the surrounding watershed. Composed of a series of underwater valleys and ridges, the sound has an average depth of 450 feet, and it gets as deep as 930 feet just north of Seattle. Altogether its four interconnected basins cover more than 16,000 square miles of land and water. In 1871 Scammon described this "peculiar arm of the Pacific with its many branches, which, in the main, trend to the south; but the direction of the bold waters are more varied than the points of the compass, sometimes running in the mountain-gorges, again meandering through the rich bottom-lands, or surrounding islands, till lost in the mazes of its own intricacies."

Gray whales are often known to get lost in these mazes. "Around the central and southern sound," says Calambokidis, "is where we're seeing a recent pattern of high numbers of grays as well as high mortality. They're animals we've never seen in a previous year. One out of every three or four that we can document there ends up dead."

Calambokidis does not think these deaths are linked to pollution. This has long been a prospective issue with gray whales, since they feed on the bottom, where contaminants like PCBs can concentrate in the sediments. In 1984, when more than a dozen gray whales washed ashore dead in Washington—over half of them in Puget Sound—a local veterinarian made front-page news citing the existence of toxic chemicals in one of these. Calambokidis himself saw scant evidence of such a reality, but his group agreed to start stockpiling tissues and examining gray whale carcasses. Comparative tests were made on grays that died in Puget Sound, California, and Alaska.

In the early 1990s the National Marine Fisheries Service (NMFS) analyses of the tissues collected by the collective found that, no matter the location, the levels of most contaminants were actually fairly low—PCBs in gray whales being in the low parts per billion range, compared with hundreds of parts per million for harbor seals. Aluminum levels did come out high in grays. Aluminum is the most abundant element in the earth's crust, so this finding was probably a by-product of their feeding on sediments. "The lack of data from apparently healthy gray whales, however, limits the assessment of whether the levels of anthropogenic contaminants found in tissues may have deleterious effects on the health of gray whales," wrote the authors of an NMFS report published in 1994.

Overall there's been a stabilization of contamination levels in Puget Sound's once-foul environment. While estuaries such as Elliott Bay and Commencement Bay remain heavily polluted, Calambokidis notes that these are not areas where gray whales feed. Additionally, the tiny amphipods, ghost shrimp, and other bottom-dwelling creatures that are the grays' prey are low on the food chain and not themselves heavily contaminated. "You end up with the highest levels in animals living in coastal waters, close to human sources of contamination, that eat high on the food chain," Calambokidis explains. That puts killer whales at the top of the list, followed by harbor porpoises and seals.

"What the contaminant studies did provide," Calambokidis adds, "was a much clearer perspective on what *was* going on with gray whales. The key finding was, most of these animals that had died in Puget Sound had virtually no oil left in their blubber. They were emaciated and starving to death." This reflected not short-term pollution exposure but problems dating back to the previous year on the feeding grounds.

It's a situation that appears to be getting worse, not better. Gray whales stranding in the middle to lower regions of the Puget Sound maze are, Calambokidis says, "clearly animals going through the motions of feeding, with mud plumes on the surface. But the fact is, there's not much in that area to eat. Divers going down to the bottom have not been able to identify any abundant prey. So I think these whales are desperate, unable to complete the migration and unfamiliar with the area, knowing they have to find food but not knowing where to do it. That's why so many of them end up dead."

During the spring of 1999 unprecedented numbers of grays started showing up and staying in Puget Sound, something also being witnessed in places along the northern migration route like Monterey Bay, Santa Barbara Channel, and San Francisco Bay. Calambokidis's collective, which responds to the majority of gray whale strandings in the sound, suddenly had its hands full. Although a few of the twenty-eight whales found dead that spring in Washington showed evidence of other causes—ship strikes, drowning in fishing nets, attacks by killer whales—by far the majority were emaciated. Something was obviously amiss.

This finding coincided with the Makah Tribe's intent to conduct a springtime hunt. And, while Calambokidis says he personally feels strongly "that they have a legitimate right to hunt gray whales both legally and morally," he does have concerns about the hunt's effects. The Makah had said they would "target" migrating gray whales. Calambokidis points out that, while the NMFS has established certain criteria to ensure that's the case, "they have not shown the criteria are effective and are not gathering the data it would take to tell if they are wrong. In fact, NMFS has agreed with the Makah *not* to photographically identify any animal they kill, even to verify after the fact whether it might in fact be a resident gray whale and not a migrant."

What difference did it make? In Calambokidis's view, possibly quite a bit. His photo-ID studies of gray whales had determined that many return to the same local waters year after year. In his studies of humpbacks Calambokidis learned that they show loyalty to particular feeding areas, a "site fidelity" that is passed along from mother to calf. This is an indication of mitochondrial DNA that is dramatically different for whales in different feeding spots.

"It's a hard concept for people to grasp," says Calambokidis, "because they say, well, aren't all these gray whales going and interbreeding in Baja? The answer is, they probably are. But if the method of recruitment to learn where to feed is maternally directed, and if you kill off the Washington or Vancouver Island resident whales, it could take thousands of years for them to recolonize those areas. You'd have violated one of the principles of the Marine Mammal Protection Act, to keep animals as a viable part of their ecosystem. You'd actually have lowered the effective carrying capacity of gray whales."

Calambokidis has gone on record about this situation, writing an article for *Sea Kayaker* magazine that did not win him rave reviews with either the Makah or the NMFS: "Depletion of these resident whales could rob this region of a significant component of its ecosystem and lead to wider economic and social consequences, given the neighboring communities where whale-watching of these animals has developed into a major activity."

The best solution, Calambokidis argues, would be to confine the tribe's hunt to the peak migration times: late December for southbound grays, March and early April for northbound grays. This, of course, would mean encountering worse weather than either earlier in the fall or later in the spring. But it would help ensure that any whale taken wouldn't be a seasonal resident, which by late spring constitute the *majority* of the population around Neah Bay. Calambokidis adds that he could foresee allowing some takes of "seasonal resident" whales—as long as it is shown that their numbers can support this.

The biologist adds: "Besides the Makah, reviving hunting is coming up among native tribes in Canada and with other whaling nations. If the United States goes on record ignoring the existence of potentially significant subpopulations of whales, that is a really bad precedent. What's the United States going to say when Canada starts taking twenty whales a year from the same group—'oh, now seasonal residents are a reality and it's a problem'? They've got to be consistent."

He concludes: "The burning questions are these: How many residents are there? How big an area do they use? What is their relationship to the larger population? Is there frequent interchange, or are they distinctive as a group? If hunted, would resident whales be replaced by others from the main group, or is the knowledge they have of these local waters something which has developed over a long period of time, and which would be lost?"

Such questions loomed large in mid-May 1999, when the Makah hunting team received a weeklong permit from the tribal council allowing the continuing pursuit of gray whales.

Chapter 12

THE KILL

The first streak of dawn is the signal for lowering the boats, all pulling for the head-waters, where the whales are expected to be found. . . . When within "darting distance" (sixteen or eighteen feet), the boat-steerer darts the harpoons, and if the whale is struck it dashes about, lashing the water into foam, oftentimes staving the boats. As soon as the boat is fast, the officer goes into the head, and watches a favorable opportunity to shoot a bomb-lance. Should this enter a vital part and explode, it kills instantly, but it is not often this good luck occurs; more frequently two or three bombs are shot, which paralyze the animal to some extent, when the boat is hauled near enough to use the hand-lance. After repeated thrusts, the whale becomes sluggish in its motions; then, going "close to," the hand-lance is set into its "life," which completes the capture. The animal rolls over on its side, with fins extended, and dies without a struggle. Sometimes it will circle around within a small compass, or take a zigzag course, heaving its head and flukes above the water, and will either roll over, "fin out," or die under water and sink to the bottom.

—CHARLES MELVILLE SCAMMON,
describing the killing of a gray whale, 1874

AROUND two o'clock on the morning of May 17, 1999, Theron Parker was awakened by a phone call. It was someone from the U.S. Coast Guard, who'd been keeping an eye on the protesters intent upon disrupting the Makah Tribe's gray whale hunt. Two days earlier the Coast Guard had seized all four Zodiacs of the Sea Defense Alliance for shooting off smoke bombs, flares, and chemical fire extinguishers in one such attempt. Ironically, one of the ten protesters arrested had been charged with violating the Marine Mammal Protection Act by disturbing the route of a migratory whale. The Sea Shepherd Conservation

Society's ship *Sirenian* had sailed to Washington's Friday Harbor to load three more boats and some volunteers. Now, Theron Parker was informed, Paul Watson's crew had just hauled anchor and was en route back to Neah Bay. But the tide was against them. It would probably be as many as five hours before they could arrive.

Parker, the Makah's designated harpooner, began dialing up the other members of the whaling team. "Wake up," he instructed. "It's time." They made rendezvous for a 4:00 A.M. breakfast at the Makah Maiden café. Then seven men drove outside town to Sooes Beach, where they'd hidden their thirty-two-foot-long cedar canoe called the *Hummingbird.* Meanwhile, a pair of motorized support boats left Neah Bay and headed west, in the direction of Tatoosh Island.

Armed with a harpoon and a cell phone, the Makah launched the *Hummingbird* from the beach. They took on water as they broke through the first waves and had to bail the canoe. Everyone started arguing. Parker called them to silence, and a prayer. After that they began paddling easily, steadily, with the current.

It was about five miles to two small islands known as Father and Son Rocks, south of the reservation, between Cape Alava and the Point of Arches. This is the westernmost point in the continental United States, where geologists have calculated some of the rock formations to be 144 million years old—more than twice the age of the oldest rocks in the nearby Olympic Mountains, and possibly remnants from an earlier North American continent.

The Makah had seen gray whales coming in here and rubbing their barnacles against the rocks. This particular stretch of coast maintained a depth of only about fifteen fathoms eight miles out into the Pacific before it started dropping off. So even if a whale sank after you'd nailed it, Theron Parker reasoned, you could probably get it back up. He cradled a harpoon he'd made in a machine shop out of three pieces of wood.

As the support boats circled Tatoosh Island and made the southbound turn to meet the canoe, Wayne Johnson, the Makah whaling captain, placed the appropriate cell phone calls to the Coast Guard and the National Marine Fisheries Service, whose own vessels were required to join them. On the bow of one Makah support boat stood Glen Johnson. He was the brother of Makah Whaling Commissioner Keith Johnson and had only recently joined the crew. He was holding one of several rifles on board—a high-caliber .577 customarily used for long-distance target shooting or by big-game hunters.

Dawn arrived gray and misty. A Coast Guard and a Fisheries Service vessel arrived in the vicinity but gave the tribe plenty of room. Several media helicopters circled overhead. There wasn't a protest vessel anywhere in sight. At 6:54 A.M., all three Seattle-based TV stations interrupted their regular programming. LIVE—BREAKING NEWS. MAKAH HUNT IS ON, the Channel 5 logo read. All across the Pacific Northwest, and as far away as Nome, Alaska, citizens sipped their morning coffee and watched the Makah canoe close in on a gray whale within the boundaries of the Olympic National Marine Sanctuary, about an eighth of a mile offshore of Father and Son Rocks. The whale was not a large one, perhaps three years old, still a juvenile. As the canoe approached, the animal paused in its swimming. It surfaced right under the bow of the canoe and rolled sideways. It turned its head and looked up at its pursuers.

Alberta Thompson rarely sleeps past first light, but she did that morning. The seventy-five-year-old Makah elder was still fast asleep when the phone rang in her trailer home. It was her firstborn daughter. She was crying but wouldn't say why. "Oh, Mom, turn your TV on" was all she could manage. Alberta did. She saw "a young whale come up and look at them, not expecting what it got. Especially when you looked at its eyes, you just knew it probably thought it was a boatload of those who like to pet them." As the scene unfolded before her, Alberta burst into tears.

Theron Parker didn't even have to throw the stainless-steel-tipped harpoon. He simply leaned forward and pushed it into the whale's arched back, about one-third of the way down its thirty-and-a-half-foot length. Momentarily, the bow of the canoe rolled right on top of the whale. "The harpoon did stick," a newscaster said. The whale thrashed its tail. The Makah crew members let go of a rope and a yellow rubber float attached to the harpoon. Parker's harpoon popped back up. He grabbed it and put it back on the bow. The rope started spooling out as the whale began to run, towing the canoe in a wide circle. Blood crested on the waters.

> *Instantly the deadly spear flies from ready hands, and plunges into the mammoth creature. The water is lashed into a pyramid of bloody foam, the boat is "fast," and the whale in vain endeavors to escape by running over the surface of the sea, then diving to the depths below; but its human pursuers still cling to the line attached to the fatal harpoon. The whale rises again to the surface, in some degree exhausted. Another boat approaches, and darts its murderous weapons, and the pursuit is continued with renewed vigor.*
>
> —SCAMMON

Now a support boat carrying six Makah revved its engine and advanced on the wounded whale. Glen Johnson aimed the high-powered rifle. The first two shots, fired inside of a minute at 6:58 A.M., missed completely. The whale dove.

The worried creature may dive deeply, but very little time elapses before the inflated seal-skins are visible again. The instant these are seen, a buoy is elevated on a pole from the nearest canoe, by way of signal; then all dash, with shout and grunt, toward the object of pursuit.

—SCAMMON

When the whale resurfaced at 7:01 A.M., six minutes after the initial harpoon thrust, the support boat moved directly in front of it. A third rifle blast at point-blank range brought a surge of foam as it shattered a ridge of the whale's skull. A second harpoon, thrown from the support boat, struck near the rear. The whale did a complete rollover just below the surface. A fourth shot pulverized the whale's braincase, followed by a third harpoon striking its right side. The water turned crimson. The whale ceased moving. A cold wind cut across the cape.

When, at last, a vital part is pierced, the animal deeply crimsons its pathway with its remaining life-blood, and lashes the sea into clouds of spray in its dying contortions. . . . Spouting its last blood from a lacerated heart, it writhes in convulsion and expires. . . . Then follow wild cheers by the crews in the boats.

—SCAMMON

Under the circling TV helicopters, the crew members waved their arms and their harpoons and their paddles in the air. It had been a ten-minute-long public death. Eric Johnson, dressed in an Oakland Raiders jacket and a wool cap with a whale on it, used a cell phone to alert Neah Bay of the success. Many of the tribe had already been watching the live TV broadcast. Over breakfast in the Makah Maiden café, people roared and applauded at the first harpoon strike. "Good shot, good shot," one man yelled when the rifle finally hit its mark. The Makah reservation closed its schools and declared a holiday.

"It was easy," the Makah crew member Darrell Markishtum would tell the press. "The whale gave up its life for us freely."

"Almost like clockwork," Eric Johnson would tell the press. "The whale came right to us."

Paul Watson's ship was approaching Tatoosh Island when the news came. Standing on deck, Jonathan Paul answered his ringing cell phone. It was someone from the media. "How do you feel about the whale being killed?" the man asked. Jonathan didn't respond. He walked up to the bridge and informed Watson and the other crew members. An eerie quiet settled over the *Sirenian*. Watson's face paled. He had assumed the Makah wouldn't try to hunt on the outgoing tide, when the danger of losing the whale was greater. The captain's gaze remained rigidly ahead as he stayed at the helm. "Paul, are you okay?" Jonathan asked. "Yeah," Watson said. By now some of the Sea Shepherds were shouting with anger. Others wept.

Arriving on scene forty-five minutes too late to try to stop the killing, the *Sirenian* continued to be held at bay by the Coast Guard. Watson merely announced over the ship's radio that this constituted "a pirate whaling action" and blasted a lonely air horn across a chill and churning sea.

> *The whale being dead, and floating, the grapnel is brought into requisition, and the animal's head is hooked and hauled up, when holes are cut through the lips, and a short warp is rove through, by which means its mouth is closed, and the tow-rope is made fast; then the prize is taken in tow to the vessel.*
>
> —SCAMMON

A distraught Watson and his crew watched as a Makah diver went overboard to try to perform the traditional sewing-up of the whale's mouth in order to prevent it from sinking. He was unsuccessful. The whale filled with water and sank ninety feet to the bottom. The crew managed to pull the whale, attached to a single line, by hand to within about twelve feet of the surface, then called for some help. Two of the tribe's longline fishing vessels came to the rescue. One of these, the *Heidi*, attached a large hydraulic winch to the line holding the whale. Slowly the dead whale was spooled upward as they turned the drum. A few times the winch overheated and had to be shut down because of the weight. Finally, after several hours, the whale's body floated to the surface. A rope was tied between the two boats to hold it in place. The tail was secured to the *Heidi*'s stern. Theron Parker got aboard the longliner, had a couple of cups of coffee, and went to sleep in a bunk below. It was 10:30 A.M. when the long tow back to the reservation began.

> *Then the whole fleet of canoes assists in towing it to the shore, where a division is made, and all the inhabitants of the village greedily feed upon the fat and flesh till their appetites are satisfied.*
>
> —SCAMMON

The whalers and their escorts reached Neah Bay around 5:30 P.M. They stopped first in the harbor, where four canoes from other Pacific Northwest tribes were waiting to greet them. In the old days they'd wait until a high tide to bring a whale ashore and tether it. But nearing the shore at low tide, this crew unfastened the whale from the longliner and retied it to the five canoes. Together they paddled and towed the whale a brief remaining distance to Front Beach. This was within a stone's throw of a memorial park and mass grave where many Makah ancestors were buried after the smallpox epidemic that swept across the region in the early 1850s. This was also right to the shoreline beyond Charles Claplanhoo's house, the same place he said the last gray had been brought, in 1909. The elder watched from his front porch while what he later described as "just a little pup" was landed almost at his doorstep.

After the kill in Neah Bay (AP/Wide World Photos)

A steady rain was falling. Along the beach hundreds of tribal members cheered and shouted. Many surrounded the whalers in the surf. Harpooner Theron Parker, bare-chested, sang a song in the Makah language. He performed a ceremonial dusting of the crew and the carcass with the down of eagle feathers. He slipped off the first time he climbed up on the whale. The crew all took turns standing on the whale, next to a harpoon stuck deep in its flesh, and pumped their fists. One fellow who wore his hair in a ponytail performed a flip off the whale's back. Kids swarmed over the dead whale, touching its flesh and bouncing on its gelatinlike mass.

It wasn't close enough to start the butchering, so a cable attached to a big Army surplus truck was hooked to the whale's tail. But the winch wasn't strong enough. Between seventy and eighty Makah men grabbed the towlines. They placed some little logs of hemlock on the sand. They pulled and shouted and hauled the carcass hand over hand, tail-first, across the logs. "Ee-heee, ee-heee," they cried in unison, until they had the whale above the high-tide mark.

"It's more than a miracle, it's a way of life," one tribal member told the press.

Offshore, the *Sirenian* sounded its siren as an accompanying dirge.

> *Then the body-blubber is cut in spiral folds . . . and rolled off down . . . behind the vent, where the whole flesh of the carcass is cut through.*
>
> —SCAMMON

The whaling captain made the first cut. The butchering began. An Alaskan Inuit whaler named Daniel had come to Neah Bay to show the Makah how to do it properly. They used knives more than a foot and a half long tied to poles taller than a man, sawing at the carcass and cutting away bloody chunks. While some men cut others sharpened knives. First they sawed the blubber: thick, white, dotted with blood. They threw the chunks onto blue plastic tarps spread on the sand. Then they sawed the meat: bright, red, lean, stripping it away with meat hooks. The whale's mouth was agape, showing its white baleen. The whale's eyes remained open. Blood from its shattered skull dripped past the eyelids. As the butchers worked other tribal members washed the blood away with a fire hose.

Tribal members of all ages pressed close to watch. Even small children reached out for samples of the fresh-cut blubber. They giggled and tugged on the chewy substance with their teeth, as if it was taffy. Some shouted "Yuck!" Some spit it out on the beach. Some washed the blubber down with Cokes and Diet Pepsis. Some grabbed for second helpings. Some put pieces in their pockets to take to elders waiting in nearby parked cars. Some let long pieces of the meat dangle from their mouths.

It got dark and the butchering continued and the rain kept falling.

At a press conference held by anti-whalers in nearby Sekiu, Paul Watson said: "They are acting totally different from their ancestors, who were sad and somber and respectful after a hunt. We'll mourn for the whale. They're acting like a bunch of redneck hunters having a good old time."

The tribal elder Alberta Thompson stood before the cameras and openly wept. She gave the whale a name—*Yabis.* In the Makah language, this means "My Love" or "Beloved."

On the evening news NBC's Tom Brokaw described "a day of ritual, death, and protest." ABC's Peter Jennings called it "a bad chapter" in the fight to protect whales as well as tradition. CBS's Dan Rather spoke of the "death of a beautiful titan of the sea."

In Seattle a vigil was attended by more than one hundred people. Outside the federal building, many of them clutching flowers and candles, they stood silently for ten minutes in memory of the whale.

Makah cutting up a whale, circa 1900 (courtesy Washington State Historical Society, Tacoma; by Curtis, Negative No. 19239)

Makah cutting up a whale, May 1999 (AP/Wide World Photos)

Back on the reservation, "I'll make a whale of a burger," Joddie Johnson, owner of the Makah Maiden café, said.

"We made history today. This is a great day for the Makah Nation," Tribal Chairman Ben Johnson said.

> *And the backbone being unjointed, the main portion of the mutilated remains of the animal floats clear of the ship, or it sinks to the depths beneath.*
>
> —SCAMMON

The crew worked by the headlights of a truck parked on the beach, the carcass steaming eerily in the beams. Dogs stole scraps of the strange new meat and gummed it briefly before laying it down again, apparently uncertain what to do with it. Around 11:30 the tide started coming in. Panic ensued on the beach. The whale was perhaps one-third butchered, and the tide seemed about to reclaim it. The Army surplus truck was recalled to try to pull the whale farther up the beach. After the flukes were cut off, chains were attached to the tail. The truck started pulling, but the chain ripped into the spinal cord and yanked the tail rigid. The truck heaved. The whale wouldn't move. There were concerns that the tail would snap and splatter the remains of the whale everywhere. Nobody knew what to do.

So most of the Makah went home. By one o'clock in the morning, only a few teenagers, two mothers with their kids, a few reporters and four fisheries scientists were still there—along with the Inuit from Alaska. Now the rising tide washed the whale's blood back into the sea. Daniel suggested if they cut out the whale's guts, the animal would be lighter and they'd be able to drag it away from the tide with the truck. The Inuit and the scientists set to work doing so. Before long they were removing the intestines.

"Do you know how to get ahold of the Makah Tribe?" Daniel asked one of the few remaining onlookers. "We need some more Makahs." Nobody responded.

"Is the fire truck still here?" he asked a few minutes later.

"Doesn't look like it," someone replied.

"We need to wash these innards out," Daniel said.

By now about fifty feet of intestines had been extricated.

"Where are you from?" a boy asked the Inuit.

"Top of Alaska," Daniel said.

"Do you do this a lot?" the boy asked again.

"Yeah, but we cut up our own whales," Daniel said.

A biologist with the National Marine Fisheries Service walked by and threw up his arms in disgust. "We shouldn't have to be doing this stuff," he said.

"I'm really tired right now, and there's no one helping us," Daniel said to no one in particular. "I'm ready to go home."

After removing about four hundred feet of intestines, they gave up in exhaustion and called for more trucks. Three vehicles working together dragged the whale up the beach. The men covered it with a tarp and left.

The tribal leadership held a morning-after press conference. Ben Johnson said he'd received thirty-two nasty telephone messages from "people calling us murderers, things like that." Hackers had broken into the tribe's Web site, turned it red with dripping blood, and changed the town's name to Death Bay. When a Seattle radio newscaster asked Johnson to respond to critics who said it was irreverent for tribal members to perform back flips off a dead whale after towing it to shore, the tribal chairman evicted the man from the news conference.

By the time the Makah's whale skinners returned to the beach early that afternoon, most of the remaining "sacred meat" had been spoiled. When all was said and done, it took front-end loaders and five men armed with gaff hooks to remove the larger organs from the beach. The lungs were as broad as a table, the spine thicker than a ceiling beam. The plan was that students would reassemble the cleaned bones, which would then go on display at the Makah Cultural and Research Center. Toward the end of the day, the whale skeleton was trucked to a fish-processing plant at the east end of town to be boiled. That was where the butchered meat and blubber were already being stored in freezers, awaiting a weekend gathering of tribes known as a potlatch.

An anonymous mourner left a wreath, a flower lei, a candle, and a note at the spot on the beach where the whale was butchered. The note read, "Baby, you didn't deserve this. We love you." The remains of the whale were cordoned off with yellow police tape, the kind customarily found at murder scenes.

Five days after the kill, more than one thousand native people came to the Makah reservation. They came from the West Coast, the Great Plains, Canada, Alaska, even Fiji and Africa. While Paul Watson's ship patrolled the outer harbor, a tribal roadblock kept any demonstrators outside the reservation's boundaries. It was a bright, sunny day as the Makah paraded their cedar canoe through the streets. They went from the museum at one end of town to a ceremonial flag raising at the other end. One of the dozen-some floats said, WHALE TO LIVE, LIVE TO WHALE. Another float proclaimed, PROUD MAKAH. They were led by Theron Parker, wearing a cedar headdress with eagle down and a bleeding harpoon tattooed over his bare chest and a harpoon rope decorated with likenesses of human skulls. An Army-surplus flatbed truck carried the whaling captain, Wayne Johnson, who wore a jacket with the inscription .50 CALIBER SHOOTER'S ASSOCIATION. A four-year-old boy, wrapped in a traditional black-and-red shawl, wore a sign declaring him a "future whaler" and waved to the crowd. Tribal Chairman Ben Johnson drove a new black Volkswagen Bug that his wife won in a tribal bingo game. "We've just started," he said. "This will inspire native people everywhere." Tribes from across the state of Washington joined the parade with bearskins and carved cedar masks and hand-painted drums. A Masai tribesman, from Tanzania and Los Angeles, was invited to join a circle of honor.

Guests entering the potlatch feast at the high school gymnasium were each handed copies of the Makah's 1855 Treaty of Neah Bay with the U.S. government, granting their right to whale in seeming perpetuity. Ben

Johnson publicly thanked the government for keeping its promise. The crowd gave the U.S. Coast Guard a standing ovation for backing up that promise. A simple lunch of braised beef, corn, mashed potatoes, fruit salad, watermelon, and fruit punch soon gave way to songs and dances.

Dinner that night featured a main course of gray whale.

> *After the feast, what oil may be extracted from the remains is put into skins or bladders, and is an article of traffic with neighboring tribes or the white traders who occasionally visit them.*
>
> —SCAMMON

After the feast, according to tribal elders, some Japanese visitors paid a call on the Makah reservation.

Within hours after the Makah killed their gray whale, *The Seattle Times* had received almost four hundred telephone calls and e-mails. These ran almost ten to one against the hunt. Over the next week the newspaper would print dozens of letters. The outrage—in many cases taking the form of blatant anti-Indian sentiments—was as striking as the Makah's deed itself. Here are some examples:

> The white man used to kill Indians and give them smallpox-infected blankets. Is this a tradition we should return to?
>
> Barbaric violence in the name of tradition must not be tolerated.
>
> This is not a hunt. It's premeditated murder. How do you preserve or revive your culture by going back to a nineteenth-century custom using twentieth-century tactics and weapons?
>
> The Makahs should honor the gray whale by becoming its caretaker, not its predator.
>
> Just saw some pictures of a bunch of testosterone-laden, high-fiving, party animals dancing on a dead whale. Whoops! I guess they were Makah tribe members spiritually reconnecting with their deeply respected tribal past.
>
> What can a victory be in killing an animal that has learned to trust people?

> I have always taken a lot of pride and enjoyment from the reputed Indian blood in my family, but what I felt this morning was shame. . . . Not exercising a right occasionally is a mark of civilization.
>
> Hey, I think we should also be able to take their land if they can take our whales.
>
> I am anxious to know where I may apply for a license to kill Indians.

In some people this event clearly brought out the worst. Political correctness—and basic human decency—had definitely gone out the window. The Seattle paper subsequently raised a large question: when are rage and racism the same thing? One scholar opined that the often vitriolic reactions revealed a particular hypocrisy in American culture. Publicly, most Americans espouse diversity and multiculturalism, even support reviving indigenous cultures. But the moment a native community acts in a way that "doesn't fit into our preconceived notions of who we want aboriginals to be," we threaten our wrath, that of the majority. Wayne Johnson, the Makah whaling captain, put it like this in an op-ed piece for *The New York Times*: "Now that whales have been elevated to near-deified status in Euro-American culture and most people think that meat comes from shrink-wrapped packages, derogatory terms are once again directed at us."

At the same time, the Makah's "success" had shattered the belief of many non-Native Americans that Indians are the trustees of a special relationship with nature. As Professor Shepard Krech III put it in his 1999 book, *The Ecological Indian: Myth and History:* "For every story about Indians being at the receiving end of environmental racism or taking actions usually associated with conservation or environmentalism is a conflicting story about them exploiting resources or endangering lands—and inevitably disappointing non-Indian environmentalists and conservationists. In Indian country as in the larger society, conservation is often sacrificed for economic security."

The annual meeting of the International Whaling Commission began in Grenada but a few days after the Makah hunt. Japan and Norway proved more vocal than ever in seeking to lift a thirteen-year-long ban on commercial whaling. They used a variety of approaches. One of these was outlined in a paper, "What Can We Do for the Coming Food Crisis in Twenty-first Century?" issued by Japan. It began: "It is estimated that over 5 hundred million tons of marine resources are consumed annually by cetaceans. This means that whales are consuming five times more fish resources than humans (the annual harvest of marine fish by humans is about 90 million tons). Thus, utilization of whales could lead to an increase of fish catches for human consumption."

The Japanese delegate to the commission accused the U.S. government of hypocrisy for endorsing the Makah hunt, even subsidizing it with a $335,000 grant, while at the same time rejecting Japan's petition to allow them "traditional" coastal whaling. "It's a double standard," Mayasuki Komatsu said. "I mean, our tribes were whaling right up until the moratorium. The Makahs hadn't killed a whale in seventy years!" Japan announced it would work to remove some whales from an international endangered species list. Toward the close of the gathering, the IWC spent several hours discussing proposed new methods from Japan and Norway to kill whales more efficiently and "humanely."

As reported by the *Los Angeles Times*, the meeting "brought into sharp focus Japan's new alliance with Grenada and several other small Caribbean states, which consistently backed every one of Tokyo's pro-whaling votes . . . as they have every year since 1992, when Japan began sending them foreign aid that now totals well in excess of $80 million. Those votes have helped Japan build a solid bloc of support after years of near-isolation on the whaling issue."

How much of a role did the Makah hunt play in bringing Japan out of such "near-isolation"? The way Micah McCarty sees it, "I think it's a bit disrespectful to scapegoat us for a preexisting agenda." Maybe so, but I wanted to find out more about the agendas of the tribe itself. On my way back to Neah Bay, I arranged to meet first in Olympia, Washington, with McCarty, who was giving a speech there. He is the twenty-nine-year-old son of former Makah Whaling Commissioner John McCarty, with whom I'd visited the previous November. And, after being the first one selected to throw the harpoon, Micah had pulled out of the hunt—for reasons I hoped to learn more about.

A talented carver whose work has been exhibited in several galleries, Micah had been raised by his mother in Olympia, then returned to Neah Bay after graduating from high school. There, while commercial salmon fishing with his father, he learned of Makah traditions and his family's whaling history. Micah is a staunch believer that "treaty rights are consistent with natural law, and must be protected." Thus, no matter what the "eco-evangelists" say, he feels every Makah has a legal entitlement to go whaling. From the start, however, he and his father were proponents of adhering strictly to the traditional methods. "I hate to say it," he says, "but some of the elders would like to have gone out themselves to do the first hunt—and the only way they saw it possible was to use motorboats, because they're too old to get into shape in a canoe."

The McCartys' insistence on a canoe prevailed, but by 1998 Micah was having second thoughts about participating. "When I was active in the canoe, I had this dream that we were being towed by a whale," he recalled. "It was careening along the coastline, dragging the canoe across and over the rocks. Then the whale actually dove *through* a rock, as if it was just part of the water. If the men were all in spiritual unison, we'd have been able to come through that rock and be with the whale. But I knew the canoe wasn't going to make it, because we weren't really together.

"That was part of the reason I decided to back out of it, because I felt there was too much ill will between the people in the crew. I know I was a part of that, too. There was a lot of jealousy with the idea that I might be the first harpooner in over seventy years. But I also felt traditional perspective wasn't being valued enough." Not long after Micah bowed out, the first whaling captain also resigned his position. The initial hunt, according to Micah, devolved into "family feuding. . . . A dark-horse family came in and basically stole the hunt away. Some of them have fabricated histories, and others don't remember their histories. Theron Parker took all the glory."

Eight years older than Micah McCarty, Theron Parker is also a carver. He has his wood shop in a trailer a couple of blocks inland from where the gray whale he harpooned was landed. I found him there one evening, and he agreed to a brief interview over a dinner of Chinese takeout. Parker pointed to a framed photograph on the wall of the trailer. It depicted a fierce-looking Makah in a cape, holding a harpoon. "That's my great-grandfather," he said. "And his grandfather was Tsekauwtl, who did the treaty. I'm a direct descendant from the chief, from what I gather. It's in our history, a living history. My family is a whaling family, and always has been."

Parker continued, "When they brought back artifacts from the [archaeological] dig at Ozette when I was younger, I wondered what it would be like to be that guy—the whaler who had the big ceremonial saddle with all the otter teeth. It was just amazing to consider how strong a man had to be to sink a harpoon. After it all became a reality—shoot, I *am* that guy."

How had he ended up chosen to be that guy? I asked. Parker replied, "I don't really know, to tell you the truth. It seemed like the first crew was only there to get their names in the papers. We called 'em the Hollywood whalers. As the beginning guys started fading out, pretty soon all us guys started getting in there. Since I do all the carving—made all the paddles and bailers, kept the maintenance on the canoes—I started learning everything I could about it. Sitting with all the old whaling gear that they'd hauled up from Ozette. How we prepare is very ceremonial. Lotta prayer. Lotta sacrifice. Very deep within the heart. So I ended up being where I was at, and that was that. It's the way it was supposed to be. You are who you are from the time you're born, is how we say it around here."

Had Parker gone through things such as his ancestors had? I wondered. Like wearing stinging nettles, living in isolation, having to swim long distances? He replied, "Lot of those things, they'll come back to you, yeah, if we're serious about what we're doing here. Like I say, prayer answers a lotta things. We've all been through that—nettles, isolation, fasting—everybody has a variation of what they do, there's a lotta things that go on."

Had Parker been a religious person before getting involved in all this? He replied, "Hell no, I was probably about the meanest bastard in this place. I didn't give a shit about anything or anybody, where I was going or what I was doing. I did some jail time, sat and thought about that. Before I even started whaling, thinking how this isn't the way I should be acting. Almost everybody on the crew has had some kind of drug and alcohol problem one time or another in their life, but this straightened all that out. I never stop anymore, as far as being an influence to the younger kids. It's a great help to this small community, because there's not much out here, you know."

Why did this happen through whaling instead of, for example, through being a dedicated crewman on a salmon fishing boat? Parker replied, "Like I say, this is very spiritual. This is something white people won't understand. We can talk till we're blue in the face, and you're never really gonna understand what it is for us."

Had he tasted gray whale before the hunt? Parker replied, "No. Oh yes I had. The one that washed ashore a few years back. We cut that up and smoked all the meat. I liked it then. It was different, gives you a kind of energy, something our bodies need. A little missing link in our lives. Like when a wino needs a fix, for example—seemed to put a calm in people's system." (Several other accounts I'd heard about the whale that became enmheshed in a Makah fishing net in 1995, including that of the elder Charles Claplanhoo, differed from Parker's. A fetid smell is said to have emanated for many months from the majority of the meat ending up in the town landfill.)

The night before the hunt, Parker went on, "everybody became really calm." An elder brought the crew together to pray, "and we knew tomorrow's the day." After being "really pumped up," he remembered a calm settling over him when he threw the harpoon. Had more than two harpoons entered the whale, it might not have sunk. "But it was probably better it sunk so everybody didn't have to see all the blood. A lotta people disagree with what we do. I wouldn't change anything."

What about departing the scene late that night, leaving the fisheries agency and an Alaskan Inuit to finish the cleaning job? Parker replied, "We were tired, man. They wanted to stay, that was their deal. They were working just on the bones and intestines; it's gonna be there tomorrow."

Parker gazed up at the photograph of the old harpooner in the cape, and I asked whether he felt the spirit of his great-grandfather had in a sense passed on to him. "Well yeah, it's in the blood."

Was his own fourteen-year-old son interested in following Parker's footsteps? "Not really. I didn't really raise him up, he just kinda came back into my life."

Parker concluded, "There's a lot of things I can't answer for you, or I won't answer. The things I do to prepare myself in my mind. When I'm old and gray, maybe I'll pass that on somewhere."

Gary Ray, the man who first put forward the whole idea to the tribal council, concurred that Theron Parker had undergone a personal metamorphosis once he became part of the whaling team. "I can honestly say Theron did an about-face," Ray said. "He's finding out who he is."

Ray was thrilled with what the hunt had brought about in Neah Bay. "Teenaged kids sitting down with their grandfathers saying, 'Tell me what you know, I want to learn my family song now, put this hip-hop music down.' This is healthy for us as a people. To get into the spiritual aspect of whaling is a whole new ball game—a disciplined life, your accountability. We can use this as a tool to begin to get back our value system and who we are. So the whale is not just something to go hunt and catch but is used to bring a broad transformation to our people."

There had been a number of other options. "We had one with big bucks that was brought to our Makah Whaling Commission," Ray said. "These people were willing to pay a million bucks per whale that we don't harvest." He paused for a moment to let the figure sink in: $4 million a year, $20 million over five years. "Think about that. So we walked down the road with them on this negotiation process. But we'd have had to sign a paper giving up forever our right to whale. They're asking to take something away that's guaranteed in our treaty. I mean, give us back all that land, we'll give up whaling."

Had this offer gone to the tribe for a vote? I wondered. "No, the proposal never made it past the heads of families on the Whaling Commission. Never made it to the tribal council. If you got approval with both of them, only then would it go to the general families."

Ray looked out the window of the Makah Maiden café at the harbor. "You know, we could blow one little pinnacle off Waadah Island out there and make that a deepwater port. A duty-free, tax-free port. I see no reason why we can't be a miniature Hong Kong. We've gotta live, too. The tribes going to gambling casinos, they're doing what they have to do for where they're at. Well, let's use what *we've* got."

Whoever made the multimillion-dollar bid to the Makah Nation to stop hunting whales, this wasn't the only offer on the table. Dr. Deborah Brosnan is a marine biologist originally from Ireland; in 1992 she formed the Sustainable Ecosystems Institute in Portland, Oregon. She'd worked with Native American tribes in the Pacific Northwest on various issues. She was deeply concerned about the Makah situation, viewing the hunt as "negating where we're going as a society in our relationship with marine mammals, and as a critical step towards opening the door to commercial whaling."

So when Brosnan received phone calls from colleagues in October 1998 suggesting that she seek a meeting with the tribal leadership to explore alternatives, she agreed. She and an assistant traveled to Neah Bay. "They were open to talking," Brosnan recalled. "We said up-front we respected their treaty rights, weren't here to tell them whether they were right or wrong, but wanted to start a dialogue. We deliberately didn't talk to any media but really wanted to go one-on-one behind the scenes. Sometimes it's the messenger as much as the message. I didn't have millions to offer, and they knew I would have to raise the money. But I'm a woman, which made a difference, and secondly I'm not American. I grew up in a culture not hugely dissimilar from theirs in Ireland.

"We said, why don't you tell us what your issues are, what you need. And they needed everything. Eventually the tribe came back with a concept of a research program that would involve the whaling team. I said fine, I've got funding to cover at least a year's research. We were looking at photo-ID, but also some biopsy work on whales that stranded—to tie this in with ongoing work but have it done by the Makah themselves. We wanted something they could be in charge of and feel good about, something which also had an element of adventure."

By early May 1999 the talks appeared to be progressing well. The tribe seemed to appreciate Brosnan's approach, and she'd raised a quarter million dollars in up-front funding for the prospective program. "Then things snowballed," according to Brosnan. "On May 10 they had a blowup with one of the protest boats on the water. I got a call that day from a tribal member who said he thought I should come up to Neah Bay. I did. He brought in a bunch of other people. There were people on the team who weren't necessarily happy even then about whether they wanted to kill a whale, or whether the time was right. But they were training, and the issue was getting tense. Suddenly it was put up or shut up time. They were saying, 'We're being attacked as a tribe, we have a point to prove.' Also, the IWC meeting was coming up. Even the night before they killed the whale, I talked to one of the trainers and he said, 'I think we still have some time.' I was planning to go up again the next day. But it was too late.

"Now I don't know how long they'll continue," Brosnan concluded. "I feel there are certain things that are going to have to play out internally within the tribe before anything changes. I don't know what those things are, but I know they're there."

One hint of "what those things are" comes from a non-Indian whaling opponent, Jeff Pantukhoff, founder of the Whaleman Foundation. Based on what a source within the tribe revealed to him, Pantukhoff says: "I believe the reason the Makah put on the big push to kill a whale right before the IWC meeting was to fulfill an earlier agreement brokered between a tribal member who worked for Crown Pacific, a Japanese-owned fishing company, and Japanese whaling interests. Several million dollars had been promised to the tribe in the form of new contracts, and certain people within the tribe were given cash payments as new trucks and hot tubs began to appear. Of course, this may seem difficult to prove, but the trail is there for anyone who cares to follow it."

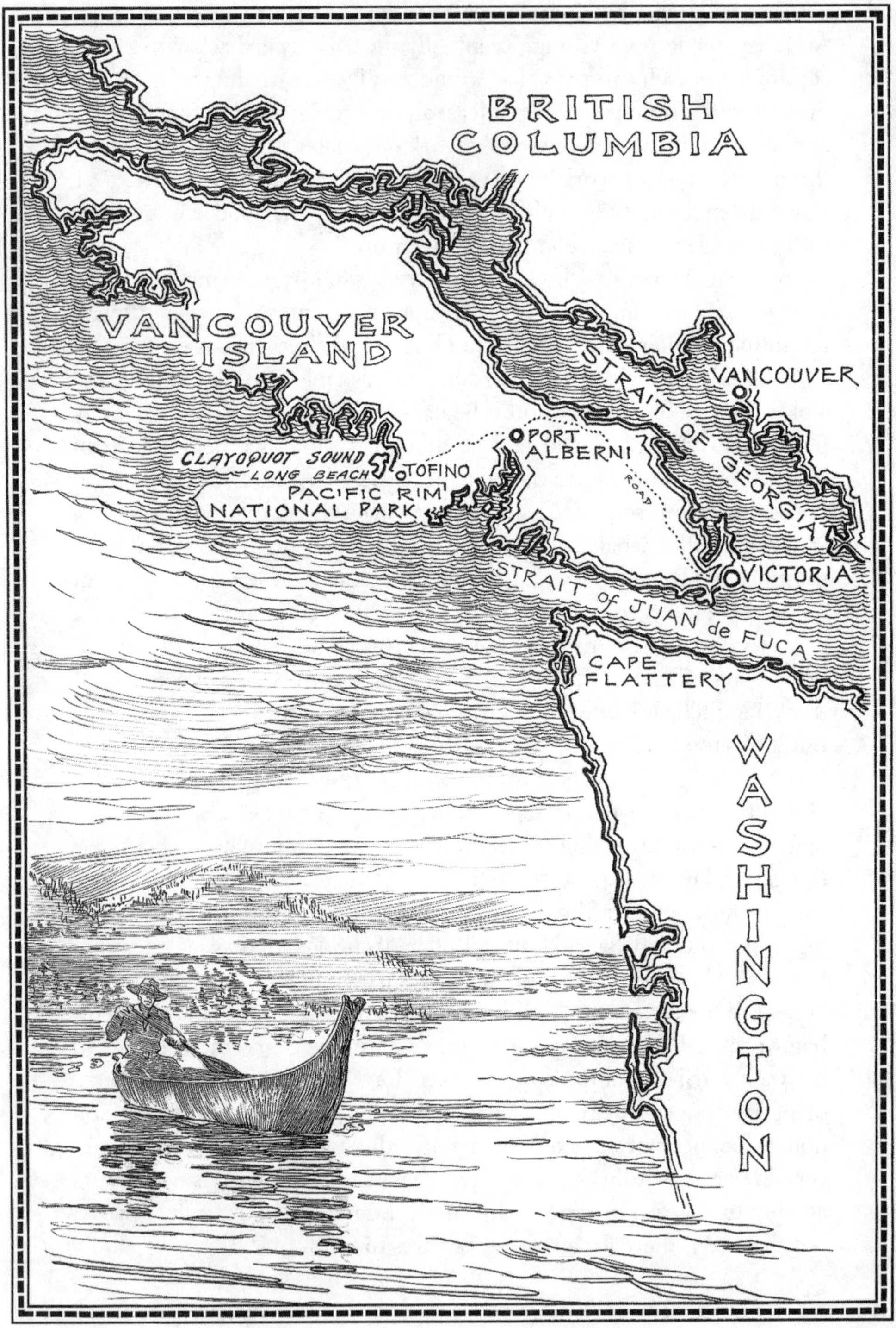
BRITISH COLUMBIA
VANCOUVER ISLAND
STRAIT OF GEORGIA
VANCOUVER
PORT ALBERNI
ROAD
CLAYOQUOT SOUND
LONG BEACH
TOFINO
PACIFIC RIM NATIONAL PARK
VICTORIA
STRAIT of JUAN de FUCA
CAPE FLATTERY
WASHINGTON

Chapter 13

WHALEMEN OF VANCOUVER ISLAND

Crossing the strait, we come to the island of Quadra y Vancouver, which is only a continuation of the high, broken, and wooded coast to the southward. Along its sea-board are many inlets, sounds, and bays. . . . The coast tribes of Vancouver and Washington territory also meet in their canoes on the same fishing-grounds. It is a fine sight to see a large flotilla of them so distant from the land as to appear veiled in deep azure. . . . It is, above all, when in chase of the whale that they show their peculiarly wild character.

—CHARLES MELVILLE SCAMMON, October 1874

FOURTEEN tribes, first called the Nootka by Captain Cook, once possessed a territory that stretched for more than three hundred miles along the west coast of Vancouver Island. Crossing the Strait of Juan de Fuca from Neah Bay, gray whales followed those Nootkan shores from Port Renfrew, through Barkley Sound and Clayoquot Sound, up past Hot Springs Cove and Esperanza Inlet, to the tribe's northern boundary at the Brooks Peninsula. One archaeological site containing the remains of whales has been estimated to be at least 9,100 years old.

Today the Nootka refer to themselves as the Nuu-chah-nulth Tribal Group. Though a part of Canadian British Columbia, they consider Nootka their anglicized name. They also consider the Makah Nation part of their original tribal family. Immediately after the Makah killed a gray whale in the spring of 1999, Francis Frank, who cochairs the Nuu-chah-nulth Tribal Council, told the media: "The Makahs' success . . . certainly lays out a blueprint for us."

Leading the charge toward revitalization of gray whaling among the six thousand–plus indigenous peoples of Vancouver Island is a forty-one-year-old former firefighter named Tom Mexsis Happynook. Traditionally his family name was Hup-n-Yook, which he says has a double meaning in his native tongue: "You can get your whale before you go full circle around the territory" as well as "Understanding life, how things are connected." Mexsis is the short form of an older name, Mauk-sis-a-noop, which Happynook says translates "Gray Whale Hunter." He is a hereditary chief of the Huu-ay-aht band and maintains that whaling can be directly traced back ten to twelve generations in his family but is undoubtedly thousands of years older. Asked if he's ready to go whaling himself, Happynook replies, "Absolutely. And my son would come with me. People need to realize that a thousand years from now, five thousand years from now, there is always going to be a Happynook whaling family."

Happynook's vision, however, goes well beyond his own lineage. He is the chairman of the World Council of Whalers (WCW), formally established in January 1997. As he announced at its first general assembly: "The WCW should be a focal point for the whalers of the world. It has been established as an organization for whalers to maintain and strengthen their customary relationships with whale resources, to promote sustainable use, and include local people in the decision-making process."

Interested in learning more about the WCW, after the Makah hunt I took a two-hour ferry trip from Port Angeles, Washington, across the strait to Vancouver Island and the British Columbian capital of Victoria. Tom Happynook and his wife, Kathy, live in suburban Brentwood Bay, in a spacious, two-story ranch-style home. They had invited me to dinner. Kathy is of Scotch-Irish descent. The couple became acquainted growing up in the coastal town of Bamfield and have been married for twenty-five years. Happynook's eighty-five-year-old grandmother, who lives with the family, joined us for salmon and halibut. Happynook explained that, after his father died prematurely, he'd been raised by his grandparents, who inculcated in him knowledge of his heritage.

"Our last traditional whale hunt was in 1928, when my great-grandfather and grandfather went out in a canoe carrying a wooden harpoon with a mussel-shell tip and sealskin floats, and dispatched a gray whale. My grandfather told me that the decision to stop hunting was made because the numbers just weren't there anymore. There are a number of principles that guide our resource extraction."

At one point Happynook brought out his family's long whale-killing

Tom Happynook holding his ancestral lance (George Peper)

lance, which had gone from being fashioned out of yew wood to this iron variety about two hundred years ago. "We grabbed on to the technology of the day," he continued, drawing a comparison with the Makah's utilization of a 770-caliber big-game weapon.

I wondered how his coming to lead the World Council of Whalers had evolved. Happynook recalled that, in 1994, after working his way to captain during sixteen years in the fire department, his tribe had asked if he'd represent them in treaty negotiations with the British Columbian government. Whaling was among the "substantive issues agenda" that Happynook placed on the table for negotiation. Around the same time ad hoc meetings on revitalizing "community-based whaling" had been taking place—first in Glasgow, Scotland (1992), and then in Kyoto, Japan (1993). The WCW was drawn up at the last of these, in 1996, held in Berkeley, California, where representatives of traditional whaling communities from the Arctic, Caribbean, North Atlantic, North Pacific, Southeast Asia, and South Pacific were in attendance. Happynook says he suggested they form an international organization. "I, of course, was immediately nominated chairman, since it was my idea," he added. During the WCW's first year the Nuu-chah-nulth Tribal Council provided office space and administrative support, but the secretariat has ended up being run out of the Happynooks' home.

"We're not rich like the protest industry," Happynook went on, "that multitude of multimillion-dollar corporations like Greenpeace and the International Fund for Animal Welfare—the list goes on and on—soliciting funds off the marine mammal issue. Quite simply, they target people's emotions, and very photogenic creatures, to raise money."

"Of course," Kathy chimed in, "their conspiracy theory is that we're really headquartered on the fifty-first floor of some skyscraper in Tokyo."

Since the subject had been raised, I remarked on having heard rumors that start-up funding for the WCW came from Japan and Norway. "That is true," Tom Happynook replied. "Not the governments, though. That needs to be made clear. We received a bit of funding from the Japan Small-Type Whaler's Association, a community-based group along their coast. The Norwegian Whaler's Union also contributed some, and so did the Nuu-chah-nulth Tribal Council. We got little bits of money from a whole lot of people to buy the computers, get the stationery set up."

By 1999, Happynook says, he'd come to realize that treaty negotiation "was not the end-all or be-all of our social issues or our poverty, but in fact only one of the tools at our disposal to bring prosperity to our community." So he'd begun putting together a tribal business corporation—fisheries, forestry, cultural tourism—and devoting more and more time to the WCW. He'd begun traveling widely on the whaling organization's behalf—to Iceland for the WCW's second general assembly, to Africa and the South Pacific, and to Japan.

The last trip had taken place a few months before the Makah's gray whale kill, when Happynook's tribe was invited to perform their songs and dances as guests of the Japan Whaling Association. "They invited us because they have a very deeply rooted cultural aspect to their whaling, just as we do," Happynook said. Fourteen Nuu-chah-nulth had entertained a crowd of three thousand at a sumo wrestling arena in Tokyo. Happynook had actually arrived a week early, dined on whale meat at a restaurant in Osaka, and all together visited four Japanese whaling towns. After having his picture taken with members of a Japanese whaler's group known as TRUE ECO NET (TEN), Happynook had written: "I believe many years from now that the photo will remind the whalers of the future that a small group of people can, and did make a difference in our fight to reestablish whaling as a time-honored profession."

What he's ultimately talking about, Happynook says, is "cultural biodiversity," or the human relationship with ecosystems and animals. His grandfather taught him that you can't manage resources, or people, but you can manage the relationship between the two. "Over millennia, we had been part of the whaling, of the gray whale's life. As the hunters, we helped maintain the balance with that population and ecosystem. We do have a role to play."

It's a role Happynook insists has nothing to do with opening the floodgates for industrial harvesting of whales again. "First of all, there are no markets for the ambergris of the whale, which was used to make perfumes; for the baleen, which was used to make corsets; for the oil, which was used to lubricate engines and have lighting. For these we have all the synthetic things we need. No, today it is about food—in Siberia, the Philippines, the Caribbean states, up through Greenland and the Faeroe Islands."

But did he really believe that whales-as-food was as vital here on Vancouver Island as it would be for the impoverished Eskimos of Siberia? "Absolutely. If you look at the scientific findings about sea mammal oil, you'll find it combats high cholesterol and helps with arthritis, diabetes, and other diseases my people are suffering from." Indeed, Happynook added, because humans and whales are both mammals, the oils are readily absorbed into the body—unlike fish oils and flaxseed, which must first undergo a chemical decomposition.

Happynook seemed to have a ready answer for everything. And he pulled no punches about his disdain for the two international regulatory bodies governing whaling practices—the International Whaling Commission (IWC) and the Convention on International Trade in Endangered Species (CITES). He called the IWC "a good old boys' club that meets in very exotic places, where everybody has a great time and nothing gets done." Its original mandate was "to manage whaling, not to be a protectionist group that stops whaling completely," and "the pro-whaling side has many times offered solutions to initiate dialogue." While his WCW "has no intention of becoming the IWC . . . we want a place at the table." What Happynook envisions as the ideal solution would be "regional management regimes incorporating science and traditional resource management knowledge. . . . You include indigenous peoples in the decision-making process."

As for CITES, which falls under the auspices of the United Nations Environment Program (UNEP), Happynook believes that "wealthy countries and environmental lobbying" are the driving forces behind it.

The CITES Convention, drafted in 1973, prohibits global trade of any species judged endangered (including nearly every type of whale, the gray among them). Happynook cites 600,000 tons of blubber being stored in Norway because the Norwegians "have no palate for it," which could otherwise be traded to Iceland, where the people do. He equates this to "the ivory issue" in Africa, where impoverished villagers are not permitted to deal in ivory stockpiled from elephants found dead or slain as rogues.

Happynook would like to see both bans overturned, so long as whale products and ivory are sold in a "sustainable way. . . . Wealth generation has always been a part of every nation's growth. We used to have barter networks all through British Columbia. Well, today it's money. . . . The whales are back now. . . . I describe the whale as our national bank."

In the next breath Happynook will talk glibly of how he was taught by his grandfather "that the Nuu-chah-nulth whalers had to suffer for the honor and responsibility of taking the life of the greatest mammal on earth." And he is ready, having grown up knowing "where the sacred places are, the secret medicines and the caves, where we got the yew tree to make our harpoon, where we gathered the mussel shells because they had to be tough and right."

As for all the so-called whale lovers, as Kathy Happynook sees it, "Why don't people have the same feeling about cows? They're a mammal with big doe eyes that prances with their babies. I think it's just years of brainwashing."

Whale, I want you to come near me so that I will get hold of your heart and deceive it.

—A Nootkan prayer

A study of the World Council of Whalers' General Assembly Reports for 1998 (Victoria, B.C.) and 1999 (Reykjavík, Iceland) proved instructive. They contained, for example, a public relations strategy for whalers outlined by Brian Roberts, a senior adviser in Canada's Department of Indian and Northern Affairs (which also provided twelve thousand dollars to help subsidize the 1999 meeting). Making reference to earlier controversies over sealing and the fur trade, Roberts stated: "The first step was to neutralize the appeal of the animal-protection lobby. To accomplish this it was necessary to mount an equally emotional powerful counterappeal. This counterappeal was based on the survival needs of aboriginal communities which depended upon the continuing taking of fur-bearing animals." This lesson, he went on, would be useful in whalers' efforts "to deal with a poorly informed and emotional public, and with politicians seeking electoral approval from such publics."

Happynook's executive board had no real surprises, its members including representatives of Norway's High North Alliance of whalers and the Japan Small-Type Whaling Association. More interesting was the WCW's acknowledgment of "the generous financial assistance" received from, among others, the International Foundation for the Conservation of Natural Resources. This organization's president is a Washington, D.C., attorney named Steven Boynton. He had attended both general assemblies and also served as first vice president/legal counsel to a group called the International Wildlife Management Consortium (IWMC)/World Conservation Trust, which one of the WCW reports noted has "an especial interest in whaling, the rights of resource users, and the sustainable use of natural resources." The IWMC's president is a French Canadian named Eugene Lapointe, who also happened to have served as secretary-general of the CITES organization between 1982 and 1990.

At the first WCW assembly, before more than a hundred delegates from nineteen countries, Boynton had opened a panel discussion about the IWC and the future of whaling. He noted that not since 1993 had the U.S. Congress passed any unanimous resolutions opposing the resumption of commercial whaling. "Nor do I think it will do so again," Boynton said. "Indeed, we now have many letters written to the U.S. administration by congressmen in leadership positions, protesting the U.S. position on whaling. Congressmen such as the chairman of the House Ways and Means Committee . . . and the chairman of the Senate Foreign Relations Committee, among many others.

"These congressional concerns," Boynton continued, "are based upon the U.S. threatening economic sanctions against whaling nations. . . . However, the U.S. is aware that if it were to apply such sanctions, and the sanctioned party were to take the U.S. before a GATT [Global Agreement on Tariffs and Trade] panel, the U.S. would lose the case.

"A recent poll carried out in the U.S. by the International Wildlife Management Consortium indicated that a representative sample of U.S. citizens would support a resumption of commercial whaling of non-endangered species used for food for human consumption. This poll result was released during the 1997 IWC meeting in Monaco."

That was when the Makah received their imprimatur from the U.S. delegation to the IWC. Boynton had gone on to discuss the IWC's recognition of aboriginal or indigenous rights to whale. "But in countries which were not conquered and colonized, such as Iceland, Japan, and Norway, it could be argued that the people are also aboriginal."

Subsequently, Boynton talked about the Marine Mammal Protection Act [MMPA], first passed by Congress in 1973. "There is a serious question concerning the need for the MMPA when the Endangered Species Act covers any species that are in need of conservation protection." This sweeping generalization failed to mention ongoing efforts by congressional Republicans to remove the teeth from the Endangered Species Act. Boynton went on to comment that the MMPA needed to be "compatible" with regulations promulgated by the World Trade Organization and GATT.

Boynton's colleague Eugene Lapointe stated that "because of gatherings such as this general assembly, the sustainable use movement is now making very substantial progress." Then Lapointe addressed himself to CITES: "The down-listing of certain nonendangered whales is the next step in reestablishing international trade in whale products."

It was as if blueprints for the future were being laid out. Indeed, at the next meeting of CITES, in Nairobi in the spring of 2000, Japan's Fishery Agency would seek to down-list three whale populations—the South Antarctic and northwestern Pacific minke whales, and the northeastern Pacific gray whale. These initial attempts were voted down by the member nations.

So just who are Tom Happynook's backers? Further research revealed that Steven Boynton, besides having worked for Japan's Institute of Cetacean Research, is a conservative activist on other fronts. According to an article in *The Washington Post*, Boynton was among a trio of initiators of the so-called Arkansas Project, which set out back in 1993 to derail the Clinton presidency. "Boynton received at least $577,000," the *Post* reported, from a foundation headed by the billionaire right-winger Richard Mellon Scaife. Boynton also authored a 1995 legal briefing paper—"United States Is Fostering an Illegal Policy to Obstruct Whaling"—for the Washington Legal Foundation, a right-wing think tank backed by the Coors family.

A new book, *The Hunting of the President: The Ten-Year Campaign to Destroy Bill and Hillary Clinton*, elaborates on Boynton's background: "His field of expertise was environmental law, with a subspecialty in curtailing legislation aimed at protecting endangered species. An avid hunter and angler himself, Boynton combated environmentalist encroachments on behalf of trade associations such as the American Fur Resources Council and the International Shooting and Hunting Alliance. Among his most prestigious clients was the Republic of Iceland, which hired him to help overturn the international ban on whaling. He also served as general counsel to the Congressional Sportsmen's Caucus Foundation, a progun group that sponsored hunting junkets for legislators."

As for Eugene Lapointe, he was removed from his post as secretary-general of CITES on November 2, 1990, by UNEP's then–executive director, Dr. Mustafa Tolba. The official reason was Lapointe's active campaigning against an ivory trade ban adopted by CITES member nations in 1989. Lapointe blamed environmental groups for his downfall and later received a settlement, based on a three-member CITES board's concluding his dismissal had been "arbitrary and capricious." He soon established the International Wildlife Management Consortium to lobby for free trade interests in whaling, fisheries, forestry, mining, and ivory. Today his IWMC is based in Lausanne, Switzerland—also the headquarters city of CITES—with additional "desks" in Japan, China, Argentina, and the United States. Besides promoting whaling, the IWMC is seeking, among other agendas, "the resumption of a limited and well-controlled ivory trade" and "sustainably using marine turtles."

An IWMC press release dated May 15, 1999—two days before the Makah harpooned and shot their gray whale—bore Lapointe's byline and was headed "A Special Tribute to Our Friends, the Makah." He wrote: "IWMC World Conservation Trust is dismayed but not surprised, that our friends the Makah whalers once again came under siege and interference with their hunt. . . . Members of a traditional culture were being unlawfully attacked by outsiders for the legal, routine manner in which Native people were trying to secure food for their community. . . . May they soon meet each whale which is meant for them, and secure it, and always return home safely in victory, under the eyes of their ancestors."

After the Makah met and secured that first whale, Lapointe immediately issued another press release. Noting that "Nature demands carefully balanced biodiversity, but includes cultural diversity," he opined that this event "spells more than the renewal of a nearly forgotten skill." He followed up the next day with a pointed jab at the IWC on the eve of its annual meeting: "Nudging the fate of ancient whaling cultures worldwide one step closer to the brink of extinction is the unstated but quite intentional agenda" of the IWC, Lapointe began.

It was a theme he would hammer home at the IWC's 1999 gathering in Grenada, where he joined Happynook, Boynton, and a representative of the Japan Whaling Association for a panel discussion on the IWC's future. In a media release headed "IWC Is Driving Itself to Extinction," Lapointe wrote that the Japanese had "waited patiently" for the IWC to acknowledge their people's "nutritional and cultural needs."

> *Though the clans peopling the coast between the Columbia and the northern boundaries of Vancouver live in Indian luxury and plenty, still they are regarded as a treacherous race, whose hands have been stained with the butchery of many a shipwrecked sailor. Hardly a trace of civilization appears on the outer borders of the island, except the deserted cabins of lumbermen and a trading-post of the Hudson's Bay Company, or perhaps the temporary shanty of some transient adventurer, who risks his life to barter for peltries. Nevertheless, during the last century, Clayoquot Sound and that of Nootka were the chief resorts of exploring and trading vessels visiting the North-west Coast.*
>
> —SCAMMON, 1874

From the region around Victoria, where the Happynooks reside, it's about a five-hour drive to British Columbia's town of Tofino on Clayoquot Sound. This is the home base of Dr. Jim Darling, who's been studying the gray whales of Vancouver Island since the early 1970s. Among marine biologists Darling has contributed to a number of "firsts"—including the first to suggest that a specific population of grays resides south of the Bering Sea in the summer; the first to utilize photo identification of individual gray whales as a basis of study; perhaps the first to be approached by a "friendly" gray on the summer feeding grounds; and even, arguably, the first to study living large whales in Japan.

Over the past quarter century Jim Darling has written popular articles for *National Geographic* and scholarly reports for scientific journals. He was originator and associate producer of *Island of Whales*, which received Canada's prestigious Gemini Award as the best TV documentary of 1991. Largely dividing his time between studying gray whales off Vancouver Island and studying humpbacks in Hawaii and around the world, Darling is also a conservationist who has written: "We are training dolphins to go to war for us. We are capturing whales so they may entertain us. We are hunting them for tradition, for food, and for profit. We are shooting them because they compete with us for food. We are drowning them incidentally in our fishing nets. We are slowly killing them with pollution. Ironically, these are animals we love."

Darling and I had made a date to go out in his boat and do some research among the summer resident grays of Vancouver Island, weather permitting, since the region receives annual rainfall of more than ten feet. To reach Darling's domain one must travel through some of British Columbia's most rugged and beautiful country. Highway 4 remains the only east–west road to cross the nearly 300-mile-long island, the largest off the west coast of North America. Not until 1956 was a dirt logging road finally completed along the 122-mile stretch from the old mill town of Port Alberni to the quiet fishing village of Tofino. The paved highway came fifteen years later, enabling young men like Jim Darling access to one of the premier surfing spots in Canada. Even today, though, beyond Cathedral Grove, one of the few roadside-accessible, surviving uncut stands of old-growth timber, few signs of civilization are visible. Two lanes snake across the Mackenzie Mountains past the shallow, rushing Taylor River and a series of deep lakes, hemmed in by reforestation projects at various stages of comeback. At the base of the Mackenzies awaits the Pacific Rim National Park, seventy-eight miles of spectacular forest trails and coastline. There at Quitsis Point, high on a rock wall, is an ancient petroglyph of what is probably a gray whale.

The port of Tofino, population 1,286, looks out over Clayoquot Sound and a Pacific Ocean that flows all the way to Japan. Less than a mile across from Tofino Harbor is Meares Island, dominated by two mountains and with a nearly sixty-mile-long shoreline of its own. Often cloaked in clouds that condense vast quantities of water, its aquifers provide Tofino with some of the purest drinking water anywhere. Continuously inhabited for more than five thousand years by two Nuu-chah-nulth tribes (the Clayoquot and the Ahousat), Meares is also host to some of the world's largest cedar, spruce, and hemlock trees. Canada's most monumental cedar, with a girth of over sixty-one feet, has probably stood on the island for more than fifteen hundred years.

Temperate rain forests like Meares's are extremely rare, covering about 0.2 percent of the planet's land area. About half of what's left grows along the west coast of North America, between here at Clayoquot Sound and Washington's Olympic Peninsula. Whereas overall only a tiny portion of Vancouver Island's rain forest is still intact, fully 80 percent of Clayoquot Sound's has been granted at least a temporary stay of execution. This is the result of over a decade of local environmental activism, culminating in 1993, when an estimated twelve thousand people joined forces to halt the clear-cutting. That became the largest peaceful civil disobedience campaign in Canadian history (almost nine hundred arrests)—and put Tofino, once and for all, on the map.

Today the tourists start arriving with the spring gray whale migration. More than a million people a year check into Tofino's resorts, bed-and-breakfasts, and campgrounds. They visit the Whale Centre, the Whale Song Gallery, and the Orca Lodge. They venture out into passages, channels, and inlets with one of more than twenty whale-watching operations; the largest, Jamey's Whaling Station, can hold forty-seven passengers on its big boat, at seventy-nine dollars per adult head for a two- to three-hour cruise. (Whale watching brings between $6 million and $8 million a year into Tofino.) They may see porpoises and sea lions, and the occasional humpback or orca—but gray whales are beyond doubt the stars of the sound's show.

Jim Darling lives at the end of a long lane overlooking Tonquin Beach—the land purchased in 1980, before Tofino got "discovered." Now it's the crown jewel off an otherwise seeming millionaires' row. It's a sprawling cedar dwelling that blends deep into rock ledges and trees; its long plate-glass windows fill a combined kitchen, dining room, and living room. Some fifty feet down a steep cliffside, gray whales generally make their appearance by April.

Approaching fifty, Darling has sandy hair that's begun showing

The marine scientist Jim Darling (George Peper)

shades of gray. But he remains a tall, attractive, soft-spoken man, whom it's still possible to imagine catching a wave at nearby Long Beach. That's where his fascination with gray whales began in 1968. He'd grown up in Victoria, where he attended the university, but he'd been surfing since high school, and in the summer there was only one place to go—that particular twelve-mile-long stretch along the western coast. Darling remembers lounging with his friends outside the surf breaks known as Tiny's and the Rock, watching for the incoming swells or paddling out with their boards, "when a whale would suddenly blow close enough to stop our hearts." At first they thought these might be killer whales, which had "a history of checking out wet-suited, seal-sized creatures and scaring them near death." Darling could scarcely believe that gray whales would purposely be inside a surf line, enveloped by clouds of sand. But they were, and he quickly grew enamored of riding the waves in their midst.

> *To our surprise, we saw numbers of these grays going through the surf where there could barely have been depth to float them. We could see in many places, by the white sand coming to the surface, that they must appear to be touching bottom. One in particular lay for half-an-hour in the breakers playing.*
>
> —SCAMMON, 1869

Scammon's was one of the first books Darling read when, in 1973, the new Pacific Rim National Park hired him to conduct a review of all the literature about gray whales. Darling had spent the previous summer in the park piloting a charter boat advertised as a sea lion cruise. "They had no idea there were whales in the middle of the park," he remembers. "But the superintendent knew of my budding interest and had four thousand dollars left over in his budget. At the time, that was more money than I'd ever dreamed of."

Years later, in his own photo-essay book about gray whales, Darling would put Scammon at the top of his "recommended reading" list. "Even though he was killing them afterward," Darling says, "Scammon was probably the last man until recent years who spent any time looking at live whales. Especially during the period of industrialized whaling that followed, you just shot them from a distance. But when hand harpoons were still being used, Scammon really had to concentrate on figuring out where the whales were, what they did, and how best to approach them. His observational ability holds up surprisingly well after almost 150 years. Of course, by contrast you could say how little we've progressed in our knowledge if Scammon is still one of our major references!"

Darling laughs. In truth he has been one of the key figures in rapidly advancing the study of gray whales well beyond Scammon. While doing his sea lion cruises, Darling remembers, "There was one gray whale with a big teardrop orange scar on its side. You couldn't miss it, and we saw it fairly often." Returning to the university—"I'd decided to study marine biology, but it never occurred to me I'd actually *be* a biologist"— he ended up visiting with a graduate student who was conducting bird and mammal surveys for the park. Dave Hatler casually asked Darling if he'd seen any whales he could recognize. When Darling mentioned Orange Scar, Hatler "literally leaped up from the supper table and rushed to a closet and pulled out his pictures. 'Is this *it*?' he yelled at me. It very clearly was. He told me he'd taken this picture in 1970. We were now in '72. He asked if I could see whether the whale came back again next season—and get a picture of it."

So Darling initiated his fieldwork with a little camera and a ten-foot-long Zodiac. Sure enough, Orange Scar returned to Tofino, and he snapped its photograph. This led to a landmark paper coauthored by Hatler and Darling in 1974. For the first time it was suggested that individual gray whales could be identified in the wild; until then scientists believed they could work only with dead specimens. For the first time, too, it was shown that *not* all gray whales made the full round-trip migration. This one, at least, had been observed over several years, stopping to set up summer seasonal residence halfway up the northbound route. And it had to be feeding on *something*. "Now these things go without saying," Darling says, "but back then it took a lot of convincing for anybody to believe it."

As it happened, essentially unknown to each other, several other marine scientists had initiated similar photo-ID studies of whales. "There was no Society of Marine Mammalogists like there is now," Darling explains, "and nobody had ever really gotten together to talk about living whales. Then a graduate student at the University of Indiana decided to invite everybody to a conference." In 1975 the biologists gathered to compare notes at A Celebration of Whales. Michael Bigg, head of Canada's marine mammal research in the Pacific, had discovered one could identify killer whales from pictures of the nicks on their dorsal fins. Roger Payne stole the show with his slides on the right whales of Patagonia. Suddenly Darling had a new mentor when Payne agreed to be part of the committee for his master's thesis on gray whales.

By the following year Darling had some 4,500 photographs—"the thirty-five-millimeter camera did more for whale research than any other single thing"—and had confirmed that at least thirty-five to fifty of these grays were "dropping out" annually. They were largely foraging along a shallow, sandy seafloor around Tofino, which was similar to that of the Bering Sea. The whales were given names like Whitepatch, Blackjack, and Squirl. One of them, Two Dot Star, TDS for short, has now been observed returning every year to Vancouver Island for more than a quarter century. Darling says its entire body is somewhat whiter, but TDS's pigmentation marks remain crystal-clear. Only a few years ago, with the use of genetic sexing techniques, was TDS determined to be a male.

A whale called Elvis, because its stripes looked like it was wearing white pants, "turned out to be a female, which surprised everybody." There was also Apache, whose pigment was reminiscent of war paint, Collage, and more. Most of these were never very sociable. This hasn't been the case, however, with some of the area's residents, one of which has been spotted here as late as mid-December before joining the southbound migration. Darling had first experienced a "friendly" encounter at San Ignacio Lagoon in 1977, when he journeyed there at the close of the initial field season of Mary Lou Jones and Steven Swartz, to pass along his photo-ID knowledge. A gray whale had leaned gently against Darling's rubber skiff, inviting him to pet it. Still, the phenomenon wasn't thought to happen beyond the lagoon.

"I was stunned when this occurred here," Darling remembers. "If I hadn't gone through it in Mexico, it would probably have scared the hell out of me." It was the summer of 1983. He was out behind Meares Island with a female graduate student, going from whale to whale taking pictures of a small group "feeding in a very calm area." Suddenly one of the grays approached their boat, surfaced tail-first, then dove underneath and began rubbing against it. Not far away the skipper of a whale-watching vessel stared with his mouth agape.

"He caught me in town the next day and said, 'What do you *do* out there to make the whales do *that*?' I'd hoped to keep it a secret. I had a pretty good idea what would happen if people realized that whales were starting to play with boats here. Well, pretty soon there were pictures on the front page of the *Vancouver Sun*. Then everything exploded. People, even dogs, were leaping off the bows of their boats onto whales' backs. It was atrocious."

Things have calmed down over the years, while every season at least one "friendly" has shown up off Tofino. Unlike at the lagoons these are not mothers with calves but lone and generally young grays. It doesn't seem to matter to these whales what size boat it is, or whether the engine is on or off. A large whale dubbed Ditto has shown a particular propensity for petting and will even chase boats that try to get away from it. On one occasion Darling passed a Native American fellow with an eight-track stereo cranked up full blast and a whale leaning on his speedboat as he patted its head.

Darling expresses wonderment at this apparently "classic learned behavior." He draws a comparison to something observed among monkeys in Japan, where mothers pass on to their offspring the washing of potatoes before they eat them. "Gray whales' skin is incredibly sensitive, and they rub as part of their daily routine," he observes. "The enthusiastic response they get every time they rub against *us* may have messed up their brains totally!"

He laughs, but somewhat ruefully. "Not many big wild mammals that come into close contact with people have benefited in the long run," Darling adds. "Especially with the forces behind this resumption of whale hunting being, I'm afraid, pretty substantial. Most of us who study whales felt we'd seen the end of industrial whaling, for lots of good reasons, not the least being that there's no real market for most of the products. I think this is going to be proven entirely wrong, that we're incredibly underestimating what's going to happen."

I ask if Darling would elaborate. He explains he was introduced to the so-called wise use or sustainable use movement during the long battle against clear-cut logging in Clayoquot Sound. He had put his whale research on the back burner for several years to run the Clayoquot Biosphere Project—a research program on the temperate coastal ecosystem. "We brushed up against the tactics of these guys: high-end advertising, gross misinformation, orchestrating community divisiveness, personal targets. The more you look into it, the more you don't want to know. It's quite alarming. Apparently—according to a few in-depth investigations—it's connected to ultra-right free-enterprise groups worldwide, with at least one of their mandates to keep resource markets wide open. There are numerous front groups and connections in world capitals. It's widespread, well-supported, and a very slick movement. And now all the same procedures and strategies are appearing on the whale front."

Darling hesitates, clearly uncomfortable about discussing this subject. "Actually, I sort of wish I didn't know about it. It can be pretty depressing. But I think in many cases a biologist is probably the only advocate these issues have." Finally, he relates something that took place at the last biennial conference of the Society of Marine Mammalogists, in Hawaii. As the scientists relaxed around the hotel pool at the end of the day, Darling noticed someone was videotaping them. "Apparently it was for an advertising program in Newfoundland: 'Here are the people telling us we can't hunt seals, lying around drinking their mai tais in Maui.' Very effective propaganda."

He continues: "If you read some of the 'wise use' manuals, one of the strategies is to pit native peoples against environmentalists. They tried it here with the logging issue. It completely takes the wind out of conservationists' sails, because they don't know how to handle something like the Makah hunt. Greenpeace, for example, is left squirming because of the cultural and native rights questions involved. So Greenpeace has effectively been eliminated from the whaling issue—which is a startling success for the big money forces that are looking to open it all up again."

Darling is scarcely paranoid and, given what I'd been finding out, did not seem to be jumping to conclusions. Like the Washington scientist John Calambokidis, he was troubled about the Makah hunt directly by the likelihood that, because it occurred after most of the migration has passed, the tribe would be taking whales from a small and distinct resident population. The same would be true if the Nuu-chah-nulth Tribe set about hunting off Vancouver Island. "If we have, say, fifty whales that live between Neah Bay and Tofino during a certain season, are these the ones being targeted rather than the larger population?" Darling asks. "Unfortunately, to date this debate has really been obfuscated by the National Marine Fisheries Service. They declared there is no proof that this is a seasonal population, and the whales should not be called seasonal residents. The bottom line is, the Makah hunt has been based on politics, not biology. I don't really know what the World Council of Whalers is up to, only that Happynook is pushing the agenda."

Darling can speak somewhat freely because he is not beholden to government agencies. His gray whale work, administered by the nonprofit West Coast Whale Research Foundation, is supported in part by local whale-watch companies, which tack an extra couple of dollars onto their ticket prices to help fund research in the region. Darling continues to divide his time between gray whale studies here and humpback research in Hawaii, Alaska, and elsewhere. The latter dates back to 1977, when Roger Payne enlisted Darling to journey to Hawaii to work on the first effort to record humpback songs through an entire winter season.

Over the years Darling has developed a hypothesis about those songs that contradicts the prevailing scientific wisdom. Most investigators have suggested that the songs are sung by males to attract mates. Darling's idea grew from combining observations of intense competition between male humpbacks, including fierce physical battles on the breeding grounds, with the characteristics of the song. His research determined that not only were singers all males but the whales that interacted with them were also males. He suggests that the song could be a display reflecting male dominance status, and as *National Geographic* wrote of his work in 1999, "even wonders if singing could be the acoustic version of carrying around impressive horns or antlers." However, Darling emphasizes that this is still just an idea, far from proven at this time.

The fact that gray whale males have a completely opposite approach to breeding is, in Darling's view, "one of the most fascinating things going." Not only are the grays noncompetitive but "there appears to be a sharing of the females." While mating among grays occurs within a very short period, either toward the end of the southern migration or in the vicinity of the lagoons, mating among humpbacks continues for as long as six months. "It's all based on the varying distribution of females at any one time," Darling says, "which is also based on the variable food supply for each whale species. Basically we've got different forces at work here. If we can gain a little more insight into this, it will probably answer a lot of questions about what's important to whales."

One of Darling's "insights" along sexual-behavior lines has raised more than a few eyebrows, if not more questions, about gray whales. It was some years ago in early April. He was out along the migratory route on a picture-perfect day when he chanced upon a trio of frolicking grays. Darling first presumed this was what the literature described as a mating group. However, after he'd been observing the scene for some time, "at one point they all rolled over, laying over each other's bellies, with these giant pink penises all wrapped around each other. What's going *on* here? It was sort of startling at the time. In reality, this turns out to be very common during the migration, especially among the younger males.

"The fact is, we still know so little about gray whale behavior that it's almost shocking," the marine scientist says. "Particularly when it comes to their social organization. But the amount of money that goes into gray whale research is minuscule, especially now that they're off the Endangered Species List. Government agencies tend to respond only to a crisis, and otherwise they are just not high on the list. But you'll hear the same complaints from people who study elephants or caribou or grizzly bears. Entire university departments involved with wildlife studies have disappeared, because more limited funds these days are going to fields that potentially make money for the school—biochemistry, microbiology, whole new departments developed for genetics."

Darling peers out his picture window, beyond the owl, orca, and gray whale emblems hanging from it, through a steady rain that's been falling all afternoon. He peers out across the Duffin Passage and Templar Channel, past the old whalers' camp at Echachis Island. Over fifty-some years of modern shore whaling in British Columbia, about 25,000 whales of various species had been killed. A joint venture between Canada and Japan didn't stop until the late 1960s, when the last of the big Japanese boats departed because of "poor markets." Just over two decades later Jim Darling had flown to Japan to undertake the first intensive study there of large living whales. The publicity surrounding his work with humpbacks off Ogasawara and Okinawa was instrumental in igniting what's become a burgeoning whale-watching industry in Japan. It's been increasing by over 37 percent a year since 1991, with more than 100,000 people taking part by 1998.

"It didn't take long to catch on," Darling says and does a brief knock on wood against the leg of his chair.

The next morning finds more clouds on the horizon, but for the moment the rain has ceased. I meet Darling at the harbor, where he keeps the eighteen-foot Boston whaler he's had since 1989. He's wearing yellow coveralls and carries his Nikon camera gear in a waterproof case. He's been told the same gray whale keeps showing up day after day in Grice Bay, and he hopes to get its picture. He revs the Mercury outboard, and we head out through swift tidal currents and into the narrow Browning Passage, which runs alongside the prehistoric forests of Meares Island. A southeasterly wind sends a chill up my spine. "This sure isn't Hawaii," Darling says.

It would take about twenty minutes to reach Grice Bay. Once you knew gray whales returned to the same areas each year, Darling says, "the questions were how many and how often and why. We're still dealing with those questions." For years he'd presumed, along with everyone else, that grays were strictly bottom feeders. Darling recalls surveying in Ahous Bay on a hot midsummer day. Inside breaking waves of maybe two or three feet, he saw a trio of whales lying on their sides, with most of their girth and one pectoral fin out of the water. They were, he later wrote, "wiggling back and forth with the waves breaking along their sides as if they were jetties." Fearing they were stranded, he was about to put on his wet suit "when, with a couple of casual snakelike slithers, they backed out of the shallows, moved along the beach a short distance and made their way back into the breakers to repeat the activity." The whales were feeding in the sand of the intertidal zone, in water about five feet deep.

While it remains clear that their predominant food, at least in the northern seas, is also these tiny benthic amphipods, "the more we looked, the more we found that they're eating a whole variety of things." Gray whales have been witnessed feeding around the sound not only in shallow sand but in mud bays, eelgrass beds, kelp beds, in the ocean water column, and at the surface. Besides amphipods, they're eating herring eggs and larvae, crab larvae, mysids, and ghost shrimp—a whole community of species—and more opportunistically than has ever been observed elsewhere. Food here was abundant, unlike the situation John Calambokidis described where gray whales couldn't find enough to eat in Puget Sound.

"Yet it appears, when you watch them feeding on plankton, that they're not real good at it," Darling is saying. "With a humpback, there's this giant mouth, sort of set as a net that closes and fills. Gray whales feeding on crab larvae at the surface are biting the water, which seems an extremely inefficient way of doing it. When they're after these swarms of shrimplike mysids, you see all kinds of weird postures. They stick their noses into crevices and, I guess, suck in. But, I mean, they'll be standing on their heads with their tails out of the water, flopping around. Maybe this is the best you can do when you're a generalist and need to have equipment for several different types of food."

As we shoot past *Leviathan II*, a whale-watching ship with about twenty-five people onboard, the back of a gray emerges just to the right of us in the passage. Another vessel, *Chinook Charters*, sees it, too, and closes in on our wake. The whale forges on. "This guy's just traveling," Darling says. It starts to rain.

We're chugging past extensive tidal mudflats lined with eelgrass and sea lettuce. Small flocks of curlews, dowitchers, plovers, and sandpipers appear to be busily feeding. Darling points out that this is part of the Pacific flyway migratory route and is in fact the second largest shorebird resting area in British Columbia. He also suggests I scour the island coastlines for black bears, just emerging from semihibernation to forage for fish in the intertidal zones. Wolves, too, are sometimes seen swimming between islands, along an archipelago that stretches for fifty miles to the tip of the farthest fjords.

The shiny black head of a hair seal rises momentarily above the surface. Among these waterways are often found harbor porpoises, with their black backs and small fins, and tawny-headed Steller's sea lions come to spawn. "I've seen sea lions chasing gray whales," Darling informs me, "harassing the hell out of them and biting at their tails."

We're passing what Darling describes as a once-great fishing spot for chinook and coho salmon in the summer. "Commercial fishing is virtually gone, but sportfishing can be quite good here." I see a trawler approaching, *Creative Salmon* emblazoned along its hull. This is one of the many fish-farming boats, a relatively recent phenomenon and one that is surrounded by controversy. "There are two big issues with farmed salmon," says Darling. "One is the Atlantic salmon escaping from their pens. Tens of thousands have done so. Potentially they could breed in the wild and displace the Pacific salmon from their habitat. The other is transmission of disease from farmed to wild stocks."

We've gone about six miles, and Grice Bay is just ahead. The sun has emerged once more. It's about an hour past low tide. Darling decides to take a route into the bay around the corner of Indian Island. "It's easier to work our way in from here," he says, "because of how shallow it is. At low tide, most of Grice Bay turns into a completely dry sandbar. A lot of the water is waist-deep, and shoulder-deep even at high tide. Really it's a lagoon, very protected from storms. And there's quite a supply of food in here for young gray whales. Some have even stayed through the winter to utilize this area."

At Indian Island there's a little campground obscured by fog hovering over the hemlock, cedar, and occasional spruce tree. "Two local characters decided to make their millions off of bringing people to this spot," Darling says. "I can't say they've had too many takers, but when the whales are here it *is* spectacular. They come in and rub about a hundred feet away from you, where that spit of land goes out along the granite outcroppings there. It happens on a regular basis, scratching themselves for fifteen or twenty minutes, then going back to feeding."

What the yearling grays are eating is mainly ghost shrimp. "You can stick your hands in the mud and pull up half a dozen." Darling brings out his binoculars and scans as we move slowly down the Grice Bay shoreline. Because of all the shadows and reflected mist, the whales are not easy to pinpoint. "You can sometimes spend an hour looking for them."

The ever-changing weather is pelting us with rain again. A bald eagle soars over our heads, coming to rest at the pinnacle of a cedar.

Even with the now-incessant rainfall, there is a serenity here unlike any I've known since San Ignacio Lagoon. I'm contemplating the diversity of habitats the gray whales choose to spend time in, pristine whether desert or rain forest, hallowed with multitudes of creatures. Twice we cruise the couple-of-miles length of Grice Bay in silence, listening to the wind grown moist, listening for that sudden whoosh of breath that would take my own breath away.

Then, as we round the point at Indian Island again, the lone gray whale we saw an hour ago is up against the shoreline in an inlet just beyond the bay. *Centurion II* and another Zodiac of watchers are keeping a respectful distance. "This whale is wandering," Darling says. "It could be a migrant that's taken a wrong turn." He kills the motor, and we drift awhile. Finally, he near-whispers: "I think it's asleep. Often these young guys will lie like this on the surface, dead-still. You have to stare to decide whether it's a whale or a rock. Sometimes this goes on for long periods of time, even more than an hour."

To the hush of a whale, Jim Darling turns the boat around and starts back toward Tofino. "Best we don't disturb him," he says.

Part Two

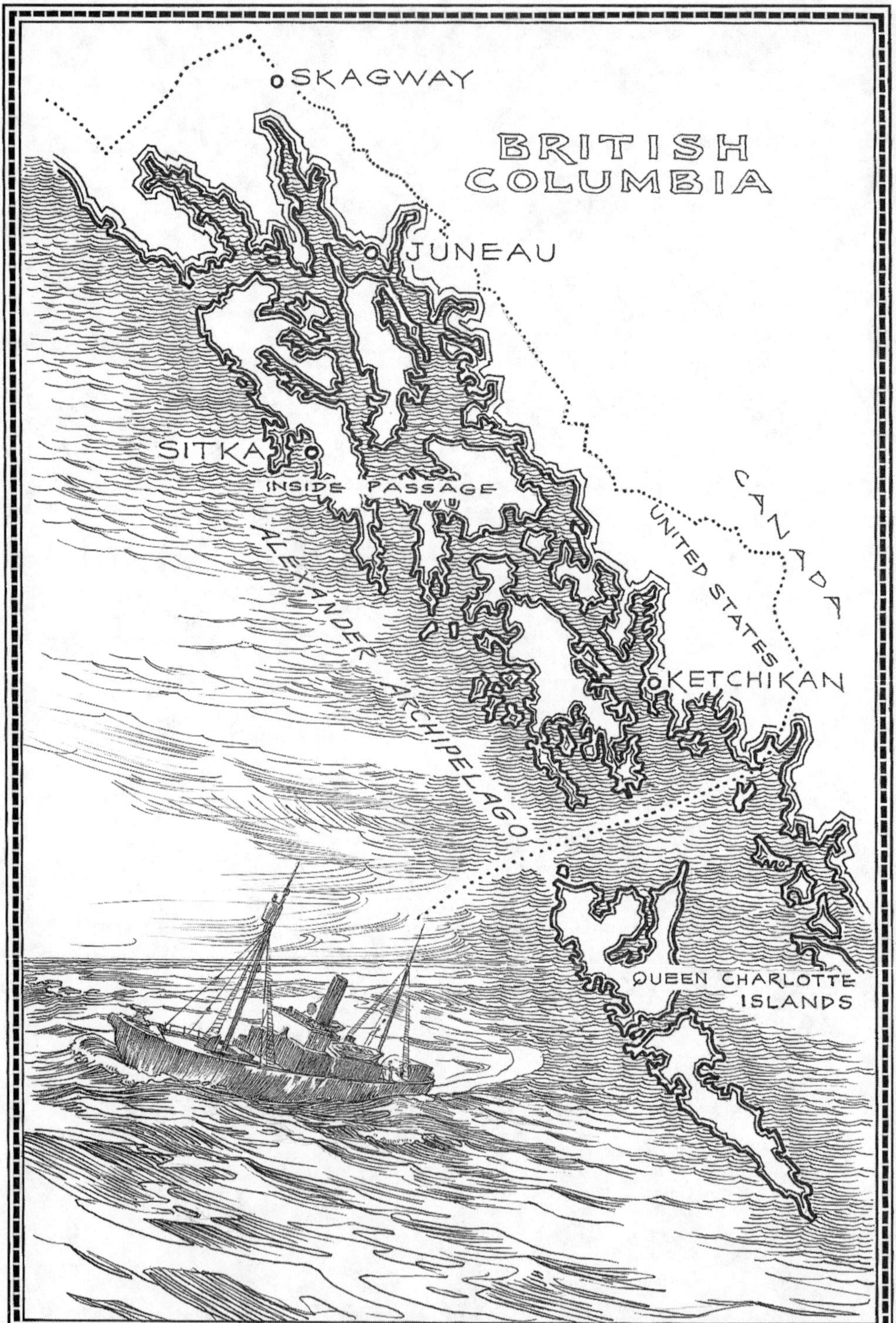

SKAGWAY
BRITISH COLUMBIA
JUNEAU
SITKA
INSIDE PASSAGE
ALEXANDER ARCHIPELAGO
CANADA
UNITED STATES
KETCHIKAN
QUEEN CHARLOTTE ISLANDS

Chapter 14

Alaskan Journey: Beginnings

Here leviathan,
Hugest of living creatures, on the deep
Stretch'd like a promontory, sleeps or swims,
And seems a moving land.

—John Milton, *Paradise Lost*, Book VII

Above Tofino nobody pays too much attention to gray whales for a while. That's because no main roads exist along the west coast across the entire upper half of Vancouver Island. At the island's apex, around Cape Scott Provincial Park, the whales make an easterly turn over to Queen Charlotte Sound. There, for some three hundred miles along British Columbia's north-central coast, they follow the same basic route northward as the ferries on the Alaska Marine Highway. No roads, and little human habitation, approach the whales' passage here either. They migrate along the largest intact temperate rain forest left on the globe, an 8-million-acre expanse known as the Great Bear Rain Forest. And not without reason: this is the domain of the spirit bear, or white Kermode, a creature found nowhere else. Actually a black bear with a double-recessive gene, which turns its coat the color of vanilla custard, it lives primarily on Princess Royal Island. Near the northern hub of Prince Rupert, named for the first governor of the Hudson's Bay Company, the grays maneuver across Hecate Strait and up and around Digby Island, peak of the Queen Charlotte chain of some 150 islands, a terrain where deer and eagles still far outnumber people and the world's largest black bears reside. Across the Dixon Entrance, marking the Canadian-U.S. border, Alaska awaits.

There the whales' pace will begin to accelerate, on the way to the Unimak Pass leading into their summer feeding area of the Bering Sea.

By land, sea, and air I intended to make my way along their route. It would be impossible to follow them directly, as much of the Alaskan coastal sector remained roadless. Nor were any research vessels operating in their proximity in the summer of 1999. Those closest to the whales, in fact, would be the native peoples of northwestern Alaska and Russia's Chukotka Peninsula, where the grays continue to be a source of sustenance for Eskimo hunters. I needed to move among them as well. I needed to learn about their relationship with the whales, whether I would feel differently among them than I had among the Makah.

I planned to be on the road for another two months. On June 10 my wife, Alice, joined me in Bellingham, Washington, where the Alaska Marine Highway begins, for the first leg of the journey along the southeastern shores of Alaska. For the next five weeks a photographer friend and I would continue on through the Aleutian Island chain and up to the Bering Strait. From there we would fly from Alaska to Sakhalin Island in the Russian Far East, where a team of U.S. and Russian scientists were studying a remnant population of the western gray whale. I'd applied for a visa to visit the impoverished region of Chukotka after that.

With the exception of Sakhalin Island, where he had earlier ventured as a whaler, this was essentially the same route followed by Charles Melville Scammon after the Civil War ended in 1865—a journey that was to shape the remainder of his days.

As he turned forty Scammon took a leave of absence from the U.S. Revenue Marine and accepted a position as chief of marine for the Western Union Telegraph Expedition. He would be in charge of a flotilla of ships that were to move men and supplies through a vast, largely uncharted territory. The idea, as conceived by a promoter named Perry McDonough Collins, was to connect North America with Europe by wire along an approximately five-thousand-mile cable. The most obvious way to hook up the two continents would seem to have been across the Atlantic, but the last of several attempts to do so had failed; in 1858, less than a month after an Atlantic cable from Ireland to Newfoundland was finished, it broke. Two years before that disaster the Russian czar had granted a charter to Collins for an international telegraph line to connect the continents by way of the Pacific. Collins merged with Western Union, and President Lincoln approved a plan to start the line in San Francisco, which was already linked telegraphically to America's East Coast. It would run up through British Columbia and on across the

Scammon at forty (courtesy John Decker)

then-Russian territory of Alaska, then along the bottom of the narrow and shallow Bering Strait. The cable would continue for eighteen hundred more miles across the steppes of Siberia to the mouth of the Amur River, where there was already a link to the Russian capital of St. Petersburg, and beyond toward Western Europe. Most of the construction would be on land and could thus be repaired. Of course, the construction would also take place under ferocious conditions of weather and terrain. It would require a small army of men, and an initial capitalization of $10 million was raised.

It was no surprise when Scammon was selected by Colonel Charles S. Bulkley—engineer in chief of the Western Union Telegraph Expedition and a stern military disciplinarian—to be flagship commander of a fleet that would eventually include seven ships. What considerable reputation Scammon had established as a whaler had only gained in stature after he joined the U.S. Revenue Marine. One San Francisco newspaper wrote of Scammon's vessel *Shubrick*: "The visitor, on going aboard, is struck with the order and neatness apparent in every department. Her commander has already so systematized everything, that each and every officer and man not only knows what he has to do, but does it promptly and quietly." Scammon was kept busy busting up mutinies in Washington's Puget Sound, routing squatters off San Francisco's Farallon Islands, and checking vessels for Confederate privateers and violations of safety and revenue laws. In 1864 he had led the rescue or relief of six ships. The most dramatic of these involved a Russian corvette, part of a squadron sent by Czar Alexander II on a friendship mission to the United States. When the *Novick* ran aground at Point Reyes, north of San Francisco, Scammon saved more than 160 crewmen and most of the ship's effects, a rescue that received publicity all the way to Washington, D.C.

So the captain already had an affinity for the Russians, whom it would be important to court in Alaska and Siberia, having even given his second son the middle name Elfsberg after a Russian captain Scammon befriended. Scammon had successfully piloted Colonel Bulkley, starting late that winter of 1865, on a preliminary three-month voyage "through almost continuous stormy weather" to New Archangel, also called Sitka, in Russian Alaska. "On arrival there," Scammon recorded, "we were visited by the then Russian Governor—Prince Maksutoff—who informed us that the *Shubrick* was the first United States Government vessel to visit Sitka."

Scammon was also known for his naturalist's propensity, and the expedition would include a six-man scientific group to explore Alaskan flora and fauna under the primary auspices of the Smithsonian Institution. One of that team's members was a young man named William Healey Dall. And the relationship that would develop between Dall and Scammon was to prove a turning point in each of their lives.

Only nineteen when the expedition began, Dall was originally from Boston. He was something of a prodigy, having begun collecting shells at twelve and been tutored in his studies of mollusks by the world-renowned Harvard naturalist Louis Agassiz. Dall had been living in Chicago, working as a railroad clerk and spending evenings at the Chicago Academy of Sciences, when the academy director, Robert Kennicott, took him under his wing. Kennicott had led a three-year expedition into central British America (British Columbia) and parts of Russian America (Alaska). When he was named director of the Scientific Department for the Western Union Telegraph Expedition, he invited Dall to join him.

An aggregation of gray whales in the breeding lagoons (© *Marilyn Kazmers/Innerspace Visions*)

A calf viewed underwater off Monterey, California (© *Phillip Cola/Innerspace Visions*)

A calf leans on its mother, underwater at San Ignacio Lagoon
(© *Michael S. Nolan/Innerspace Visions*)

Stirring up the bottom, feeding on small shrimp that live in the mud
(© *Flip Nicklin/Minden Pictures*)

Passing through California kelp beds (© *Bob Cranston/Innerspace Visions*)

A baby breaks the surface in San Ignacio Lagoon (*© Dr. Nita Lewis Miller 2000, courtesy Heidi Tiura*)

Marine scientist Steven Swartz preparing to radio-tag gray whales off California's Channel Islands (*Mike Bursk, courtesy Steven Swartz*)

A killer whale attacks a gray whale calf (© *Sue Flood, while working with Nancy Black under NMFS permit GA no. 8, file GA15, Monterey Bay, California, April 1998*)

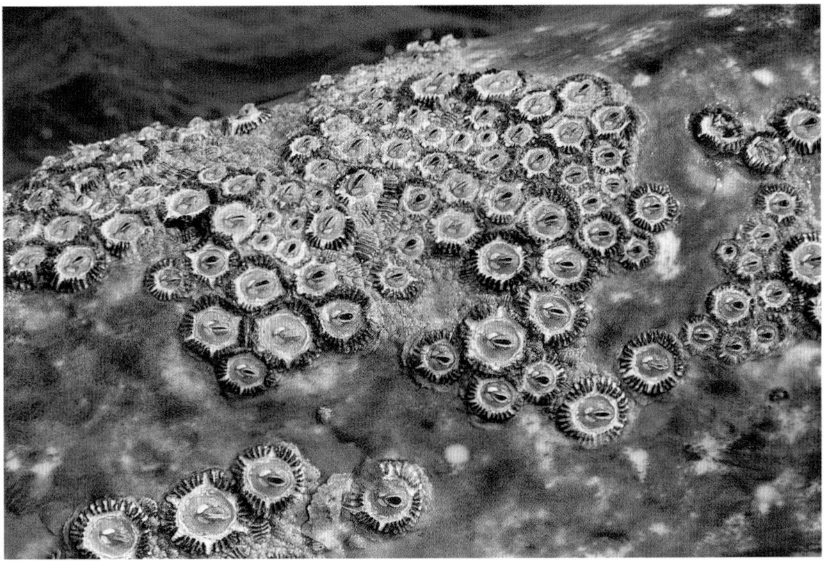

Grays are the most heavily barnacled of whales (*Steven Swartz*)

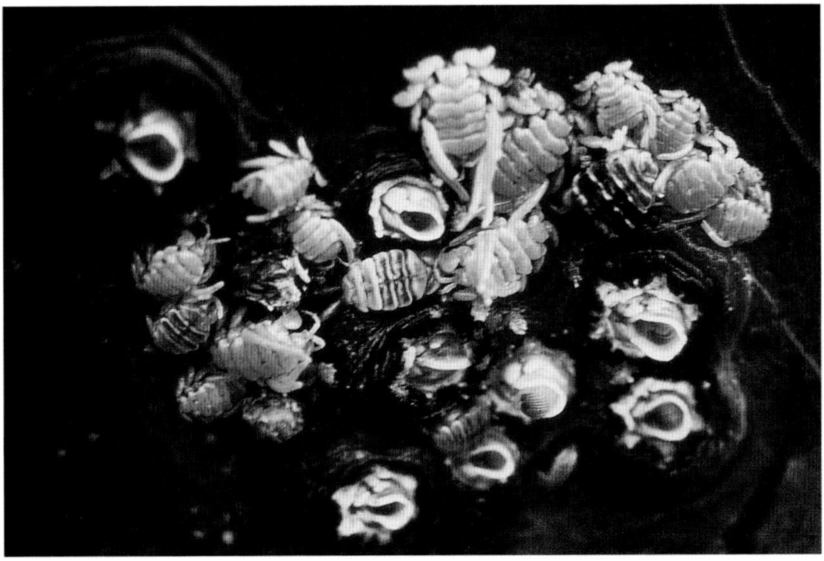

Cyamus scammoni, whale lice (*Steven Swartz*)

A rare moment of simultaneous "spy-hops," San Ignacio Lagoon (© *Michael S. Nolan/Innerspace Visions*)

Flukes up off Tofino, Vancouver Island (© *Jim Darling*)

Coming up for a "spy-hop," San Ignacio Lagoon (© *Jeff Pantukhoff/Whaleman Foundation*)

Coming in for a "close encounter," San Ignacio Lagoon (*George Peper*)

Surfacing at San Ignacio (*Richie Guerin*)

Awed by a whale (*Shari Bondy*)

John Spencer and friend *(courtesy John Spencer)*

Scientist Mary Lou Jones (foreground) at San Ignacio Lagoon (*Steven Swartz*)

A gray shows off its baleen (*Jim Sumich*)

Mother and calf greet their guests, San Ignacio Lagoon (*Brian Keating*)

Making a motorboat look minuscule (© *Jeff Pantukhoff/Whaleman Foundation*)

A meeting of two worlds (© *Michael S. Nolan/Innerspace Visions*)

William Healey Dall at nineteen
(courtesy Bancroft Library, University of California, Berkeley)

For William Healey Dall this would mark the beginning of a career as one of America's leading naturalists. From aboard Scammon's ship Dall would be credited with making the first coastal surveys of Alaska. He would soon become the leading authority on the natural features of the territory, after its acquisition by the United States in 1867. Altogether he would author more than fifteen hundred publications, including the classification sections for Scammon's *Marine Mammals of the North-Western Coast of North America*. In Alaska a major river, a large island, a nine-thousand-foot mountain, and a promontory all bear Dall's name. So do wildlife species, including the Dall porpoise *(Phocoenoides dalli)*, Dall white mountain sheep *(Ovis dalli)*, Yakutat brown bear *(Ursus dalli)*, and more than one thousand mollusks. World War II had a Liberty ship named the *William H. Dall*. Today south of St. Michael, Alaska, and facing the Bering Sea, is a prominent headland called Point Dall. Thirty miles to the north is the coastal Eskimo village of Scammon Bay.

The first mention of a relationship between Scammon and Dall appears in a letter written early in 1865 by the younger man to another of his mentors, Professor Spencer Baird, assistant secretary of the Smithsonian Institution. Dall had just spent two weeks in Monterey, California, adding "some ten or twelve species to the fauna of this part of the coast that have never been found so far north before." He continued: "With regard to the whale jaws they were eighteen feet long too big to send and have been corralled by a newspaper man for gate posts. I have made some interesting notes of the California gray whale at Monterey which I will try to copy & enclose. I obtained some baleen from a dead one which I shall send by the first opportunity."

Dall added that he thought Scammon could be relied upon for "biographical information and exterior measurements" but that he couldn't get the captain to "describe and name the whale . . . separately" from men Scammon apparently considered of more scientific repute than he was. At this juncture the gray whale had yet to receive a scientific classification; indeed, naturalists knew little about it. Dall went on to describe how Scammon had been furnishing a member of the California Academy of Sciences "with specimens for some years, supposing that when described they would be sent to the S.I. [Smithsonian Institution]." However, the fellow "has sent some to the Brit. Mus. [British Museum] and kept the rest." Dall said he was making it a point to make sure the Smithsonian would not be left out in the future.

From this piece of correspondence, we learn that rivalry was brewing within scientific circles over Scammon's collections. We also learn that Dall and Scammon had somehow become acquainted before the telegraph expedition. It's possible Dall had gained his initial awareness of the captain through Jonathan Young Scammon, the elder brother and Chicago entrepreneur who had been among the founders of that city's Academy of Sciences, where Dall spent his evenings. In a letter to Baird not long before the telegraph expedition was to set sail, in the summer of 1865, Robert Kennicott described the captain as "a brother of our Chicago Scammon" who is "deeply interested in Nat. History," planned to publish a paper on cetaceans, and "can aid us enormously in marine zoology." Kennicott reasoned that, while Scammon knew "nothing of scientific zoology, as yet," he would be an "apt scholar."

He concluded:

> This Capt. Scammon is [a] splendid man; a little hard headed where whales are concerned, but excusable for this, for his having been an old whaler. I propose, with Col. B. [Bulkley]'s consent, that Dall should accompany Captain S. in charge of marine zoology, under the direction of Capt. S.—Dall will work well himself, but will not induce others to work as readily as will Capt. Scammon.
>
> I do not think J. Y. Scammon appreciates his brother or knows how much of a man he is. Col. Bulkley chose him in preference to any Naval officer or Capt. here. It might be well for you to put a spoke in his wheel when ever you write to Chicago.

So on July 12, 1865, with "weather thick, heavy seas," as Scammon recorded in his logbook, his new 450-ton bark *Golden Gate* departed San Francisco "toward New Archangel, R[ussian] America." The captain hadn't gotten a steamer for his command vessel, but most of his other conditions had been met by Colonel Bulkley. Scammon had requested a surgeon, a paymaster, and a distinct chain of command by which all the other vessel masters would receive their expedition orders only through himself. In the absence of a regular trained doctor, who was to meet the ship at Sitka, Dall had been detailed to report to Scammon as acting surgeon. Dall wasn't too pleased with that responsibility, since his chief mission would be to coordinate the collections made on the expedition's scientific forays and to send everything back to the Smithsonian. But he knew it would provide an opportunity for him to get closer to Scammon, and for this the young commissioned Army lieutenant seemed grateful.

Dall's own father was a minister who had seen his family only at long intervals since becoming the first Unitarian missionary to India ten years before. Scammon does not seem to have been especially close to his own firstborn son and namesake, who was then fifteen. So it may have been that Scammon and Dall each filled a father-son void in the other. The two quickly found common ground, and Dall wrote at some length about Scammon to his father. "I had seen a good deal of him in Frisco, and was rather a favorite with him," he noted. "He is a self-made man, having worked his way up from a sailor . . . [and] has for some years been preparing careful accounts of the biography & habits of the N. Pacific whales."

When the ship departed, Dall wrote, "Capt. Scammon was quite worn out with previous work in fitting out his vessels, and being the only one with any knowledge of medicine on board, I went back to the city to obtain proper drugs, as the medicine chest was perfectly inadequate." The *Golden Gate* docked first at Victoria, British Columbia, dropping off a land party to explore the Fraser River region. There an Englishman named Frederick Whymper, also to become a close friend of Scammon's, joined the expedition as staff artist. Whymper would write the first account of it, *Travel and Adventure in the Territory of Alaska.*

In another letter, to the Smithsonian's Baird, Dall described how on the first leg to Sitka, "Capt. Scammon was very low, being almost worn out with overwork and dyspepsia. I was able however to cause a decided improvement, which has gained me the lasting good will of the Captain." In an article for the *Alta California* newspaper, Dall would add that the initial voyage "was made very pleasant by the uniform kindness and courtesy of Capt. C. M. Scammon." Meantime, Dall himself "found some new and very curious creatures on the surface of the ocean, three hundred miles from any land. I caught them by means of a towing net. Capt. S recovered a good deal and we had very little sickness . . . a pleasant voyage and arrived at Sitka Aug. 9th." The journey from San Francisco had taken almost a month.

The modern vessel *Columbia*, upon which Alice and I embarked for our summer voyage to Sitka, is the largest of the Alaska Marine Highway's nine ferries—418 feet long, 17.3 knots service speed, and carrying some six hundred other passengers. It took approximately two days and eight hours to get to Sitka from Bellingham, Washington, with several intermediate stops along what's called the Inside Passage. Charted by George Vancouver in the late 1700s, this was the same basic route taken by Scammon on his initial sojourn. The first leg is nonstop from a six o'clock departure on Friday evening to first light on Sunday morning. You cruise for twelve hours between Vancouver Island to the west and the mainland of British Columbia to the east, then onward into the wider Hecate Strait and past the Queen Charlotte Islands. It's a whole new world: multitudes of islands, no habitation, nothing but greenery and sea. Waterfalls cascade from snowy peaks embraced by mist. Fir- and spruce-covered mountain islands kiss the gray sky. In some places the pines appear to be growing right out of the water. There is a sense of timelessness, voyaging for mile after mile past the same vistas. In that sense it often feels as though you're on a grand river.

Oddly enough, as we look around our ship, there are comparatively few peers of our middle age. Most passengers are either in their twenties, camped in tents out on deck, or in their sixties, settled comfortably into their staterooms. Too late to obtain a cabin ourselves, we spend the nights in sleeping bags on the floor of what's called the observatory. When sunset approaches around 9:00 P.M., awaited by a pewter sky and a tranquil, silver waterway, the two predominant age-groups mingle in lounge chairs. All sit together in a respectful silence, watching an opalescence that lasts for more than two hours before giving way to an onyx, star-filled sky. We are left to ponder another meaning of "inside passage," of youth into adulthood, of old age unto death. Our ship moves on, past Cape Caution, past Egg Island.

The timber town of Ketchikan, Alaska, is the first port of call. It was originally a Tlingit Indian fishing camp, then by 1930 regarded as the salmon capital of the world. It's known today as the rain capital of Alaska—Ketchikan gets an average of half an inch per day, we're informed—and Alice and I don't feel like disembarking during a short layover to travel fifteen drizzling miles into town and look at totem poles.

Instead, we visit with the ferry's resident naturalist. He tells us that last week, for the first time in anyone's memory, a gray whale was seen here. They rarely travel the Inside Passage at all, preferring to migrate along the westerly Pacific side. "That's not a humpback, it's a gray whale!" a local crane operator had exclaimed to dozens of visitors watching the whale in the harbor. It had seemed to be resting, floating quietly near the surface, only the arched back showing. The head would slowly rise, take a deep breath, and sink again. It resembled a huge log bobbing at the surface, and indeed this behavior is sometimes called logging by marine scientists. But what the lone gray whale was doing in Ketchikan was anybody's guess. Had it somehow lost its bearings? Might it be looking for something to eat? The gray was gone as rapidly as it appeared.

Our ship moves on across Alaska's panhandle, through the coastal fjords and past the majestic stands of spruce, cedar, and hemlock of the Tongass National Forest. With 17 million acres of lush temperate rain forest, the Tongass is more than three times larger than any other designated national forest. It's also America's most heavily subsidized timber industry, with over four thousand miles of logging roads. Within the remote reaches of the Tongass are the most sizable concentrations of bald eagles and grizzly bears to be found anywhere. Stopping at the pretty little port of Wrangell, Alice and I hike a mile to beach-side petroglyphs that predate the Tlingit tribe. Like the Baja cave paintings, the images in these rocks are of unknown origin and age.

The town is not visible from the sea, being snugly nestled among the hills and islands, and the entrance to the harbor would not be easily found, were it not for Mt. Edgeconder [Edgecumbe], which towers some 2,300 feet above the sea. The channel is narrow, winding and intricate, but finally widens into a snug harbor dotted with innumerable islands.

—WILLIAM HEALEY DALL, describing his first glimpse of Sitka

On August 10, 1865, according to Scammon's journal, the steamer *George S. Wright* towed his ship into safe harbor at Sitka, on the western shore of Baranof Island, and fired a twenty-gun salute that was immediately returned by a Russian shore battery. This was the capital of Russian America. With Prince Maksutoff away inspecting various trading posts, Scammon called upon the acting governor, who "manifested a desire to assist the expedition in every possible way."

With the local Indians, however, things did not start out on a pleasant footing. One of the officers on watch, annoyed at so many canoes constantly circling their ships in the harbor, ordered the use of pumps to spray the natives with water. There was already a long history of bad blood between the Tlingits and the white man. The Kiksadi Clan had lived in and around Sitka centuries before Russians or Americans set foot on the island's rocky shores. The Indians called Baranof Island "Shee" and, in choosing the seaward side for a settlement, dubbed it Shee Atika, or "people on the outside of Shee"; Sitka is merely a contraction. In 1802 the tribe had attacked and destroyed the original Russian fort, founded as a fur-trading outpost under a charter by the czar three years earlier. Two years later Alexander Baranof returned with 120 soldiers and 800 enlisted Aleut tribesmen in 300 *baidarka* canoes. The combined Russian-Indian forces defeated the Tlingits in what was to be the last major resistance by any of the northwestern coastal tribes.

How much Colonel Bulkley had heard about this history is unknown, but he quickly set out to make amends by bringing the chiefs together with his officers aboard Scammon's *Golden Gate* flagship. Scammon decked himself out in full regalia, epaulettes and sidearms. There were speeches, tours of the ship, and "a scrumptious repast," as one crew member wrote in his diary. Scammon described "a grand Pow-wow—with Kaokhan and two other Russian Indian chiefs whose tribes are represented when in force to number some 800 to 1,000 warriors. . . . Their huts and cabins are separated from the port of New Archangel by a stockade separating the races." The natives offered entertainment for

Watercolor of Sitka by Frederick Whymper, 1865 (courtesy Bancroft Library, University of California, Berkeley)

the benefit of Scammon, Bulkley, Kennicott, Dall, and the others. "We were much surprised at the gentlemanly manner of the Chiefs at the table," Scammon recorded. "They handle knives, forks and spoons at least not awkwardly, in fact they deported themselves as well as any half civilized men and their bearing throughout was far above any savages I have ever seen, especially the war chief of the tribe called Sitkouns or Kaloshos, his general appearance was dignified and clearly was comparatively well dressed in good fitting English costume and evidently a very superior man of his race." In the margin of his journal, Scammon noted that the expedition had received presents of blankets and tobacco.

For some days thereafter William Dall wandered about the town and, "comparatively little being known of this part of the world," set down his observations for readers of the *Alta California* newspaper and in letters. Sitka's European population of Russians, Germans, Swedes, and Finlanders stood at about eight hundred, all employed by the Russian-American Company and including a garrison of soldiers. "The Russian-American Company is a joint-stock arrangement," Dall wrote, "on a plan quite similar to the East India and Hudson Bay companies." The governor was appointed by the Russian government, and his house, standing in the midst of the fortifications, "is quite a large mansion, spacious and elegantly furnished." Beyond lay "a neat counting-house, and still further on, on each side of the street, log houses, shops, etc. line the way."

In each house a guest was immediately presented with a glass of tea, which Dall found "vastly different from that insipid compound known by the same name with us. Tea here is hardly steeped, is very strong and all its original aroma is preserved." Of another Sitka custom Dall wrote: "Everywhere you go they present you, on entering, with the inevitable 'fifteen drops,' or vodka, a liquor between brandy and rum, and tasting like liquid fire. You have to take a slice of raw turnip to prevent choking which is always at hand."

The Russian-American Company's revenues, Dall continued, "seem to be derived from collecting furs, their trade with San Francisco, and the fisheries. The salmon fisheries commence about the middle of May and extend to the latter part of September. From 100,000 to 150,000 salmon are annually exported from this place to Hamburg and the Sandwich Islands. . . . It is quite a novel sight to see the long fish-boats, manned with from fifteen to twenty oars, returning at night laden to the gunwhale." As for the natives, some three thousand of whom lived in town and on various islands in the Alexander Archipelago, Dall noted they supported themselves by fishing and hunting and small gardens. Russian-American Company policy was to furnish them as much employment as possible "and to conciliate them in every way." Each worker was provided a daily loaf of bread and soup, and charged nothing for fresh fish.

Dall wrote of having "made some very good collections at Sitka and had a pleasant time there. . . . The scenery about the town is romantic and picturesque in the extreme. On every side of it, save towards the sea, high mountains lift up their snow-clad peaks, while the heavy timber and verdure, with which their sides are covered, relieve their otherwise rough appearance. The islands in and about the harbor are like emerald gems, the constant rain and dampness preserving the fresh appearance of the verdure."

Beyond Sitka the various companies of the telegraph expedition would start to separate. Some would proceed north to Fort St. Michael, preparatory to land explorations of the Yukon. Others would head for the mouth of the Anadyr River along the coast of Eastern Siberia. Dall himself was named purser of the *Golden Gate* by Scammon and wrote of him: "Capt. Scammon is the only one who really cares for the work, and he runs red tape like a chancery clerk." Added the expedition member William Ennis in his diary: "The Captain amused us by relating many of his yarns both by land and sea."

When the expedition's vessels departed on August 23, the entire population of Sitka turned out. A Russian battery at the harbor entrance gave a full but irregular salute that nearly brought down the old wharf. The *Golden Gate* returned the gesture. They were off into the Alexander Archipelago, bound across the Gulf of Alaska toward Kodiak Island, the Aleutians, and the Bering Sea.

Still no roads lead into Sitka. You can get there only by ship or plane. It's the most remote ferry stop in southeast Alaska, and the only major one to front the Pacific Ocean. To reach Sitka you've got to take a long detour from the northernmost Inside Passage juncture at the Alaskan capital of Juneau, down again and across the scenic but treacherous Peril Strait, separating Baranof and Chichagof Islands. The ferry must time its crossing to coincide with a high or low slack tide, since at one point the passage narrows to only three hundred feet wide (twenty-four feet deep). You know you've made it into Sitka's outer waters when you spot a snow-capped volcano, Mount Edgecumbe. The town, with a population slightly over nine thousand, is right in the middle of wedge-shaped, hundred-mile-long Baranof Island. Because the city/borough consists not only of this island but of smaller isles as well, Sitka lays claim to being "the biggest city in America" in terms of area. (New York City covers 301 square miles; Sitka sprawls across fully 4,710, most of it composed of trees.) Since its timber-processing mills closed in recent years, fishing and tourism are Sitka's main businesses. Several cruise ships arrive every month during the summer, each bearing almost two thousand visitors. (Passengers aboard cruise ships as tall as ten stories spend more than $160 million a year in southeast Alaska; tourism has tripled since 1980 to become the state's second biggest industry, after oil and gas.)

Atop Castle Hill a pair of bronze cannons marks the spot where first the Tlingits had their lookout post and then Baranof built his colonial governor's mansion. The house burned down long ago. But as I stand on a stone promontory looking out to sea, it's not difficult to imagine Scammon and his company making their way up here. They had been entertained in a two-story dwelling (forty-five by eighty-seven feet) built of heavy hewn logs painted yellow and crowned with a red sheet-iron roof, with an octagonal cupola on top housing a lantern to help guide vessels into the harbor. Inside the tall glass doors thick carpets surrounded a grand piano, portraits, and mirrors. A large room on the upper floor was used for receptions, plays, concerts, balls, and billiards. Other rooms housed a library, a collection of clothing from Northwest Coast native peoples, and a display of Russian and American animals.

Atop Castle Hill, on October 18, 1867, the first American flag was raised to signify the transfer of Alaska from czarist Russia. Atop Castle Hill, on July 4, 1959, the first forty-nine-star American flag was raised to signify Alaskan statehood.

Just below Castle Hill, in April 1999, underneath a bridge that crosses the Sitka Channel between Halibut Point and Crescent Harbor, a gray whale swam in—and decided not to go any farther. Between mid-March and mid-June, the majority of northbound gray whales cruise past Sitka's Cape Edgecumbe and the surrounding islands. Almost never had one simply hung around the town beaches for nearly a month as this one did, often roaming within ten feet of the shoreline. The local *Daily Sentinel* headlined a lead article GRAY WHALE FROLICS OFF SITKA.

"This whale was feeding right off people's porches," Jan Straley told me. She is a marine mammal biologist with the University of Alaska Southeast and has lived in Sitka for twenty years. "I was getting all these calls from people afraid the whale was going to beach itself. But it looked plenty healthy to me. It was about twenty-five feet long, not a very large gray, and it could move! You'd see it go under the bridge, through the channel, everywhere. The water was clear enough, you could watch it spiraling and blowing a few bubbles and coming up. It appeared to be clockwise feeding, making circles on the ocean floor as it collected food."

One of the residents observed the whale stripping something off the kelp. The herring spawn coincides with the grays' migration, and most likely, Straley said, that's what the whale was after. Last year she had done some diving from a submersible when gray whales were feeding out by West Chichagof Island. There the feasting was on cumacea, tiny invertebrates that live in burrows or mucous tubes in bottom muds or sand. "With this swarm of cumaceans, it was like diving through tapioca, and the gray whales were just hoggin' out," as Straley put it.

Walt Cunningham had taken his twenty-six-foot charter boat, *Olympic*, out for a closer look at Sitka's porch-side whale this spring. "This was like three weeks after the herring spawned," he is saying, "so they should have been hatched and they'd look like wire or lead pencils down in that kelp. Any eggs left by then would be dead, but this whale was just rolling on its side and picking up that kelp and stripping out clouds of herring fry! All the way between here and clear up toward the ferry dock, along that whole area. Some people said they'd seen *two* whales at one point. Anyway, the one gray I saw, where it was and the fact it was eating like this, was certainly *not* normal."

A gray whale observed in the Inside Passage at Ketchikan. A gray whale dining on herring fry in Sitka's inner harbor. What we were hearing about was definitely out of kilter, even though these were two isolated instances among a population estimated at more than 26,000. Still, might they be part of the trend being observed by scientists such as Jorge Urbán, Bruce Mate, John Calambokidis, and Jim Darling in other sectors of the migration?

I made a 6:00 A.M. date for Alice and me to go out looking for gray whales with Walt Cunningham. He is an electrical engineer who's spent thirty-seven years running boats instead. Indeed, behind his well-trimmed salt-and-pepper beard, his glasses, and his baseball cap, he looks like a professor disguised as a fisherman. And one with an academician's knowledge on numerous subjects, especially those relating to Alaskan natural history.

In fact, Cunningham turned out to be the fellow who re-revised the scientific conjecture on the gray whale's migration patterns—and proved that a much earlier observer named Scammon had been right all along. Here's what Scammon had to say in *Marine Mammals*: "Occasionally a male is seen in the lagoons with the cows at the last of the season, and soon after both male and female, with their young, will be seen working their way northward, following the shore so near that they often pass through the kelp near the beach. It is seldom they are seen far out to sea. This habit of resorting to shoal bays is one in which they differ strikingly from other whales. . . . The mother, with her young grown to half the size of maturity, but wanting in strength, makes the best of her way along the shores, avoiding the rough sea by passing between or near the rocks and islets that stud the points and capes."

Scammon's view was largely accepted until Dr. Raymond Gilmore came up with a different notion in the mid-1950s. "We know little of their exact course through the Bering Sea and across the Gulf of Alaska," Gilmore said. He suggested that, rather than continue to hug the interminably roundabout and lengthy coast of Alaska along the continental shelf, most gray whales abandoned their coastwise nature at Vancouver Island and made a straight-line westerly run across the Gulf of Alaska's deep-ocean waters, following the currents while passing through the Aleutian Island chain into the Bering waters. Walt Cunningham was working for the Alaska Department of Fish and Game when he came, unexpectedly, to challenge Gilmore's theory.

As we pull out of Sitka harbor, fog hanging low on a misty morning, Cunningham tells the story: "My wife, Susan, and I spent a number of years on Kayak Island, a ways northwest of here along the coast. It's just this side of Prince William Sound, where Steller and Bering first hit the continent. We were studying seabirds and sea lions around Cape St. Elias, near this 150-meter-high, pyramid-shaped rock called Pinnacle Rock on the nautical charts. When we went there the first time in 1977 and came back after three months, we said we'd also counted twelve hundred gray whales going by. All the scientists said, Bullshit! At that time, according to the latest literature, gray whales came out of Cape Flattery and went straight across the Pacific until they hit Unimak Pass way out there, and right on into the Bering Sea. Well, we went back the next year to Kayak Island and set up a core study. That time, we counted twenty-one hundred grays! So people began to take notice. It's just that nobody had ever really collected the data. Over time it's been realized that gray whales do *not* go across the Pacific but literally follow the fifty-fathom curve all the way up Alaska."

As he later wrote in a report, Cunningham also observed three types of behavior "that appeared feeding oriented." Most prominent of these was "surface gulping," along with less frequent "surface skimming" off the reef extending from Pinnacle Rock. These findings, too, flew in the face of then-gospel knowledge that gray whales didn't feed along the migration, and certainly not on the surface. Additionally, Cunningham saw what he termed "paired lateral rolling," "fluke display," and "spy-hopping" that seemed attached to sexual behavior. This marked the first time such had ever been reported in Alaskan waters. It peaked toward the end of April, during a third pulse of migrants consisting mainly of younger, smaller whales of both sexes. Cunningham suggested—now proven correctly—that "it may be merely sexual experimentation by subadults and other immature animals." Among other varieties of fore-play, he recounted: "For twenty minutes, two whales repeatedly rose and sank in the spy-hopping mode. Suddenly, with their ventral surfaces opposed, they rose vertically out of the water beyond the level of their flippers and fell backward, away from each other."

Now, standing at the helm, Walt Cunningham has his map folded out in front of him. He points and says, "Going south, starting here around

Peisar Island, is where a bunch of grays apparently spent last fall. They may even have been there over the winter, it's just nobody goes down there. Sometimes we see young humpies here all winter long, even one by itself. *Resident* may not be the right word, but patterns and distributions are changing."

We cruise past the Sitka Sound Seafoods processing plant and dozens of trollers and seiners and long-liners lying at anchor. Sitka has more dock space than it does road system; the docks would stretch for twenty miles end to end. Soon we leave the harbor and enter a world of small, tree-filled islands where the mist rests on the tips of the Sitka spruce in feathery billows.

"I mean, think about the changes we've seen with the eagles," Walt continues. "There are more bald eagles than there have probably ever been in Alaska. One of the most obvious reasons is, they've gotten the DDT worked out of their systems. But they've also learned to live with *us*. Twenty years ago, if you saw an eagle down along the shore and started toward him, he flew away. Today he's just as happy staying right there. We're now feeding them, too. Year-round our two fish processors here take the heads and guts of the fish brought in by the commercial fleet, grind it all up, and pump it out in the channel, where it bubbles up, and the eagles have learned to come in for it."

You just never know, Walt says, as slowly the fog begins to lift passing Makhnati Island. While studying sea lions back in the 1960s, he was astounded to find that they often swallowed—and later regurgitated—as much as eight pounds of rocks they'd use as ballast on their dives for fish. He slows his engine as we approach what looked at first like a floating log but turns out to be a lone sea otter. It's lying on its back, large black flippers sticking up out of the water like oversized shoes.

Walt says the otters generally sleep for a couple of hours like this on their backs. Often you'll see them in little colonies called rafts, all wrapped up in kelp and snoozing together. They segregate themselves by sex. Males are about a foot longer and twenty to thirty pounds heavier than females. Sea otters can't sleep for long, because they have an incredibly fast metabolism. With an average body temperature of 104 degrees, they consume about one-quarter of their body weight every day, mostly shellfish off the bottom. "They're eating machines," as Walt puts it, sometimes using rocks as tools to break open their sea urchins, abalone, clams, or crab. They rely completely on their fur to keep themselves warm. Fortunately nature has endowed them with the thickest fur on the planet—as many hairs in one square inch as a German shepherd possesses on its entire body. This same soft, smooth fur was why the Russians first settled in Sitka; one ship's hold of those pelts would fetch a year's pay. By 1911 only about two thousand sea otters remained at widely scattered sites in the North Pacific. Restrictions on hunting, and reintroduction programs, have restored the population to about 150,000; around 8,000 inhabit southeast Alaska.

The fog, wind, and accompanying heavy waves have subsided. At the mouth of Sitka Sound, we approach a small island of breathtaking beauty. This is the sixty-five-acre St. Lazaria National Wildlife Refuge, the island created when lava pushed up through the seafloor in a process much the same as that which created the Hawaiian Islands. As the clouds part, St. Lazaria reveals itself lying near the volcano crowning Kruzof Island just to the north. It probably last blew ten thousand years ago, Walt says, and is technically referred to as dormant. Ancient lava has formed arches, tunnels, and grottoes in elegant, compelling shapes on nearby St. Lazaria. The sheer cliffs are lined with salmonberry bushes and crested by spruce shaped like brushstrokes in a Japanese painting. These cliffs are a nesting ground for up to half a million storm petrels, murres, auklets, and puffins. It is a bird-watcher's paradise. "Magic," Walt pronounces in a reverential whisper.

He cuts the engine, and we stand silently watching thousands of birds, clinging to cliffsides, diving for prey. Some species will fly around for days before coming in for a landing. Walt points out all the ledges and cracks that provide different types of habitat for burrowers and for sitters. There is the rhinoceros auklet, which digs burrows and fishes the same way as its tufted puffin "cousins," but "instead of having the cuticle on his beak, he's got that little fleshy horn above it, hence the name." There are the sooty shearwaters, known as mutton birds in their nesting area in New Zealand. The common murres are sometimes referred to as the Exxon birds—"because that was the guy covered with oil in all the pictures you saw." There are the ancient murrelets: "The chicks hatch with most of their down and a lot of their feathers. A few days after they're born, the parents come in from the sea and call for them. And these tiny balls of down leave the nest, cross the beach, go out and join Mom and Dad, and you won't see them again until the following spring."

As we round the Pacific side of St. Lazaria, the sea turns rough and the lava formations loom suddenly craggy and violent. Walt brings out the map again and points south, toward Peisar Island. "At first I didn't think those were gray whales there last fall—the spouts looked strange. But then I started looking more closely—yep, two adults and a smaller one. They were still here the last time I looked, at the end of *December.* This was very atypical. Except, when you think about it, these inner lagoons begin to look like where they go down in Baja. These whales were in a sort of lagoon, except way north of where they should've been. I don't know what it means, if anything. Anyway, we could go take a look."

We reach an area around Peisar Island called Frosty Reef when I see a spout ahead. "Blow!" I call out. Walt closes in. It turns out to be not a gray whale but a humpback. A younger one is also in the vicinity. They're feeding amid the kelp very near the shoreline. It's a pleasure to see them. Still, I hunger to see gray whales in their northern territory. They feel just out of reach.

"That narrow slot where those whales are feeding is locally called the Keyhole," Walt says. "That means a point where you turn around and go back."

And so we did, six hours after we set out, as the noon hour approached. It took nearly an hour to return to Sitka harbor. As we came up on Castle Hill, Walt described how extreme high tides long ago would cover its sheer rock face. Several years back, when the National Park Service began making a handicapped access ramp, they came across "all kinds of archaeological stuff" in what's now claimed to be the most significant "first contact" find in Alaska. "They came upon an old Russian foundry and evidence of a tanning shop, parts of women's dresses, complete fur sealskins that were headed for the market, and one of the largest pieces of what's called raven's tail weaving ever seen. Sixty thousand artifacts the first summer, seventy thousand the next. The whole community was over there daily, watching to see what they'd discover next."

I thought of Scammon and Dall wandering over Castle Hill, and of the gray whale that spent a month this spring feeding where the *Golden Gate* had once docked. Now the whale had moved on, and, presuming it was continuing north, I calculated it might well have reached Kodiak Island by now. That was some four hundred miles from Sitka—and my next stop.

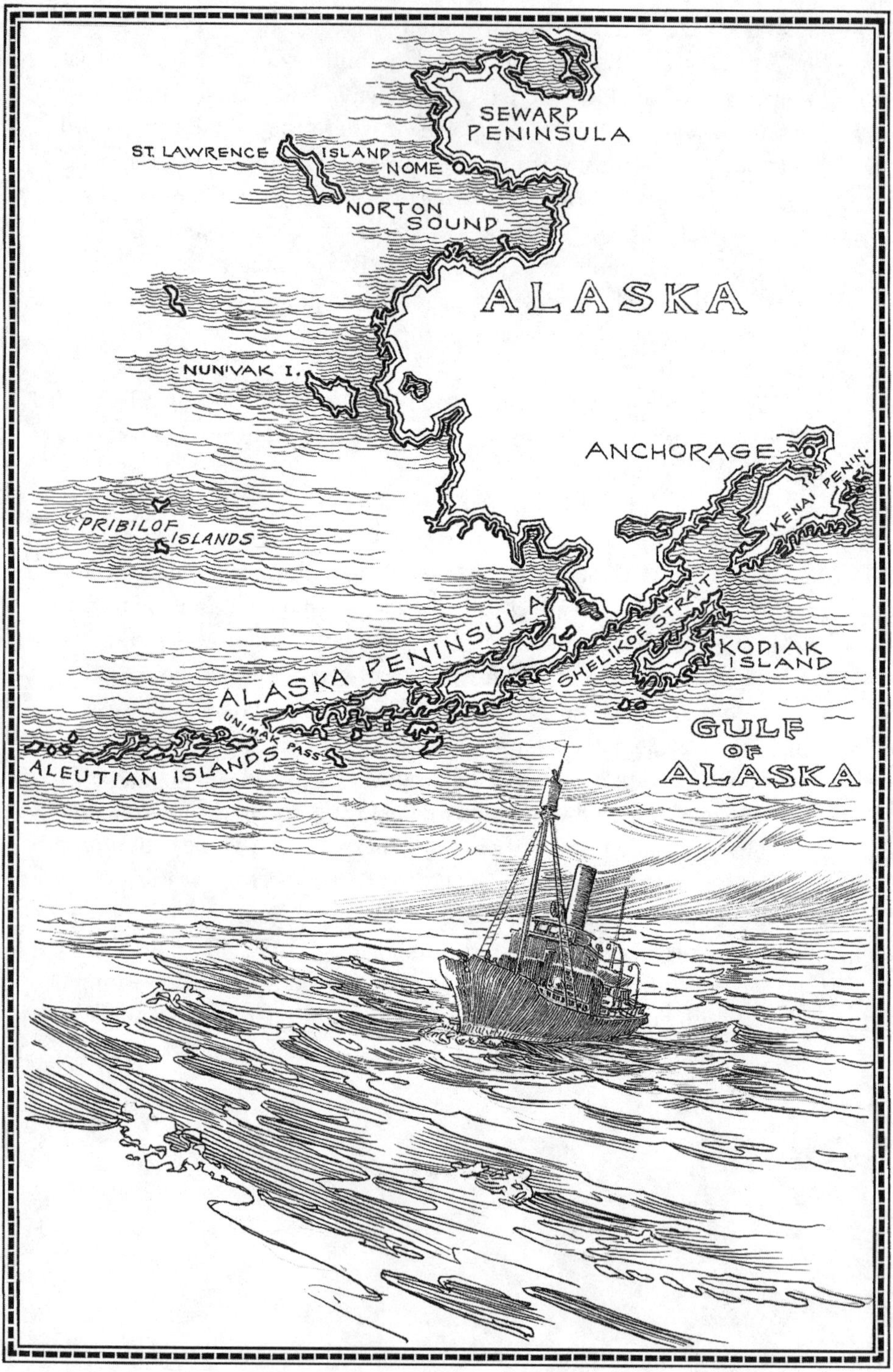
SEWARD
PENINSULA
ST. LAWRENCE ISLAND
NOME
NORTON
SOUND
ALASKA
NUNIVAK I.
ANCHORAGE
KENAI PENIN.
PRIBILOF
ISLANDS
ALASKA PENINSULA
SHELIKOF STRAIT
KODIAK
ISLAND
UNIMAK PASS
ALEUTIAN ISLANDS
GULF
OF
ALASKA

Chapter 15

WHALES IN STRANGE PLACES: KENAI, KODIAK, AND UNIMAK PASS

Kodiak—or the Great Island, as the natives call it—is about seventy miles long and fifty wide, and is separated from the main-land by the Straits of Chelikoff. . . . The mountains crossing the central portion, from north-east to southwest, are covered with perpetual snow; foot-hills, high and precipitous, cover the remainder; and nowhere on the island can any great extent of level land be seen. Surrounding Kodiak are numerous smaller islands, many of which are nothing but immense bowlders, elevated above the water. . . . The navigation is extremely dangerous. The tides, which sometimes rise to the height of thirty feet, rush with great velocity and irresistible force through the numerous channels, creating "tide rips," which the sailors dread. The south face of Kodiak is exposed to the swell of the Pacific, unresisted for thousands of miles; and, during the storms that frequently rage, the thunder of the surf can be heard far inland.

—"Kodiak and Southern Alaska," *Overland Monthly*, June 1872, found among the papers of Charles Melville Scammon

AMID the rugged shorelines of Alaska, gray whales move largely incognito. Walt Cunningham's landmark mapping of their northbound migration had observed their turning northwest after passing Cape St. Elias, paralleling the coast toward the Copper River Delta. Avoiding direct entry into Prince William Sound—site of the 1989 *Exxon Valdez* oil spill—they pursue a southwesterly course from Cape Hinchinbrook across to Montague Island and down into Blying Sound. They then follow the coast of the Kenai Peninsula.

That is where the marine mammalogist John Hall picked them up heading north as he was making an aerial survey during the spring of

1977. He recalls flying parallel to the coast, west to east about two miles offshore, and seeing nothing for fifteen or twenty miles. When he thought he might have glimpsed some blows way inshore, Hall turned around. "All of a sudden, gray whales were everywhere," he remembers. "This was around Gore Point, one of the worst stretches on the planet in terms of bad ocean. It gets hideously big swells, twenty to thirty feet high, rolling and crashing against these huge, very blocky cliffs. And what we saw were gray whales, at the last second, rolling out of the backs of these swells just before they broke on the cliffs. The whales were literally on the beach!"

Probably no one has stayed more abreast of the gray whale's migrations in the years since than Dr. David Rugh. He is a reddish-haired, soft-spoken wildlife biologist who, since 1976, has worked out of the National Marine Mammal Laboratory's complex of offices on the shores of Seattle's Lake Washington. Rugh (pronounced "Roo") has been as influential in the methodology of counting whales as Bruce Mate has in tagging them. It's Rugh's theory that the grays' unique migratory pattern each fall is a primary reason this population has increased so dramatically when other whale populations have not.

"Think of their migration as being a kind of funnel, which draws them together on a tight coastal corridor," Rugh says. "This increases their chances of finding each other when the timing is right for mating to occur. Most females become pregnant during the early part of the southbound migration, generally sometime in December, and they give birth about thirteen months later, late in the following southbound migration. Compare the grays' situation to the North Pacific right whales, which migrate on parallel lines north-to-south through the open sea. That's the saddest case, because the opportunities for them to come across each other are pretty low. Their population numbers reflect that. There are so few right whales in the North Pacific Ocean they are probably on their way to extinction, whereas gray whales have been removed from the Endangered Species List and are not even considered threatened anymore."

Unlike other whale species he's watched on the move, Rugh says, gray whales have a unique type of pulsing. This is a phenomenon wherein there might not be *any* whales for an hour, then they're "peppering the water" during the *next* hour. "For a while group after group seem to be coming along in a row. Then it trickles off, and there isn't much for a while. It's something other than random. This pulsing indicates to me that there's some gregarious sense of interaction among them."

Such as extended families that like to travel together? I wonder. "Yes, those types of associations seem like a possibility," Rugh continues. "In social terms, grays are often in pairs, trios, or groups up to about a dozen whales. A few or many groups are seen in a pulse. The proximity between grays is much closer than among bowheads, for example, who rarely travel in tactile proximity; however, often bowheads seem to stay at least within acoustic range of each other—which could be several kilometers. Belugas sometimes migrate in huge pulses of hundreds or even thousands of whales, but with few or none for days between pulses. Other whale species tend to be more loners, with the exception of mating periods. Whereas with gray whales, it's almost tactile. As you watch them moving, they look as if they could be rubbing together."

Another marine scientist, Dr. Michael Poole, spent seven years observing the migration during the 1980s and noticed "that individual grays will stop migrating and wait for another individual or pod to catch up and then move on together, or individuals will leave a pod and increase their speed to catch up with a pod that is further ahead, and then will travel together."

This pulsing—these rows of grays that appear to be tracking one another on a parallel course—is mysterious. "Enough time has gone by between whale groups, you wonder, how they can follow each other?" as Rugh asks. Poole likewise detected "remarkable similarities in routes taken by pods," but if *too* much time had passed, the trail seemed to grow cold.

It's been speculated that grays are leaving a scent for other whales to follow. This phenomenon is known by scientists as chemoreception. Unique to gray whales among cetaceans—indeed, not known to occur on any other mammal—is a peculiar thick-walled, saclike structure embedded in the blubber of the tail behind the anus. It contains a fluid and was called the stink sac by flensers in the whaling days. Termed a postanal gland by biologists, the organ has been found in both adults and near-term fetuses of both sexes, always in the same general anatomical location.

Poole and a microbiologist, Jack Beierle, tried collecting water samples in the "footprints" left by passing whales and analyzing these for the presence of the fluid, but either none was present or the quantities were too weak to be detected. They even placed collected samples of the fluid into the water in advance of approaching whales and sought to monitor their responses; again, the results were not clear.

However gray whales maintain contact, equally fascinating to Dave Rugh are the often treacherous storms and currents through which they find their coastal route. Other scientists have wondered whether they might be following the earth's magnetic field. Experiments with homing pigeons have shown an ability to do this. Many birds and some whales carry tiny particles of magnetite, a black oxide of iron, in their brains. Like other iron oxides, magnetite particles are strongly affected by magnetic "clues." Fin whales in particular possess these particles and follow magnetic contours. It's not known if the grays have the same cranial material.

However, looking at a map of the migration, Rugh points out that the grays travel in numerous directions— "north, north-northwest, northwest, west, here almost headed back southwest. So the theory that they navigate according to magnetic orientation would mean they have to know where they are in latitude and longitude, to know which route to take and which magnetic bearing they should be on. It would be hard for them to be able to reorient their magnetic compass as they travel the full migration route. Whereas, a theory that says they're following bathymetry [bottom contours] makes more sense.

"Gray whales must have a navigation device other than just following the surf line," Rugh reasons, "because it would take several *lifetimes* to follow the surf all across British Columbia and Alaska. It must be that they are oriented to depth contours. This would enable them to avoid circling islands and the convoluted course around rocks, shoals, and all the fjords. Gray whales do a series of surface dives, where they catch their breath as they swim along, and then they do a deep dive. It's in the deep dive that they probably get down close enough to the bottom that they're aware of it."

Rugh has seen young grays wandering in and out of inlets, in very shallow water, as though they're still trying to sort out the route. Whales that go from headland to headland, crossing canyons and skipping some of the contours, are the larger, older ones, who've apparently learned from experience more efficient ways of navigating through a migration. Grays often take shortcuts in California while coming north, and they make shortcuts in Alaska on the way south. Rugh says, "They seem to have remembered where they can take these shortcuts."

One such shortcut, which takes the grays across open water rather than following a coastal inland route up and around Cook Inlet, brings them farther south from the base of the Kenai Peninsula. Crossing the Kennedy Entrance, which leads down through the Barren Islands, it's about fifty miles altogether to reach Point Banks, at the uppermost portion of the Kodiak Archipelago.

Situated in the Gulf of Alaska about 250 miles southwest of Anchorage, the archipelago encompasses more than five thousand square miles. It's a wild and stunning landscape, especially when first witnessed from the air: jagged snow-covered peaks, alpine lakes, fjordlike bays, coastal wetlands and meadows, some four hundred rivers and streams, and wide U-shaped valleys left by glaciers that covered most of its islands ten thousand years ago. Kodiak Island, largest of sixteen major islands, is also the second-biggest island in the United States after Hawaii. Kodiak was the first capital of Russian America, preceding Sitka, and a major center for the fur trade. Today a full two-thirds of this island's 3,588 square miles is composed of the Kodiak National Wildlife Refuge, for the most part accessible only by float plane. It was established by Franklin D. Roosevelt in 1941 to protect, foremost, the Kodiak brown bear—some three thousand of which forage on the archipelago along with about sixteen thousand humans. More than half of those people live either in the town of Kodiak or on Alaska's largest Coast Guard base, both at the northeastern edge of the island. That's also where, at the base of Pillar Mountain, lies the third most profitable commercial fishing port in America—nearly eight hundred vessels that in 1999 brought in 331.6 million pounds, worth $100.8 million, most lucrative being the salmon harvest and the deepwater trawling fishery for pollock and cod.

Located on the edge of the Aleutian Trench, aptly nicknamed Chain of Fire, the archipelago is heavily influenced by both volcanic and seismic activity. In 1912, when Mount Novarupta erupted on the Alaska Peninsula just across the narrow Shelikof Strait, the town of Kodiak was engulfed by nearly two feet of windblown ash. The massive Alaskan earthquake on Good Friday of 1964 triggered tidal waves that sent residents scurrying to the top of Pillar Mountain. Still, Kodiak's reputation as the Emerald Isle of the North has remained intact. The volcanic ash, in fact, created a lime-filled soil that now enhances the greenery. Kodiak, I would find, has many such dichotomies.

The winter of 1998–99 had been the coldest in decades here on the island, and sea surface temperatures in its bays were the chilliest in a hundred years. As the second week of June began, so did the rain. Kodiak receives an average of about sixty-eight inches of rain a year, but never so incessantly. This cloudburst wouldn't let up for twenty days. It was pouring when Alice's and my two-hour flight from Anchorage touched down, along a runway so close to the water that the landing gear seemed almost to be riding the surf in. We came to a stop beneath Barometer Mountain, so named because you can see its nearly 2,500-foot peak only in good weather. Not today.

From our bed-and-breakfast I placed a call to Eric Stirrup, the first charter fishing skipper on Kodiak to promote whale watching as part of his trips. Miraculously, the sun decided to make an appearance the morning we ventured out in his fifty-foot boat. *Ten Bears* was heading west, in the direction of Whale Island, with twin 455-horsepower engines that Eric said could top out around twenty-three knots if needed. He predicted it would be about a two-hour run to Whale Pass, adding: "I am of the old, not bold, pilot school. I've spent my fair share of nights in over a hundred knots of wind and lived to tell about it. And I get more and more chicken as each year goes by." Now in his mid-forties, Eric wears a light beard on a broad, cherubic face. He's a salt-of-the-earth type ready with incisive and thoughtful commentary on the plight of the ocean. He's got a degree in marine biology but reminds you that there's more than that to science—that men like Scammon, who live and love the sea, are often the best observers of nature.

Eric makes his own rules on what size halibut you can keep on his boat, which he said had earned him the designation Kodiak's black sheep charter skipper—because customers all want to kill a three-hundred-pound fish. "With me, you get one any bigger than 150 pounds—here's the camera, let it go. Because these are the big breeders, thirty million eggs a year. My boat rule is, you can't keep a fish thirty-two inches or under either. I'd rather go home with nothing."

Eric figured we might run into some minkes, fin whales, or humpbacks around Whale Island. He wasn't so sanguine about grays, although a couple of weeks ago he ran into several off Raspberry Cape. "Normally we don't see any here past Mother's Day, but this year there's still a few straggling by. You can't hardly get close to a gray whale here in Kodiak, though. Fifty yards at best, and you've gotta be real sneaky and figure out where they're moving, then move out in front of 'em and get everything shut down and quiet. Then they'll come up for a minute and be gone. They'll usually go down and you won't see 'em again for a quarter or half mile. They just don't like being around boats that much. I don't know where those 'people-conditioned' whales from Baja are, but they're not here."

Could their keeping a distance be some kind of holdover from whaling days? According to records kept by Kodiak's American-run whaling station at Akutan between 1912 and 1939, a total of 6,188 whales taken included 2,498 fin whales, 1,510 humpbacks, 835 blue whales, and 482 sperm whales. Surprisingly, not a single gray whale, though I wondered if that was merely an indication of how few remained at that time. "I imagine," Eric said, "it's also because when the grays were going by Kodiak, the weather was too rough to run those boats." The more likely reason for the grays' nonchalance about Kodiak whale watchers, Eric thinks, is that the whales are simply "all business" when migrating.

During the spring of 1999, somewhere between fifty and one hundred gray whales hung around the two-mile radius between Kodiak Island's Narrow Cape and Ugak Bay. Clouds of little krill, known as euphausiids, were in abundance, and the whales definitely seemed to be eating. More were thought to be bottom feeding around Fog Neck Island. I'd been told that they'd even shown up around the mainland town of Homer for the first time, also apparently to feed.

As we swung around Citrus Rock (named after the Coast Guard cutter *Citrus*, which hit it and almost sank), Eric began speculating on the reasons for some gray whales' altering their habits. "Maybe as the gray whale population grows and returns to a more normalized level—or maybe the normal level has been exceeded and they're on a peak—they're starting to use fringe environments. Maybe it's too crowded in the Bering Sea, so they're being forced to expand their range. We don't know historically how many gray whales existed and what their real range was. But look, with climate change like we're seeing here, everything's out of whack. Last year in April the water temperature was thirty-nine degrees Fahrenheit, and it was *thirty* this year—a nine-degree difference! Last year we didn't have much phytoplankton, with fog and rain all spring. This year we had a lot of sun, but it was cold and very windy, and phytoplankton was so rich and thick that it would plug fishing nets. Why are we seeing more and more glaucous gulls around Kodiak? This is an Arctic gull that spends its life up near the ice edge and never used to come south. Well, are they starting to push their range because of food demands? Bering Sea plankton has not been healthy the last few years with warmer waters."

The glaucous gull, I realized, is a seabird that is largely dependent for food on what gray whales leave behind after surfacing in the Bering Sea.

Eric said, "You can see Whale Island over the top of that peninsula there."

It wasn't long before I called out to Alice to come up to the wheelhouse, there were whales ahead. Eric was as excited as if he were seeing this spectacle for the first time. "Oh, we got a big bunch up in here! Different kinds, I think. See how tall and straight those blows are? Like cannon fire. I just saw a humpback tail! They come up with kind of a rounded, teardrop spout. Oh, that's classic humpback, you can see the curls at the top and they're getting that nice roll in their backs. I've heard 'em sing in the Marmot Straits—killer whales are also real chatty—yeah, there's whales all the way down the edge of Whale Island there. They're feeding over deep, deep water. We'll work our way over quietly, right past those puffins. See if we can figure out who-all's here."

As *Ten Bears* moved in closer, Eric put a hand in the air and seemed to be holding his breath. "I *swear* that's a sei whale!" Alaskan whale scientists had been telling him sei whales didn't come into this particular region, so he had to be seeing fin whales. The two are similar in many respects. "You can't tell the difference in the blows." Eric was almost whispering. "You've gotta have good light to distinguish the markings. Seis are blotchy, brassy, real golden in color, whereas fins are dark slate with clear white tummies. Seis are like motorboats. Fins are more gentle and lethargic. Look at that!"

Yes indeed, two of the whales were showing their arched backs, very much the mottled brown color of the sei whale that Eric described. Slowly the arches decreased as the whales submerged and disappeared into the mist. Eric reflected, "See how they came up and didn't really lift the fluke, but angled it up so gracefully? Now when a fin whale goes down, it just disappears, goes forever like a giant submarine. Those sei whales are feeding down there at forty fathoms, then coming up and resting."

Eric said he had a few books in the cabin below, where we could make some additional comparisons. As the whales faded from our sight, I walked down to see what I could find. There on a shelf was a reissued paperback edition of Scammon's *Marine Mammals of the North-Western Coast of North America.* I began leafing through the pages, as I had so many times before. Oddly, I could find no reference to sei whales. I did, however, come across Scammon's description of "the Finback, or Finner," as the fin whale was then called. "Color of back and sides, black or blackish-brown . . . belly, a milky white." I knew that was *not* what we'd just witnessed off Whale Island. Here in the North Country, the variegated species mix in a crucible.

> *The fisheries around Kodiac* [sic] *are the most prolific and valuable on the northwest coast of America, as the Island occupies about the central position of the salmon run and of the codfish grounds. . . . The population consists of Russian creoles and Aleuts, who are principally gathered in five villages and towns. . . . whale fishing is a general occupation. The flesh of the whale is used for food, and the oil for culinary purposes, being preferred even to butter. The creoles eat the oil in moderation; but the Aleuts are so fond of it, that they will dip it up and eat it like soup.*
>
> —From an 1865 newpaper article pasted by Scammon into a scrapbook

About 2,200 Aleuts still live in six villages scattered around the Kodiak Archipelago. In the basement of a Russian Orthodox seminary in town, a little lady of about seventy-five is immersed in old manuscripts about their ancestors' whaling rites of passage. Lydia Black, originally from Kiev, Russia, prefers to describe herself as a theoretician whose area of expertise is symbolic systems. She is the author of *Glory Remembered*, an analysis of the wooden headgear that Alaskan native whalers and warriors wore at sea. She has also written a lengthy study of whaling in the Aleutians. Black has concluded that the majority of the Aleut peoples haven't whaled since prehistoric times, although they did utilize whale carcasses that drifted ashore.

It was a different story on Kodiak Island. As one of Black's students, Dominique Desson, put it in a doctoral thesis titled "Masked Rituals of the Kodiak Archipelago": "Whaling for the Alutiiq speakers of the Kodiak Archipelago was much more than a search for food. It was an exhilarating, frightening, awesome enterprise intertwined with solitary ritual activity, and replete with danger, which placed the whaler in immediate contact with the spiritual world."

According to Desson's and Black's research, the Kodiak whaler often went out alone after his prey, using a harpoon tipped with a vegetatively derived poison called aconite. He would seek to strike the whale on or near the side fin or flipper, where the poison interfered with the animal's control of motion. If successful in landing the whale, the islander would then excise the area of the body surrounding the wound. Either this portion would be discarded or the hunter himself would eat it. If he suffered no ill effects, distribution of the meat would commence. (Other scholars believe that the poisoned carcasses that eventually drifted ashore south of Kodiak actually discouraged further native whaling between the Gulf of Alaska and Vancouver Island.)

"These whalers were very much like shamans," Black explains. "Before a hunt, they would live apart in remote and inaccessible places, at the top of a high cliff or in a forest. Sometimes, wearing the costume of a crab, they ritually disinterred the bodies of important people. They placed these bodies in a secret cave, obtained oil from them, and kept the mummified remains or skulls as their talisman. The spear point to be used during the hunt was smeared with human blood and then left to stand by itself on a shelf for a while. During each phase of his preparation, the whaler pronounced incantations. Before going out in his *baidarka*, he had to stay awake for four days; otherwise, the whale would get away. In isolation he was cleansing himself of any possible human interference or 'vapor,' which would be unattractive to the whale."

At last the hunter placed a magnificent high-crown bentwood hat upon his head and "transformed" himself into a killer whale to be able to face his mighty prey. Once on the sea, he first had to symbolically trap his whale within a defined area. Desson writes: "This was done by pouring human oil across the mouth of the bay in which one wanted to entrap a whale or the whale was ensnared by a circle drawn on the cover of the *baidarka* with a piece of human fat." There were more incantations while the whaler threw his harpoon, again after his quarry had been struck. A spiritual entity was called upon to help tow the whale to shore. Entering his home, the whaler made five circles around a fire and would isolate himself without food for another three days.

"It could be," Black says, "that the introduction of whaling techniques to southwestern Alaska is relatively ancient—if, as some recent scholarship indicates, the Alaskan Norton Culture and the Old Koriak Culture of Northeast Asia are related, and other links to the Sakhalin-Hokkaido Cultures dating to around the first millennium B.C. hold up. The use of poison-tipped harpoons has a precedent in the North Pacific."

I am left to consider certain similarities to the Makah's ancient whaling practices, such as the ritualistic use of corpses. I am left wondering, too, why whaling apparently took root only in Kodiak and not in the nearby Aleutian Islands, and about what other comparisons I might make among the native peoples of the Arctic.

•

It's raining again as Alice and I head out of town, past the airport and the Coast Guard base, along the pothole-filled Chiniak Highway. We stop briefly at Kalsin Bay, where gray whales are said to have been recently spotted. The two of us stand on a beach that feels similar to ones we are familiar with along Baja's East Cape—except here there is no visibility through the fog. We drive on toward the end of the road, fifty long miles to Narrow Cape, where gray whales pass by the hundreds on their migrations—and where the Kodiak Launch Complex sent up its first ballistic missile the previous November as part of the Air Force's new "atmospheric interceptor technology" program.

Customarily, gray whales would have been in the vicinity when the Minuteman II rocket went off. But the southbound migration was late, so late that the bird-watchers who come to Narrow Cape every Christmas season had seen droves of whales going by on December 27. Alaskan newspapers reported this "very abnormal" occurrence, the same late start that had so concerned Homero Aridjis and Bruce Mate. The gray whales had eluded the Makah, and the U.S. Air Force, during that winter of '98. But as more and more scientists were theorizing, the reason was that they'd had to stay longer in the Bering Sea to try to get enough to eat.

The potholes get worse and the rain keeps falling as we turn onto Pasagshak Bay Road. Even in such weather the verdant hills of Kodiak are lovely, reminiscent of Ireland. Then, abruptly, the road is graded and the terrain flattens as if bulldozed. A white car comes racing up from behind and speeds around us, going at least sixty. Out of a barren landscape, a three-story concrete bunker with no windows and a huge metal door looms on the left side. There are several other buildings, but no sign of human presence. Around a sharp curve just ahead of us is a roadblock, where the white car now awaits. I brake and roll down my window.

"Sorry, this is a restricted area," a man wearing a dark suit informs me.

"I'm trying to get to Narrow Cape," I reply, "and see some whales."

"Sorry, you're going to have to turn around," the man says, and points me back down the road. As we pull away I notice the white car is staying about a quarter mile behind me, like an escort. It's eerie at the edge of Kodiak Island.

As the days went by I would come to learn more about the origins of what was going on at Narrow Cape. Kodiak had become a military staging area during America's North Pacific operations in World War II, when the population of a formerly tiny village soared to more than 25,000. A submarine base, an air station, and an army outpost had all been built. And the legacy had remained, in more ways than one.

Kate Wynne, a marine mammal specialist with the University of Alaska's Sea Grant program, has lived on Kodiak since 1993 and explained it to me like this: "You have the pristine side of Alaska—and then there's the other side. We're paying the price today for all the World War II stuff. Dump sites are being cleaned up now that are just nasty, noxious things. Some houses were built right on top of these. Mercury levels off the scale. People have gotten so sick they had to leave the island. And I'm pretty sure there's a lot more we've never heard of, like nerve gas and mustard gas storage areas. On other islands in the archipelago, you can see big piles of fifty-five-year-old barrels rotting. On Chirikof Island it looks like a moonscape, completely barren, and then these little rivulets of orange stuff going into the ocean. When I moved here, and everyone said how the sea lions have declined from 141 degrees westward, I looked at the NOAA charts and noticed all these designated dump sites offshore of Kodiak and Seward. I asked people, what was dumped there? It was the World War II munitions, solvents, fuels, barrels, batteries, you name it. All the sea lion declines are downstream of those sites."

When Eric Stirrup came out twenty-one years ago from New Jersey, he figured he'd "seen paradise." Now he's not so sure. He explained that "the Air Force, within the last five years, spent the whole summer with a crew doing some sampling up on top of Pillar Mountain. That was a White Alice station, microwave communication, World War II and post–World War II. Their solution to dealing with the transformers was, dump the oils down a pipe and into a tank in the ground. Well, Pillar Mountain is right on top of this watershed that drains into Pillar Lake, which is the water supply for Kodiak. What toxins have accumulated in our drinking water over years and years of leaching of these PCBs down through the mountain? Hello? People wonder why the cancer rate here appears to be higher than the norm, although the town officials say it isn't."

Now came the 3,100-acre rocket launching site on Kodiak's remote eastern shore at Narrow Cape. The island's citizens were initially told that the facility would be used strictly for sending up communications satellites. Only later did it come out that it was really to evaluate existing radar systems or, in the Air Force's words, "target launch capability to realistically simulate inbound missile threat trajectories from potential Pacific Basin adversaries." Starting in early 2002 the Strategic Target System (STARS) booster would be launched, carrying a specially configured threat-representative payload known as Generic Rest of World (GROW), as part of the National Missile Defense flight test program. Kate Wynne puts it all into lay language: "The first launch was on a path that would simulate a nuclear attack by North Korea. The Air Force is planning to send the rest in drones over the North Pacific, which are going to be shot down like clay pigeons from Hawaii and Vandenberg, to prove they can in fact protect us."

Since gray whales are known to migrate along both the east and west sides of Kodiak Island, how much of the population actually passes Narrow Cape is not known. A hastily done environmental assessment by the Air Force in 1997 concluded there was "no significant impact" for any wildlife in the area; thus, no follow-up environmental impact statement was required under the law. Potential effects upon gray whales and other marine mammals depend, of course, on when and how many launches there are. The Air Force is talking about nine rockets annually, but the blueprints indicate room for four launch pads. So they could be sending up three rockets a month. "Launch noise will be audible on Kodiak for a distance of twelve miles for approximately one minute," the Air Force said in the environmental assessment. "Sonic booms will be heard only on the open ocean." What nobody has looked at, Wynne says, "is the effect of the reverberations underwater, what that sound is doing as it bounces across Pasagshak Bay. If a launch happens during the gray whale migration, it could really push the whales offshore."

To say nothing of the aftermath, when substances such as hydrochloric acid drain into the bay. "Basically all that rocket propellant becomes localized acid rain," Eric Stirrup told me. "It's either going to end up in the groundwater or eventually leaching out into the intertidal zone." Susan Payne, who works on Kodiak for the National Marine Fisheries Service and initiated the island's annual spring Whale Fests, added: "There's going to be a lot of runoff, and people have seen gray whales bottom-feeding off Narrow Cape. This may even be a staging area for the remainder of the migration."

Local opposition to the launch facility, however, had been minimal. "That's the frustrating thing about living in Alaska," said Kate Wynne. "This state is so pro-development generally, whether it's mining or fishing or oil or whatever. The politicians always push economic value to the nth degree." Or, as Eric Stirrup wistfully described it: "Over the last fifteen years, Kodiak has become very much a government town. Fishing is no longer the engine that keeps things going, though people want to think it is. Government, and the industries that thrive off government, represents about 60 percent plus of the workforce."

The next day we left Kodiak, flew back to Anchorage, and drove south again to the coastal town of Seward on the Kenai Peninsula. It's at the head of Resurrection Bay, discovered and named by Russian sailors who took refuge here during a Sunday storm. Snowcapped Mount Marathon provides its backdrop, and extending down the coast from Seward is the Harding Ice Field, with its eight tidewater glaciers left over from the last Ice Age. We'd been told that the Kenai Fjords boat tour, which offered a fifty-mile cruise out through the bay to the Holgate Glacier, was well worth rising early for.

Our guide was an old salt named Frank Hansen, who sat on the upper deck and talked to several hundred passengers over a loudspeaker as if we were all friends at his dinner table. I went up to ask him about the chances of our encountering a gray whale. "It's so late in the year," Frank said, "the bulk of them came through almost two months ago. Most of the stuff they like to feed on, we don't have out here. There's no muddy bottom. But you never know, there've been a few stragglers. Another of our boats claimed to have seen a mother and calf over by the Chiswell Islands four days back. We'll be heading right near there. I'll keep my fingers crossed—and my eyes out—for you."

The weather was sunny and mild, temperature in the sixties. The passage through the fjords was as beautiful as any in Alaska—past Cheval Island and Chat Island, alongside vertical rock rises lined with spruce and covered with snow. Around 2:00 P.M. we approached the Holgate Glacier. In this enormous turquoise mass, we bore witness to antiquity. For ten minutes the entire boatload stood at the railings and listened to the glacier "talk," listened to the deep moanings as ice cracked and fell into the sea, a geological wonder known as calving. All around us floated icebergs in miniature. The wind and the fresh scent of clean air wafted through us. A feeling of serenity washed over us.

Reluctantly we watched Captain Hansen put the ship into reverse and start back along Holgate Arm. On the return leg we were to follow the edges of the Harris Peninsula out beyond Harbor Island, where the Alaska Maritime National Wildlife Refuge begins. This would be an excellent area, our guide said, to spot all kinds of creatures. Sure enough, before long we were within whisker-counting distance of a sea otter, floating on its back, rolling, grooming itself, and diving for shellfish. Frank informed us they're sometimes called the "old men of the sea," and in fact they are the largest members of the weasel family in North America. Next came a small, exquisite island that hosted a colony of Steller's sea lions. Beneath a volcanic arch similar to the one on St. Lazaria Island off Sitka, bulls weighing up to a ton hauled out on the rocks with their yearlings, all sunning themselves and bellowing boisterously. Inside a grotto with an interior pool, reminiscent of a Roman atrium, females gathered with their pups. Then came the tufted puffins, suddenly ubiquitous, with their parrotlike orange beaks and dressed in their full regalia summer plumage. Flapping their wings, they dove deep into the water pursuing small fish to bring to their young.

We were passing along the sheer cliff face of Harbor Island, where the pines come right down to the water, when it happened. "I think that's a gray whale, folks, off to the right!" More than fifty people, including ourselves, charged the railing. The boat leaned with the weight. We scanned the craggy shoreline. Nothing. The minutes went by. The captain doubted himself. "I don't know, folks, I might have given you a false alarm there. My eyes aren't what they used to be."

Then a gray whale calf surfaced and spouted. A collective sigh seemed to emanate from our ship. It was a small whale, clearly born within the last few months. It was swimming within feet of the shore, in very shallow water. Its mother was nowhere in sight. I glanced at my watch: 3:20 P.M. It stayed on the surface, cruising steadily, for close to five minutes. As we slowly moved on, the whale dove and vanished from sight.

I went up top again to see the captain. "I was really hoping you'd see that!" he exclaimed. "It wasn't much of a blow at first, and I only caught it out of the corner of my eye. Then I kept looking and looking. That's gotta be the calf of the pair our other boat saw earlier. So the mother has to be around here someplace. She doesn't need to breathe as often, but she'll be close. She's probably right behind us now, because the little one was making tracks. Wasn't that great?"

It sure was. That evening Alice and I sat feasting on halibut cheeks and king salmon, looking out across Resurrection Bay. We'd been just in their wake as the gray whales themselves ventured into certain new domains for the first time. The image of the baby gray we finally saw, swimming so close to shore in the shadow of the cliffs, seemed implanted in our consciousness, as had happened to us that first time at San Ignacio Lagoon. We mused that the gray whale's blow is heart-shaped because it traverses the wild heart of the planet, the stunning coastline from Baja through Alaska, from desert to ice. I tried to envision what it would be like to catch up with the whales again all the way in the Arctic. Would it, we imagined, require something of a purification ritual?

Once, not all that far from Resurrection Bay, it had certainly been so for Scammon and company on the Western Union Telegraph Expedition.

In tandem with the steamer *Wright*, Scammon's bark *Golden Gate* had pulled away from Sitka on August 20, 1865. The captain recorded in his log: "The change from the almost incessant rainy weather . . . to the calm clear sea is pleasant and all on board seem to realize its benefits." Scammon knew this might not last for long. They were bound northwest across the Gulf of Alaska, toward the same narrow waters that the gray whales utilized to cross to and from the Bering Sea. In those days it was called the Ounamak Passage; today it's known as the Unimak Pass. It was tricky for large ships that had to dart between many islands in what seasoned skippers often called the worst weather in the world. Scammon had brought along a Russian pilot, knowledgeable of the passage, to help them make their way through it.

They passed south of Kodiak Island, within sight of Chirikof Island and on by the Shumagin Islands before dropping anchor at Ounga Harbor off the Alaska Peninsula. It had taken ten days to get here from Sitka, a trip described in elegiac terms by the young naturalist William Healey Dall: "It was a delightful sail among green islands dotted with volcanic peaks of surprising height and ruggedness, and in many cases silver cascades fed by the eternal snows, leaped from some rocky crag in one misty veil." Upon reaching Ounga, Scammon wrote: "Its pleasant green appearance as the sun shone between the overhanging clouds gave one a desire to visit it. Especially as we were looking for a barren mass of rocks. Our Russian pilot informed us that it is inhabited by about twenty souls, men, women and children, whose employment is to collect furs for the Russian American Co."

The expedition leaders had heard that Ounga Harbor might possess a decent vein of coal for setting up a refueling station. According to Dall, "The coal didn't amount to much, being more properly lignite or brown coal." However, he added, "We discovered here large quantities of petroleum. Won't some broker organize an Ounga Oil Company?" (As far as is known, no one ever did.) The small island, with its lofty volcanic mountains crowned with snow, was breathtaking. "All hands rushed ashore on a glorious day," Dall recorded, into "meadows covered with a luxuriant growth of grass and flowers, I counted 43 species, and preserved specimens of 15 sp. I never saw richer grass on an Illinois prairie." Scammon stayed onboard ship, noting that the reconnaissance of the harbor was "very satisfactory and the scientific [team] made good collections."

The next day, after Dall went out with a small boat's crew and dredged up more shells, the respite at Ounga Harbor ended. As they had done leaving Sitka, the steamer set off with the *Golden Gate* in tow. Then, early on the morning of September 1, a gale struck.

Scammon would write: "At half past two o'clock A.M. the wind increased so much that signal was made to the steamer to cast off the hawser, which was obeyed. For some time after she kept ahead of us but finally fell astern, and at half past five A.M. in a squall we lost sight of her. Sail was immediately taken in, but she did not make her appearance. I had before cautioned the *[Wright's]* captain to be careful to always keep company, and in the written orders his attention is called particularly to the matter. There can be no excuse for his parting company even if the engine gave out, as there was a strong breeze at the time and the vessel under sail was able to keep up with us with a fair wind."

Dall would write: "We were going nearly nine knots an hour when it became so foggy that we could not see the steamer or anything else, and we went under double reefed topsails. About 8 A.M. the lookout discovered breakers directly in front and on each side of us . . . in close proximity to a very dangerous reef." They were near Halibut Shoals. Scammon quickly realized they were in considerable trouble. The sea was breaking furiously over a huge mass of rocks about a half mile away. The captain ordered all hands, "landlubbers and all," as Dall put it, to have the helm put over and to reset the sails. Still the *Golden Gate* rushed toward the rocks. No one knew whether the ship would "answer the helm" or the sails be capable of pulling them out of harm's way in time. Then, several hundred yards from certain destruction, the ship turned itself with a shudder—and shot past the shoals.

"It was as near wreck and certain death as I care to come," Dall set down. "The coolness and intrepidity of Capt. C. M. Scammon, and the unbounded confidence of all in his thorough seamanship, prevented any panic. We escaped with a mere 'scratch.' " Scammon called all officers and crew together and praised them for the "manner that all strove to save the ship and the lives of those on board. I have never, in my thirty years of experience as a sailor, seen a ship so near the breakers without going on them."

What followed we gather from Scammon's log:

> At noon the wind moderated but the weather was extremely foggy. We fired signal guns but could see or hear nothing of the steamer during the night.

> September 2 Saturday. From 12 m to 4 o'clock AM we fired signal guns every half hour hoping the steamer might come within hearing distance. The weather being quite calm, but [we] were forced to proceed on our course without her. This I regretted very much as the bad season was fast approaching and a few days delay might put us to serious inconvenience.

And so, alone, they came to the Ounamak Passage, or Unimak Pass.

> September 4th Monday. Commenced with a strong gale from the W.N.W. preventing us from getting through the passage. At 12 m the wind increasing so much that we were unable to make any progress. Came to anchor under the south shore of Ounimak Island in 45 fathoms water, about three miles from the land. A swell from the S.W. made our anchorage rather uncomfortable, causing the vessel to roll a great deal when the tide was running to the westward. At this anchorage we caught thirty five cod fish and one hallibut *[sic]*. Several casts of the patent sounder shows the bottom to be composed of volcanic rock, with occasional patches of coarse dark greenish or black sand.

Against the powerful headwinds, they were going nowhere fast. The following day, "at sundown within six miles of land we observed Mt. Shishaldenski in a slight state of eruption. I had anticipated this pleasure, as our Russian pilot informed us that there were several active volcanoes on Ounamak Island." It was, Dall wrote, "about 9,000 feet high and smoking; a beautiful white peak" that beckoned them finally into the pass after three days "beating through." Scammon recorded the final hours of this harrowing voyage matter-of-factly on September 6: "During the night entered Ounamak passage and by 9 o'clock a m we were in Behring Sea."

•

For gray whales, all told it's 4,125 miles between San Ignacio Lagoon and Unimak Pass. Below Kodiak Island some are known to "island-hop" among the Trinity, Semidi, and Shumagin Islands, it being speculated that this maximizes their traveling in shallow water. Desolate and volcano-shrouded Unimak Island, with 267 miles of shoreline, is nearly contiguous with the long, narrow, roadless Alaska Peninsula, separated by only a narrow strip of water known as False Pass. West of Unimak Pass the Aleutian Island chain begins, continuing in a broken line all the way to Russia's Kamchatka Peninsula.

The grays' winter migration southbound from the Bering Sea takes place in the darkest, bleakest, stormiest months. It was then, in November 1977, that Dave Rugh first came to Cape Sarichef, a promontory at the northwestern edge of Unimak Island. A Coast Guard station was operable there at the time, and during this same season back in 1961, an officer had alerted scientists about hundreds of gray whales streaming by his watch. In the spring of 1977, the biologist John Hall had also discovered Unimak Pass to be "standing-room-only for gray whales." This belied the then-current theory that the grays zipped across the deep offshore Gulf of Alaska between Washington and somewhere in the Aleutians. These were also the first pieces of evidence to back up Scammon's assertion that this whale "makes regular migrations from the hot southern latitudes to beyond the Arctic Circle; and in its passages between the extremes of climate it follows the general trend of an irregular coast." So the National Marine Mammal Laboratory decided to conduct a winter census. Rugh, recently arrived from graduate school at the Institute of Polar Studies in Columbus, Ohio, was dispatched.

He flew from Anchorage to King Salmon on a commercial plane, then hopped a chartered Cessna out to windswept Cape Sarichef. At thirty miles between there and Akun Island, easternmost of the Aleutians, this is the widest point of Unimak Pass (eleven miles is the narrowest). Except for "the Coasties," as he called the U.S. Coast Guard personnel, Rugh was alone. The accoutrements at the base were decent enough—a warm building, showers, three square meals a day; however, watching for gray whales posed quite a challenge. The average temperature hovered near zero Celsius, with horizontal rain, sleet, or snow the daily norm.

"The winds were awesome," Rugh remembers. "There were times, probably in the seventy-mile-an-hour range, when the whole place would roar. Even with those concrete buildings, you could hear the wind

Dave Rugh below Cape Sarichef, fall 1978 (courtesy Dave Rugh)

churning and moaning across the pipes and wires. Outside you could hardly stand upright and could only get around pretty much by leaning down almost on your hands and knees. I would somehow maneuver a hundred meters to the edge of the sea cliff and then climb about halfway down a steep, eroding bluff—because the Coasties had a foghorn up top that was so loud it could be damaging. My perch was a little over thirty meters above the sea. I hid behind some pieces of a broken compressor damaged in the '64 earthquake, and used some building scraps for an additional windbreak. I'd look out to sea and watch huge, wind-chopped waves rolling toward me in slow motion before they exploded up and over the sea cliffs. Just the sound of the surf hitting the shore was amazing; the deep booming was spine-chilling. The only other thing I had to worry about was grizzly bears, a daily nuisance even in the dark of winter. I'd stay at my perch on the cliff recording whale sightings until I got hypothermic and could no longer hold on to my pencil, then I'd go back inside and warm up."

Because he was so far north, Rugh never had more than seven or eight hours of daylight, and the sky was hardly more than dusky at its brightest. Nonetheless, during 82.5 hours of systematic observations over a seventeen-day period, he recorded 2,055 gray whales swimming south. "Through mathematical approximations, I tried to account for the beginning and end of the migration," he later wrote, "giving me a grand estimate of 15,100 gray whales. This was surprisingly high in light of the fact that 11,000 was the commonly quoted population figure to that date." (Rugh noted that the 11,000 figure came from an uncorrected estimate based on data collected at the standard counting station in central California; later, when appropriate mathematical procedures were applied, the abundance proved to be over 17,000.)

Because in 1977 he'd arrived in midmigration, in 1978 Rugh decided to start much earlier and conduct a full-season census with a team of four observers. This time the Coast Guard built him a wooden structure to provide protection from the elements. Rugh also did some aerial observations along the north coast of the Alaska Peninsula. However, in 1978 the Coast Guard abandoned the station in an attempt to cut costs. During Rugh's third and final winter season with his crew, in 1979, they shared the landscape with only a bird biologist and his family. To obtain more precise calculations of gray whale locations, Rugh installed a theodolite in the glass shell formerly used to house the Coast Guard's light beacon. To see if any whales rest at night, he and the others did tests with Army night-vision goggles. The goggles amplify reflected visible light in a manner similar to a TV screen but without the magnification. They found that the whales traveled as much after dark as during daylight. They also detected no evidence that the whales traveled more slowly as storms worsened. Over the course of those three observational winters, Rugh wrote, "Most whales (79 percent of 10,223 sightings) were seen within one kilometer of shore, which meant they passed in water less than twenty meters deep."

With the Coast Guard (and their foghorn) gone, Rugh was also able to station his assistants at both higher and lower elevations to test for differences in sightability. So he experimented with paired observation teams who did not interact with each other. They found an approximately 85 percent concurrence. "That meant, as good as we thought the record was, there were still whales being missed within the viewing range of observers watching in good conditions," Rugh says. In later years, when he took the same independent observer test to Granite Canyon, California—now the primary location periodically used to determine gray whale abundance—that 85 percent figure held true. Soon this became a standard correction in the abundance estimates.

Rugh learned something else about counting whales at Cape Sarichef.

Logic would seem to dictate that the stormier the sea conditions, the harder it would be to see them. Conversely, on a dead-calm day one would think the whales would be perfectly visible. Not so. In fact, the reverse is generally true. "At Unimak, I've watched in thirty-foot seas, where, as the whales come up and the waves go out from under them, they're *very* distinct. High counts of grays in Beaufort 8 conditions. I've also watched in Carmel, California, flat sea, cloudless sky, no wind—and the counts are way *down*. That's because the water is glazed and gray. Only in certain light conditions are you realizing that the whales are there; in other zones you can't detect them even with the best observers."

Without the logistical support the Coast Guard provided, 1979 was the last time anyone braved the winter elements to track gray whales moving south through Unimak Pass. Another researcher, Pauline Hessing, did watch the northbound migration from there between March 23 and June 17, 1981, as a follow-up for the work she did with Rugh in the winter studies. What she described as the "mercurial weather" wasn't a whole lot better. She noted that, early in the season, the grays passed farther offshore but during the last several weeks, when mother-calf pairs predominated, 90 percent came within 110 yards of the offshore rocks. The calves seemed to huddle behind their mothers' flippers, which Hessing thought probably served to lessen water resistance. During aerial surveys, whales were "sometimes seen loitering" at what she speculated might be "rendezvous areas where the migrating whales regroup before continuing north."

These studies occurred at a time when the Outer Continental Shelf's environmental assessment program for oil was especially interested in the ocean around Alaska. In 1971 Unimak Pass had received one of the highest impact ratings for modeled oil spills in a study conducted at seventeen sites along the Alaskan coast. Hessing noted that gray whales migrate near two of three areas leased for possible oil drilling in the southern Bering Sea (St. George Basin and the north Aleutian Shelf). She also warned, in her conclusion, about oil's potential to impair whales' metabolism. To the industry's credit—and perhaps thanks to the work of Rugh and Hessing—nothing has yet transpired along these lines.

Beyond Unimak Pass, covering the coastal regions of two continents, lies one of earth's largest semienclosed seas: the fecund feeding ground of the gray whale.

Chapter 16

INTO THE BERING SEA

Bering Sea itself is probably the richest of all the northern seas, as rich as Chesapeake Bay or the Grand Banks at the time of their discovery. Its bounty of crabs, pollock, cod, sole, herring, clams, and salmon is set down in wild numbers, the rambling digits of guesswork. The numbers of birds and marine mammals feeding here, to a person familiar with anything but the Serengeti or life at the Antarctic convergence, are magical. At the height of migration in the spring, the testament of life in Bering Sea is absolutely stilling in its dimensions.

—BARRY LOPEZ, *Arctic Dreams*

As soon as the winds falter, and the thirty- to fifty-foot seas cease their tumultuous welcome against the cliffs of Cape Sarichef, the latest group of northbound gray whales streams to the right along the north shore of the Alaska Peninsula in their first taste of the Bering Sea this season. There remains a considerable way to go to the prime feeding areas in the Chirikof Basin, south of the Bering Strait. In a spring like that of 1999, the grays run into a barricade of ice a couple of hundred miles up the peninsula. They stack up for two or even three weeks, waiting for the ice to retreat from Bristol Bay. Some prefer to spend the summer in bays along the Alaska Peninsula, such as in Nelson Lagoon, feeding in a narrow channel on little purple shrimp that live in the gravel bottom. The majority of the grays will ultimately pass through Bristol Bay, site of the world's largest sockeye salmon fishery. Sometimes the grays will go right into the river estuaries and spend periods of time in only a few feet of water.

Terry Johnson is a marine extension agent for the University of Alaska Sea Grant program who runs a summer tourism business in northern Bristol Bay. "From the air you sometimes can see their feeding plumes," he reports. "When gray whales occasionally die here, their carcasses draw brown bears from miles around, although sometimes the bears eat only the tongues and leave the rest. . . . In the Togiak Bay area the grays play with the sea lions. Late in the spring migration, small groups of sea lions join them and just wrap around them, writhing and spinning and clearly enjoying themselves."

Past Walrus Island and Togiak Bay, around Cape Newenham, the grays continue on up through Kuskokwim Bay. Above the strait that runs between Nunivak Island and the mainland, the tight migration starts to disperse. Some whales will turn almost due north into open water, heading toward their goal around St. Lawrence Island. For much of this shortcut, they are still on the continental shelf and thus in shallow water; the hundred-fathom line cuts right through the center of the Bering Sea, starting from Unimak Pass and running on a northwest line toward Russia. Other pods will stay coastal, moving past the bay—and a small Yupik Eskimo village (population 326)—each called Scammon Bay.

The native people's name for this village is Mariak, and it's believed to have been first settled in the 1700s because it had high ground and good water. The name was changed by the Alaskan legislature in 1967, to honor the centennial of Captain Scammon's service as chief of marine for the Western Union Telegraph Expedition. Located a mile from the Bering Sea on the banks of the Kun River, and within the 19-million-acre Yukon Delta National Wildlife Refuge, Scammon Bay residents continue to live a largely subsistence lifestyle. They hunt beluga whale, seal, geese, swans, cranes, ducks, loons, and ptarmigan. Fishing yields salmon, whitefish, blackfish, needlefish, herring, smelt, and tomcod. From the vicinity of Scammon Bay, more gray whales veer northwest into the open sea and toward the Southeast Cape of St. Lawrence Island.

I entered the realm of the Yupik and the Aleut, the Inupiat and the Chukchi, with a photographer companion, George Peper. Friends for more than twenty-five years, we had worked jointly on the campaign to save the Atlantic striped bass. This was to be a voyage that would ultimately take us not only to gray whale feeding areas but into Arctic native hunting villages. We would be venturing with our sleeping bags and provisions into some of the most remote regions remaining on the planet.

We met up late in June in Alaska's largest city, Anchorage, initially to take part in a media tour designed to raise awareness about problems facing the Bering Sea. This was organized by the World Wildlife Fund

(WWF), which had recently identified the Bering's as a Global 200 ecoregion, meaning it is among the two hundred most important areas for conserving life on Earth. *Time*, the *Los Angeles Times*, and *The Washington Post* had all dispatched reporters to cover the WWF's weeklong trek into the Aleutians and the Pribilof Islands.

The event began with a daytime gathering in a hotel conference room of scientists, fisheries officials, native peoples, and environmentalists. One participant recounted how a Yupik elder, asked to describe the changes she'd been seeing in recent years, replied: "The world is turning faster." Asked what she meant, the woman noted that her people used to be able to predict the weather accurately three days in advance. Now it was down to a day and a half, maybe two days at best. Alan Springer, of the University of Alaska's Institute of Marine Science, elaborated that the summers of 1997 and 1998 were "extremely anomalous in Alaska, and particularly the Bering Sea. Air and water temperatures were far higher than usual, and this seemed to bring about a number of biological changes that hadn't been seen before."

An estimated 80 percent of Alaska's glaciers were receding, while the extent of ice in the Bering Sea had shrunk over the last thirty years by up to 3 percent. This sea ice, as a WWF report later described, "influences a wide range of physical activity from atmospheric events to oceanic mixing and sea bottom temperatures. It provides essential hunting and breeding sites as well as protective habitat for polar bears, seals, walrus, and other wildlife. It is an essential environment for the growth of micro-algae that are the base of the entire Bering Sea food web. And the spring phytoplankton blooms along its extensive ice edge boosts productivity into early summer."

Already the walrus, a benthic feeder like the gray whale, was seeing a decreased rate of reproduction. At the same time unusually high numbers of seabirds were dying, including nearly two hundred thousand short-tailed shearwaters, which washed ashore on the Russian and Alaskan coasts. Nesting in southeastern Australia but migrating to forage each summer in the usually rich Bering Sea waters, they appeared to have starved to death. Deep-diving common murres were having similar difficulties.

Could whatever was affecting these creatures be part of the same situation driving more gray whales to visit locales they'd never frequented before as well as to strand themselves? When our WWF tour landed in Dutch Harbor, on the mountainous Aleutian island of Unalaska, Mayor Frank Kelty informed me that a few gray whales had started coming right into the harbor chasing herring. This meant they were traveling *away* from their accustomed path toward the Bering Sea feeding areas, roaming fifty miles to the *west* of Unimak Pass along the Aleutian chain—yet another indication that weather or productivity factors were forcing a shift in the whales' migratory patterns. If this wasn't anomaly enough, the mayor said that killer whales had started to eat black cod, also known as sablefish. Local offshore fishermen reported that no sooner would they start hauling their long-lines than the orcas would race in to pick off the black cod, commercial catches of which were down significantly.

Dutch Harbor and the neighboring community of Unalaska are the most lucrative fishing port in America, with 678.3 million pounds of seafood landed in 1999, valued at $140.8 million. Seven fish-processing plants ring the harbor, their massive, multiunit complexes churning out pollock, salmon, cod, king crab, snow crab, herring roe, and halibut (sportfishing's world record halibut, a 459-pounder, was taken here in 1996). About 60 percent of the commercial catch is exported to Japan, which once struck Dutch Harbor with two days of aerial bombardment during World War II.

I'm at dockside aboard the *Hamilton*, a 378-foot U.S. Coast Guard cutter. It started out in the mid-1960s collecting oceanographic data in the Atlantic, got deployed two years to Vietnam, went on to intercept vessels carrying illegal drugs and illegal immigrants in the Caribbean, and now has a new mission. The *Hamilton*'s crew of approximately 160 is charged with enforcing U.S. fisheries laws throughout Alaska. They've been focused lately on the Bering Sea, this 885,000-square-mile wilderness with at least 525 species of marine mammals, fish, shellfish, and seabirds. Here lies the world's most extensive bed of eelgrass and its largest salmon grounds. Here more than half of America's seafood is produced.

Here, too, the massive nets of factory trawlers sweep across miles of ocean floor, tearing up the bottom, including corals, mollusks, plants, and a host of microorganisms that whales and other marine animals require to survive. Most of those nets are targeting walleye pollock, a small bottom-dwelling species whose catch—some 4 billion pounds annually—constitutes the greatest fish harvest on the planet. You'll find these pollock in most of your packaged fish sticks, your fish fillets at Long John Silver's and McDonald's, and your *surimi* paste or imitation crabmeat. Staggering as it seems, the pollock catch today is half what it was in 1985.

The fish's decline in abundance is directly tied in to the food webs of at least twenty other declining marine species.

So the Coast Guard's Operation Ocean Guardian sends C-130 spotter aircraft and cutters like the *Hamilton* out into the Bering Sea. They patrol the pollock grounds to prevent our maritime boundary from being infringed upon by foreign factory trawlers. They also look for illegal high-seas drift-net vessels, banned worldwide since 1993 under a United Nations moratorium because of their capacity to overfish species like salmon while hauling in hundreds of marine mammals, sea turtles, and seabirds as unwanted "bycatch." Captain Vincent O'Shea, a former commercial fisherman who now supervises the fisheries law enforcement branch of the District 17 Coast Guard operation in Alaska, had this story to relate:

"About four years ago, we had a drift-netter that refused to stop. Our cutter ended up chasing it for four days, even through a typhoon. Finally the commanding officer launched a small boat in six- to eight-foot seas. It cut immediately in front of the drift-net vessel and dropped a line. This passed under the hull, was sucked up into the propeller, and stalled the engine with a very dramatic black cloud of smoke. Our cutter then sent out another small boat with a boarding party, climbed aboard the disabled ship, went up to the bridge, and took over control. The crew was Chinese, but both China and Taiwan disavowed any knowledge of it. So we seized it as a 'stateless' vessel."

Listening to the tales of Captain O'Shea and his knowledgeable explanation of the Bering Sea's problems, I couldn't help but think of Scammon. For thirty-three years he'd served as captain of seven ships with the Revenue Marine, forerunner of the Coast Guard, his last command coming at the age of fifty-eight. He'd performed numerous "search and rescue" missions. And he'd spent considerable time here in the Aleutians and also at the Pribilof Islands, where his duties had included bringing supplies to the natives and protecting the seal fishery. As an article in the *Overland Monthly* reported in 1872, "Congress had passed a law forbidding the killing of any fur-bearing animals in the waters of the [Alaska] Territory, and the cutter *Wayanda* [Scammon's ship] was sent to cruise near the islands and prevent all vessels from landing."

The Fur Seals have so wide a geographical range—extending nearly to the highest navigable latitudes in both the northern and southern hemispheres—and are found assembled in such countless numbers at their favorite resorts, that they become at once a source of great commercial wealth; and, among marine mammals, they are the most interesting we have met with. . . . In Behring Sea, the islands of St. Paul and St. George are now the main resting places of the Fur Seals. . . . The Aleutians, under the direction of officers of the Russian-American company, were employed in taking the seals. . . . It was the custom to drive thousands of them inland, that their capture might be more easily accomplished. The loud moanings of the animals when the work of slaughtering is going on beggars description; in fact, they manifest vividly to any observing eye a tenderness of feeling not to be mistaken. Even the simple-hearted Aleutians say that "the seals shed tears."

—SCAMMON, 1874

Two hundred miles north of Dutch Harbor, the Pribilof Islands are the first, and only, inhabited place one comes across for a substantial sector of the Bering Sea. St. Paul, largest of the five islands, is thirteen miles long by six miles wide. There are no trees, and the winds can be so ferocious that it's not hard to imagine any having been long ago whisked out to sea. These islands are sometimes called the cradle of storms, with winds accompanied by "swirling fog, thick mists, winter ice packs, blizzards—sudden, erratic change," as one resident has described the "volatile, capricious weather." Yet in June, as we arrived, St. Paul still bears the strange charm recorded by Scammon: "The interior is irregular, and near the centre of the island rises an extinct volcano to the elevation of a thousand feet. In summer, when approaching, it presents a green appearance, and on landing one finds a luxuriant growth of grass waving over hill and valley, intermixed with field flowers in full bloom."

The Pribilofs also have probably the largest concentration of mammals and seabirds anywhere on earth. St. Paul is a haven for northern fur seals, nearly a million of which return every summer to breed and give birth. St. George, forty-five miles away, boasts what may be North America's most prolific bird rookery on a sheer cliff rising to almost a thousand feet above the sea. More than 210 bird species have been identified on the Pribilofs. Rarely, but not without precedent, gray whales have shown up along these isolated shores as well.

It's early morning on St. Paul. George, myself, and a travel writer from *The Washington Post* are heading out of town past a "bachelor beach" of northern fur seals, all subadult males. Terry Spraker, a veterinarian from Colorado State University, is at the wheel of a white pickup.

We're bound for a point known as Reef Rookery, the largest fur seal haul-out on the island. Terry is looking for dead pups, which he will take back to his laboratory to try to determine the cause of death. He's been doing this for the last fourteen summers.

We come to a stop beside a gate and a sign that says: FUR SEAL ROOKERY. DO NOT PASS BEYOND THIS POINT. Terry, bespectacled and decked out in rain gear and rubber boots, hands out knee pads to the three of us. The vet is carrying a long pole with a noose on the end. Sliding open the gate, we set out hiking through a meadow of purple lupines, sprinkled with bright yellow Arctic poppies. It's deceptively beautiful. Coming around a bend, there they are: thousands of fur seals spread amid the rocks and endlessly down the beach. Their sounds—grunts, bellows, groans, bleats, growls, baahs—an entire language, and not a friendly one at that, fill the air. We're going to walk right through some bull seals, the ones Terry calls "the wishers" because they wish they were out there with the females. We'll hope, he says, they won't consider us competition. I glance around in vain for some place of refuge.

"The ones to watch for," Terry continues, "are animals making circles. That's your clue there might be some trouble. Just watch them out of the corner of your eye as you go by. Otherwise, it's considered a challenge. Walk kinda slow. I'll go first."

In silence we wind our way past the wishers, a few of which bare their teeth. I remind myself they only *have* four teeth, but trying to avoid eye contact with six hundred pounds of snarl isn't easy. I'm grateful the first catwalk isn't far. We ascend a ten-foot-high ladder and emerge onto a narrow scaffolding consisting of a pair of two-by-six planks. "It's sometimes slippery," Terry warns, "and can be kind of dangerous. So scoot your foot to test out the catwalk. I'm not sure how it's gonna be with this mist. See these studs about every eight or ten feet? It's best to keep one stud between us as we're walking because of the amount of weight. Once we reach the females, we'll crawl a little bit out to a big T. That's a good place to just sit and watch the animals."

The wind is picking up—it rarely blows less than fifteen miles an hour here—and the rickety, weather-beaten beams start to sway ominously. Only a couple of feet below me, I can see what seems an interminable series of doglike snouts, all pointing up and looking *very* annoyed.

"Don't worry," Terry says as he begins to walk the planks. "But just remember, one false step and you're history!" I listen, to no avail, for at least a chuckle. Pretty soon I'm down on my knees—and remembering the last discussion I had with my wife before she left Alaska, something about a "purification ritual."

After what seems an interminable time, I reach the T. The scene I contemplate is like nothing else I've ever witnessed. Huge bull seals stand guard over harems of as many as a hundred females, ready to fight off any intruder. The blond-mustached bulls are at least five times heavier than their pale-bellied mates. Once they establish their territory, the males won't leave for as long as two months, during which time they'll neither eat nor drink. Little black pups lick themselves and huddle up against their tan-coated mothers. You wonder how either can possibly survive all the stomping around and the incessant belches, honks, snores, barks, coughs, splutters, and bleats. I'm no longer afraid but mesmerized by thousands of flapping flippers. The seals appear to be fanning themselves.

Through binoculars Terry Spraker has spotted a dead female over by the next catwalk. He says we can either stay here or follow him, but be aware that we'll need to traverse a literal minefield of bulls en route—and he's forgotten the pole he usually uses as a prod. The *Washington Post* reporter elects to stay put. George and I accompany the vet back down the ladder and off across the boulder-strewn grass. Terry picks up a stick and bangs it against the rocks as we go. "Come on, guys," he says threateningly at intervals, not to us but to the seals. He points to a very long, flat rock, where a lone seal sits magisterially. "That's a special rock in their territory," Terry says, "so be sure and steer clear of it."

We bluff our way to another cacophonous catwalk. Below me the bickering of a bull trio suddenly turns into a raging battle. Terry motions us to our knees. As if on cat's paws, or so I'm hoping, I trundle on. My knees feel gelatinous under the pads. Terry is way ahead of us. He's leaning over the edge, trying to lasso a dead female that's hemmed in against a piece of driftwood. Every time he gets close to putting the noose around the seal's neck, a bull rushes in and bites at the pole. George moves past Terry and down to the other end of the catwalk, where he can get a better angle to photograph the operation. Terry doesn't seem to notice until my friend starts back. "Be careful there," he says, "because one of the legs is missing." George peers down a definite gap in the catwalk, into the vapory breath of a waiting bull. "Thanks for telling me now," he says.

It's more than half an hour before Terry manages to elude the bull, get the noose around the female's neck, and pull her 135-pound body to the catwalk. She has white whiskers and two deep bites on her hide, one

Terry Spraker carrying a dead fur seal, author in background (George Peper)

below the neck and another near a flipper. Terry explains she was an older female, and apparently got caught between two males fighting over her. Often he's seen a male toss a straying female around like a beach toy.

Terry drags the seal back to where we climbed onto the catwalk and, with it finally hoisted onto his back, we begin maneuvering our way back to the truck. "One female, but no dead pups today," Terry says. "Which is good." On the return to town I ask about his analyses of fur seal mortality. "The main issue I see is starvation of the pups," the veterinarian replies. "Where is that coming from? From something happening to the females. The mother could be killed, like the one we've seen here, or could be sick. There's something going on out at sea. Many of these animals are not coming back to the island. Whether they're not getting enough food here, or going other places, I don't know. And the scientists don't seem to know either."

By way of analogy Terry tells a story, one that he believes "really has implication for wildlife diseases." It's about a dog swimming across a river in the middle of winter. One at a time you add one-pound rocks to a little pack on the dog's back. Each rock slows the dog down a bit more. The fourth one sinks it. "So," Terry questions socratically, "which rock drowned the dog? They all weigh the same, right? Well, obviously it's a cumulative effect. But this is the same thing that happens to animals in the wild. Scientific researchers focus on their particular rocks of expertise. All of them confess that a bigger picture is real, but they don't act like it and they don't manage like it. This is a problem."

> *The Island of Saint Paul, in Behring Sea, a barren rock frequented only by sea-birds and seal, has for years past been of more value to the Russian-American (Fur) Company than all the rest of its fur-producing country. At certain seasons the fur-seal resorts to it for breeding, and the valuable skins obtained here are numbered by the hundred thousand. Previously, indiscriminate slaughter of these animals nearly exterminated them; but, by the wise policy of the Russian-American Company of late years, only the old males have been killed and their numbers have wonderfully augmented. If Government allows free-traders indiscriminately to kill these valuable animals, they will soon, as in the past, become so rare as to destroy a valuable source of revenue.*
>
> —WILLIAM HEALEY DALL, 1867

Not long after Gavril Pribylov discovered these then-uninhabited islands in 1786, the Russians began forcibly bringing native peoples from the Aleutians as their "slaves of the seal harvest"—a labor force to kill and skin hundreds of thousands of fur seals for the animals' lucrative pelts. Like the gray whales in the Baja lagoons, the seals were basically sitting ducks on their breeding grounds. Their demise would eventually result in the first international conservation agreement, late in the nineteenth century—but the practice of sending the Aleuts onto the seal-killing fields continued under U.S. domination until commercial sealing was finally outlawed altogether in 1983.

When Scammon visited the "Seal Islands of Alaska," as he termed them in an article in 1870, the Aleuts living on St. Paul numbered about 250. They impressed him greatly. "The extreme conscientiousness which governs their varied domestic vocations is without a parallel," he wrote. At that time, in addition to hunting sea otters and fur seals, Scammon tells us, the Pribilof Aleuts practiced whaling. "The whalemen put out from the shore whenever a favorable time offers, while the old men and children watch intently from the hills for the lance to be hurled into some vital part of the animal, when all who are able hasten in their boats to assist in towing it to shore. This having been done, the *Tyone*, who is elected by the villagers as their Chief, attends to the division of the spoil. The man who makes the capture gets the choice pieces, which are the flukes, lips, and heart, and the rest is distributed among the people of the whole settlement."

Today St. Paul's population of about eight hundred is the world's largest Aleut community. Nearly all continue to live where Scammon observed them: "Their village is situated on the south-east side, upon a gentle slope, with terraces facing the northeast, commanding a pleasing view of inland scenery." Still standing as well is "a wooden church, erected by the Russians, with red roof surmounted with white crosses . . . on the highest terrace." This church is located just above the small hotel where we stayed; most of the townspeople attend Russian Orthodox services here.

While they no longer whale, during the summer months when the fur seals abide on St. Paul, the Aleuts are allowed to take as many as one thousand juvenile males for food. Groups of men round them up on the beaches and club the seals to death. The meat is distributed among the people. But in recent years the allowable "subsistence harvest" quota has not been filled. It's often hard to gather enough of a team, especially during halibut fishing season. Many of the elders, the ones who grew up dining on seal meat, are dying off. It's a tradition that may not survive long into the new century.

Something else, though, is being introduced. It's part of a stewardship program funded by the U.S. Fish and Wildlife Service. Known as the disentanglement team, it's young Aleuts trained to remove debris such as plastic packing bands, fishing nets, and tangled rope that's wound itself around the necks of fur seals. Fish-processing vessels often ignore an international ban on tossing plastic overboard. Around the Pribilofs strong tides wash ashore pretty much anything thrown into the water within a ten-mile radius. Curious seal pups play with such detritus and end up getting caught in it. As the seals grow the stuff becomes lacerating necklaces that can eventually kill. Since the disentanglement effort began a couple of years ago, several hundred animals have been saved. And some of the program's "graduates" have gone on to major in marine biology and related fields at Alaskan universities.

It's late on a foggy afternoon. A seal has been spotted through binoculars with a white packing band around its neck and what looks to be a 360-degree wound. Eric Galaktionoff and Candace Stepetin, both twenty-one-year-old disentanglement team veterans, are preparing to move into the bachelor rookery. They're dressed almost identically, in caps, sweatshirts, jeans, and black boots. Each holds a long pole with a noose attached by gaffer's tape, and a razor blade at the tip. The seal they're after is about a quarter mile away, conveniently sprawled in front of a large rock. Eric and Candace will try to sneak up on it from behind. They set off through the tall grass toward a nearby knoll, from which they'll slowly descend upon the colony. Unless the wind changes abruptly, from that angle of approach the seals probably won't pick up their scent.

A group of us are watching from the road. For some twenty minutes we lose sight of Eric and Candace altogether, but the entangled seal hasn't moved. Dave Cormany, with the National Marine Fisheries Service, is ready to join the two young people if necessary. "Fur seals can move as fast as you can run for a short distance," he informs. "That's a big danger if you happen to fall down." In fact, Candace had once been bitten when she stumbled.

"The seals are starting to move down below," someone observes. Still no sign of Eric and Candace, but they must be closing in along the bank. Yes, there they are, hunkered low and tiptoeing around the back of the rock where the entangled seal remains. Now Eric charges, the long pole held fast at his side. The sand churns into a cloud of dust as the entire herd seems to be running in many directions. Eric gets close enough to thrust the noose around the racing seal, but in less than ten seconds the animal shakes loose. The chase is on again.

Suddenly and astonishingly, the seal stops in its tracks. Young man and young beast take a long look into each other's eyes. The seal sits down. Eric walks over and gently slips the noose over its neck a second time. It's with permission, as if the seal somehow knows that Eric is here to help him. Then, as the animal starts to move again, Eric removes his cap and drops to his knees. The seal tugs hard at the end of the pole. Candace advances, trying to attach the hooked end of her pole to the plastic band around the seal's neck. She can't quite do it. Eric is lying on his stomach, the pole extended as far as his arms can reach, when Candace signals for help. Dave Cormany, also holding a pole, starts off at a run.

Within minutes Cormany has managed to affix the blade and cut away the plastic band. The noose is lifted. Eric stands up. The seal waddles off to join its brethren. All this time, from a safe distance toward the water, the entire bachelor herd seems to have been watching what's transpired, scarcely making a sound.

The trio returns to where we're standing, as wide-eyed as the seals.

"That was courageous," one of the reporters tells Eric. "Truly something to see."

I'd been introduced to a new version of an ancient culture. In the way he stalked and herded and "communicated" with that fur seal, Eric Galaktionoff seemed to my outsider's eye to be drawing upon centuries of precedent. Except he was not hunting, he was rescuing.

The northern fur seals in the Bering Sea have not shown signs of much recovery since two hundred years of killing ended almost two decades ago; the population remains designated as "depleted." In 1997 and 1998, possibly related to the warmer sea temperatures, came the first recorded outbreak of a southern fungal infection. One of the observed results was a loss of guard hairs, the coarse outer layer of seals' dense fur coats that helps the animals retain body heat and provides protection against the cold waters.

Some species of marine mammals are faring far worse. Steller's sea lions suffering an 80 percent decline since the 1970s, now listed as an endangered species; sea otters in the western Aleutians, down 90 percent since the early 1990s; harbor seals losing 3.5 percent of their population every year.

For gray whales the first signs of trouble had occurred in 1999, when, by midsummer, more than 170 had been found dead along the northern migration route: 60 in Mexico, another 60 in California, 24 in Washington State, 7 along the coast of British Columbia, and 20 more in Alaska. For every dead whale detected, it's been estimated that possibly 10 or even 20 others never wash up onshore. If true, that would bring the count to well over a thousand. Even the known figure was a record high number, although some scientists hastened to add that perhaps we simply hadn't paid as close attention in the past (more observation equals more dead whales, the logic went). These were the same scientists saying that the gray whale population might have reached its natural limit, or what they term sustainable carrying capacity. Hence, they believed the unusual numbers of strandings were worthy of study, but not cause for undue alarm as far as the overall health of the population was concerned.

But what might be going on in the "bigger picture," the one Terry Spraker said scientists often bury their heads in the sand about? To begin to consider that question, let's pick up the grays at the primary destination of their northbound journey. This is a region south of the Bering Strait and north of St. Lawrence Island, in the northern Bering Sea between Alaska's Seward Peninsula and Russia's Chukchi Peninsula. It's known as the Chirikof Basin, a shallow area of depths usually between 100 and 130 feet. Dr. Sue Moore, director of the Cetacean Assessment and Ecology Program for the National Marine Mammal Laboratory, has made several summertime trips to this region. "In some years," she says, "it seemed you could almost walk across gray whales in the Chirikof area. They were just thick. You had the feeling of watching a big grazing herd. You'd see these huge mud plumes from the air, animals turning back and seemingly working the same area again." The lab's Marilyn Dahlheim, who also conducted studies in the Chirikof Basin in the 1980s, recalls: "At times I remember ten or fifteen up breaching simultaneously. No matter where you looked, there were gray whales. You could smell them from a distance, just like the whalers said."

It may well have been here that Scammon first observed: "They are known to descend to soft bottoms in search of food, and when returning to the surface, they have been seen with head and lips besmeared with the dark ooze from the depths below." This is what sets the gray apart from other baleen whales. It's the only one that feeds mainly upon benthic, or bottom-dwelling, organisms. In this niche it has few competitors. The unique foraging strategy is, in Sue Moore's view, a primary reason why the grays have come back so well. Right whales and bowheads, by contrast, must compete with seals, cod, and other creatures for the copepods they eat. Right whales skim along the surface, and humpbacks swim up from beneath their prey. While some opportunistic gray whales have been observed feeding similarly off Vancouver Island and elsewhere, during the five or so months that most spend in the North Country, they feast predominantly from that "dark ooze."

Gray whales start arriving at the Chirikof Basin in May and are present in large numbers by June. By this point there are nearly twenty-four hours of daylight. Only about 20 percent of the whales will stay here all feeding season, into mid- or late October; the majority eventually move farther north, at least as far as the also fecund Bering Strait. In the feeding regions pregnant females are the first to arrive and the longest to remain; commonly they feed individually. Other grays feed in aggregations of between twenty and thirty.

If we followed the whales down to the bottom, what might we see? First of all, what repast do they enjoy? Many nineteenth-century whalers believed that gray whales were eating buried shellfish. Hence, the derivation of one of their many nicknames: mussel digger. Scammon wrote: "From the testimony of several whaling-men whom we regard as interested and careful observers, together with our own investigations, we are convinced that mussels have been found in the maws of California grays; but as yet, from our own observations, we have not been able to establish the fact of what their principal sustenance consists."

Today, based largely on marine scientists' analyses of the stomach contents of gray whales, we do know. Ninety percent of what they ingest are very small crustaceans, resembling shrimp, known as amphipods. One of the world's densest communities of amphipods can be found in the seabed around St. Lawrence Island. Around ten to fifteen thousand years ago, when melting ice covered what was left of the Bering Land Bridge, the amphipod habitat was deposited: first beach sand and gravel over the silty tundra peat of the land bridge, then a thin sheet of fine sand. Amphipods have continued to thrive on massive amounts of phytoplankton that fall to the bottom during the winter. This is a particularly nutrient-rich area, bathed by productive cold water from the Gulf of Anadyr and the Bering Shelf, at the same time relatively protected from wave-generated bottom currents and too deep for extensive gouging by grounding sea ice, which is present for almost half the year.

The approximately 3,600 known species of amphipods range between a quarter inch and slightly over a full inch in length. (About the only ones that humans are slightly familiar with are sand hoppers, also known as beach fleas when encountered along coastal shores.) Here the quarter-inch *Ampelisca macrocephela* is an especially popular food item among gray whales. These construct mucous tubes out of grains of sand that penetrate several inches into the seafloor. From such dwellings amphipods snare the microscopic plankton and other tidbits. Between 3,000 and 14,000 amphipods per square meter reside in colonies or sedimentary "mats."

An average-sized gray whale is believed to consume between one and two metric tons *a day* of *Ampelisca macrocephela.* Not to mention the isopods, gastropods, bivalve mollusks, hydrozoans, and worms that creep in there. (By way of comparison, imagine us opening wide to engulf a dinner of dust mites.) Gray whales are known to spit out things they don't like, but their stomachs have also been observed to contain such inedible items as sand and pebbles.

Marine biologists long debated the mechanism by which this feeding occurs. Do the gray whales first stir the sediments with their snouts before filtering suspended food items into their mouths? Or do they plow the bottom, using their lower jaws as scoops? A breakthrough came from observing a gray whale calf captured at Scammon's Lagoon and taken into captivity at San Diego's SeaWorld during the early 1970s for a year of scientific research. Gigi, as the staff named her, wouldn't eat for the first two weeks. She had to be force-fed a concoction of cod-liver oil, vitamins, yeast, cream, water, fish, and ground-up squid. Then, as the marine mammalogist John Hall remembers it, suddenly "Gigi began eating them out of house and home. They would back a flatbed truck up to her tank and throw in these fifty-pound blocks of squid. We called them squidsicles, because they were frozen with freshwater so they'd float. Gigi would come up underneath and start vacuuming off the squid. Well, as they thawed they'd hang down like icicles on a Christmas tree and finally sink. Gigi would go down and mop up the bottom. That's where we first really saw how gray whales eat. Gigi rolled onto her side and created a suction through the right side of her mouth."

This "sucker" stratagem is unique to the grays. Swimming parallel to the seafloor, they turn slowly and bring one side of the mouth to within a few inches of the bottom. Scuba divers have reported that gray whales take in vast patches of the sea bottom, leaving shallow, elliptical depressions. Biologists have also deduced gray whales' propensity to right-sided eating, because most have fewer barnacles as well as more wear and tear on the baleen plates and abrasions where their heads touch down.

How does a gray whale ingest its food? As it opens its mouth slightly, the gray's throat grooves expand and then contract while the tongue retracts. This tongue weighs at least a thousand pounds; that's literally as big as a horse, and it has the effect of a giant vacuum cleaner. Or, from the perspective of an amphipod, of a tornado that funnels up the hometown, all its residents, and several neighboring villages besides. All of this material gets jetted over to the other side of the mouth and the awaiting baleen filters. The gray has 160 pairs of these short, stiff plates, grayish yellow in color, each up to fifteen inches long and ten inches wide. Made of keratin, the same material as our fingernails, they hang in rows down both sides of the upper jaw, looking not unlike venetian blinds. The blades run crosswise, and the inside of each plate is a fibrous fringe that compacts into a porous wall and allows water to vent back out. At its anchor point in the jaw, the top of each baleen plate is as solid as a rock. The leading edges feather into extremely coarse bristles, akin to a very stiff toothbrush.

It's an ideal system for separating out amphipods from mud. Sometimes gray whales will make two passes along the bottom, the first to stir up the top inch or so of sediment and the second to suck in the resulting cloud of floating organisms. This tactic permits them, to some degree, to avoid heavy-duty indigestion. Having gouged out say a seven-foot-long pit, a gray will rise off the bottom, dispersing a mouthful of sand and mud through the comblike baleen in a vast plume. Then the tongue will swing into action again, scraping off the food that's trapped on the baleen and pushing a few billion more members of the Ampeliscidae family down a narrow throat.

And so it goes, about 67 tons' worth of intake over these Arctic months. That's somewhere between 16 and 30 percent of a mature gray whale's total body weight, blubber-restored and ready for the next migration. If such consumption sounds gluttonous, in fact gray whales are extremely efficient feeders with very low metabolic rates. They burn about ten times fewer calories a day, per pound of body weight, than humans do.

These whales possess an uncanny ability to stay down long enough to obtain plenty of amphipod calories. Consider that they've generally been observed to take between three and four breaths on the surface—no matter whether they're submerging for two minutes in the Baja lagoons, three to five minutes along the migration, or nine to twelve minutes on the feeding grounds. To comprehend how they can do that, we need to make a few more interspecies comparisons. Gray whales store 41 percent of their oxygen in their muscles, compared with 13 percent for humans. They have two to three times more blood per unit of body weight than we do, another boon for maintaining oxygen. Their lungs have more air cells and a double layer of capillaries (versus one for us), which strongly enhance air exchange. The lungs of a whale are capable of taking about twice as much oxygen from the air as ours. They expel between 80 percent and 90 percent of their oxygen-depleted air; we top out around 15 percent. The heart of a gray whale weighs, on average, 285 pounds. During a deep dive its heart rate slows to half of its normal eight to ten beats a minute. This restricts blood flow, so the supply of oxygen depletes very slowly.

Yet none of these factors accounts for the gray whale's capacity to withstand extremely chilly water temperatures. Especially when we consider that the normal internal temperature of a gray whale is much like that of a human being, between 96 and 99 degrees Fahrenheit, but that the summertime water temperature on those feeding grounds usually ranges between 37 and 39 degrees Fahrenheit. It would seem to make sense that the blubber—five inches of exterior insulation—is somehow a key factor involved in maintaining body heat. Well, that's true enough for swimming. But it never could account for how gray whales endure with their cavernous, and blubberless, mouths open to Arctic seas for as many as twenty hours a day. How do they avoid succumbing to hypothermia after a few meals?

Recently a scientist came up with an answer. It's connected to that wondrously large tongue, which represents 5 percent of the entire surface area of a gray whale's body.

Dr. John Heyning, a thin, mustached fellow in his early forties, is deputy director of research and collections at the Los Angeles County Museum of Natural History. All of his work for more than two decades has centered on whales and dolphins. As part of a new program to study the anatomy of large baleen whales, Heyning accompanied a Smithsonian Institution colleague, Dr. James Mead, to New Zealand a few years ago. The men were looking into how whale feeding patterns are affected by ocean currents when they discovered, in the tongue of a pygmy right whale, little networks of arteries and veins that appeared to be tied in to temperature conservation. Previously, scientists had detected these so-called countercurrent heat exchangers in flukes and fins. But nobody had ever thought to look at the tongue.

Heyning and Mead brought their research to bear on two gray whale calves that stranded and died along the Southern California shoreline in 1996. Dissecting the vascular system of the tongues, they found more than fifty "heat exchangers" in parallel arrangement. The scientists labeled these the lingual retia, or tongue network. But, as Heyning later wrote, "we lacked the physiological proof that they actually worked to conserve body heat in gray whales. And proof seemed impossible to get: no baleen whales were held in captivity, and free-swimming gray whales were not likely to have much interest in allowing us to take temperature readings from inside of their mouths."

Over lunch during a conference break in Long Beach, Heyning is saying: "Well, the neat thing about science is, you never know what you're going to find. When you learn science in school, it's usually a kind of cookbook—you just add this chemical and that chemical and, like with cooking, it's usually not as tasty when you're done. The reality of science is, you often blunder into something new that takes you off in a different direction."

In this instance Heyning "blundered" into J.J., an orphaned baby gray whale found off Marina Del Rey, California, in January 1997. Heyning had considerable experience working with stranded and entangled whales and, when lifeguards noticed a small whale floundering around not far from shore, he was the first to be alerted. "I looked out and realized this was a dependent calf with no female adult around," he remembers. "The week before, there were ten-foot waves, and the whale would undoubtedly have been dead before anybody could get to it. But this was an absolutely flat, calm day. I got on the lifeguard boat, and we started following it around. It got stuck on the rocks in the jetty, and I jumped down from the bow of the boat and was able to push the animal back into the water. I'd like to say I had superhuman strength, but in reality the whale is already neutrally buoyant and was just so weak. It was clearly emaciated and sick. None of us thought she was going to make it."

By the next day dozens of news stations and thousands of people were lining the beach. The whale was still there offshore. Heyning and a host of others—personnel from the National Marine Fisheries Service, a representative from a local marine mammal "rehabilitation center," the lifeguard team, the police and fire departments—began mapping out a rescue plan. "This wasn't like a baby bird you can put back into its nest," says Heyning, "but a 1,670-pound whale! We had to drive it ashore but really had no idea how to do that. But we had a small flotilla and thought we could maybe corral it and move it in. Well, that worked for about twenty yards before it took off. The person running the *Baywatch* lifeguard boat took a water cannon and sprayed it, and the baby seemed to like that and followed along for a while. Like all kids, the attention span isn't very long. The whale started swimming toward the Marina Del Rey harbor, which we didn't want because of all the big boats in there on the weekend. At one point we jumped in the water and splashed, tried to make noise to distract it. Then, for reasons that are still totally unclear, the whale turned and went straight into the shallowest, most protected part of the beach—right where we had all of our equipment set up, and where it would be easiest for us to work."

The *Baywatch* followed the whale in. As soon as the baby touched bottom, Heyning and others jumped into the surf and tackled it. "Had we tried this the day before, the whale would have been thrashing all over the place," he remembers, "but it had basically gone into a coma; its blood sugar was all gone. Yet it happened to strand in the only place where a museum had a truck that could pick it up off the beach. And only 120 miles from SeaWorld, the only institution that had ever housed a gray whale in the past."

Even in shallow water the crew of men couldn't lift the baby in the wet sand. So the winch of a large four-wheel-drive truck was hooked by cables to a stretcher, which had been placed to carry the whale to the truck's tilt bed. As the baby gray was finally hoisted into the back of the truck, the beachful of whale watchers erupted in applause. Another group of volunteers had gone to find a closed truck, since it was unlikely the whale could survive the trip to San Diego in an open flatbed. "When we backed the two trucks together on the street, they happened to match perfectly," says Heyning. "Everybody who lived around there came with buckets of water to help keep the whale cool. After we moved it into the enclosed vehicle, several of us followed behind in a kind of parade. The police closed streets off ahead and gave us an escort to the freeway, where the California Highway Patrol was waiting to accompany the whale safely the rest of the way to San Diego."

Heyning broke away to attend a surprise fortieth birthday party planned by his wife—but this was not the last he would see of J.J. (The whale was named in honor of Judy Jones, the late director of operations of Friends of the Sea Lion Marine Mammal Center in Laguna Beach, who had recently died of cancer.) Emergency medical care at SeaWorld and several months of pampering found J.J. slowly beginning to recover. That's when Heyning paid the first of several visits. "One of my interests with baleen whales was, how do they control their tongues to get just the right flow of water coming in? Do they back-flush to get whatever they're eating off of the baleen? Gray whales really have a tremendously dexterous tongue. It looks literally like a loaf of bread in their mouth, and some of the literature talks about it not being very mobile. I quickly realized with J.J. that the tongue is quite muscular and can move all over the place. Then I also realized: I could test my hypothesis about thermoregulation while she was still being fed from a tube."

As J.J. ate and gathered strength that spring and summer, Heyning leaned over the tank with an infrared thermometer and took some measurements. Later, in a magazine article titled "Cold Tongues, Warm Hearts," the scientist would write:

> As she suckled on her feeding tube, J.J. needed to open her mouth to the water only a little way. After about one minute of feeding, the temperature of the surface of her tongue was a mere one degree Fahrenheit above that of the surrounding water, and one to four degrees F below the temperature of her head and neck region.
>
> Just as I had predicted, J.J. was losing little heat through her

Keeping J.J. alive after the rescue (Alisa Schulman-Janiger, courtesy John Heyning)

> tongue. Furthermore, her tongue remained a vivid pinkish color, indicating that the blood vessels there were not simply constricting to conserve heat. Our observations suggest that the lingual retia are extremely efficient: gray whales appear to suffer more heat loss through their body-encasing blubber layer than through the tongue, in spite of the fact that the tongue has far more blood vessels and possesses significantly less insulation.

In the evolution of baleen whales, Heyning concluded, heat exchangers in the mouth are probably as important as the development of the baleen itself—and "the mouth is a crucial site for the regulation of body temperature during filtration of their prey."

Having offered that new knowledge to whale science, fourteen months after being rescued J.J. was lifted by crane onto a truck specially equipped with a foam-rubber bed, provided another police escort to the deck of a 180-foot vessel, and lowered gently into calm seas two miles off the San Diego coast. When released J.J. was thirty-one feet long and weighed 19,200 pounds. As far as anyone knows, the now-healthy juvenile joined the northbound migration to the Bering Sea.

•

One might wonder what effect the gray whale's voracious appetite has on the ecosystem in its feeding grounds. Every year a single gray will remove amphipods from more than fifty acres of bottom sediments. Considering that populations of amphipods are pretty widely dispersed—one square foot of Bering Sea sediment might yield less than half an ounce of the critters in certain sections—grays also spend a lot of time searching. Over a feeding season one whale will sift through a hundred acres or more of seafloor. In fact, gray whales "disturb" their feeding grounds more than any other mammals, including elephants. In some portions of the Bering and Chukchi Seas, over 40 percent of the seafloor is scarred by huge, recently formed pits they've left behind. One researcher has described these as looking like "mortar-pocked battlefields from an old John Wayne movie."

All this activity, it turns out, is not detrimental at all. Rather, it plays a crucial role in maintaining the larger bottom-dwelling community. Scientists first realized this after a 1980 study in the Chirikof Basin, where they took core samples from the depressions left by the feeding grays. The amphipods' only limiting factor is space, and the whales were clearing room for more amphipods to build their tubes and attach. They were also "seeding" areas with juvenile amphipods small enough to escape through their baleen sieves. Other types of amphipods, those of the genera *Atylus* and *Anonyx*, are known to feed on pieces of worms, sponges, and other prey that the grays don't choose to ingest. So the gray whales actually help the very colony that sustains them, and more, to run at a maximum reproductive rate.

A 1991 study published by the Monterey Bay Aquarium elaborates:

> Feeding gray whales also tend to stir up nutrients that have previously settled to the bottom. Suspended again in the water column, the nutrients stimulate the growth of plankton, a primary food source for tube-dwelling amphipods and other invertebrates. Accidental farmers, the gray whales tend to their invertebrate "crops" in one final way: they prevent mud, naturally discharged into the Bering Sea by the Yukon and other large Alaskan rivers, from smothering the sandy habitat of the amphipods. Whenever the feeding whales spew sediment into the water, heavy sand and grit soon sinks to the bottom, while lighter weight particles of clay and fine silt are carried by water currents away from the feed-

> ing grounds. Whale researchers estimate that a minimum of 156 million tons of sediment is sifted by gray whales in the Bering Sea every year. This is nearly three times the amount of mud-bearing sediment that the Yukon River discharges each year into the same body of water.

And this still isn't *all* the gray whales are doing. Even the deepest-diving of birds can't get down to where the whales forage. So on the surface flocks of glaucous gulls, northern fulmars, black-legged kittiwakes, red phalaropes, thick-billed murres, and various types of auklets wait to converge upon the multiple mud patches that contain food "spilled" by the whales. Although several of these bird varieties have vastly different body sizes and bill dimensions, they select amphipods of similar size to eat. As they hover around whale slicks, it's a scene not unlike gulls following a fishing boat that's cleaning a catch and throwing morsels overboard. Some of these birds are simultaneously nesting and flying the grays' leftovers back to their young. The ornithologist Craig S. Harrison has written: "The benthic organisms brought to the surface by these whales may support several hundred thousand birds, including migrants, non-breeding juveniles from nearby rookeries, breeding adults, and fledglings."

Finally, in an ironic kind of reciprocal relationship, three relatives of the amphipods eaten by the gray whales live as ectoparasites on its throat grooves, blowholes, skin folds, and barnacle clusters. Two of these, *Cyamus scammoni* and *Cyamus kessleri*, are found hitchhiking on no other kind of whale. The largest, *Cyamus scammoni*, is also the most abundant—with one hundred thousand having been found on the skin of a single gray whale.

It was William Healey Dall who would name *Cyamus scammoni* after the captain who knew the gray whales so intimately. The Western Union Telegraph Expedition's flagship, after the near disaster not far from Unimak Pass, had maintained a northbound course along the Alaskan coast. In what Scammon described in his journal as "a hard gale" on September 9, 1865, "I this day gave orders that the men on duty on deck during rainy and bad weather should be permitted to secure a dry place aft, or near the mainmast, so that they would not be exposed unnecessarily to the spray and sea that is continually deluging the forward part of the ship."

They reached Norton Sound by the thirteenth and dropped anchor off Fort St. Michael, a trading post of the Russian-American Company. Here they were greeted by the advance land party of the scientific team led by Robert Kennicott, who joyfully came out to meet Scammon's ship. Kennicott informed them that Colonel Bulkley's steamer *Wright*, which had parted company from the *Golden Gate* much to Scammon's dismay, had already come and gone. It was heading farther north to explore Grantley Harbor and Port Clarence, planning to rendezvous again on the Siberian side. The *Wright* had left Kennicott poorly supplied, according to Dall, who wrote that many things intended for the expedition's Yukon explorations "had been used on shipboard, many were not to be found. . . . A miserable leaky dingy, or small boat, was left by the *Wright* for K's use." This situation found both Dall and Scammon enraged at the expedition overseer, who it was becoming clear wasn't much of a team player.

Scammon set about supplying Kennicott's party from his own stores, being "indefatigable in his exertion to fit him out with everything in his power," according to Dall, including "a good sail etc." for the dingy. Dall elaborated, in another letter to the Smithsonian's Spencer Baird, "Capt. Scammon has been the main stay of the Scientific Corps this season, and it was only owing to his earnest cooperation that Mr. Kennicott obtained a passable outfit." He added, "I have made some very curious observations on floating mollusca and have, thanks to Capt. Scammon, been able to keep my trawl constantly going." Despite Dall's describing the year's work so far as "one fight against ill will, prejudice, and petty spite, ever since we started," the relationship between the young naturalist and the captain was growing stronger through adversity.

Meantime, Scammon was setting down his first impressions of the Northern Alaskan Eskimos: "The Indians at first seemed rather shy of us as they paddled about the ship in their bidarka's but gradually gaining confidence ventured alongside and finally some few of them came on board. Their appearance was certainly favorable . . . well formed, tall and athletic. The bidarka is a skin boat the same as the kayaks used by the Greenlanders, that is a wooden boat frame with the skin of the seal, sea lion or some such animal stretched over it. . . . The reindeer furnished them to a great extent with both food and clothing. They are well dressed in furs and skins. . . . Their pipes are sometimes made of wood but generally of walrus tusks and like a Japanese pipe contains only a pinch of tobacco."

Scammon, Dall, and company planned now to sail west across Norton Sound, passing along the gray whale feeding grounds near St. Lawrence Island, and docking next at Plover Bay on the Russian side. Dall sent his letter to Baird describing the latest developments with another of the expedition's schooners, which was bound for St. Paul in the Pribilof Islands; from there he hoped the letter would be taken to Kodiak Island on one of the Russian-American Company's vessels, and "thence by ice-vessel to San Francisco." Dall reported that they left Kennicott "in good spirits and with a good party, our best wishes going with him. To me it was like parting with a brother. I hope next spring to return and find him, his work accomplished and pitching into Natural History. God speed him. . . . I am now left in charge of the Natural History operations for the next year. It is a responsible position and one which may prove too much for me; but I shall do my best. . . . Capt. Scammon is frankly in favor of doing everything he can to help Natural History."

Scammon recorded, "It was with sincere regret that we bade adieu to . . . Fort St. Michael." Neither he nor Dall had any idea this was the last they would ever see of Robert Kennicott. Nor could they imagine what awaited others of the telegraph expedition's party in the far northern reaches of Alaska—the territory my own journey was about to enter.

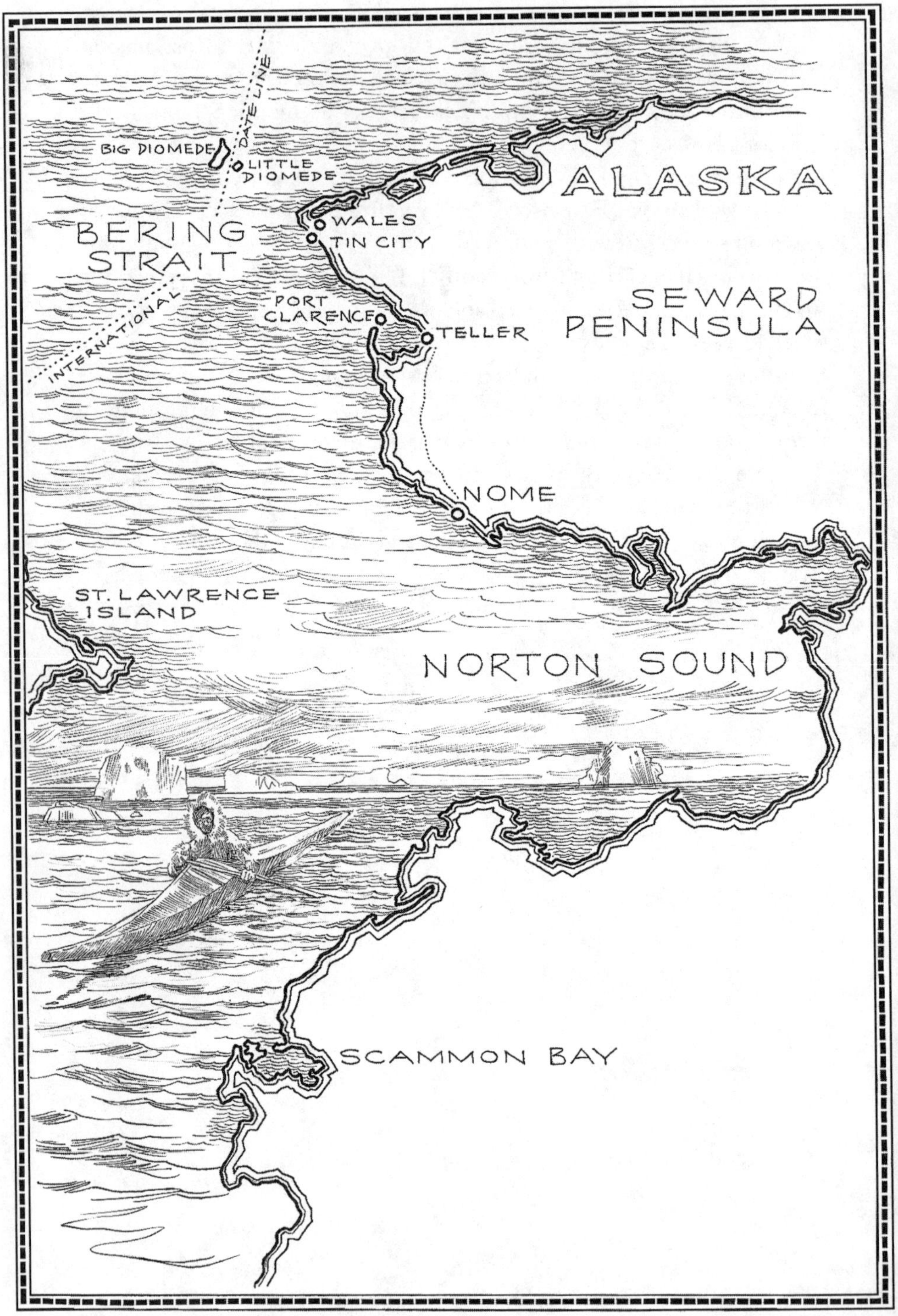
ALASKA
BIG DIOMEDE
LITTLE DIOMEDE
DATE LINE
INTERNATIONAL
BERING STRAIT
WALES
TIN CITY
PORT CLARENCE
TELLER
SEWARD PENINSULA
NOME
ST. LAWRENCE ISLAND
NORTON SOUND
SCAMMON BAY

Chapter 17

AMONG THE HUNTERS AT BERING STRAIT

In a few minutes the whale skin started to move as if it were alive. As they poured on more water it began to grow. Each morning it would need more water. It grew and grew until they could see a mouth, then fins, and finally a tail. . . . The whale grew and grew until it filled the tent.

—From an oral tradition among the Eskimo of Wales, Alaska

As the Arctic days grow longer and the winter ice begins to recede in Northern Alaska, the whale procession begins. The acolyte bowheads and belugas start moving through in early April. The pontiff grays generally don't get to the Bering Strait until early June, using the same leads in the ice. None of these three types of whales possesses a dorsal fin, so it is easier for them to swim through and sometimes under heavy ice floes. The belugas, or white whales as they are also known, are associated with the deepest water and the heaviest ice cover (they can swim a mile or more beneath the ice). Gray whales prefer the shallowest water and the lightest ice cover, with bowheads choosing the intermediate range. Unlike the other two whales, the grays occupy the same regional habitat throughout the summer and the autumn.

From the vicinity of St. Lawrence Island and the amphipod-rich Chirikof Basin, hundreds of gray whales forage farther north. They can be seen near the shores of Nome at the apex of Norton Sound, about two hundred miles east of St. Lawrence. They continue on past King Island, ninety miles northwest of Nome, where birds in the millions nest along the rocky crags of sheer cliffs towering seven hundred feet above sea level. They journey by the remote Alaskan coastal settlements of York and Tin City and, at the easternmost point of the Seward Peninsula as well as the narrowest sector of the Bering Strait, they hug the often fog-shrouded shoreline along Cape Prince of Wales.

From the rocky headlands crowning this cape, on a clear day similar terrain can be glimpsed fifty-five miles across the strait at Cape Dezhnev in the northeastern extremity of Russia's Far East. This is the closest point between the mainlands of Asia and America. Almost precisely halfway between, separated by 2.5 miles and the International Date Line, are two islands known as the Diomedes. Big Diomede belongs to Russia, Little Diomede to the United States. We look across at yesterday, they peer into tomorrow. Both islands rise vertically from the sea for more than a thousand feet, black and flat-topped, looking from a distance not unlike rocky whales themselves.

Gray whales will forage around the Diomedes as well, once the spring currents have transported the pack ice farther north. This shallow sector of the continental shelf, never any deeper than 180 feet at a given point across the strait, is the submerged Bering Land Bridge. The bridge

Scammon's California Grays Among the Ice

disappeared around ten thousand years ago, during the last warming period of the Pleistocene, when tremendous volumes of water were released from melting glaciers. Over the centuries since, cold, upwelling water from the Gulf of Anadyr has carried rich concentrations of nutrients up into the Bering Strait. Grays will eagerly seek out the "hot spots" along Cape Prince of Wales and the Diomedes, where high-quality organic carbon has settled into the sediments of the shallow shelves. Walruses and bearded seals often follow them down to feed along the same rich benthos.

The presence of all these marine mammals has long been of keen interest to the human beings who reside along the Bering Strait, and who for centuries have had to sustain themselves by hunting amid strong currents and fierce storms, here along the "bottleneck" where the Bering and Chukchi Seas collide.

At the airport in Nome, Jim Stimpfle readily picked George and me out of a crowd of several dozen men wearing caps labeled GPA—Gold Prospectors of America. "For three thousand dollars," Jim would explain, "they come to Nome for two weeks and live up the coast in an authentic gold-rush mining community made out of plywood, where you can actually go panning. All the food you can eat, all the mosquitoes you can swat, and all the lore and excitement of yesteryear. The guy who thought it up died of a heart attack trying to spend the millions he was making."

I'd been exchanging phone calls and e-mails with Jim Stimpfle for a couple of months. He is a Nome real estate broker who's taken part in several boat expeditions in the Bering Strait. Jim is an energetic fellow in his early fifties, a transplanted New Yorker who looks a little out of place in Nome's frontier-town atmosphere. He'd ended up here after Vietnam, fallen in love with a pretty Inupiaq Eskimo woman, and been on the Arctic coast ever since.

George and I were hoping that the weather might be decent enough to entice Jim to make a boat trip he'd written me about, to a little native village called Wales. This was said to be where the first land bridge–crossing migrants from Asia into North America had arrived. It was also where one of the exploring parties from Scammon's Western Union Telegraph Expedition had visited. Today it was where the Inupiat Eskimos continued to hunt whales, along with walrus, seal, and polar bear, as they had since prehistoric times.

We checked in beneath the moose antlers at the Nome Nugget Inn, checked out an unpaved Main Street with more bars per block than probably any other street in America, then set off in Jim's Chevy pickup to explore the outskirts. Jim explained that the derivation of Nome harked back to a map made in the late 1700s by Captain Cook. Next to a cape the explorer had written: "Name?" A British gentleman later failed to properly decipher Cook's handwriting and designated the spot Cape Nome. So, as Jim says, "it was a misnomer."

On a hilltop about ten miles outside town, we arrived at what remained of a White Alice site: now-deactivated twin radar towers of steel, arcing skyward and looking slippery and rather prehistoric. These were part of the DEW Line, or Distant Early Warning system, which stretched across much of Alaska during the Cold War. We were, after all, in the closest American town of any size (population, 3,500) to Russia, where a Soviet bomber could make it across the strait within six minutes. Over the next horizon we came upon the last two gold-mining dredges built in the 1930s by the Yuba Dredge Company. The dredges and the DEW Line mingled with willow and alder bushes, willow greens being, according to Jim, "one reason why the Eskimo people have survived 25,000 years, because they always had a fresh source of vitamin C. When you mix these greens in with seal oil and put them in a sealskin poke, they stay fresh all winter long."

If all this seemed an incongruous brew to a newcomer, perhaps it was appropriate to a locale where, as Jim's eldest son, Sean, described, "it's so cold in winter you can throw a glass of water into the air and it'll evaporate before it hits the ground." Right now, it was in the low sixties, about as balmy as Nome ever gets.

Over dinner that night at the Polar Cub Café with Jim and his wife, Bernadette, he invited us along to Teller, eighty miles to the north, where the road comes to an end. He'd be trailering his nineteen-foot open fiberglass boat on the back of the pickup. If George and I wanted to come, in exchange for groceries and gas, weather permitting, we could try to take the boat on down the coast another seventy miles to Wales. Jim had old friends there but hadn't been to visit in seven years. He warned us it could be a risky, even treacherous passage in any season. Fog could close in quickly. So could floating icebergs. Wales was about as isolated a place as one could imagine. Maybe we'd rather fly there on the eight-seater Cape Smythe plane that left Nome pretty much every day?

The last words of one of Jim's e-mails were simply too tempting: "As a stranger in a strange land . . . you will be following Vitus Bering, Capt. Cook, Amundsen, Stefansson, Harriman, and other explorers making the quest . . . looking . . . for whales." So we set out across the treeless tundra, with his wife driving a second car accompanied by three of their children.

As we crossed the continental divide, Jim spoke of his dream of starting a ferry service between the United States and Russia across the Bering Strait. He'd already been instrumental in raising the so-called Ice Curtain during the late 1980s. As Jim explained, the Cold War had really started in this area in September 1948, when first the Americans and then the Soviets closed their borders. "Before that," Jim said, "the Eskimo people had been traveling back and forth forever, trading and intermarrying. Not only between Big and Little Diomede Islands, but between St. Lawrence Island and the Russian mainland, even by walrus-hide skin boat between Wales and Cape Dezhnev. I used to listen to Eskimo elders talk about how much they wanted to see their relatives again."

So, while visiting some in-laws on Little Diomede in 1987, Jim had painted the word *mir* (Russian for peace) onto three pieces of plywood and tacked these onto a storefront that faces Big Diomede. Someone from the Soviet military snapped a telephoto of the sign, and somehow it ended up on the front page of a Moscow newspaper. The next year Jim began besieging the State Department and Alaska's congressional delegation with cards and letters. The result was the first "friendship flight" across the Bering Strait in 1988. After the Soviet Union collapsed, in 1992 Jim had sailed the fifty-two miles across the strait from Wales to Siberia in a successful search for some of his wife's long-lost relations.

We stop to gather some freshwater in buckets, where Gold Creek intersects with the Bluestone and Cobblestone Rivers, all three disgorging into this large inland lake called the Imuruk Basin. Around the next bend is the onetime gold-mining town of Teller, which consists of two long rows of wooden houses along a gravel spit. This is the farthest you can go by road in North America; from here on it's either by boat or on foot. We launched Jim's boat, piled in with his wife and family, and headed across a channel to their fish camp. Jim pointed toward the open Bering Sea nearby and said: "This is a protected harbor. The whaling fleets would come around these points, between those two sandbars. Or behind Port Clarence, which you can't quite see from here but we'll pass by it tomorrow. That's where the Western Union Telegraph men came in."

•

In September 1865, as the San Francisco *Evening Bulletin* reported, Colonel Bulkley and the *Wright* had "steamed into Port Clarence, where the Colonel took soundings" to determine if the depth might be right for the proposed telegraph line across the Bering Strait to Siberia. The *Wright* then proceeded under sail to take further soundings, "but meeting a large field of ice at the mouth of St. Lawrence Bay, she was prevented from entering, and steered her course for Plover Bay [Russia]." That was where Scammon's ship was to make rendezvous, a story we shall return to.

Eventually, Port Clarence would be deemed a suitable location to commence a section of the telegraph line that would continue for several hundred miles inland, across the unexplored wilderness that is today known as the Seward Peninsula. In mid-September 1866 thirty-nine men went ashore at Port Clarence under the command of Daniel B. Libby. He was a young Civil War veteran from Maine who named the settlement Libbysville. The supply ships departed. While finishing the initial part of the telegraph line to the head of Grantley Harbor, "we experienced terrible cold and stormy weather," Libby recorded, "and the thermometer fell to fifty-five degrees below zero. Great was my anxiety for my men camping in common tents as some of them were, but fortunately, none were frozen."

To amuse themselves the men of Libbysville published a monthly handwritten newspaper they dubbed the *Esquimaux*. By April they'd run out of food. All work stopped. Libby dispatched his men to live with the local Eskimos and learn to hunt and fish for themselves. One elder, Darby Gougan, probably saved their lives with his "exertions to supply [them] with food." Seal and walrus became part of the expedition team's regular repast.

By the following summer, having laid twenty-three miles of telegraph poles through the bone-chilling winter, they would learn that the success of an Atlantic cable had made their labors moot. They sailed away from Libbysville on July 2, 1867, leaving behind lumber and other supplies for the Eskimos who'd helped them.

Daniel Libby settled down in San Francisco, but he held on to the notes and maps he'd made in the Arctic. On a wilderness foray he'd seen traces of gold in one of the river valleys. Thirty years later, when word came of the discovery of millions of dollars' worth of gold in the Klondike, Libby organized another expedition—to seek out that stream where a bonanza might yet await. On April 23, 1898, his small prospecting party would make the first big gold strike on the Seward Peninsula. Council City, near the mouth of Melsing Creek on the Niukluk River, became the first permanently settled mining camp in the region. It would produce almost $12 million in gold over the ensuing generation, and the founder of Libbysville would live out his years a wealthy man.

Inside a plywood shack near the shoreline where Libbysville once stood, we sit down with Jim's family for a dinner of reindeer stew, willow greens, and dried seal. The wind gnaws at what's left of a screen door and lashes the flaps of two pup tents we've pitched outside. Jim's wife tells us that her ancestral home was King Island, fifty miles offshore but visible from here on a clear day. Her parents had lived in a wooden house suspended on stilts and moored to the rocks with plaited thongs of walrus hide. The villagers used handlines to pull themselves up from the water.

According to Eskimo legend, Bernie said, King Island was once a fish. Its first resident was a woman who came ashore and built a grass hut. One night the woman awoke to see a man, who'd felt she must be lonely and had brought her some meat. As time passed she realized that her new husband, while usually in human form, was really Polar Bear. Neither one of them, however, knew the identity of someone else who left many seals at their doorstep.

One night an angry suitor came to confront Polar Bear. This was Black Whale. The two shifted from their human to their animal forms and fought claw to tail on the beach. Finally Black Whale cried out, "I can fight no longer," and dove back into the sea. Polar Bear, victorious in battle, transformed himself for the last time into a man.

As far as she knew, Bernie concluded, the people of King Island had never hunted whales.

The next day's weather couldn't have dawned more splendidly. "If we're going to see gray whales," Jim said, "this would be the perfect day." Around 11:00 A.M. he motored Bernie and the kids across to the car. Bernie made sure we had extra foodstuffs in case of unforeseen delays. Then George and I boarded the boat with Jim and his twelve-year-old son, Ian.

We cruised past a small Coast Guard station at Port Clarence, after which the coast bore few signs of habitation. Mounds of driftwood lined the beaches. Jim said they'd washed down the Yukon River as far as 250 miles, becoming a strange source of firewood in a treeless landscape.

Nature's provisions. These looked sometimes like totems, having tumbled into place in the form of a human sentinel or a great bird hovering on the shoreline. Inanimate objects here possess such presence. A rock of weeping moss shields black cormorants. The jagged, ocher-colored spires of a cliff seem to envelop in tentacles the white-striped auklets. We pulled ashore briefly at Lost River Valley, an old gold-mining site where Jim has taken a snowmobile through the gorge in wintertime.

Up ahead a first sight of floating Arctic pack ice is unforgettable. It's maybe a couple of hundred yards offshore, stretching as far as the eye can see, and we'll be passing right alongside it. I remind myself again that it's late June. Jim says that, only two weeks earlier, this was where he'd gotten socked in for three days. I realize if we did have to stop because of the weather, we could wake up iced in and unable to move on. Images of polar bears surface from my subconscious . . . and black whales, too. But no spouts appear on the horizon.

"All the native hunting is from the ice ledge," Jim says. "Even now when it's broken, they still attach their boats to the icebergs and float with them when they're hunting seals or walrus. Every day they check the strait to see if it's frozen in."

Even from the ice edge ten miles away, Cape Prince of Wales is a stunning sight, rising abruptly to over two thousand feet. The wind is picking up again. I notice a shining white radar dome glistening from a hillside above a rocky, barren shore. "That's Tin City," Jim says, "built in '52 to watch for Soviet bombers, and still a working Air Force base. Also, until recently, the only working tin mine in North America."

If you see a crystallized vapor trail flowing around the dome, he adds, that means you'd better put in for shore here because the wind won't let you land around the bend. So far, so good. Around the bend the village of Wales opens onto a rich shoreline that stretches north to emerald green marshes and barrier islands. We've made a trip known to take as long as two weeks in less than four hours.

The ancient trading center of Wales had been noted on the earliest Russian maps, when North America was called simply Greatland by the Russians. Northern Alaska was first sighted, near Wales, by the Cossacks in their little ship *Gabriel* in August 1732. Almost a half century passed before the shoreline was seen again, when Captain James Cook sailed by on his final voyage seeking the Northwest Passage and called Wales "the western extremity of all America hitherto known." On the rim of the continent there existed probably the largest Eskimo settlement in all Alaska. Wales was known then as Kingigan, which means "High Bluff," and it held upward of four hundred people.

In 1848, ten years before Scammon's discovery of the gray whale's Mexican lagoons, a whaling captain named Thomas Roys had made an equally pivotal find here in the Bering Strait. Few vessels had ever plied these dangerous and yet-uncharted waters, let alone tried to harvest whales. Captain Roys had heard rumors of "strange whales" in the vicinity but concealed the course of the bark *Superior* even from his crew. When they'd found out where they really were that July, there had been a near mutiny. Now his small ship lay becalmed in the strait, with a strong current carrying it north in the direction of the Diomedes. From the mainland at Wales seven walrus-skin canoes were moving in the *Superior's* direction carrying large numbers of natives. Captain Roys stood on the deck with a pistol, having no idea how hostile they might be and knowing his crew was vastly outnumbered. But a sudden southeasterly breeze moved his ship out of their reach. A heavy fog soon enshrouded the whaler. When it lifted after twenty-four hours, the *Superior* found itself surrounded by "strange whales."

This whale came to be called bowhead because, as Scammon would describe, the most striking feature "is its ponderous head, forming, as far as our observations go, more than one-third of the whole creature." As strictly an Arctic species, bowheads had never before been targets of men like Roys. These massive animals, growing to lengths of seventy feet and weights of sixty tons, possessed blubber coats as thick as a foot and a half. When this blubber was rendered into oil, it yielded vast amounts compared with other species'. After Roys's ship filled its hold with 1,800 barrels of oil that summer, it didn't take long for the word to spread. Within four years more than two hundred vessels would be operating in these waters. The great polar whale, as the bowhead was also termed, became "by far the most valuable, in a commercial point of view," Scammon wrote. And, just as happened to the grays once Scammon penetrated their nursery areas, the bowheads feeding in the Arctic would soon be driven close to extinction.

Scammon whaled twice in the Arctic, the last time in 1862, and would recall: "The Esquimaux about the north-western shores of Behring Sea speak about the *Balaena mysticetus* [bowheads] resorting to the bays when the 'small ice comes,' and they look forward to that season as a time of plenty, and reap a kind of marine harvest by catching numbers of them, thus securing an abundant supply of food for winter store."

As for the gray whales, "In those far northern regions, the animals are rarely pursued by the whale-ship's boats; hence they rest in some degree of security; yet even there, the watchful Esquimaux steal upon them, and to their primitive weapons and rude processes the whale at last succumbs, and supplies food and substance for its captors."

Initially those "watchful Esquimaux" did not bear ill will toward Western newcomers, as Captain Roys had feared. Early in January 1854, Captain Henry Trollope made plans to cross the Bering Strait to Siberia in one last effort to learn something about the fate of the famed British explorer Sir John Franklin. In the course of searching for a passage between the Atlantic and Pacific, Franklin's two ships had last been seen almost nine years earlier in Baffin Bay, Canada. The largest search party in the history of sail—eight Royal Navy ships and a private yacht—had gone looking for Sir John. Captain Trollope left Port Clarence by dogsled for Wales, from where he and his men hoped to reach Siberia by way of the Diomede Islands. When they rounded the cape sequestering Wales, the rugged, broken ice suddenly turned smooth, "studded with innumerable holes, each surrounded by a snow wall, within which people were fishing."

Trollope went on to record that "our arrival seemed to attract a good deal of attention, and as we neared the village, the whole population turned out to meet us; men, women, and children, the latter shrieking and shouting, wrestling and tumbling one over the other with great glee; the anglers left their rods and lines, which they very dexterously haul up and wind on a short rod about two feet long, and accompanied us up the steep bank on which the upper village stands."

The captain was ushered into a belowground dwelling with a long, low, narrow entryway that trapped the cold air outside. He found the interior "neat, tidy . . . the floor smooth, clean swept and polished, two cheerful rows of lamps or lights, burning almost with the brilliancy of gas, gave the place a most comfortable and warm feel, most grateful to us cold and wet as we were." As the master of the house was out hunting seal, Trollope wandered around with a large crowd in his wake, being warned that the neighbors of another village on a sand-spit below were "great thieves." (Trollope noted, however, that he was "treated with equal civility" down there.)

That night his Eskimo host's wife repaired their shoes and "even a woollen sock was mended, with a piece of fawn skin." He dined upon chunks of hot seal meat in a rich gravy. When Trollope's crew departed almost a week later, all of Wales came to see them off. "They are intelligent and ingenious in a very high degree," he concluded, "displayed in their habitations, in their boats, their sledges, and their weapons."

When Daniel Libby and his Western Union Telegraph team ventured to Wales a dozen years later, they partook of a feast amid some 150 Eskimos where the dancers "stripped to the waist, with their fancy knee-breeches and decorated boots, their heads set off with beads and feathers" and "pow-wowed in a frantic manner." The next morning Libby saw an "immense crowd, assembled on raised ground about one-fourth of a mile distant, and who seemed to be attending a mass meeting, or listening to a stump oration. On enquiry we learned that the gathering was on account of bad weather and lack of *cow-cow* [a whaling term for food], and for the purposes of making offerings and prayers for more food and temperate *selami* [Eskimo for weather]."

As we approach Wales a long, low sandbar causes several rows of surf to begin breaking well out from the shore. "Get ready to jump and bring the boat through the last of these breakers," Jim shouts, "but it can be tricky." Scarcely are the words out of his mouth than a deserted beach is suddenly lined with motorcycles and their larger, sturdier, four-wheeled counterparts, arriving from different directions over the dunes. Half a dozen men wade out to help pull us ashore. Waiting there is a petite Eskimo woman straddling a four-wheeler. She greets Jim with a warm hug and introduces herself to me as Ellen Richard. Her husband is out grading the road but ought to be back soon, and we should all come up to the house.

Today about 175 people live in Wales. Behind the beach, down a few sandy lanes and a little ways up a steep hillside, the village stretches for all of a mile. The majority of the wooden buildings are old and in serious decay. Housing consists of many prefab government dwellings, interspersed among some older homes with steep Cape Cod roofs. There's a school, a church, and Wales Native Store. In the town center stands a white, three-story geodesic dome: post office and clinic on the ground floor, bingo hall on the second floor, town offices on the top floor. Jim explains it was built on the site of the ancient *qualqi*, or ceremonial village meeting place, by Ellen's husband, Dan Richard, who'd gotten the design idea from reading Buckminster Fuller.

The Richards' seaside house at the end of the road is the largest one in Wales. They started building it twenty years ago, and it remains in a perpetual state of renovation. Nonetheless, it's so homey that we scarcely notice the open two-by-fours attached to the long ceiling. Beyond the entryway is a spacious living and dining area: an alcove lined with computers, a woodstove that's drying some wet socks and boots, a multitude of hanging plants, shelves lined with a collection of local history books behind a couch that faces a satellite TV, potbellied kitchen stove opening onto a round dinner table, bedrooms and bathroom beyond.

A bulldozer chugs to a halt outside, and Dan Richard walks in. He's a Bunyanesque figure, as massive as his wife is diminutive. He wears a bandanna around his neck, a carpenter's tool belt, and overalls over a plaid shirt. Under the suspenders are a pair of shoulder holsters, each bearing a pistol.

Later I'd learn how Dan came to be here. We'd be walking along the tundra when he spoke of having grown up on a farm in California. He'd been part of a flight crew during the Vietnam War, shot down over Cambodia, spending months in a POW camp where the Khmer Rouge applied bamboo stakes to his body and set them on fire. They'd especially enjoyed torturing him because of his size. One night he'd been rescued by Australian commandos. In a malarial fever he'd walked to safety from the Killing Fields.

Upon his return to America, Dan had been assigned to an Air Force base in Las Vegas. After all he'd been through, the neon-drenched casino capital proved too much for his psyche. He'd asked to get stationed the farthest possible distance from civilization. Tin City was the loneliest place they could find. Dan arrived in the mid-1970s. Soon he took to hiking seven miles around the cape mountain to Wales. There he fell in love with Ellen Oxereok, the sister of an Eskimo friend.

Now, over a dinner of one of Ellen's specialties—pasta with a moose sauce—Dan nodded his head as I explained that I'd come to Wales to learn what I could about Eskimo subsistence hunting, especially for gray whales. "When I first got close to a gray whale—this was many years ago—I was out in a boat with Ellen's dad," he said. "We were standing on the ice talking when we heard this real loud whooshing noise. Right next to us, the ice started moving apart, and this big old gray whale wiggled his way up through it. You could see the barnacles, the little crabs on its back—right there!

"It's a brine shrimp over around the point at Tin [City] that the gray whales here like a lot," Dan continued. "Usually they're just thick this time of year." Then, in an echo of what I'd been hearing all along the migration route, he added: "But you know, the weather's changed quite a bit since I first came up here. Just seems like more severe winters and no

Dan Richard with his wife, Ellen, and son Sherman (George Peper)

good springs and the summers have been real late, too. Something's going on. Here it is almost July, and the gray whales are only now starting to move through."

He reached for a salad bowl heaped with willow greens and added, "But when there's a gray whale comes by, you can bet that Raymond and them'll be out after it. About a week or so ago, his crew was chasing one, or trying to. They don't look to get too close to 'em, because there's been a few knock their boats out of the water, and it's kinda deadly."

Dan was referring to a hunting captain named Raymond Seetook, and after dinner I put a call in to him. "The gray whales haven't been traveling," he confirmed. "Still too much ice out there. We're waiting."

Dan offered to put us all up. Jim and his son Ian could spread their sleeping bags out beside the woodstove. George and I could have the bedroom of the Richards' teenage son, who was away for a few days taking a computer class. Ellen announced she was "going picking" in a greens-filled meadow not far away, below Razorback Ridge. The rest of us set off in Dan's pickup for a look at the surrounding area, driving past the dirt field that serves as Wales' landing strip for planes, into the outskirts of town.

Dan, who maintains three mechanic's shops attached to his home, runs all the heavy equipment in the area. "This is a mighty fine road," Jim complimented him. "Not too bad," Dan replied. "All I did was four passes today, didn't do any real work on it yet."

"This is the road to Tin City?" George asked.

"This is the road to nowhere," Dan said as we rounded a smoothly graveled bend. "We argued with the Bureau of Indian Affairs forever on it. Their excuse was, they can only build a certain amount of mileage for any given project. It's 11.3 miles from Wales to Tin City. They'd only contract to build to 4.1. So the road costs $535 a foot, and goes to nowhere. The worst part is, originally we started at the other end, pioneered a bypass road through. When we first brought the heavy equipment over to here, we came through a valley and right up over the tops of those mountains."

We passed Loch Lagoon, where Dan pointed out a local herd of reindeer penned behind some wooden corrals. "Except the wild caribou have started showing up these last few years, and taken some of the reindeer off into the hills with 'em." Down these hills a considerable amount of snow still cascaded toward the road. There had been more snowfall this past winter than Dan could ever remember. Typically, they just had the cold—somewhere between sixty and ninety below—and the wind, whipping along at an average speed of twenty-two knots. We stopped beside a dry streambed where Dan said his family went canoeing over what had been a raging river—only a week ago. Things change quickly in the North Country.

Along the edges of the hills, blackberries and salmonberries grew amid the volcanic rock. We climbed the tundra to pick a few, savoring the wild taste as we sought sure footing on the spongy earth. *Tundra* is a Russian word for an Arctic land area covered by lichens, moss, and grass. It's a rich, rust-colored layer of plant life above the frozen permafrost. "Most people don't realize that tundra grows," Dan said. "In the wintertime, the ice could be six to eight feet deep underneath it." Now, in summer, the tundra also blooms with wildflowers—forget-me-nots, cotton flowers, and wild lilacs.

A raven flew overhead, and Dan said: "Ravens are like your best friend in the country. They always tell you where game is. Where you see them circling, head that way. Dad told me about that years ago."

Dad was Ellen's late father, Frank Oxereok, Sr. "I spent my first few years up here with him to learn how best to survive up North. Another thing Dad said: Herders in the old days and even still today, whenever they kill deer, they always face the antlers to home. So if you're ever lost in the country and you run across some, follow the antlers."

We came to where the road ended at nowhere, and Dan backed up a steep grade until he could find a place to turn around. It was well past midnight. The sun cast mauve shadows over the horseshoe-shaped valleys. Dan swept one of his huge hands across the horizon and said, "Sometimes it's empty, but other days I've seen this land so thick with cranes, it looks like a concert with a hundred thousand people standing around." On the way back he pointed up a rise. "There's a lotta neat geometry out here," Dan said. "That rock structure up there looks almost like a sled." Indeed, it seemed to coax a passerby to trek up to its limestone ledge, but we kept moving.

"One thing you learn about the Arctic," Dan said. "Mother Nature only gives you one chance." This soft-spoken bear of a man started telling stories. About polar bears. "I used to run a trapline down the coast, and I finally gave up because there was too many standoffs between me and the polar bears. They'd get between me and the snow machine, and it was a matter of, okay who's moving first? They're pretty tricky, too. They can sneak right up on you before you even know they're there."

It wasn't difficult to picture Dan almost nose to nose with a polar bear in a staring contest. All polar bears are left-handed, he was saying, so if ever one swings at you, move to your left. "They're the most amazing animal, I'll tell you. I've watched them poised for hours on the ice above a seal's hole, waiting there with one paw across their nose. Because they know their nose is black, and they're actually covering it!

"Polar bears aren't really like bears at all. The way they move, their head and shoulders are more like a snake. Most bears will track you and maul you. Polar bears are the only bears in the world that'll track you and *eat* you. If they're hungry, no matter where you go they're gonna follow."

He'd had some close calls. Some years back he and his eldest son were about forty miles "back up the hills" cruising along on a snow machine when Dan glimpsed something moving across a hillside. Thinking it might be a moose, he headed in that direction. "All of a sudden, the snow just kinda leaped out at me," Dan said quietly.

"From how far away?" George asked.

"About as far as from me to Jim there." Jim was sitting on the other side of the backseat. "It was about a twelve-footer. Somebody'd wounded it, because there was a big festering hole in its side." Dan said he had floored the snow machine, spun around, and shot the polar bear "with some specially made guns that I carry with me just for those reasons, made for my hands."

Those hands loomed on the steering wheel like massive paws. "Polar bear claws don't retract," he added.

Ellen came into view along Razorback Ridge. Greens overflowed the top of her basket. She placed it on the back of her four-wheeler. Dan waved, pointed back at the cliff side and said: "Many generations of Wales people are buried on Razorback. The higher on the hill you were, the higher status in town you had. Walk around up in there, you'll see old caskets and bones. You might see a lot of things."

On a walking trail that winds across Cape Prince of Wales, I'm with Luther Komenseak, captain of one of the five Inupiat Eskimo whaling crews in Wales. I'd been told to expect a reticent man, since I am a white outsider. In fact, Luther turned out to be engaging and loquacious. He is in his forties, tall and muscular, and sports a black cap with a whale embroidered on it. His grandparents had moved to Wales from a now-abandoned village across the strait in Chukotka. They brought their belongings in a skin boat and built their first sod house around the point at Tin City.

The wind here feels omnipotent, but I'm grateful for its choke hold on the equally omnipresent mosquitoes. Luther sweeps a hand toward the sea below and says that, in other years on any given June day from this spot, you'd see at least a hundred gray whales going by. You'd see blows about every five minutes from as many as half a dozen at a time. So far this year very few had been sighted. The ice never used to be this thick this late. You could count on being able to start hunting at a certain time.

"Well, right now we have the wrong wind direction anyway," Luther continues. "Being on a point, any kind of south wind—especially with the power of the wind—it's too bouncy for us to go out." They didn't generally launch their aluminum skiffs into a northwest wind either, because the area near the beach is so shallow. They needed a calm offshore breeze from the north or northeast.

So the crews are "on standby" right now. The village prefers to hunt and eat bowhead whales and has a quota of two bowheads per year granted through the Alaska Eskimo Whaling Commission. But no bowheads had been landed during their migration past Wales this spring or, in fact, since 1996. So, according to Luther, permission for taking of a gray whale had been obtained through their tribal council.

He lights a cigarette and points to something on the seaward face of the mountainside. It's a stone figure, distinctly shaped out of an assortment of rocks. For a moment I could swear it's an Eskimo wearing a parka. The figure is hunkered down with arms outstretched toward the Bering Strait. Vigilant. Ominous.

"This is one of the stone men," Luther says. "The Inukshuk."

Long ago the people of Wales had fashioned the rocks into the Inukshuk. They were designed to be silent sentinels, guarding the passageway into North America against invading Siberians. I thought to myself, They were the Cold War centuries before there *was* a Cold War. They stood watch beside craggy hideouts, perhaps five feet long and six feet in diameter, room enough for a couple to take refuge with their furs and ivory.

The Inukshuk stood watch, too, over the ancestors buried aboveground on the adjacent Razorback Ridge, their gravestones marked by whale bones. "A lot of them were buried with their possessions," Luther says. "There's even a whole skin boat, and another person with all their polar bear skulls." And there was a certain rock on this burial slope, a white rock known as the Door. It was said to lead into the realm of the spirits, because a great Eskimo shaman was buried there. Sometimes this Unutkoot would appear in the form of a comet flashing its tail above the heads of the hunters out at sea.

Across a bridge at the base of the mountain, not far from Dan's house, is what Luther calls "the historical site." This was an area of town separated by an inlet during the time when Wales consisted of three sectors, none of which got along. The so-called local island people had their own chief and clan. Some say it was a murder, others a rape that precipitated their fearing revenge from their neighbors and pulling up stakes. Early one spring morning their Unutkoot is supposed to have summoned a thick fog. The people silently paddled away in their skin boats. "They migrated as far north as Barrow," according to Luther, "so we've got lots of relations along the coast. The people up there always call us, in our native tongue, the Unexpected."

Luther and I have descended the mountain and are standing on the seashore. Behind us, in a juxtaposition that diffuses time and space, a satellite dish stands not far from where a walrus-hide skin boat, or *umiak*, has been put upside-down on a driftwood rack. Luther says skin boats were still used for hunting when he was growing up, because they were light and handled well in variable ice conditions. Usually they were between twenty-six and twenty-eight feet long, but his grandfather's stepdad had a thirty-eight-footer. That was the one Luther had learned in, "all paddling in sequence and so quiet too." Skin boats are still used on Little Diomede. But here in Wales, Luther figures, 1985 was probably the last time.

> *The Esquimaux whaling-boat, although to all appearance simple in its construction, will be found, after careful investigation, to be admirably adapted to the purpose, as well as for all other uses necessity demands. It is not only used to accomplish the more important undertaking, but in it they hunt the walrus, shoot game, and make their long summer-voyages about the coast, up the deep bays and the long rivers, where they traffic with the interior tribes. When prepared for whaling, the boat is cleared of all passengers and useless incumbrances, nothing being allowed but the whaling-gear. Eight picked men make the crew. Their boats are twenty-five to thirty feet long, and are flat on the bottom, with flaring sides and tapering ends. The framework is of wood, lashed together with the fibres of baleen and thongs of walrus-hide, the latter article being the covering, or planking, to the boat. The implements are one or more harpoons, made of ivory, with a point of slate-stone or iron; a boat-mast, that serves the triple purpose of spreading the sail and furnishing the staff for the harpoon and lance; a large knife, and eight paddles. The knife lashed to the mast constitutes the lance.*
>
> —SCAMMON, 1874

By the 1870s the Yankee whaling fleet was enlisting the villagers of Wales to assist in its catches. And Yankee technology came to supplant the time-honored hunting methods of the Inupiat that Scammon had observed. The lances were shelved. The harpoons were now fixed with triggering devices that fired black powder bombs into the whales. The darting gun came to be used for bowheads, the heavier shoulder gun for gray whales.

At the same time schooners from San Francisco and Hawaii were making yearly runs to northern Alaska, trading their rotgut whiskey and contraband firearms to the native peoples for furs and walrus ivory. Amid these forays of the Western world into their territory, new attitudes apparently arose in Wales. One U.S. Revenue Marine captain described these Cape Eskimos as "great bullies . . . the worst on the coast," who traveled "in large numbers compelling smaller bands to trade with them at their own terms."

In what is still remembered here as the Massacre of 1877, thirteen Wales men were killed after they allegedly tried to pirate a trading brig from Honololu that lay surrounded by fog in the Bering Strait. Not surprisingly, there are different accounts of what happened. The ship's captain, George Gilley, would later tell a reporter that the Eskimo chief was "about six feet five inches tall, by far the most powerful native we had ever seen," and a "murderous villain" who demanded ammunition and rum. According to Gilley, each Eskimo drew muzzle-loading pistols and tried to shoot the crew, which managed to overpower them. "Toward the last we hauled them out with gaff hooks," the captain added icily.

Ellen Richard's uncle, Ernest Oxereok, heard quite another version from his own grandfather. The Wales men had been accused of cheating the trader, but the actual culprits were from a different village. A great Eskimo warrior, a master of Kung Fu–type tactics, had almost won the battle single-handed. In the end he was shot and thrown overboard with the others. The bodies are said to have been recovered and are buried where the whale ribs emerge from the cotton grass in back of the village.

Not long after the Massacre of 1877, Christianity had come to Wales. It was brought by a missionary named Harrison Thornton. Luther Komenseak had pointed out to me a white gravestone on a bluff leading away from the village. A GOOD SOLDIER OF CHRIST JESUS, read the epitaph from 1893. One night that year three teenage boys had rigged one of the new whaling darting guns on the preacher's door. Then they'd knocked. When Thornton answered he was blown away. Dan Richard explained what ensued: "In those days, if you did something wrong, you went before the Council of Elders and you paid for the crime. Everybody in the village understood that laws had to be upheld. Two of the boys were taken to a clearing and killed by their uncle. The third was shot in the leg."

Next to devastate Wales was the Spanish influenza epidemic that swept the globe in 1918. The lethal flu was carried to the village by a visitor stricken with the virus. Within four days the majority of its population was dead. The survivors were too weak to carry all the bodies to Razorback Ridge. Between three and four hundred people, including all the elders, were buried in a mass grave in the dunes.

Where there had been twenty-three whaling boats, there weren't enough villagers left who even knew the rudiments. Years of occasional, halfhearted attempts went by until, in the late 1960s, a Mormon teacher from Utah named Charles Christiansen came to live in Wales. He saw resuming whaling as a way of revitalizing a community tradition. Nearly all the equipment had been sold off by the villagers; only two darting guns remained. Christiansen set about hand-crafting a harpoon shaft and building skin boats.

Luther Komenseak, then in the eighth grade, assisted Christiansen. Luther also remembered editing his school newsletter, which contained a section called "Whale's Tales." He commenced taking a "whole series of lessons, different stages," starting out in the stern of one of the five skin boats that Christiansen would complete within a four-year period. In 1969, and again in 1970, both a bowhead and a gray whale would be landed by the people of Wales.

As Luther told the story, I felt almost assaulted by ironies and contradictions. The first missionary had been killed by a darting gun. The last missionary had put the darting gun back in the villagers' hands. I thought of something else Dan had said: "Today's elders, when they were growing up, were tormented by the missionaries. To the point where they couldn't Eskimo-dance and, if they spoke their native language, they were in trouble. So when *they* had kids, they wanted them to have an easier time and so didn't teach them these things and weren't strict with them either. This next one became like a lost generation."

George and I are sitting now in Luther's kitchen. It's very simple and functional and neat, with family and religious photographs adorning the walls. His wife, Christine, is serving us coffee. Luther is rustling through a stack of handwritten pages. These are notes and correspondence once kept by Christiansen. The teacher, when he left Wales, had given it all to Luther's father, who had passed it along to Luther. There were some photographs of whaling gear, sent from a veteran whaler who lived on St. Lawrence Island. There were letters from other Eskimo whalers at Barrow and Point Hope, describing how to load the bomb and push the shell, and how to divide up the prize among the village. There were several pages of specific "instructions." Luther held these up and started reading aloud.

Best targets and proper order for the striker: First a neck shot, aimed forward and down. Then a body shot from the side, also aimed forward and down. If you're within eight or ten feet and only the head is out of the water, try to put the bomb in the eye, aimed in and down. If the head is pointing up and the back is down, he'll go backwards and then forward close to the surface, when you've maybe a chance to strike with the shoulder gun. If the whale is coming at you head-on, aim two feet down into the water.

> *The boat being in readiness, the chase begins. As soon as the whale is seen and its course ascertained, all get behind it: not a word is spoken, nor will they take notice of a passing ship or boat, when once excited in the chase. All is silent and motionless until the spout is seen, when they instantly paddle toward it. The spouting over, every paddle is raised; again the spout is seen or heard through the fog, and again they spring to their paddles. In this manner the animal is approached near enough to throw the harpoon, when all shout at the top of their voices. This is said to have the effect of checking the animal's way through the water, thus giving an opportunity to plant the spear in its body, with line and buoys attached. The chase continues in this wise until a number of weapons are firmly fixed, causing the animal much effort to get under water, and still more to remain down; so it soon rises again, and is attacked with renewed vigor.*
>
> —Scammon, 1874

In some ways, as Luther paints the picture, today's chase is not so dissimilar. The captain steers the boat, with bowman and harpooner up front. Primary and secondary "float men" ensure that all the lines are in order, that the rope won't slip or the float snag, because, once a whale is hit with the harpoon, there will be no time for corrections. You have to know how to pick up your paddle properly, without its rubbing against the boat. You look to approach whales from behind, where they don't seem to sense you as easily. As you near the whale, there's a certain way to coil the harpoon. After a successful strike, when the whale surfaces you want to be on the right side and away from the flukes.

We walked outside onto the slope where Luther lives, where we could still see the foundation holes of the prehistoric sod houses. Occasionally squirrels unearth them, and archaeologists have come to examine their remains. Many of these are delineated by whale bones jutting out of the ground at odd angles. Luther says if you look closely you can tell the difference between the gray whales' and the bowheads'. "What puzzles me," he says, "is why my ancestors always cut a section out of the whale bone, always in the same shape. That's true on the Siberian side, too."

In a wooden shack down the hillside, Luther keeps various whaling implements: An old-style harpoon with a rounded blade. A lance his cousin made for him. Even though these have long since gone out of fashion, Luther always carries the lance at sea. He even used it on his first whale, when it was floating upside down after being harpooned and "darted." That was a bowhead taken in 1991, four years after he started his own crew.

When hunting bowheads in the spring, the men of Wales set up a camp on the pack ice. The whaling itself takes place in open water corridors, or "leads," that the bowheads utilize to move through. They're easier to go after than gray whales, because the ice hasn't yet melted when they're passing, so they're more confined. Bowheads are also "very gentle whales" compared with grays, which are faster-moving and especially aggressive when a calf is around. "There's a certain time of year that mothers and calves go by, we all know that from our elders and our observations," Luther says. "You notice when there's a calf, because the mother will come up for air more often. We leave the mothers and calves alone. That's mine and everybody else's respect for whales."

Luther estimates probably five or six gray whales have been taken since whaling started up again in Wales, mostly in shallow water north of the village. They watched for a distinctive ring left by the tail, which I knew as a fluke print and here is called *cameliok*. "Old folks always say, you never get scared of the whale because the whale will sense this right away," Luther says. The "devil-fish" tales were alive and well in the Arctic. A crew would generally stand well off and shoot, using the fifty-pound shoulder gun to fire the harpoon penetration bomb with its timed fuse.

There was this story: One time a crew split apart six gray whales and pursued only the medium-sized ones. Closing in on a trio, the gunner fired and hit one, but the bomb failed to go off. The other two whales slowed their pace, turned, and came around to join the first. All three whales then moved synchronously under the boat. They lifted the boat out of the water with their tails. The one that had been shot broke the motor mount. The three whales all vanished. Another crew towed the boat back to shore.

There was also this story: One of the whaling captains had shot a thirty-foot gray whale, killing it instantly, even before the bomb exploded. Two other whales were in the vicinity, separated by perhaps a hundred yards. They closed ranks. They came toward the boat. The dead whale was already floating belly-up. The other two whales came up from below. They "wrapped their arms" around the dead whale, "under the whiskers," one of the crew would remember. And then they dove and carried it to the bottom of the sea.

Around the same period, in 1985, a thirty-five-foot-long gray was made fast to a twenty-four-foot-long skin boat. The whale dragged the crew more than two miles out to sea on a "Nantucket sleigh ride." It took twice as long as the sleigh ride to bring the whale ashore.

> *It is the established custom with these simple natives, that the man who first effectually throws his harpoon, takes command of the whole party: accordingly, as soon as the animal becomes much exhausted, his baidarka is paddled near, and with surprising quickness he cuts a hole in its side sufficiently large to admit the knife and mast to which it is attached; then follows a course of cutting and piercing till death ensues, after which the treasure is towed to the beach in front of their huts, where it is divided, each member of the party receiving two "slabs of bone," and a like proportion of the blubber and entrails; the owners of the canoes claiming what remains.*
>
> —SCAMMON, 1874

The gray whale caught in 1985 had fed the village all winter. "We have a distribution format that's been handed down to us," Luther says. When a whale comes ashore, "first somebody cuts a piece of the *muktuk* [blubber] and runs it up to the village, for evidence—that's what we call the runners." Luther's last whale, a bowhead taken in 1994, had required sixteen hours to "dress," although with considerable help the process can happen in half that time. The captain gets what Luther calls "the choicy parts," the tender and thicker *muktuk* on the bottom of the belly around to the tail. The striker's piece is from underneath the chin, it's the captain's choice how big or small. One side of the lips goes to the captain and crew. Those from the second and third helper boats get a cut down the back. Much of the meat from the shoulder blades forward goes to those who assist in pulling the whale onto the beach with block and tackle and with the butchering. The community makes its own tools for the job, long-handled knives akin to spades.

After Luther landed his first bowhead in 1991, a celebration was attended by three hundred people from all the outlying villages. Diomeders and King Islanders came and danced. On April 19, 1995, when Wales's last bowhead was taken by Raymond Seetook along with his two brothers and his nephew, another huge feast was held at the school gym.

Raymond helped his wife prepare the *muktuk* and meat from the sixty-foot-long "big one." Eskimos came from as far away as Nome and even Anchorage. And there was a contingent from Shishmaref, an island seventy-some miles north where many are related to the folks of Wales. They'd been given their fair share. But the night after the feast, about half of what remained of the whale disappeared from a "meat hole" covered over with canvas tarps. One of the Shishmaref citizens is said to have piled so much onto his dogsled it broke. Now, when the Shishmaref people journey to Wales, it is said that "*Novoks* are coming." *Novoks*, in Inupiat, means "seagulls."

> *The choice pieces for a dainty repast, with them, are the flukes, lips, and fins. . . . The entrails are made into a kind of souse, by pickling them in a liquid extracted from a root that imparts an acrid taste: this preparation is a savory dish, as well as a preventive of the scurvy.*
>
> —SCAMMON, 1874

When he landed his first whale, Luther remembers, "we ate it for one week straight, all different ways—fried, boiled, dried, raw—and my kids got really healthy, no bad colds, rosy cheeks." Sometimes his family will have spaghetti or a hamburger, but, he says, "if I don't eat my native food, I'm not gonna be too happy."

Luther says everything gets eaten—stomach, heart, liver. They even scoop out the brain to make a kind of pea soup, and the intestines are very good, too. They slice off the cartilage and eat what's in between the vertebrae. Different parts of the whale have varying tastes.

At the opposite end of town, Dan Richard has respect for utilizing every edible part of the whale—"that's nice to see"—but to him whale is assuredly an acquired taste. The *mam-mok* between the baleen "tastes basically like bubble gum." *Muktuk*, slices of skin and blubber, reminds both Dan and Jim of Crisco. "They boil just about everything," according to Dan, "so basically if you've tried walrus or seal, you've probably tried whale. Because once you boil it, it's pretty much all the same."

To the Eskimos of Wales, a culinary discussion of whale meat is easily as refined as comparisons, of, say, filet mignon, rib eye, and flank steak. In the first place, the textures of bowhead and gray whale are distinctly different. Bowheads have far more fat, both an inner and an outer layer, which is why they seldom sink after being caught. Thus their *muktuk* and meat are thinner than gray whales', and not as difficult to chew. Raymond Seetook glowingly calls bowhead "better than president's dish, nothing can beat it." He would prefer to cook the gray's *muktuk* rather than eat it raw, as with bowhead. Neither variety *needs* to be cooked, though; there's no danger of trichinosis as with walrus or polar bear, which must be prepared like pork. The *muktuk* can also be pickled.

The villagers say they prefer to eat younger grays, which don't have as many *kumuk*, the sea lice and barnacles that need to be scraped off or cut out. (Bowheads don't pose this problem.) "Gray whales are a real delicacy dried," Luther believes, the meat being stored in whale oil and becoming like jerky. When his wife, Christine, opines that she doesn't care for gray whale, Luther shrugs and says, "Well, it's like shrimp tastes better than clams to some people."

Dan keeps what's called an old Eskimo lamp out in one of his shops. It's basically just a stone, similar to Indian grinding stones for cornmeal, though not as deep. "They'd bring the blubber in and hang it above," Dan explains. "As it would heat up, the lamp would take more oil down into it and would keep the fire going."

Late in the nineteenth century seal-oil lamps had replaced the whale-oil variety. Then came gas lamps and Coleman stoves. Electricity didn't make it to Wales until the late 1960s. For nearly all the residents, there's still no running water. Without plumbing or a sewerage system, everyone uses what they call "honey buckets" for their toilets, which must be emptied daily. But you can shower in the Kaingikling Washateria and do laundry there as well. And a makeshift water line from a creek on the other side of Razorback Ridge provides crystal-clear water, which people haul to their homes by bucket.

In today's Wales there are TV and telephones. "Stove oil" is the heating fuel, and prices are going up. Groceries are expensive in the Wales Native Store. No alcohol is sold, or technically even allowed, in the village. But, of course, people have their stashes. There are plenty of dogs but no dog teams anymore. With the advent of snowmobiles and four-wheelers, the need to "mush" across the hills no longer exists.

Raymond Seetook is a thin, wiry man who in 1999 was fifty-three years old and looked considerably older. He lives across a bridge about a half mile from the Richards, in a small new housing complex. You can't see the ocean from here, so whenever Luther sees something from his higher vantage point, he'll give Raymond a call. Besides running one of the whaling crews, Raymond works as an agent for Cape Smythe Air and is seasonally employed loading mail bound for Little Diomede Island onto Evergreen Helicopter, which flies back and forth all day Wednesdays. Last week he and his wife, Debbie, loaded ten thousand pounds of mail. She sometimes makes beaver hats, or cuts up polar bear hides, to sell. Welfare helps with electricity and phone bills.

"Today," as Dan puts it, "subsistence includes your hunting, your welfare, and your food stamps."

Raymond and I are sitting at his kitchen table. Through the picture window we can see the fog rolling in. Hunting's been slow this summer. He's talking about his growing-up years. Raymond was raised in a house on Main Street with twelve brothers and sisters. "No kind of heating except a woodstove. Yet we survived." In those days the native store received a once-yearly shipment of supplies by way of the North Star Freighter that moved along the coast all the way up to Barrow. By October or November generally the store would be empty. Raymond's family maintained two wintertime food caches, one with seal oil and meat, the other with precooked walrus. His father caught seals in winter with an "Eskimo net" made out of raw sealskin, thick and with a big rock for a sinker. They'd check it late at night. When May came, "we used to be very happy. That was when fresh everything from the sea," after being icebound since November. "That sea," Raymond continues, stretching out his hand toward it, "that was our garden."

Even when his father was dying of stomach cancer in 1981, he still went out walrus hunting. Raymond remembers him falling down while pulling a skin boat in over the shore ice, "but he still had the strength to hunt."

Raymond is silent for a long time, and then says: "That's how the Eskimos were in those days. They don't give up."

Chapter 18

TO THE DIOMEDES: LIFE ON THE EDGE

There was a telepathic connection, Eskimos believed, by which the whales could understand them. So it was that hunters once sang songs calling the harpooned whale back to them with promises of kind treatment after they had killed it. Each crew's song was both secret and sacred, and some cast a magical spell.

—DAVID BOERI, *People of the Ice Whale*

THE gray whales were passing by the island. Inside a stone house under a skylight filtered through walrus membrane, a small crew of men worked at sewing a sealskin float. Present were an old man and a woman. On Little Diomede she symbolized the *inua*, or whale spirit.

Once everything was in readiness, the woman took a position in the middle of the room. The old man sang and beat slowly on a drum. They carried the whaling gear outside. The villagers helped bring the walrus-skin boat down from a rack. All the accoutrements for the hunt were put in their rightful places in the boat. The crew members dressed in brand-new clothing, made for them by their wives, adding mittens and waterproof parkas.

The old man took the lead. The woman brought up the rear as all climbed to the roof of the steersman's house. She brought a water bucket and a dish laden with reindeer fat. Part of the fat had been shaped in the form of a whale. All circled the roof in a clockwise formation. They went back to the boat. The woman put her bucket and dish inside.

They unloaded the boat and turned it upside down. The woman sang and washed down the harpoon with a grass mat soaked in urine. A fire was passed beneath the boat. The woman concluded her song. The villager who would throw the harpoon shoveled the remains of the fire onto the shoulder blade of a walrus. He carried this to the shore. He gave it to the sea.

They reloaded the boat. The crew assembled in their rightful places. The woman picked up her water bucket and dish and stared at them. She began to dance and sing above the whale molded in fat. Four times the men lifted their paddles. They uttered a howling sound like wolves. The crew got out of the boat. The old man strode around it, tapping the bow with the steering paddle, singing and four times raising the stern.

The whalers then took hold of the boat. They raced to the left of the woman to the shoreline and back again to where she stood. She leaned and put her dish of fat in the boat again. She picked up a short piece of walrus intestine that was filled with ashes. As the old man called out behind her, she scattered the ashes while the crew fell in line carrying the skin boat. The children of Diomede followed as well. At the edge of the sea, the old man sang. The woman sliced the reindeer fat and gave the pieces to the children.

So the boat was launched, the crew pausing after each made four paddle strokes. Once more the woman sang. She turned her back as the men paddled toward her. The harpooner raised his weapon and pantomimed thrusting it into the woman. Once more the boat was brought ashore and turned over. With the old man leading the way, all returned to the village for a feast. A shaman came and conferred his blessing that the rituals had been properly bestowed upon the whale that waited to give itself to the people of Little Diomede.

The next daybreak, while the gray whales moved past Fairway Rock and between Little Diomede and Big Diomede just to the west, the men paddled stealthily but silently in the direction of the spouts. Not until they returned would the old man and the woman emerge from the steersman's house.

The Diomede Islands weren't named until early in the eighteenth century, when the Russian explorer Vitus Bering sailed by and labeled them in honor of St. Diomede. There is a legend about how the islands came to be. In olden times there was a sand spit over which people could walk between the narrowly separated islands. Perhaps this was when the Bering Land Bridge yet existed, when humans and animals still crossed between Asia and North America at the apex of the gray whale's migration. One day a young orphan boy went hunting and caught a seal. He tied it to his skin boat. The seal wasn't dead. The boy never saw the seal come up from behind and grab him. The boy's grandmother had come down to the beach to wait for him. When she saw that he'd been killed, she was so angry that she skinned the seal alive and put it back out to sea.

Another old woman, one who lived "under the sea," also grew angry when she saw what had been done to one of her creatures. She caused the water to rise and cover the sand spit. Since that time, according to the legend, the people have lived on separate islands.

When America purchased Alaska from Russia in 1867, the 2.5 miles between the two isles became the dividing line between the two nations. When the Iron Curtain came down eighty years later, all of Big Diomede's native residents were removed to the Soviet mainland and the island became a military base. Anyone from Little Diomede who happened to stray into Soviet waters was taken prisoner. Hence, the traditional hunting grounds for the Little Diomeders were reduced substantially. Sometimes they'd observe helicopters rising from their neighbor's turf and briefly heading east before coming to an abrupt halt—as though they'd run smack into a glass partition—then flying north or south along an invisible line of demarcation.

Even today Big Diomede remains strictly a Russian military outpost. In the winter small planes from the Alaskan mainland are able to land on the pack ice between the Diomedes and, depending on where the ice is thickest, set down first either in Russia or in America. Once the ice melts air arrivals on Little Diomede are limited to Evergreen Helicopter's weekly flights directly into the village. The fare from Wales, only twenty-eight miles distant, is one hundred dollars each way.

Additionally, for non-Eskimos, the Diomede Native Corporation charges a one-hundred-dollar debarkation fee. For any film or photographic reproductions, you are supposed to negotiate a contract in advance. There are strict rules on what you can and can't take pictures of, lest your camera be confiscated. But at least an outsider could go there. When I'd contacted the native officials at Gambell on St. Lawrence Island about paying a visit and conducting interviews, they'd turned me down flat. The Eskimos there permitted only quick, supervised package tours.

Dan Richard said the suspicions were understandable. The Diomede islanders had once allowed in a Jacques Cousteau film crew and ended up being portrayed as marine mammal butchers. Then, in 1991, came a federal government "sting operation" called Operation Whiteout. Some of the Diomede hunters were induced to trade walrus ivory for marijuana and found themselves behind bars. In 1998 more hunters were indicted for "wasteful taking of marine mammals," which meant discarding everything but the ivory tusks of ten walruses.

Little Diomede, as Dan put it, was "life on the edge." He thought it would be a shame to come all this way and not at least take a day trip over there. With all the mail and other supplies being sent from the mainland, there wasn't room for both George and me on the helicopter, so we'd have to go separately. Shortly before noon I strapped myself into the cockpit, cargo piled high behind me. "It'll take us about ten minutes," the pilot, Eric Penttila, said. Momentarily we were soaring above turquoise floating ice that etched an assortment of picturesque patterns across the Bering Strait. I scrutinized the sea in vain for whale spouts.

On Little Diomede Island (George Peper)

The two Diomedes are so close together that until you get near you can't quite tell them apart. Little Diomede is lined by sheer cliffs that rise steeply on all sides, nothing but shattered granite and limestone boulders, among which it's impossible to imagine habitation. I can see what seem to be hundreds of birds drifting in and out of the ledges. The pilot says that murres nest on the cliffs, their pear-shaped eggs clinging to the narrow ledges, where the Diomeders also climb to gather them. The chopper rounds a corner, where the northwest slope takes a gentler grade down to a narrow strip of stony beach. There, at maybe a forty-degree angle, a cluster of shacks climbs the rocky hillside: the village the Eskimos call Inalik, which means "the other one" or "the one over there." It's known continuous residency since before Christ. Today about two hundred Inupiat live here. As the helicopter hovers my eyes scan the steeply angled rows of houses, and I remember Dan's saying that, when Diomeders visit Wales, they take great delight in a simple straight-line walk along the beach, where the land is level.

The helicopter touches down just above the beach, on a makeshift landing pad that's no more than a leftover piece of metal from a tugboat that once got torn from the shores of the island while unloading fuel and supplies. Children come running over to greet the chopper, racing like little land crabs over the rocky paths that wind through, around, and behind the houses. My first impressions are of a large water storage tank, a crane and backhoe, and a satellite dish perched beside a couple of upside-down skin boats. The narrow wooden houses, some of two or even three stories, are built along five tiers up the mountainside. A few of them are perched on wooden stilts. Life on the edge indeed.

I make my way up the rocks to the Diomede Native Store, where several people are standing on a walkway looking out to sea and passing around binoculars. "They're already shooting out there," one says. Walrus is the order of the day. One six-man hunting crew is at sea, another is getting ready to go. From not far away I hear the loud report of gunfire. Dogs bark. An acrid smell fills my nostrils, and I realize it's the seal meat hanging on a wooden pole above a clothesline.

I'm told I must first report to the offices of the local IRA (Indian Reorganization Act Council) to pay my landing fee. Sidestepping a series of what look like otherwise man-eating dogs straining against their chains, I slip and slide along the path and up two sets of steps to a side door entryway. Only the secretary is in.

After waiting a half hour for a town official to arrive and receive my hundred bucks, I'm provided with a husky young escort. Most of the men I should talk to, I'm informed, are already out hunting. Down below, though, one crew is just getting ready to launch. This is the crew of Tom Menadelook, or Tommy Junior as everyone calls him. I already knew a little about him from Jim Stimpfle, and Dan and Ellen. One of his grandfathers had been a great whale hunter in Siberia. Tom was the son of the former mayor of Little Diomede. He was eleven when he shot his first seal in 1975 and so was welcomed into manhood. On May 4, 1999, he fired a darting gun that brought home the Diomeders' first bowhead whale in over sixty years.

This had occurred only four days before the Makah Tribe shot their gray whale, and I'd read about the event in an Associated Press story. It reported that the Diomeders had landed a twenty-eight-foot bowhead, their first since 1938. According to spokesman Patrick Omiak, "bad luck and sea ice over the decades" were the reasons it took so long. Learning of this had first piqued my interest in comparing the lifestyles of the Alaskan Eskimos with those of the Makah.

Menadelook, a stocky Inupiaq in his thirties, agrees to talk until the rest of his men get there. He tells of having divided the twenty-eight-foot bowhead equally among about thirty households. He tells, too, of how he'd immediately afterward flown to Anchorage, where his wife was undergoing chemotherapy. "She said she had good news. She'd beat the cancer. So I missed our feast for the whale—but I brought my wife home!"

Why did he think it had taken so many years between bowheads? I ask. Menadelook replies it's a combination of strong currents, unpredictable and ever-changing ice conditions, clear water, or sometimes not enough open water. He hastens to add, though, that gray whales had sometimes been landed on Little Diomede. In 1985 a large gray had required a backhoe to haul it ashore. "Last year we tried for a gray whale. We had the harpoons in it, but we lost it. The weather got too bad and it sank. Gray whales are more dangerous than the bowhead, because once in a while they'll turn around and attack you. I've seen that happen.

"That's the skin boat we used to land the bowhead," Menadelook adds, pointing to a rack beside us where it's undergoing some repairs. I lean an arm of my waterproof rain jacket on the edge of the boat, then settle my hand into perhaps the stickiest substance I've ever felt. "Oh, watch out for that," Menadelook says, too late, "that's seal oil!" It's being used to loosen the walrus skin covering. My jacket is covered with it. A trio of teenage Diomeders who've been eavesdropping on my interview find that quite amusing.

"It's not gonna come off with Clorox," one chides me. "*Nothing* will take it off!" another chirps. "You can wash and wash that jacket, forget it. But it might make it more waterproof!" a third exclaims. All laugh heartily, and I feel sort of . . . initiated.

I see the helicopter coming in for another landing. George emerges. By now Menadelook's crew has all assembled. George joins me to watch the men push their boat out over the rocks. Five hunters with high-powered rifles are equipped to stay out several days if necessary. Their faces are earnest, unsmiling, hard-bitten. "This is a serious hunting crew," George says admiringly. But our escort will allow us only so far—no close-up photographs. He suggests we go visit with a couple of the elders.

Peter Oscar Akhinga, now seventy-three, is a carver who holds up a special flutelike instrument that he fashioned out of ivory, bone, and wood. The last of thirteen children, he lives in a two-room house on the third tier of the village. He is reed thin and wearing only a white T-shirt with light-colored slacks. As a young man, he tells us, he was visiting on the Russian side when the Iron Curtain fell and he got stranded over there. He'd been held captive fifty-two days, surviving on rations of black bread. After the old schoolhouse burned down on Little Diomede, he'd never finished the second grade. Peter figures he was born to be a whaler. He says he's been in on the taking of sixteen whales. As he reminisces Peter gets so excited that he acts out the scenes. He's holding the paddle again, dipping it high and silent into an invisible sea.

"Don't put your paddle hard in the water, the whale is going to see it. . . . Don't smoke too much in the boat, the whale can smell it. . . . When the boat leader stands and crosses his hands, that means the whale is coming up. . . . Gray whale and bowhead whale, same way."

Sometimes, Peter says, they would tie an amulet to the boat. A likeness of the whale. Especially the flipper. The whalers' wives always "try to make certain that everything is new, that the men wear white clothing. If you wear black, the whale can see you."

A man whose wife is pregnant is not allowed in the bow of the boat, or to shoot the harpoon gun. Otherwise he might be killing his baby. The old people said not to sweep the floor in the house until a whale is taken. Neither should a wife sew during a hunt, "because the line might get tied up." Nor should a wife do *anything* while the sun is setting and the men are at sea.

If you lost a whale out there, you did nothing when you returned home, for as long as a week.

How the people would dance after a whale was landed! If you brought home a female whale, you couldn't make love to your wife for five days. If you brought home a male whale, you had to wait only four days. Same with putting a whale on the ice after it was cut up: female whale for five days and male whale, four days.

"When the people firmly believed in our ways," says Peter, "there was great shaman among us. In those days, the men gathered mud from the corners of the houses and rubbed the mixture into the boat.

"They never do this no more. They don't know much around here, the younger people, about the whale."

Just down the way Moses Milligrock's wife looks at my seal-oil-stained jacket, shakes her head, says something in Inupiat, then carts the jacket away to the porch for a little scrubbing. Moses, who is seventy-six, is sitting by his kitchen window, binoculars at the ready for whatever might transpire down below. He's a white-haired, round-faced gentleman with a cherubic smile. Having grown up speaking his native tongue, he speaks English with a touch of a foreign accent. He says he's always lived on "Diomede cliffs" and carves ivory for sale. The puffin on a rock here is only thirty-five dollars. He holds the pretty little piece up for display.

"Gray whales," Moses says, "are different than the big ones"—bowheads—"I mean the taste. I like those big ones' *muktuk.*" He makes about a four-inch half-moon between his thumb and index finger to demonstrate how thick it is. "But the gray whales, not so much blubber."

I ask if the blubber is still boiled and used for preserving greens. "In the old days, long ago. Right now there are too many things to eat. In the old days, there was a real small store—just tea, flour, sugar, milk for the whole island. So people had to save everything they get."

Moses points out the window at some nearby discarded remnants of last April's bowhead. "They don't save all the blubber no more, because they got stove oil now. Before, when we have no electric, we use seal oil for Eskimo lamp. But now, electric all over. Look at the wires out there."

Does he feel those days were better, the days when everything was used? "Eskimo way is better to us. We grew up like that. These young girls, they don't know how to cook Eskimo food. These young boys, they don't know nothing. Except that boy there, he's learning already."

He points at Stanley Milligrock, age seventeen, who has walked in and headed right for some cookies on a small table. Moses's grandson was twelve when he first went out with a hunting crew pursuing gray whales. Yes, Stanley says, they got one.

"If it sinks to the bottom, they have to pull it up by the boat," Moses says. "That's what happened," Stanley says. "Hard work," Moses says. "It took about forty-five minutes," Stanley says, "to bring the whale back up and another two hours to get it in to shore. There's some gray whales out there now, but they're hard to see when there's ice around."

Out the window there's a flurry of commotion at the Diomede dock. A blond-haired lady Jet Skier has shown up, come over from Wales accompanied by a smaller chase boat with a camera crew. The banner says they are en route TO RUSSIA WITH LOVE. This is part of a Bering Strait phenomenon in which, over the past fifteen years, adventurists have tried to swim, dogsled, and kayak across. Even bathtubs and surfboards have been attempted. All ended in failure. This is the first such effort by personal watercraft, and it's been in all the Alaska papers.

Moses turns away from the scene, puts the binoculars back on the kitchen table, and continues: "Right now everything is different. TV just came a few years ago. That changed [things]. Lots of Diomeders, now they move to Teller, Unalkleet, Nome, even Anchorage."

He points to a stone house down below, where whale and walrus meat have customarily been stored for the long winter. "Young woman don't put away the meat right, like in the old days," Moses says dismissively, arching his arm to utilize a homemade back scratcher. His wife returns, having done the best she could with what she calls my "poor jacket." I buy Moses's carving, and George and I depart.

One of the boat crews has returned with a walrus kill. We can see tusks being carried to a storage area. But our "tour guide," who's waited outside during our meetings with the elders, won't let us anywhere near. "No pictures," he says matter-of-factly. "No pictures allowed."

The helicopter has landed again, and we're told this is the last trip of the day. In fact, of the week. Or maybe for as long as a month, if the weather turns lousy. We've been here only a few hours but have little choice but to head back to Wales. There's much less cargo now, so

George and I can both fit. Once more a sizable contingent of children gathers to see us off.

"You know what one of those little kids asked me?" George says after we've boarded. " 'Are you white man? Are you spy?' " There's a sadness in his voice. We both know it would've taken a long time to gain the trust of these island people, given their past experiences with our culture, these hardy souls who for centuries have clung to this snow- and windswept slope and survived by hunting the marine mammals that swim along their icy shores.

Strangers in a strange land, we lift off and silently watch a thick fog begin to settle in over the Bering Strait.

When we arrived back at Dan and Ellen's place in the late afternoon, Jim Stimpfle and his son had already launched their boat in the direction of Nome. Otherwise, according to the weather forecast, they might have been locked in here for days. Strong wave action had propelled the floating ice astoundingly close to shore. I'd never been in a place where people were so much at the mercy of the elements. Reuben Nichols, one of Ellen's relatives and a member of Raymond Seetook's whaling team, sat on the couch in the Richards' living room and told me: "Last time I went walrus hunting on the ice was two years ago, maybe three. The wind's been too bad to go out and, by the time decent weather comes, the ice is starting to rot." Then Reuben added: "But I'm sure if it was our ancestors, they would have gone out anyway. They didn't have no choice. They didn't have a store you could go to and buy some food if you didn't."

So is whale hunting still a necessity for these people? It's a question fraught with dichotomies, and no easy answers. Among the papers bequeathed to Luther's father by the Mormon teacher who resurrected the whaling tradition in Wales, there was a letter from a director at the New York City Museum of Natural History. It was dated 1969. "The hunting of whales you propose to do seems to be illegal," the letter began. "The gray whale is protected from hunting by international law. The hunting of bowheads is likely forbidden by international agreement except for aboriginal hunting, which would not permit the use of a harpoon bomb but would have to be done with merely lance harpoons. I would urge you to reconsider your plans for reviving the hunting of these whales by your villagers. Aside from the questionable legality of the hunting itself, some species of whales are in great danger of extinction."

Charles Christiansen ignored the advice, and in the years since, things had changed. In 1977 the International Whaling Commission (IWC) had promulgated a complete ban on bowhead whaling as a result of biologists' estimates that only somewhere between 600 and 2,000 bowheads existed in the Bering, Chukchi, and Beaufort Sea region. The Alaskan native communities, which until then had continued to hunt bowheads, maintained these figures were too low and protested vehemently. They formed the Alaska Eskimo Whaling Commission (AEWC) to unite ten native communities along the Arctic coast. Eventually the IWC Scientific Committee revised its population estimates (today, bowhead numbers are said to be more than 8,000 and increasing by about 3 percent a year), and a subsistence quota was established by the IWC. Between 1995 and 1998, 204 bowhead whales were landed, with up to 280 additional bowheads allowed to be taken by the native communities between 1999 and 2003. The AEWC sets the allotment among its members, with 160 crews from Barrow constituting by far the largest whaling community, while Wales and Little Diomede are at the bottom of the list.

So the bowheads landed once a year at Wales in 1987, 1991, 1994, and 1995, and by the Diomeders in 1999, were all perfectly legal. Not so, however, gray whales. I didn't fully grasp this at the time I was in northern Alaska, because the Wales hunters seemed quite sincere about their village's having the permission of the AEWC to pursue a gray since they'd not landed a bowhead in the spring. Later I'd find out that the AEWC doesn't have such unilateral authority. It must have sanction from the International Whaling Commission. The IWC quota on gray whales, as worked out by the Russians and Americans, is granted only to Siberian natives—and now to the Makah Tribe. Until the early 1990s the IWC had allowed up to ten gray whales a year for Alaskan native subsistence—but the U.S. delegation had voluntarily surrendered this quota, right around the time the Makah started making noises about wanting to resume the same practice.

Raymond Seetook was quite open with me about what transpired in 1996. He said he'd first called all around, trying to determine whether it was permissible for his crew to get a gray whale. He was told there would be a thousand-dollar fine but not to worry because the AEWC would pay it. So he'd brought in a gray, the last one that had been landed in the village.

When I checked the IWC records later, it seemed this was one of two gray whales taken "illegally" that year by native Alaskans. The United States had reported the infractions at the next IWC meeting. But the IWC Infractions Committee didn't regard the transgressions as a violation of its regulations. Its rationale was, it awards quotas not to specific countries but rather on the basis of management "stocks" of a particular species. In this instance the IWC had allocated 140 gray whales to be taken annually by the peoples of Russian Chukotka. In 1996 the Chukotkans took only between 80 and 85 gray whales, so there was a "leftover" quota of 55 to 60. Thus, it hadn't mattered that Alaskan Eskimos took 2.

All of this "whale politics" left me pondering the big question: Was the Makah's claim more, or less, legitimate than that of the Wales and Diomede Eskimos? The Makah asserted a nineteenth-century treaty right but hadn't hunted gray whales in more than seventy years. The Diomeders had apparently never stopped trying, and Wales had resumed hunting in the 1960s after about a forty-year hiatus. Their extreme geographical locations—especially the Diomeders'—certainly, in my view, made their subsistence needs more real than the Makah's. That was what the IWC's definition of allowable gray whale hunting was *supposed* to be about. But are even those needs still vital *enough*? Along the Bering Strait as well, the modern world had made definite inroads. As Luther Komenseak said, "One of the main distractions is TV right now." Two of his three boys "really enjoy going out on the ocean," while the third prefers to stay home and watch the tube.

Above all, it was the striving for intergenerational continuity that I found most admirable about the marine mammal hunters of Wales. Raymond, whose father started him out when he was about ten, today hunts with his two brothers and a fourteen-year-old son. His eight-year-old fired at two seals but missed during the time I was there. "He's getting good, though," Raymond said, "gonna be a decent bowman when he gets older." Luther's crew consists mostly of his uncles, since his three brothers moved away to Nome. Luther provides native food to his elderly parents, his sisters, and, he says with a laugh, "to hungry Eskimos in Anchorage."

When Dan and Ellen's fifteen-year-old son, Sherman, returned home from his computer seminar, the first thing the boy did was venture out on an all-night seal hunt. Dan has taught Sherman the ways of the tundra, how to hunt moose and muskox. It's Ellen's Inupiaq cousin Gilbert,

Dan's closest friend, who instructs Sherman in the ways of the sea. "He just becomes his father during springtime," Dan says simply. The next day Sherman proudly came home with a six-hundred-pound seal.

It's a land where ice and wind, polar bears and orcas, command respect. When killer whales were sighted recently in the area, Raymond Seetook bemoaned the fact that they would scare away walrus, gray whale, "whatever's out there." On our last day in Wales, George and I were sitting around a conference table in the Wales community center with a young hunter named Mike Akhinga. He'd grown up on Little Diomede, a relative of Peter Oscar Akhinga. He joked around until the subject of killer whales arose. We asked if native people ever hunted them. An ominous silence descended over the room. One of Ellen's cousins, who was filling out some forms across from us, stopped writing. A sullen older man, who'd been sitting quietly in a corner, blurted out: "You leave 'em alone!"

Mike Akhinga nodded and said, "Those killer whales, we stop our boat and let them move through." Ellen's cousin said, "Some of the elders say, if you ever shoot at them, they won't forget. We don't touch a killer whale."

"Otherwise, they will hunt you down until they find you," Akhinga said. "They will sink your boat and drown you."

It was said that when a whaling crew at Shishmaref broke the taboo and killed an orca, every member died within three years. In 1889 the whaling chronicler Herbert L. Aldrich wrote:

> To destroy one is to cause the death of many people as a punishment. I was told of an instance when a great many St. Lawrence Island natives died because one of their number caused the death of a killer. The natives believe that the killers live in the mountains in the winter, and that the various warm springs there are made for them to do their cooking in. Whether the killers are supposed to go overland, or through subterranean passages, or only in spirit to the springs, I could not learn. The St. Lawrence Bay natives believe that the killers have a house back in the mountains where they live winters. To keep in favor with the killers, the natives make knives of whalebone or ivory, and throw them into the water to aid the killers in killing the whales.

It's a land where, as with the influenza epidemic that struck Wales, people never knew what to expect. On St. Lawrence Island in the early 1880s, an estimated 1,000 people out of a population of 1,500 starved to death; six villages were abandoned. Here is how the ethnographer Edward William Nelson described the scene: "In July I landed at a place on the northern shore where two houses were standing, in which, wrapped in their fur blankets on the sleeping platforms lay about twenty-five dead bodies of adults, and upon the ground and outside were a few others. Some miles to the eastward, along the coast, was another village, where there were two hundred dead people . . . In the houses all the wooden and clay food vessels were found turned bottom upward and put away in one corner—mute evidences of the famine. Scattered about the houses on the outside were various tools and implements . . . among these articles were the skulls of walrus and of many dogs."

In that case the scourge could be traced directly to depletion of one of the islanders' staple foods: walrus. The Civil War had stimulated the market for oils. Most of the whale populations had been decimated, and walrus oil was easier to refine than whale oil. Scammon estimated that, over a four-year period starting in 1868, at least sixty thousand walrus were killed by the whaling fleet. This wanton destruction, and its impact upon the native peoples, inspired one of the strongest conservationist statements in his book:

> Among the numerous enemies of the Walrus, it is to be regretted that the whalers are included, they having been driven to the necessity of pursuing them on account of the scarcity of Cetaceans. Already the animals have suffered so great a slaughter at their hands that their numbers have been materially diminished, and they have become wild and shy, making it difficult for the Esquimaux to successfully hunt them, in order to obtain a necessary supply of food. It is stated that there has been much suffering among those harmless people of the far north, on account of this source for supplying themselves with an indispensable article of sustenance being to an alarming extent cut off.

Today in northern Alaska walrus are hunted not only for their tusks but for what the women make into "meat rolls." Especially by the Diomedes, thousands can sometimes be seen. Bottom feeders like the gray whales, walrus are often in the same area. In the spring the bulls come first, followed by females with pups. The last to arrive are called green flippers because of the green algae they carry on their feet from down south. They're also known as water boys because the hunters pursue them on the water and not on the ice.

Dan told us a story about walrus that epitomized life on the edge. "Ellen, her brother, her dad, and me were out boating around one afternoon, having a good time. Dad thought he saw some ducks in the water, so we started heading that way. It was a herd of walrus. They were asleep in the water. We woke 'em up—and they were *pissed.*

"They attacked the boat. Charging at us. What they do is try to get their tusks up over the rail of the boat and grab you. Dad was trying to get the motor in the water enough to move us out, but every time he'd put that motor down, there'd be a walrus there. All you can do is just keep firing as fast as you can to try to keep 'em away. We shot probably a hundred rounds before we could get out of there."

Our last night here, returning around 2:00 A.M. from a trek into the countryside on four-wheelers in a fog so dense we could barely see a few feet in front of us, Dan pulled in beside his garage and loaded a motorcycle into the back of his truck. He said George and I could come along if we wanted. He drove again to the outskirts of Wales, where a mound of earth was built up about eight feet off the ground. He unloaded the bike and started the engine. He gunned it to the top of the mound, hopped off and flipped the bike over on its side, and pocketed the ignition key. The fog covered the mound like a shroud.

Returning to the truck, Dan said quietly: "It's Eskimo tradition to bring your weapons to your grave. Friend of mine died not long ago, we buried him there. He was always on that bike. It was his most prized possession. I figure he'd like to have it with him."

The next day there was a lot of speculation in town about how that motorcycle ended up on top of the mound. There was also talk about how Raymond Seetook's crew had gotten one *ugruk* the night before. A little over six feet and about eight hundred pounds. They'd divided the bearded seal meat among themselves, and Raymond was sending some by plane today to his older sister in Nome.

When the fog rolls in over the runway, sometimes you can't get out of Wales for days. George and I decided we'd better take advantage of a late afternoon break in the weather. Dan, Ellen, and Sherman drove us to the landing field. They said there would always be a place for us—here at the edge of the world.

On the flight to Nome I was still wrestling with the necessity for native peoples to hunt gray whales.

"Well, the Chumash Indians in Baja never hunted them, but why?" George wondered. "Was it because they had no cold storage place to put the meat? Or because of the abundance of other fish? Maybe geography plays a big part."

"I know," I said. "These are legitimate questions to ask, as far as the past goes. But another thing I wonder about, are we in a sense today taming these whales in Baja, only to see them be killed in the North? That doesn't sit right with me. But am I being sentimental? Or are *they*—I mean the Eskimos—being nostalgic? Where's the line between their being absorbed into the twenty-first-century world and their holding on to what sustained them in centuries past?"

George shook his head and said he didn't know. I talked then about how different I felt here, how much less alien, or perhaps the word was "alienated," than on the Makah reservation. "There does seem to be a spiritual dimension to what Eskimo whale hunters accomplish," I said. "When the hunting is truly part of their religion, then it's a real relationship with the whale, one that perhaps becomes not only forgivable—but enviable.

"I'd be surprised if there wasn't at least a subconscious surviving conviction that in hunting whales they contribute to the appropriate movement of souls between this world and the next. And that the sharing of wild meat, which a whale provides in such abundance, is the sine qua non of the human social order. The Makah are decades removed from that, and at *best* can be charged with indulgent nostalgia."

George responded with a telling point. "In Wales they don't have to *prove* themselves to an alien culture."

He paused before continuing: "Of course, as far as hunting goes, with government welfare checks coming in, they don't have the same basic need as they used to. Then again, as Luther said, groceries cost so much to bring into a place like Wales or Diomede. So people can't live up here on *just* those welfare checks, and maybe that's served to not only keep these people hunting but to keep them proud of having so long and so *well* proved themselves, in the face of so much adversity, right up to the days of satellite dishes and Sea-Doos."

Then George wondered, "So do you think maybe the real tragedy, as far as their lives are concerned today, is that they're not hunting *more*?" His words hung in the air, a conundrum, stretched taut across a Fourth of July, where Asia first merged with America, on the continental divide.

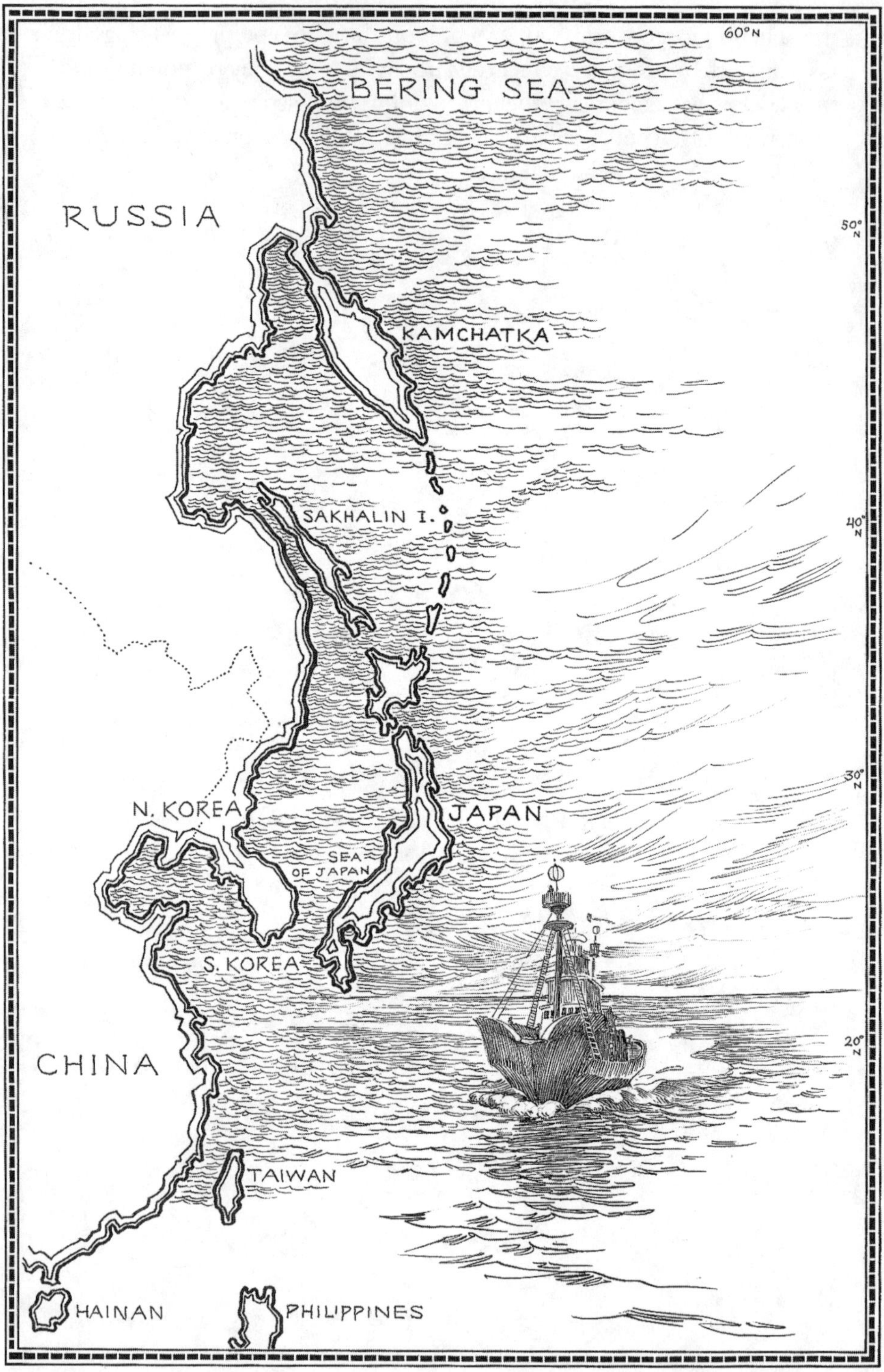

60° N
BERING SEA
RUSSIA
50° N
KAMCHATKA
SAKHALIN I.
40° N
30° N
N. KOREA
JAPAN
SEA OF JAPAN
S. KOREA
20° N
CHINA
TAIWAN
HAINAN
PHILIPPINES

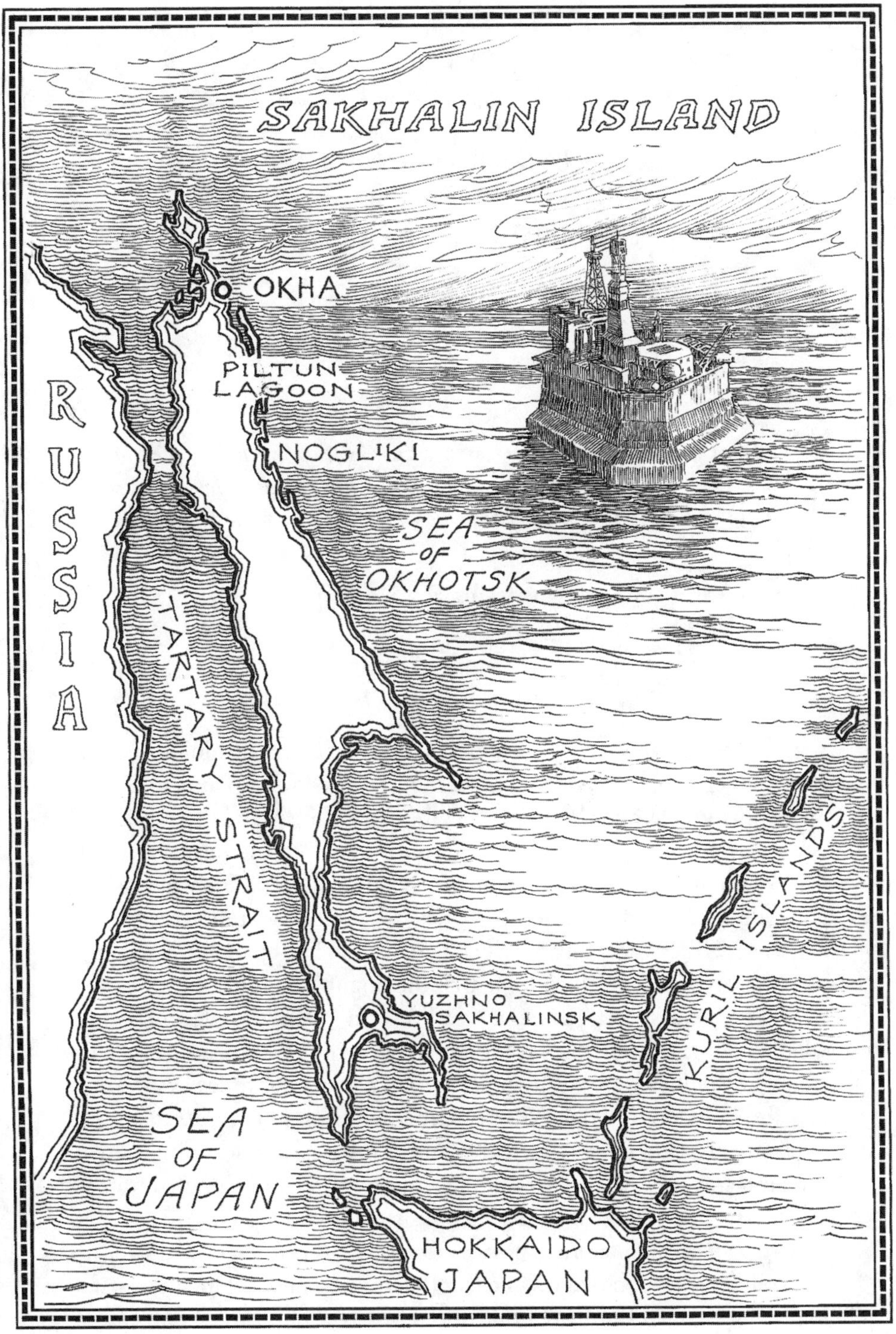
SAKHALIN ISLAND
OKHA
PILTUN LAGOON
NOGLIKI
RUSSIA
SEA of OKHOTSK
TARTARY STRAIT
KURIL ISLANDS
YUZHNO SAKHALINSK
SEA OF JAPAN
HOKKAIDO JAPAN

Chapter 19

Interregnum

Sakhalin Island: Last of the Western Grays

Here on these coasts you are gripped not by thoughts but by meditations. It is terrible, but at the same time I want to stand there forever and grieve at the monotonous waves and listen to their thunderous roar.

—Anton Chekhov, *The Island*

On September 10, 1983, a group of Russian marine biologists conducting fisheries research in the Sea of Okhotsk happened upon an astonishing sight. From the bow of their ship *Neva,* they observed twenty gray whales "moving slowly southwards" in the near-shore waters off the northeastern coastline of Sakhalin Island. "The whales were generally quiet and ignored the vessel," the scientists reported. "Some animals revealed their flukes (which had distinct light patches) on diving. . . . This gives at least some hope that the population is not extinct."

These whales were remnants of a Western Pacific population of gray whales, a separate stock from the Eastern Pacific's California grays who frequent the Bering Sea waters more than a thousand miles away. Once heavily exploited by whalers, especially those from Japan and Korea, the western grays had largely been given up as lost. Of course, nobody knew for sure, since the waters around Sakhalin Island had been off-limits to outsiders since World War II. This was the most remote outpost in the then-USSR's Far Eastern empire, closed to all foreigners and even to nonresident Soviet citizens. The fact that gray whales might yet be feeding off Sakhalin's Piltun Lagoon was scarcely a blip on anyone's radar screen. The region, as Russia's famed writer Anton Chekhov described it in 1890, remained "the end of the world . . . there is nowhere else to go."

With dusk approaching in the island's capital city of Yuzhno Sakhalinsk, my train rumbles slowly out of the station. It's a diesel dinosaur, creeping along at ten miles an hour and still barely able to cling to the rails. Yet there is the veneer of former elegance. Our sleeping car is Oriental-carpeted, the windows draped with white cotton half curtains. I peer through them at shabby gray buildings and potholed streets. George's and my belongings are hidden in a back compartment near the ceiling, lest someone break in and try to steal them in the night. An American pop tune from the early 1960s is playing over a radio in the next compartment. A female porter arrives with fresh linens and asks if we'd care for tea. I continue to gaze out at fishing boats setting salmon nets, at valleys of pine and white birch and bright purple wildflowers, at ditches lined with junked cars, at bright multicolored wooden houses surrounded by cabbage and potato gardens and wild strawberries. I feel enmeshed in a series of time warps. A fog closes in.

We'd arrived yesterday, after a nine-hour flight from Anchorage that touched down briefly in the yet heavily militarized, once-forbidden zone of Kamchatka on the Russian mainland. Now we're on our way to Sakhalin's Piltun Lagoon, the only known feeding habitat of the western gray whale. A team of U.S. and Russian scientists are there doing boat and aerial surveys. We'll be staying at their base camp. My only previous knowledge of Sakhalin Island was its "infamy" as the spot where the Korean Air Lines 007 jumbo jet strayed over Soviet territory and got shot down in September 1983, killing all 269 passengers. Geographically we are actually closer to Washington, D.C., than to Moscow, which is eight time zones and almost four thousand miles away. Situated only about twenty-five miles above Japan, Sakhalin Island is nearly six hundred miles long, with a maximum width of one hundred miles, a pointing finger of land not unlike that of the Baja Peninsula. Do gray whales, I wonder, seek out the same types of coastlines no matter where they travel?

Water, water everywhere: More than 16,000 lakes, and over 60,000 rivers and brooks flow along Sakhalin, which is surrounded by the cold Sea of Okhotsk to the north and the warm Sea of Japan to the south and west. About 620,000 people live on the island today. Early Paleolithic man is believed to have come here chasing mammoths across a land bridge. Indeed, archaeologists have speculated that in the time between the two Ice Ages, Sakhalin was part of one big land bridge, which served as a corridor for early man to cross the Bering Strait and enter North America. One of the original native peoples of Sakhalin, the Ainu, supposed that the island is a great beast sleeping in the ocean, with the northern coast where we're bound its head. The tail is a southern bay called Aniva, where in 1805 the Russian explorer I. F. von Krursenstern "hesitated to send longboats ashore for fear that they would be overturned by masses of whales that were churning up the waters."

The translator accompanying us is Natasha Barannikova, a twenty-two-year-old who works with Sakhalin Environment Watch. This small group of activists, based in the capital, monitors the island's timber and oil and gas industries and maintains liaison with U.S.-based groups such as the Pacific Environment Resource Center and Earth Island Institute. Natasha is a brunette with a kind of wistful beauty. She's never been north on a train before. When George asks where she would most dream of traveling, she says without hesitation, "To Piltun, to see the gray whales."

Morning dawns with an orange fireball of a sun over a charred landscape devastated by forest fires. "They burn every summer," Natasha tells us. "Ninety percent of them are caused by people. There isn't the equipment to contain them."

The region we're passing through was, between 1869 and 1906, a Russian penal colony. More than a thousand prisoners were shipped here every year from the mainland. The convicts called it Falcon Island. Chekhov, whose primary purpose in traveling to Sakhalin was to conduct a medical-statistical census, would tell of one who murdered a sadistic guard by suffocating him in fermenting bread dough. "Migrant coal workers ate candles and rotten wood while the czar's ministers sold the island's salmon and caviar abroad," Chekhov later wrote. The first place the convicts were sent was the southern coastal town of Alexandrovsk. Chekhov wrote that, when his ship was approaching, "I saw many whales swimming and frolicking in pairs in the Strait."

After Japan seized the southern half of Sakhalin during the 1905 Russo-Japanese War, they'd turned it into a penal colony of their own. The Japanese forcibly brought some 43,000 Koreans here for virtual enslavement in the coal mines. After World War II, Japan reluctantly returned its portion of the island to the Soviets.

We emerge from the train at 8:30 in Nogliki, end of the line on the old Japanese-built railroad. Natasha finds a driver to take us onward by dirt road. For miles the scenery is little but dead tree stumps and gnarled branches. We stop briefly in Val, a collective farm village in Soviet days where a number of the island's native Nivkhi peoples had been relocated from the coast. It was pretty much obliterated in a May 1995 earthquake that registered as high as 9.0 on the Richter scale. The quake knocked down every building in nearby Neftegorsk and killed two thousand people.

We stop again at a hillside town Natasha says is called Thirtieth. Except for an old man pushing a wheelbarrow, it too is largely deserted. There used to be oil rigs here, even a lovely hotel and shops. "This was a beautiful place, but now . . . there are many such villages on Sakhalin," Natasha says. As we finally come to a paved road outside the northern oil and gas center of Okha, our car gets a flat tire.

Mid-afternoon arrival in Okha. A statue of Lenin still stands in the park, as does one in Yuzhno's main square. There are only a few paved streets in this city of 36,000. We check into adjoining suites in a large co-op apartment building with one section converted into a hotel. In George's and my room, holes remain in the wall where a family once hung its photographs, and children's stickers adorn the bathroom walls. We take a bath by painstakingly turning on the water, gas, and heater in the old kitchen.

Yuri Shvetsov shows up. He's a gray-haired ex–Soviet police investigator with a kindly smile. For a hundred dollars cash, the whale scientists have arranged for him to drive us in a four-wheel-drive jeep south again along the Sea of Okhotsk to Piltun Lagoon. He's come to show us maps and discuss the trip. Yuri says we should plan on departing by 8:00 A.M., because it will get hot as the day progresses. If there is wind we should expect smoke from the latest fires. But wind will drive away the mosquitoes. It would be good to bring along some vodka. Yuri takes us to pick up some provisions at a nearby market. Definitely have dinner in the apartment tonight, he instructs. It's dangerous, especially for Americans, to be on the streets of Okha after dark. The whale scientists, last time they were here, got mugged.

We're having guests for dinner, both named Andrey. One is a local journalist, Andrey Kolomeets, who has been to Piltun Lagoon. His friend Andrey Bendjak does P.R. consulting for oil companies on the island and speaks fluent English. We start off talking about the fires, which recently burned through a small settlement nearby and shrouded Okha in smoke. Sometimes people could see only thirty feet in front of them.

Generally speaking, though, they say Okha is cleaner than many other places on Sakhalin because the city uses natural gas instead of coal. So they get white snow here, instead of its turning black as snow quickly does elsewhere.

As the Slavonskiya beer and the vodka warm our conversation, we turn to the subject of gray whales. Kolomeets is remembering seeing a surfacing "gray whale fountain" at Piltun Lagoon. "It's magic, when you see eye to eye with the whale," he adds, and goes on: "In Russia, people consider Sakhalin one of the unusual places. Those who are not into traditional science consider Sakhalin one of the sacred places. The fact the gray whales came here long ago, and stayed, makes this a mystical place. We, of course, don't know the place gray whales ultimately come from—maybe deep in the sea. Maybe the ones who approach closer to the seashore are a kind of expedition."

By the end of the evening, the two Andreys are singing Beatles songs and a chorus from "We Shall Overcome." What is the biggest change since Communism ended here? I ask. "We can communicate with Americans, make company all together, and say what we want without worrying about going to prison," Bendjak replies. "Just eight years ago, if we somehow managed to be here, we would have been watched."

Yuri Shevtsov, we learn from them, was trained as a hydrogeologist in St. Petersburg and knows everything about the ecology of northeastern Sakhalin. He's also considered the finest off-road driver around. "No danger to have trips with him." We raise a toast to "bad roads and good drivers." And say good night.

Yuri shows up early the next morning, wearing green fatigues, a blue shirt, and a black cap. It's only some sixty-five miles from Okha to Piltun, but he expects the trip to take about six hours, with a couple of stops along the way. Yuri has his spinning rod in the front seat of the jeep. He's director of an organization of hunters and fishers in the region.

Not far out of Okha we leave the paved road and take a dirt track across an open, flat countryside dominated by stubble from scorched earth. The fires are a terrible problem, Yuri says, with as many as a hundred burning every year and consuming as much as 25,000 acres at a time. Born in Okha, Yuri has no desire to leave. He'd been secretary of the local Party organization but doesn't think the Communist system was better, even though "life is difficult now because the economics of our country is destroyed. Ten years ago, there were no 'Yeltsin boys' in the street asking for food. People robbed very little. But now, everyone in government is stealing, so why not?"

We cross a mountainous region of undersized larch forests alongside thickets of dwarf cedar, then continue to bounce along a tundralike landscape. Soon we are on a marshy plain closing in on the Sea of Okhotsk. Oil derricks and a gas pipeline loom over the horizon. The onshore reserves of northeastern Sakhalin have been exploited since the 1920s. More than four thousand wells have been drilled and largely depleted, although about forty fields are currently in production. Natasha gets out to take some pictures at a particularly inky, vile-smelling reservoir.

Yet to the east the coastline rises as jagged and beautiful as that of Northern California. This is the ninth largest body of water on the planet, and one of the most productive seas. It teems with numerous varieties of salmon—pink, chum, coho, and more—as well as herring, cod, saury, flatfish, pollock, shrimp, squid, crab, and sea urchin. Along this shoreline, too, are large rookeries of fur seals, sea lions, and true seals. This coast stays frozen for six months of the year, and sometimes ice floes can be seen as late as July.

Just beyond the oil fields, on a bluff above the sea, we come to a locked gate and a wooden cabin. Here live Yuri's friends Nikolai and Svetlana, whom he calls "keepers of the gate." Beyond lies the *zapovednik*, a Russian nature reserve stretching the remaining distance to Piltun. No one goes past this point without a permit. The four of us are invited inside to take seats on wooden benches at a kitchen table. Svetlana slices some black rye bread. Nikolai passes around a bottle of kvass, a fermented drink of sugar and water.

Yuri wanders outside again, and I hear him shouting something. "He says there are whales passing!" Natasha exclaims. She leads the charge to the door. Perhaps a couple of hundred yards offshore, a trio of gray whales are surfacing and spouting. Yuri hands me a pair of binoculars. Yes, there are the familiar barnacled backs. They look—and, at first glance, they feel—exactly like those I know from Baja.

For centuries the chief residents along this northeastern coastline were the seminomadic Nivkhi, whose summer dwellings consisted of large one-room wooden cabins perched on posts about five feet above the ground. They used narrow, carved-log boats to navigate the treacherous coastal waters. They lived primarily on fresh or dried salmon, wild berries, and occasional products obtained from the Japanese and Manchu traders who'd been visiting for about a thousand years.

The Nivkhi regarded the region's brown bears as their kin, often capturing them to keep in pens outside their homes. Their primary legend revolved around a battle between the god of land, the bear, and the god of sea, the whale. With the help of man, the bear god had won. It was, in fact, something of a technical knockout, in that man and bear had escaped from the whale through a fork in a tree. Being too large to fit through, the whale god perished. Nivkhi tradition forbade the eating of whale meat, because it represents what is dead. They referred to the Sea of Okhotsk as the Sea of Death.

The first Americans ever to set foot on Sakhalin Island were whalers, beginning sometime in the late 1840s or early 1850s, when the Sea of Okhotsk became the region of choice for the Yankee fleet. At that time the Russians had only begun to colonize small portions of the island and wouldn't officially claim it until 1853. The following year some 160 whaling vessels headed to the Okhotsk, compared with only 45 choosing the Bering Strait. In 1855 over 130 ships sailed to the Okhotsk and only 8 to the Bering. Despite catches already going into decline, almost 150 ships returned to Sakhalin's offshore waters in 1856.

Bowhead whales, with their large yields of oil, were the favored quarry. But when they weren't around, it wasn't uncommon for a single vessel to take six or seven gray whales. Multiply that by the number of whalers and the catch would have been substantial. These were the very same Yankee whaling ships making an annual circuit from the Okhotsk by summer to the Baja coast in winter, where they began destroying the other stock of gray whales.

Scammon, in several of his articles for the *Overland Monthly*, made reference to his memories of "Saghalien Island." He described how the inlets and rivers "swarm[ed] with salmon," how the seals pursued the fish and the native peoples trapped the seals in nets and dispatched them "with their rude, bone implements." Scammon also wrote of the origins of an American cod fishery in the Sea of Okhotsk.

The captain had spent his last summer of whaling in the vicinity of Sakhalin, passing into the Sea of Okhotsk on June 6, 1862, and staying into September off Iona Island and around Shantar Bay. His ship was the *William C. Nye*, a well-known New Bedford whaler recently sold to Tubbs & Company in San Francisco. Though he described gray whales "congregating" here that summer, Scammon was also in search of bowheads. One of his log entries recorded "seeing whales during the day (6 or 8) but very wild." But persistent heavy fog made even this many sightings a rare occurrence, and only three whales were taken. The voyage was, in fact, a disaster in more ways than one. As the captain's log tells us: "Saturday June 21st. Light airs from the westward and clear at SW saw bowhead whales off Rocky Pt. Lowered two boats L. B. struck. When the whale went in his flurry, he stove the boat and killed the 4th mate and one man. The other boat came to the ship with the rest of the men got the ship under way but could not find the whale. At sunset came back for the anchorage. Came to anchor at 9 P.M. in 12 fathoms water." On September 4 another crewman drowned when one of Scammon's whaleboats was smashed by a whale.

By the time of this final and most ill-fated of Scammon's whaling voyages, both the eastern and western populations of gray whales were at very low levels. At that juncture it wasn't clear that these whales constituted separate stocks. After all, there is no physical difference between the two populations. One of the first published references to whales in this part of the world appeared in *Explorations of Kamchatka: Report of a journey made to explore eastern Siberia in 1735–1741, by order of the Russian Imperial Government*, by Stepan Petrovich Krasheninnikov. "There are a great many whales both in the ocean and in the Sea of Okhotsk," he wrote. "They often come so close to shore that it would be possible to kill them with a gun. Sometimes they come in to scrape their bodies, perhaps to rub off the shells which are so numerous all over their bodies; the creatures inside the shells cause the whales great discomfort. One of the reasons for believing this is that when they keep their backs out of water for some time, they patiently allow great flocks of gulls to light on their backs and peck at these shells." The report could only have been referring to shore-hugging, barnacle-covered gray whales.

Scammon did not differentiate the populations either, noting only that grays frequented both the Arctic Ocean and the Sea of Okhotsk. Roy Chapman Andrews, who roamed the globe collecting cetacean specimens for the American Museum of Natural History, addressed the matter in his book *Whale Hunting with Gun and Camera*. In a chapter titled "Rediscovering a Supposedly Extinct Whale," Andrews wrote: "In 1910, while in Japan, I learned from the whaling company of the existence of an animal known as the *koku kujira*, or 'devilfish,' which formed the basis of their winter fishery upon the southeastern coast of Korea. The descriptions indicated that the *koku kujira* would prove to be none other than the lost California gray whale, and I determined to investigate it at the earliest opportunity. . . . For over twenty years the species had been lost to science and naturalists believed it to be extinct."

At a Korean whaling station, 1912 (Neg. No. 218411, Photo. Andrews; Courtesy Dept. of Library Services, American Museum of Natural History)

Andrews spent two months at a whaling station in Ulsan, on the southeastern coast of Korea, in the winter of 1911–12. When he saw a pair of great flukes hanging from the deck of a whaling ship, he excitedly examined the "huge black body" and concluded that this "really was the long-lost gray whale." By the time he put together a monograph about gray whales in 1914, however, Andrews became the first naturalist to begin to place things in better perspective. He wrote: "As yet it is impossible to state whether or not the Korea and California herds mingle in the north during the summer. Information gathered from the whalers tends to show that a large part of the former herd summers in the Okhotsk Sea and the latter in Bering Sea and further north. Individuals of the two herds may mingle and interbreed during their sojourn in the north, but it is probable that whales which have been born near either the Korea or California coasts will find mates among the members of their own herd during the southward migration and return annually to their birth place."

When whaling started up with renewed vigor and larger ships in the aftermath of World War I, again both the eastern and western gray whales were thought to be virtually extinct by the 1930s. The former population has, of course, made a dramatic resurgence; the latter remains on the brink of oblivion. And to this day the mystery of just where the western gray whales go to breed and bear their young is unsolved.

It's a long thirty-five miles to Piltun Lagoon from the gatekeeper's home on the sea. Yuri alternates among forging a road across the sand, straddling a cliff above the seashore, and carving our way through the tall beach grass. He's rarely able to go faster than five miles per hour. Mosquitoes swarm incessantly against the jeep windows, and he turns on a little fan to try to make the ride more palatable. Birds that have migrated to these wetlands from as far away as Australia nest amid the creeping cedars. A rare Steller's sea eagle soars above the waves, larger than a golden eagle, with a bright orange-yellow bill and a long, white, wedge-shaped tail. Yuri stops to show us the roadside tracks of a brown bear. More whale blows are visible in the near distance.

At first glimpse Piltun Lagoon is strikingly reminiscent of portions of the Baja Peninsula, with sparse clumps of vegetation sprinkled among the dunes. Yuri parks along a sand spit on the low-lying shoreline. Less than a mile across the narrow channel, just north of the lagoon mouth, a steel lighthouse painted red and white is visible. We wait. Within a few minutes I hear the sound of an outboard. A light gray rubber Zodiac comes into view.

Dr. David Weller, the American marine biologist in charge of the camp, hops onto the beach and greets Yuri with a bear hug. He's in his midthirties, with a chestnut beard and warm smile. Accompanying him at the inflatable's helm is a young Russian who introduces himself as Grisha. "For you, Dave," Yuri says, handing Weller several loaves of freshly baked bread. He shakes hands and says he'll return for us in five days. As he drives away Dave says, "Yuri is our Far Eastern ambassador. Last year during a gas shortage, he'd bring us down six gallons at a time to help keep our boat going."

As we motor across the shallow lagoon, Dave continues: "It was a colder winter, with thicker ice than in many years here. The ice was still around when we came in the middle of June, and the sea is only now starting to warm up. The gray whales showed up—from wherever they come from—about two weeks later than normal."

This is the third summer for Dave and several of his crew at Piltun Lagoon. He explains that he's been working at sea for the past fifteen years, mostly in the Pacific and the Gulf of Mexico, "and after a while you can read the weather. But here, there's no way. Not even the locals can predict it. It can be foggy for five days in a row, then clear, then fog again the next day, or high winds that blow from all directions." Maybe only 40 percent of the days are they able to venture out among the whales at all. This July, however, had turned into the best yet weather-wise, with something like fifteen trips by the twenty-first.

I see something furry pop its head up out of the water, look around suspiciously, then quickly dive again. "We call them KGB seals," Dave says with a smile. "They're always checking us out. They're actually larga, or spotted seals, closely related to the harbor seal."

As we pull up to the shore again, another young Russian is skinning a larga seal killed for food by a Nivkhi hunter, who brought the hide to him earlier today. Alexey Vladimirov says he plans to take the hide back to Moscow with him in the fall, for his university's scientific collection. We pull the boat in and, beginning to cart our gear up a hill of white sand, are met by the remainder of the seven-person camp. Over the rise, this side of the lighthouse, sits a rectangular compound of ramshackle wooden buildings. One of these is for the lighthouse keeper's storage of diesel fuel, logs, and food obtained from a supply boat that stops here once or twice a year. For a summer rental fee of a hundred dollars, the scientific team lives inside half of a long-beamed house, the other half being occupied by the lighthouse keeper, his wife, and her grandmother.

Dave Weller says they're going to take advantage of a break in the weather to see if they can find any whales, and they have room for one more passenger. I've scarcely time to cross over a threshold of coal dust and stacked firewood, stash my backpack in the men's "bunk room" off the dining room, and make a quick trip to the outhouse before donning a heavy orange Mustang survival suit and waddling through the ubiquitous mosquitoes back to the beach. George and Natasha stay behind to accompany the two whale "spotters," Amanda Bradford and Yulia Ivashchenko, to the top of the lighthouse. They'll be communicating with our Zodiac by handheld VHF radio.

Dave Weller and Susan Reeve off Piltun Lagoon (George Peper)

"Today, unfortunately, there's so much smoke that it's hard for them to find the blows. You can see how shrouded it is out there," Dave says, pointing to where the lagoon channel meets the sea. This is the result of a new series of fires that have been burning for a week. "But we don't even know exactly where they are, or how intense," he adds.

We climb into the Zodiac. Dave is carrying a black waterproof case containing a Nikon F5 35-millimeter camera with a zoom telephoto lens. His research partner for the past three seasons, Susan Reeve, has the radio and the data sheets. Grigory Tsidulko—Grisha—is driving. He maneuvers carefully along the channel, which has shoaled up in new places this year and forced the biologists to refigure how to get in and out.

This nearly sixty-mile-long lagoon is the largest on the Sakhalin coast, fed by at least three rivers and a number of seasonal streams, but this single channel provides the lagoon the only entry and exit point to the Sea of Okhotsk. "A strong tidal current moves a lot of water through the channel, which keeps it naturally dredged," Dave points out. "Our pet hypothesis right now is that the lagoon itself is what causes the gray whales to come here year after year to feed. Because there's so much nutrient-rich freshwater effluent, and all these nutrients get sucked right out, mostly to the north, enriching the benthic community offshore of the channel mouth."

That's where the grays have often been seen densely clustered. Their food source is bottom-dwelling invertebrates, mostly ampeliscid amphipods, just what their eastern-roaming brethren favor in the Bering Sea. Here at Piltun, though, the water is quite a bit shallower, mostly between sixteen and eighty feet; whales have even been observed in depths as scant as ten feet.

We're within a mile of shore when word comes down from the lighthouse about an "M-C," short for mother-calf pair, moving slowly north. One of two pairs identified so far this season, these whales were observed by the boat team in the same general vicinity early this morning. The calf they'd previously named Katya. The mother is Chut-Chut, which means "Just a Little" in Russian, so designated because, Susan says, "she only shows a very small part of her body when she surfaces."

Katya and Chut-Chut are one of eleven mother-calf pairs observed here since research began in 1994. Altogether sixty-nine gray whales had been identified through 1998. Gray whales are the most dramatically marked of any whale species. Barnacles, parasites, and pigmentation all provide definite coloration patterns—big spots, vertical lines, a sudden blaze of color. So does scarring resulting from the bites of killer whales, which Dave says appears at different places on the bodies of about half of the gray whales at Piltun. With calves less scarring is evident, and fewer barnacles; but as they grow older over the summer period, the pigment becomes easier to see and photograph. The whales the scientists see repeatedly and recognize have been named Bud, Svetlana, Dima, Speedy, Bob, Blaze, Pirate, Hooligan, Mellow, and many more. There's even one the other biologists named Dave, since every time this particular whale showed up on their videotapes Dave Weller would say, "I know that guy."

In 1998 Dave took ninety-one rolls of film. When photographing whales for ID, he always aims first to obtain a right-flank shot, then one of the left flank and, if possible, one of the flukes. But he's also able to identify many of the whales by eye, with a quick examination of their flank markings. Such is the case now as we close in on a couple of spouts a few hundred yards away. Dave has his camera out and ready, but there's no need to shoot. "That was Katya and Chut-Chut again. I'm wondering what you may have next, over?" Susan radios to the lighthouse.

This particular afternoon, not much. The seas are calm, but the visibility is poor. Where any other whales are, nobody has a clue. In late afternoon Grisha brings the Zodiac in relatively close to shore, then makes the turn toward the channel. A couple more larga seals make furtive appearances. Dave says tonight we can look at the gray whale "photo album," where the best shot of each identified whale appears. Then I'll learn a little more of what's meant by their distinguishing characteristics.

After this year, Dave adds, they'll do an official population estimate. Up to now whale scientists had figured the endangered western gray population to be somewhere in the neighborhood of 250. If Piltun Lagoon is indeed their sole feeding ground—and nobody'd glimpsed them anywhere but here in more than isolated instances—then did a total of 69 identified whales suggest even a far smaller population?

Dave nods his head. "The thing is, 87 percent of the whales we photographed last year, we already knew from previous years. Only six new adults and seven calves were identified in 1998. And in the last twenty-six surveys we made, only two new individuals were sighted. So, yeah, I would say that's a red flag for caution about this population."

And *that* is something the sponsors of Dave Weller's research would rather not hear about.

The sponsors who provide funding for the Piltun scientists are a consortium known as the Sakhalin Energy Investment Company, Inc. (SEIC). Established in 1994, three years after the collapse of the Soviet Union, SEIC includes some of the biggest companies in the world: Texaco, Marathon, Mobil, Shell, Mitsui, and Mitsubishi. They are sinking over \$15 billion, with direct capital and operating expenditures anticipated at around \$40 billion, into offshore oil and gas development along northeastern Sakhalin. It's a production-sharing agreement with the Russian Federation, which retains ownership of the fields. The plan is for massive tankers to load crude oil destined for the burgeoning economies in the region, particularly Japan and South Korea. Long term they're looking at a trans-Sakhalin pipeline akin to the one in Alaska.

The first mobile drilling and production facility, called Piltun-Astokhskoye, is approximately twelve miles southeast of the lagoon channel. This and the nearby Lunskoye field have been calculated to hold more than 750 million barrels of oil and 14 trillion cubic feet of natural gas. The overall scale of projected development off Sakhalin Island may well be equivalent to that of Alaska's Prudhoe Bay. This region's natural conditions, however, are far more hostile than any other place oil companies have drilled before, including Alaska. Production, in fact, is possible only during the six ice-free months of the year. Piltun is a high-risk area for seismic activity, ice shears, and wave and wind action related to typhoons and tsunamis.

How then did Sakhalin Energy come to subsidize studies of the western gray whale, an endangered species that seems to be utterly dependent upon this same area for its sustenance? To understand the genesis of this curious arrangement, we must examine the history of a politically savvy marine mammalogist, Dr. Robert L. (Bob) Brownell, Jr. His career had gotten off to an auspicious start when, with the whale expert Raymond Gilmore, from the San Diego Museum of Natural History, he'd discovered the breeding grounds of the right whale in Patagonia in 1969. Brownell had gone on to study river dolphins in the Río de la Plata of South America and in India's Ganges River. Then he'd finished his doctoral studies at the University of Tokyo. While there he read a 1974 paper by a colleague concurring with a 1951 report that the western gray whale was probably extinct. Brownell had a hunch this wasn't true. He traveled to Korea and located an elderly biologist who was able to document catches of at least sixty-seven grays by South Korean whalers between 1948 and 1966, as well as incidental sightings since then off Japan. While "the continued small catches in this century have probably prevented any recovery," Brownell and Chan-il Chun wrote in a landmark publication in 1977, this Western Pacific stock of gray whales "probably still exist[ed] in small numbers" but was in desperate need of "meaningful international protection."

Brownell had gone to work for the U.S. Fish and Wildlife Service, housed in the Smithsonian Institution, and by the early 1990s became the State Department's ocean science policy adviser. His primary thrust there was working toward a United Nations worldwide moratorium on the use of high-seas drift nets, which were ensnaring thousands of marine mammals. Then, soon after the breakup of the Soviet Union in 1991, he began discussions with Russian colleagues about starting a long-term research project on the status of the western gray whale.

Brownell had been part of a marine mammal project under a U.S.-Russian environmental agreement that dated back to the Nixon era. Even at such low points in Cold War relations as the Soviet war in Afghanistan, scientists from the two nations continued their annual meetings. Once the USSR collapsed Brownell convinced higher-ups in the State Department to provide funds to help keep the Russian marine mammalogists afloat. He'd long worked closely with Alexey Yablokov, a whale biologist who was named Boris Yeltsin's adviser on ecology and public health.

It was Yablokov who came to Brownell with the revelation that the Soviets had inaccurately reported the numbers of whales they were taking. Now it became clear why they'd always blocked an international observer program called for by the International Whaling Commission. Biologists from four of their own factory ships blew the whistle; one had stolen the true data sheets and hidden them under his potato cellar. Brownell got money from the State Department to salvage and computerize the information. The results were horrifying. The figures reported to the IWC since the 1960s were off by 100,000 whales—almost 50,000 humpbacks, more than 10,000 blue whales, even 3,000 already near-extinct right whales had been killed by the Soviet factory ships. Little wonder the few remaining populations have only just started to show signs of recovery.

So Brownell was also something of a hero among whale conservationists when he left State in 1993 to become director of the Marine Mammal Division at the government's Southwest Fisheries Center in La Jolla, California. Over the years he had closely followed renewed scientific interest in the western gray whale sparked by his dramatic news in 1977—including aerial surveys conducted by Russian biologists in the late 1980s showing "an increase in the numbers," especially in the summer months around Piltun. In June 1994, Brownell attended a meeting of a scientific group primarily focused on North Pacific fisheries issues. He approached a representative from Exxon, which he'd heard might be considering offshore drilling in the Piltun region. "I asked why they weren't concerned about the gray whales and the right whales in the area off Sakhalin," Brownell later recalled. "They said there weren't any. I said, 'Says who?' They said, 'Well, our consultants preparing our environmental impact assessment.' I said, 'Well, they're wrong. These are some of the most endangered populations of large whales in the world.' They didn't want to hear about it. I tried to make contact with them afterward, and nothing happened."

That September a Japanese photographer documented nine gray whales feeding off Piltun. The following June, 1995, Brownell learned of a joint venture to initiate the Sakhalin-2 Project offshore of the lagoon, announced during meetings between Vice President Gore and Russian Prime Minister Chernomyrdin in Washington, D.C. Two months later Brownell paid his first reconnaissance visit to Sakhalin Island with some Russian colleagues. By the end of August they had photo IDs on thirty grays around Piltun. Late the following year Brownell went to his friend Yablokov. "We need to stir the pot," he told him. The U.S. vice president was pushing the Sakhalin energy development, via the Gore-Chernomyrdin Commission, as a boon to the Russian economy in terms of infrastructure. "Gore's got to be concerned about the whales; there have to be some safeguards," Brownell said. The U.S.'s IWC commissioner, James Baker, received a tacit agreement from Gore's staff. Brownell started working with a small team from the Environmental Protection Agency and the Department of Energy, drafting a one-page statement about biodiversity and the whales around Sakhalin. The statement was signed by the two leaders at the February 1997 Gore-Chernomyrdin meeting.

Then Brownell flew to Houston to meet with officials from the Sakhalin Energy consortium and Exxon Neftegas. He laid on the table a five-year proposal to study the western gray whale population. According to Brownell, "They didn't have much choice."

The initial outlay from the oil companies was $350,000 for 1997. Brownell went to Bernd Würsig, who at Texas A&M chairs one of the largest laboratories in the world where graduate students can earn master's and doctoral degrees studying marine mammalogy, and asked him to round up a research team. Würsig thought immediately of Dave Weller, then completing his Ph.D. work. When Weller jumped at the opportunity, Brownell ended up providing him an office at the Southwest Fisheries Center as part of a postdoctoral fellowship. Meantime Brownell contacted Alexander (Sasha) Burdin, chief of laboratory at the Kamchatka Institute of Ecology and Nature Management, to find Russian counterparts and start the wheels rolling on Sakhalin.

An entirely different kind of joint venture—one that might determine the fate of the western gray whales as well as a multibillion-dollar oil-and-gas consortium—was under way.

It's all happening at what must surely be one of earth's more vestigial locations. In our new abode there's no running water, no refrigerator, no electricity except between 9:00 P.M. and midnight, when the lighthouse keeper turns on a diesel generator in an outer building. In the beginning they had to cart all their water in by bucket from an outdoor well. Now the researchers use a well that probably dates back to when the log house was built. It runs under the floorboards of the dining table, with the water pumped from a hose into a couple of barrels next to the kitchen sink. This sink is where everyone's face and dishes get washed, with the dirty water draining through a pipe into a metal pail below.

The hot shower is another innovation. It's in the generator shed. Water is pumped up to the roof into a big tank, from which it runs down to cool the diesel engines when they're running at night, then gets siphoned off into a hose back up top that fuels the shower. It's available only at night, and to reach it you walk a set of loose planks to the shed; once inside you must make sure to shine your flashlight into a gaping hole in the floor. Then simply keep turning the faucet in any direction until the water gets warm. "Being at Piltun," Dave says, "really gives you a whole new appreciation for life back in the States. Here everything is a major production."

He'd spent his early years in California, having been a surfer whose interest in marine biology was kindled by "seeing gray whales and bottlenose dolphins go by." His undergraduate work was in animal cognition at the University of Hawaii, studying bottlenose dolphins in captivity. There Dr. Louis Herman was trying to assess how dolphins might "think" or problem-solve in their natural environment. Dave assisted the professor in teaching two dolphins sign language. The pair learned a vocabulary of about fifty words—nouns, verbs, adjectives—and other cognitive tasks. Dave moved to San Diego State University for his master's work on the behavioral biology of bottlenose dolphins in the wild. He went on to spend years studying humpback whales off the Hawaiian Islands. This was his first time working on gray whales.

His Piltum co-worker, Susan Reeve, was born and raised in Hawaii. Her work didn't overlap with Dave's, but she got her master's doing cognitive studies with bottlenose dolphins at the University of Hawaii. Grisha, the third member of the Zodiac crew, is embarked on a doctoral program at Moscow State University studying distribution and behavior of the western gray whale. "Even after two summers I already spent here, every day happens something new," he says. "New feelings, new information. Sometimes birds you never saw before, sometimes the engine problem you have to figure out."

Amanda Bradford, a twenty-three-year-old Texas A&M graduate working on her master's in marine mammal population dynamics at the University of Washington, opens the catalog containing images of whales identified in previous years. Dave points to "some familiar faces," including that of Chut-Chut. At least three age classes inhabit the Piltun region, from fifty-foot-long adults through not quite full-grown yearlings to calves. This range leads Dave to postulate that "there's no segregation going on in the migration," unlike among the eastern gray whales, whose younger animals sometimes stop and hang out for a summer around places like Vancouver Island. It may be, he adds, that the western population is simply too small for such segregation to occur.

From biopsy samples taken over these last two years, the scientists hope soon to know the ratio of males to females at Piltun. Dave brings out a small dart, which is fired from a crossbow into the blubber to collect some DNA. "It's no more than a pinprick to the whale," he explains. "And, just like with humans, a tiny piece of skin will contain the entire mitochondrial DNA fingerprint of a given individual." The samples can be stored indefinitely in a saturated salt solution. Bob Brownell took a few in 1995, another fifteen were obtained in 1998, and Dave would like to get up to thirty before doing some comparative analysis. This will include DNA testing to try to further determine the relationship between the western and eastern gray populations.

"Are they truly isolated, or might there be some mixing?" Dave wonders. "Or were they one large population that's now separate? We think that, at least geographically, they're isolated from each other. Whether genetically they're related, we don't know for certain."

Bob Brownell had told me that he's sure the DNA results will show the two populations to be "closely related, and at some time in the not too distant past, something caused them to split off like this. You could speculate on how it might have happened. It could have been something to do with interglacial periods or Ice Ages. Or the ice was way south and, when the gray whales were up north, some then went west looking for a place to feed and ended up off Sakhalin."

In Brownell's view that would have been at least ten thousand years ago. Some marine scientists, however, have conjectured that, as the eastern population recovered, it spilled over from the Bering Sea and now migrates down the Kamchatka peninsula coastline to Sakhalin before rejoining the others for the trip back to Baja. Since we don't have evidence on where the western grays *do* go in the winter, it's tempting to think about. Brownell, however, finds this a "superficial argument" and adds: "Gray whales are so amazing in terms of how they can migrate and navigate, why would they take such a big wrong turn basically? Think about it. That would be a pretty major thing, to be migrating south with everything to your right side instead of your left side. If you're using stars or sun for navigation, it's not gonna work. The sun would be rising and setting on the wrong side of you. You'd be completely out of whack."

Dave Weller cites another reason that the two stocks are unlikely to mingle: the geographical isolation of the Sea of Okhotsk, which is almost completely enclosed by the Kuril Island archipelago and the Kamchatka peninsula.

Where, then, might the western grays journey after they leave the Piltun area? In 1998 this happened on November 11. Dave says he "would love to go south and see if we can find them. Or better yet, put some satellite tags on them and hope they carry them all the way to the breeding grounds."

In 1874 Scammon wrote: "It has been said that this species of whale has been found on the coast of China and about the shores of the island of Formosa, but the report needs confirmation." Almost forty years later Roy Chapman Andrews discounted Scammon's hypothesis, saying he doubted "if the gray whales migrate far south of the peninsula of Korea," where Andrews had seen them. Once again Scammon appears likely to be the one who stands the test of time. Predating him by a generation, an 1844 edition of a whalemen's newspaper out of Honololu called *The Friend* related: "During the months of January and February, whales and their young resort to the coast of China, to the south of Hailing Shan, in great numbers; and during those months are pursued by the Chinese belonging to Hainan and the neighboring islands with considerable success. The fish generally seem to be in bad condition, and were covered with barnacles." The whales were observed "to roll on the numerous sand banks on the coast, in order to clear their skin of barnacles and other animals which torment them. They are often seen leaping more than their whole length out of the water, and coming down perpendicularly so as to strike hard against the bottom."

The species wasn't identified, but the description—barnacles, shallow-water habitat, breaching behavior—closely parallels the one Scammon would soon provide for the gray. If those were grays off Hainan Island, this description would place them in the South China Sea, around the corner from the Gulf of Tonkin and not far from Vietnam.

A whale skeleton in a Chinese museum, bits and pieces from old whaling logs, as well as twentieth-century strandings and scientific data, have pinpointed gray whales at numerous localities down the long Far Eastern coastline. They apparently roam from the Sea of Okhotsk down into the Sea of Japan, the Yellow Sea, the East China Sea, and the South China Sea, and along the coastlines of Russia, Japan, the two Koreas, and China. Over the years calving grounds have been surmised off Japan, along the southern coast of South Korea, and at two spots around central China.

However, five of the most recent scientific papers on western gray whales, written between 1984 and 1997, all suggest that the calving area is likely to be in the waters around Hainan Island—precisely where that first apparent reference to western grays in 1844 happened to be. That's a latitude of 20 degrees—not far below the Tropic of Cancer, which also runs across the southern Baja Peninsula.

Grays would need to cross some relatively deep water to get to the warm, remote Hainan Island. But if this *is* the chosen birthing spot for the western grays, they would have a coastline of more than nine hundred miles to select from. It's considered China's "shining pearl," unpolluted and pristine. Like the Baja lagoons, part of the area consists of natural salt flats that produce 270,000 tons of raw salt a year. Hainan's coastline includes the Monkey Island of Nanwan at South Bay, where the rare macaques feed amid the evergreens and wild mangoes. And Yalong Bay, with its calm, clear sea free from stone and minerals. And it's in the shadow of the Horse Ridge Mountain, Tian-Ya-Hai-Jao or "the Ends of the World." This was the place reserved for officials condemned for treason in the ancient dynasties. The name was given because the sea beyond is so vast.

Later, in attempting to solve the mystery of the western gray's migration, I communicated by e-mail with one Xichen Zhao of the Hainan International Travel Service. The response, in English, was as follows:

> Yes, Mr. Russell, whales ever be seen around Hainan Island.
>
> In February, a bigger whale was stranded near Haikou, the capital of Hainan. The policemen and fishermen want to help it to return the sea. But it was so big and they couldn't be success at once. Until the mid night, they tugged it to the sea with a bigger ship. The next day, it was found in the beach and died. Because the time was the Chinese Spring Festival, Hainan people hadn't a merry holiday.
>
> In last month, a little whale was found in a small bay in Wenchang country. This time, the whale was not so bigger and people was mate [sic] to help it return the sea.

> We welcome you visit Hainan. I don't know when the whale will travel to Hainan Island and whether the whale is the gray whales.

At Piltun it's raining when we arise at 6:00 A.M. Dave is hoping the moisture will clear the air. I walk with him across the compound, around a corner past a dilapidated *banya* (Russian bathing house), where two huge pigs currently reside, to the entryway to the lighthouse. In operation since 1964, it's 115 feet high, and Amanda says she's counted 166 steps, which is a decent enough StairMaster if you make the climb several times a day. I'm breathing heavily by the time we reach the top. The view is spectacular. We can see three lakes, including the camp's "swimming lake" about a mile out into the creeping-cedar-covered dunes, and a pine forest off in the distance. In front of us are miles of rough sea with abundant whitecaps. "Probably a Beaufort 5," Dave says, according to the mariner's scale for rating wind, "compared to our whale-counting trip yesterday, which was probably a 1 or a 2." Even in the drizzle, a couple of gray whale spouts are visible not far from the mouth of the lagoon. And there to the southeast is the massive oil drilling platform known as the Molikpaq. To its right across a gangway is a smaller rig, the Sakhalinskaya, basically a floating motel.

The Molikpaq has been out there for almost a year, since late August 1998, positioned on the seabed in water about a hundred feet deep. It can accommodate more than one hundred people. To install it about 5 million cubic feet of earth were removed and dumped in a deeper part of the Sea of Okhotsk. Also dumped were millions of cubic feet of sand used to stabilize the Molikpaq's foundation and build a protective wall around the platform. At sea level the platform was reinforced with steel plating for ice resistance. During the week the Molikpaq and more than a dozen support vessels arrived, particularly low numbers of gray whales had been observed by the marine scientists at Piltun. Speculating this may have been because of the increased noise, their 1998 season's report recommended that industry-related aircraft and vessel traffic maintain a minimum distance from shore of between five and six miles.

On July 5, 1999, the Sakhalin Energy consortium ignited the flare as "First Oil" was pumped after ten months of drilling. The oil was transported through a 1.2-mile-long underwater pipeline to a floating storage and offloading unit moored nearby. Every week or two some of the largest tankers in the world would be arriving to shuttle the oil to Southeast Asian buyers. By the spring of 2001, as more wells come online, the Vityaz Production Complex at Piltun-Astokhskoye is expected to be producing as much as ninety thousand barrels a day.

What about the possibility of a catastrophic spill? The 1995 earthquake in northern Sakhalin had ruptured land oil pipelines in more than fifty places, with at least 110 tons spewing into several rivers. The currents of the Sea of Okhotsk have been poorly studied. The *Exxon Valdez* disaster off Alaska's coast ultimately spread over 10,000 square miles of ocean and some 1,500 miles of shoreline. That's over twice the length of Sakhalin Island. An oil spill under ice conditions, of course, would pose the most difficult cleanup task. The gray whales would be poisoned if they ingested toxic hydrocarbons or contaminated food. At the very least their feeding area would doubtless be destroyed.

So the whale scientists had hoped that, once the Molikpaq started pumping, their own research efforts might be expanded if the oil companies were really interested in considering all potential problems. No such luck. The funding for Dave Weller's team had been cut more than in half in 1999 (from $350,000 to $140,000). Exxon, which had been splitting the whale research budget with the Sakhalin Energy consortium, resisted believing the scientists' suspicions that the western gray whale population might be less than the 250 they'd initially conjectured. So Exxon refused to work with the researchers any longer, leaving Marathon Oil their only sponsor.

Now Dave points to the deck surrounding the top of the lighthouse, where the first two years they'd set up a theodolite with a thirty-power binocular at the eyepiece. This amazing piece of equipment not only measures precisely where whales are but also reveals their swimming speeds and movement patterns, and behavior observations allow dive sequences to be assessed. In 1998 Dave's team of four—theodolite operator, two behavior observers, and a data recorder—would stay up here all day. Their work was considered just as important as getting photographic and acoustic information on the gray whales, which took place alternately from the boat. It had required seven months for Dave, Susan, and Amanda to reduce and analyze everything they collected by theodolite last summer, then write it up.

"It was worth it," Dave says, "because it's the most powerful technique we have for showing behavioral changes. It's also been the sore spot in our reports, what I think the oil companies have been most concerned about. Because from these detailed focal follows, where we'll watch a group of whales for three to four hours at a time, we were saying it appears the whales are changing their dive behavior or swim speed relative to noise disturbance."

The scientists' draft report had noted: "Tolerance of noise does not necessarily mean it has no deleterious effects. . . . Noise, seismic surveys, and other industrial activities can displace whales from critical feeding and migratory habitat. For example, strong acoustic sources like the open water seismic exploration conducted in the P-A [Piltun-Astokhskoye] field during 1997 was probably audible to gray whales at distances exceeding 100 kilometers [62 miles]."

When vessels bounced those low-frequency pulses off the seabed to map the likely areas for gas and oil, the gray whales were observed to swim faster and straighter over larger areas. "These findings were hypothesized to reflect behavior changes indicative of disturbance to feeding behavior." The whales also had a more rapid breathing rate, indicating "a fundamental physiological change which may or may not have proven deleterious to individual whales." Additionally, the whales tended to remain close to shore during seismic surveying, then shift to more offshore waters later in the summer, after it was terminated.

Unlike the Navy, which at least sought to take into account whale scientists' findings on their LFA sonar tests, the SEIC oil consortium simply didn't want to hear about this. Their response to this part of Dave Weller's draft report read: "The conclusions that implicate industrial activity as the probable basis for statistically significant differences observed during the study period are not well founded."

Dave had lacked enough personnel to do what he considered crucial—make the theodolite observations simultaneously with the boat photo-ID and acoustics work. "With full-time recording equipment going, and the observations, you'd know what kind of sound the whales are being exposed to, how intense and how far away it is, and really nail down whether the effects we're seeing are from the noise or just seasonal or natural variation in their behavior."

Now, not only had funding vanished for any theodolite work but the effort to record industrial noise and whale sounds—to see if there might be a correlation—had been taken away from the Weller crew as well. Instead, the acoustical studies, as well as aerial surveys, were being done by an all-Russian scientific contingent based two and a half miles to the north—which hadn't raised any oil company hackles—and both SEIC and Exxon were contributing funds.

Dave has heard via shortwave radio that the Russians' acoustical expert there wants him to pay a visit. Given today's inclement weather, this seems as good a time as any. Just about everyone from the camp sets out at mid-morning. As Dave and I commence the long beach hike to the North Village, we come upon abandoned boats, a jeep buried in the sand, even the remnants of a railroad track running alongside some rusting barrels. "Apparently the rail line once went as far as the North Village," Dave says, "but nobody we've spoken to knows why. It must have been strategic in some way. It's pretty amazing that we're at essentially an old military outpost that was shrouded in mystery until just a few years ago."

This stretch of coast is so rich in fish, birds, and other wildlife, it's likely to have been inhabited for many centuries. But the Nivkhi who'd lived here had long ago been relocated by the Soviets, to work in small cities in subsidized fishing and fish-processing collectives. Now a few are returning to their ancestral villages, fishing and hunting again out of necessity. Two Nivkhi, I'm told, live year-round at the North Village. It must be as strange for them to retransition themselves as for us Westerners to venture into what's been called the Uppermost East.

I ask Dave to describe the most frightening situation he's faced here. He replies, "When you're quite a ways from home out there, and all of a sudden you see this giant bank of fog coming along that essentially engulfs you. We have GPS [Global Positioning System] to navigate, and a compass, but you become kind of disoriented and it's unnerving. You've got to trust your equipment because you can't see."

He also recalls a moment in 1997, when his mentor Bernd Würsig came from Texas A&M to visit. "This was when we decided we'd better have a satellite phone," Dave says quietly. A Japanese marine scientist was also present, and this called for a celebration. Several people went picking wild mushrooms in the tundra, where many tasty varieties grow, and cooked them up. Everyone ate them, and a few had queasy stomachs that night. But nobody—except Bernd—had any real problems (interestingly, he was the only one to have eaten mushrooms and *not* drunk any vodka).

Dave recalls: "He woke up the following morning and told me he was feeling really under the weather and wasn't sure what was going on. Within a few hours he had a very high fever, lost function of his legs, and his vision started to deteriorate. We induced vomiting, kept pumping him with water and getting him to throw up, and in the interim tried to contact a helicopter in Okha by radio."

Dima Golovenkova, the lighthouse keeper, probably saved Bernd's life. He stayed at the radio, calling every other lighthouse keeper he could think of, who in turn radioed still more. Finally, a rescue helicopter arrived. "Had there been fog on that day, who knows what would have happened, because they wouldn't have been able to land," Dave continues. He and Sasha Burdin flew the critically ill Bernd to a hospital in Okha, where they would remain with him for four days. The diagnosis was mushroom poisoning. "The medical supplies available to them were limited," Dave remembers. "Really all they could do for Bernd was rehydrate him, provide a saline IV, and keep watch. But they did whatever they could, and he recovered."

Now ducks and gulls lift themselves from the water, flapping their wings and soaring above our heads. Some ten thousand waterfowl are said to visit the wetlands of Piltun during the summer months. As we near the North Village, a Steller's sea eagle is dining on a fish along the shoreline. George sneaks up within a hundred yards with his cameras, until the lighthouse keeper's dog charges in to scare the magnificent bird aloft.

Just ahead above the lagoon is an old military observation tower, beside a wooden building that was once a barracks, renovated last summer by Nivkhi fishermen and currently the abode of the Russian scientists. Beyond these is probably the biggest pole barn I've ever seen, formerly used for a fish-processing operation. About twenty feet into the lagoon, holding a fly rod, is a Russian wearing waders, a Yankees cap, and a T-shirt inscribed INTERNATIONAL SOCIETY OF CHAR FANATICS. Dave says, "He's fishing for Arctic char; they get up to fifteen-pounders here."

The Russian team, all from Vladivostok, consists of four acousticians from the Pacific Institute of Fisheries and Oceanography and six ichthyologists from the Institute of Biology of the Sea. They arrived ten days ago. Their second in command is Dr. Alexander Rutenko, who's wearing a gray sweatshirt depicting a surfer above the caption FREESTYLE BOARDING. He's a specialist in radio physics who wrote his dissertation on acoustics but who admittedly knows little about the gray whales he's come here to record. Grisha serves as translator as our small group gathers outside the main cabin.

"The gray whales here are very quiet," Dave tells Alexander, "vocalizing only occasionally, perhaps once an hour." He demonstrates a knocking sound, tapping his hand against the wood paneling. The Russian wonders, Do they use sound for echolocation or communication or what? "There's no morphological structure or direct evidence to indicate that gray whales can echolocate at all," Dave replies, "but it may be for social facilitation or communication between individuals in a feeding area."

Are their vocal mechanisms similar to those of odontocetes, such as dolphins? Alexander asks. "It's not currently known how that sound is actually created, but probably with air sacs squeezing together, opening and closing underneath the air passages," Dave says. "But it's not known whether it's the same as delphinid vocalizations or not."

Does Dave know anything about the grays' means of orienting themselves while they migrate? Is it the magnetic field of the earth? "This is also unknown, but it could be magnetic navigation: using the sunlight and orientation to the coastline, then surveying the bottom as well with acoustic sounds, along with an understanding about the underwater topography."

We enter the Russians' dwelling, where several scientists are sitting around a long wooden table having some lunch—including red caviar gleaned from the local pink salmon. Alexander ushers us into a side room lined with computers and recording equipment. "They're trying to find out," Dave says, "how far noise from the Molikpaq travels and then correlate that with how those sounds might affect the gray whales, in terms of where the whales are." Alexander explains how they have placed three buoys at varying intervals—one very close to the Molikpaq, another about halfway from there to the lagoon, and a third a few miles straight out from the mouth. Each buoy has a hydrophone platform anchored to the bottom, leading to an antenna on the surface. Each can record automatically for up to forty-eight hours before they need to change the batteries. The signals are transmitted back to a receiving antenna here at the camp. A journal is kept of all the sounds, which are recorded onto either a tape recorder or a computer.

Alexander turns on one of the tape players. The loud whirring sound I'm hearing, he says, is the Molikpaq. Is he concerned about potential detrimental effects on the whales? I ask. "It's too early to say something particular," the scientist responds, "but probably the main effect on gray whales will be that infrasound produced by the pump of the Molikpaq, a powerful low-frequency noise because it's coming from about a one-kilometer-deep hole in the earth and all that slice of sea bottom will vibrate in the same manner and produce some kind of seismic wave."

What should be done if your studies find such negative impacts? I ask. Should the drilling be stopped? Alexander lifts his shoulders and says, "It's a necessary technological process; they wouldn't be able to change it. Oil companies wouldn't change anything anyway." In his personal opinion, though, "This area is very important for these gray whales, and they should drill somewhere else where they do not disturb endangered wildlife." Since first oil was struck only a few weeks ago, "we will see if the whales are gone from this area or there are fewer animals than before. This would indicate they are affected by the sound and it will be time to start to do something."

The walk back to our camp seems interminable. By mid-afternoon, though, the rain has ceased and Dave thinks we could go look for whales around the lagoon. It takes seven of us to push and pull the inflatable over the rise and downhill again to the beach. Dave calls this "the Piltun workout."

It's still very rough as we head toward the open sea, jostling against one another in the little boat. Dave, a more experienced motorman, takes over momentarily from Grisha to maneuver us through the high swells at the channel opening. Just ahead I hear that instantly recognizable whooshing sound, almost too close for comfort. There, immediately in front of us, rolling in the surf, is a thirty-foot gray whale.

Dave cuts back on the engine. It's about twenty-five feet away, tumbling in the big waves right off our bow. Our fifteen-foot rubber boat is considerably smaller than a metal *panga*, and I feel suddenly very vulnerable. Now the whale turns and gazes over at us with that intense, that immense and impeccable, eye. Then it rolls again, offering us a dramatic gesture of pectoral fin and tail fluke, which pounds the water and sends spray our way. The whale appears to be having a fabulous time, enjoying being pummeled and moved around by the breakers. Surf-riding, no doubt! And showing off for us.

I focus on the pigment, the barnacles, the peeling skin, seeing clearly for the first time the unique markings that a scientist's trained eye sees. Susan says in a near whisper, "We're in only 3.3 meters of water here." Might the whale be rubbing itself on the shallow bottom? It's impossible to say. It just keeps rolling along as we silently ride its wake. This is every bit as good as a Baja show, I think to myself. I feel like staying here forever, churning tumultuously wherever the whale would take us.

Finally Dave says, "Let's leave it alone." Outside the tide rip, where the waves are breaking, the Sea of Okhotsk has become very calm. "Sometimes it's the opposite," Susan says.

A fish flies out of the water as we begin to motor north. "Pink salmon," Grisha says. The lighthouse watchers have honed in on a lone whale up ahead. Dave brings out his camera as we approach. "I don't recognize this one," he says. "Two large white spots on the front right flank, and on the front left flank there's a very distinctive vertical line. I think we'd know if we had him already this year. That doesn't mean we didn't identify him last year or the year before; we'll have to check the catalogs tonight."

Dave calls out the roll and frame number, which Susan records on her Piltun photo-ID form along with the time, "temporary individual name" of the whale, its body aspect (left, right fluke), and location. Temperature and salinity, which the crew stops periodically to check, are also recorded. Grisha leans over the side with a refractometer. "That's twenty-five," Dave says, reading the salinity measurement, meaning twenty-five parts per thousand. "So it's not very salty right now in this location. Thirty-five is typical seawater." Grisha then brings up a long white thermometer and announces a temperature reading of 15.6 centigrade. "Very warm," Dave says, "almost 64 degrees [Fahrenheit]. It was 1.8 C when we got here this summer."

Susan radios back to the lighthouse: "We're just finishing our station here and wondering if you still have that M/C or anything else, over?" Amanda replies that whale pods are visible in both directions, with a probable mother and calf to the north. She adds excitedly that she briefly saw through the spotting scope ten unidentified delphinids, possibly killer whales, come to the surface off in the distance.

The Zodiac takes the waves beautifully. In smooth water it could reach a speed of almost forty miles an hour if need be. Right now we're cruising north at about ten when Grisha calls out: "Defecation!" He's spotted a trail left by a whale on the surface. "Well, Grisha, get your net," Susan says as he slows the boat to idle and prepares to dip. A minute later I'm trying to ferret out the net's rather fresh and malodorous contents. "What you got there, amphipods?" I ask.

"Let's take a look," Dave says. "It's mostly digested." A closer look. "Quite digested actually." A still closer look. "Quite *heavily* digested already." He reaches into the net and plucks out a minuscule something. "Now here's the creature, maybe three-quarters of it, an ampeliscid amphipod. I thought we might've scooped a few whole ones. Oh well, let's put it in a jar."

As Dave rinses off his hands, he looks over at me and shrugs. "Close encounter . . . with gray whale shit."

We're off again. There's another whiff of something ahead. Whale breath. Amphipod-induced halitosis, one might say. Dave, however, likens the pungent smell to cauliflower. Susan prefers a comparison to cabbage.

Standing on the bow, Dave instructs Grisha to slow down. He motions with his hand to let Grisha know which direction to turn. Dave seems to possess a keenly instinctive sense about which way a whale will go next. "Left side!" he exclaims as one surfaces to starboard, then disappears. "Try the right for the next surfacing," Dave tells Grisha. Sure enough, that's where it comes up, but not long enough for Dave to photograph it. This gray is feeding and proving elusive. After we chase it around for a while, Dave decides, "We'll just stop and wait until we spot it again."

He tells us, "I think this might be a whale named Svetlana, who we already know. Some days she's very easy to approach and photograph. Last year she had a pretty fresh gouge on her back, and this year when we've seen her it seems to have healed and become a V-shaped white scar." Maybe from a killer whale? I ask. "No, more like a propeller or something that would have run across the top of her back."

Her fountain of spray unfolds in the air about a hundred feet away. "Right side would be better, not as much glare," Dave says as Grisha guns the boat. We take a wave high and head-on and can see her entire body. The scar is plainly visible. It's Svetlana all right.

More conversation with the lighthouse. We need to move inshore, toward where the spotters saw three blows about a mile beyond the South Teepee. That's a trio of poles lined up along the beach, one of a number of landmarks the scientists designate with their own code. Another is Sleeping Log, a large driftwood tree whose root center is inshore to the lagoon. Two Cans is straight off the lighthouse, a pair of large barrels on a spit of beach.

A trio of whales are off to port, moving in the company of black-headed and black-tailed gulls, who soar and dive as if part of the same dance. Not far beyond is another whale triumvirate. Each group appears to be composed of both small and larger whales. They seem to be diving synchronously, with a specific breathing rhythm. As we advance closer, Susan points out the plumes of mud at the surface in the forty-five-foot depth where the whales are eating.

Dave says, "I can ID Bud, just by his looks, he's so handsome."

Susan says, "He winked at me last time."

Dave explains, "There's a brown spot on his side we recognize."

Grisha stops to let the whales figure out where they want to go. "Right side partial fluke, frames twenty-five through twenty-eight," Dave signals. Grisha accelerates. Dave aims and shoots. On their feeding grounds the whales move much faster than I'd observed at San Ignacio Lagoon. This is also true when any approach our boat, a moment that's heart-stopping in an entirely different way.

"Do you think we got them?" Grisha asks.

"I'd like to get the last one a little better," Dave says. "Left pass on these guys, then we can go on to the next group."

And so we do. Like that of the whales they follow, the teamwork among this trio is equally striking. There's never any sense of competitiveness or domination, and I find myself wondering if the whales themselves can sense this. I can hear more blows in the distance, and I flash back to standing silently listening with Homero Aridjis that night on San Ignacio Lagoon.

It's becoming dusk when we approach the entry to Piltun Lagoon again around 8:00 P.M. As we leave the open ocean, I see a pair of gray whales swimming ahead of us and then, like a farewell, each makes a splendid full-length breach.

At the shore the others come down to help with the haul-out. Altogether today they'd counted twenty-eight gray whales from the lighthouse. This summer's previous biggest count was twenty-one. The highest ever in a day, Grisha remembers, was thirty-seven.

Over dinner Dave says, "You know, that whale rolling in the surf today on our way out has to rank as one of our top ten close encounters. Only three times last year did we see that kind of surface activity."

Have you noticed, I ask, whether the gray whales here seem to become more "friendly" as they get used to your presence?

Everyone around the table nods affirmatively. "We've seen the mothers and calves become friendlier, to the point where they're a few feet away from the boat," Dave says. "For example, Katya and Chut-Chut, when we first saw them this year, were quite nervous about us. In the past month they've become much easier to approach. What's been really neat is to watch the curiosity of the calves gradually pull their mothers closer and closer. Frequently Mom will come up above the water and check us out."

After a pause he continues: "Also, we watch the calves become independent. All of a sudden one day they're on their own, usually when they're between six and eight months old. You'll see them swimming around doing calflike things, but with no mother around. We see the transition, and in fact for me it's a bit of a worry—how do these little guys make it on their own, back to where they came from? Sometimes you see two new-of-the-year calves together, looking for support and company, I suppose. It's pretty rewarding to watch them grow up."

Susan says, "With such a small population, it's like coming back to check up on your buddies every summer. You know, 'Are you gonna turn out to be a mother this year?' "

Dave goes into the kitchen and returns with the third, and last, bottle of vodka that George and I had brought with us. Around the table, here where there is "nowhere else to go," five Americans and five Russians lift their glasses for a toast.

"To the western gray whale!"

I'm remembering what the first oil company representative to visit the camp told me upon our arrival in the capital. We'd been talking for about a half hour, going over territory I'd largely expected, when Jamie Robinson took me aside after I turned off my tape recorder. "Your chances of seeing gray whales at Piltun are a hundred percent," he'd said. "And it's a religious experience."

George dreams about swimming with the whales that night. We awake to a strong wind and high seas. No whale counting today. Dave says that's all right, because they're starting to run low on gas for the boat. Only one fifty-gallon drum left. They hope, when Sasha Burdin arrives in the next week or so by helicopter from the Russian mainland, he'll bring some more. Dave expects they'll probably be here into mid-October if they're not "weathered out" by the big autumn storms.

Mid-morning. We've spied Yuri's jeep on the other side of the lagoon. George lines everyone up for a group photograph alongside the Zodiac, after which he ends up with everyone else's cameras around his neck to take some shots for them. The mosquitoes are on the attack as always, and we certainly won't miss the hard beds or the carbo-filled cuisine. But it's not easy to leave.

Dave and Grisha motor the three of us across to where Yuri waits with "a Russian tradition." He's brought some sliced reindeer meat to put on black bread, and a little vodka to wash it down. We raise our last toasts.

At the top of the first rise, Yuri stops the jeep and beckons us to get out. We stand there watching a whale fountain unfold from the curl of a wave. Yuri tips his cap. "You can say good-bye to Piltun," he says.

Chapter 20

Breakthrough Across Troubled Waters

The freeing of the whales, whatever else it demonstrated, was a basic fable in human generosity.

—An observer quoted in *Connection on the Ice*, by Patti H. Clayton

An Inupiaq Eskimo hunter named Roy Ahmaogak was exploring an ice-covered area of the Chukchi Sea off Point Barrow, Alaska, looking for signs of bowheads, when he noticed whale spouts in the distance. It was October 7, 1988. At the northernmost geographic point in the United States, the thermometer read −10 Fahrenheit. Ahmaogak moved stealthily across a relatively thin patch of ice, getting as close as he could to the spouts. Instead of bowheads, he saw three young gray whales. They were between twenty-five and thirty-five feet in length. They were crowded into two twenty-by-twenty-foot holes in the ice, maybe a hundred yards offshore, in about forty-five feet of water. They'd probably been feeding near the Plover Islands, Ahmaogak reasoned, when the temperatures fell and the ice closed around them. He didn't hunt gray whales, but he momentarily considered putting them out of their misery. There didn't seem to be any way they could escape from this trap. The nearest open leads in the ice lay about five miles to the northwest.

It didn't take long for word to spread across America's largest Eskimo community. Two biologists from the North Slope Borough soon visited the site and immediately realized the peril the whale trio faced. Gray whales lack the specialized cartilage that bowheads have to support the blowholes atop their heads and enable them to break through thick ice from underneath. And, while bowheads are able to stay underwater as long as thirty minutes, grays must surface much more frequently to breathe. These whales were clearly struggling for enough air in their confined quarters.

A week after their discovery, the first news story broke in Anchorage, seven hundred thirty miles south of Barrow. The whales remained alive, but an attempt to break through the ice and create a path into open water by a helicopter-towed oil company barge had come to naught; the hover barge was locked in at Prudhoe Bay. Then something mysterious ensued. A flood of media calls started pouring into the North Slope Borough. Reporters from the major TV networks boarded airplanes. Only a few weeks before the 1988 presidential election, the plight of these three gray whales was about to dominate the world consciousness. And a million-dollar rescue effort would come to involve an alliance among Eskimo whalers, the U.S. military, oil companies, government agencies, Greenpeace, eventually even the Soviet Union.

Soon the whales had names. The biologists called them Bonnet, Crossbeak, and Bone after the markings on their heads. The Eskimos called them Putu, Siku, and Kanik, which translated from Inupiat as Ice Hole, Ice, and Snowflake.

Mark Fraker, a scientist who came to enlarge the holes for the whales, would recall: "We found it remarkable that the whales moved readily between the two openings, even in darkness. How they navigated over the approximately hundred-yard distance is unclear. Maybe they simply memorized the spatial relationship, or perhaps they somehow used underwater sound to help them find their way."

The temperature dropped still further, to around −20 F, and ice accumulated faster than it could be shoveled out in the two holes where the whales were struggling to survive. What to do? As it happened, two men from Minnesota had paid their own way to Barrow, bringing along some deicing machines. People found them generators and extension cords, transporting everything to the site by helicopter. And, when they installed the deicers, basically electric motors with propellers enclosed by protective wire cages, the whales seemed to find the machines fascinating. The smallest, Bone, played in the bubbles the deicer produced. One of the scientists, who'd studied gray whales at the San Ignacio Lagoon, knew that in Baja they often appeared attracted to the sound of a *panga*'s motor. Might the whales now follow the sounds of the deicing machines into new holes?

At this point Arnold Brower, Jr., entered the picture. He was one of the most successful of the younger whale hunters in Barrow. He was also the direct descendant of a Yankee whaler, Charles Brower, who'd settled in Barrow in the latter part of the nineteenth century. Charles had been the first to combine Yankee whaling techniques and Eskimo ritualistic practices. In 1897, when five American whaling vessels became trapped in the ice off the point, he'd saved the lives of the men aboard. Now his great-grandson was about to be influential in saving the lives of the gray whales his people called *agvigluaqs*.

Brower's idea was to use a method the Inupiat sometimes employed to tow a whale carcass back to shore. They'd use chain saws to cut a series of rectangular holes in the ice, stationing groups of men with poles on each side to force the slabs under the ice sheet. Doing this they could create an opening thirty feet long and as much as eight feet wide in about an hour. By the end of the first night using this tactic, new holes stretched for about one-third of a mile toward open water, six of them equipped with deicers. Still, it was difficult to get the whales to move into the new holes. The "trick"—an idea conceived by Marnie Albert, the wife of Senior Scientist Tom Albert with the North Slope Borough—was to shine floodlights into the water of the next hole. Apparently the whales could see the shaft of light, and they swam over to it.

One of the trapped whales, 1988 (Michio Hoshino/Minden Pictures)

Eventually the whales learned to do this on their own. Some observers believed they may have been following the sound of the deicing machines. Jim Nollman, the whale acoustician and pioneer in interspecies communication, reveals another possibility in his 1999 book, *The Charged Border*. Nollman had come to Barrow with an underwater sound system, which he hoped to use to entice the whales to discover the channel. Late at night, in the company of several Inupiat, he'd piped a cassette of Ladysmith Black Mambazo—a South African a capella group with a "lot of whispering voices singing exquisite harmonies"—underwater to the whales. They responded, apparently, by moving a quarter-mile closer to freedom.

However, at the twenty-fourth air hole, Bone failed to resurface. The youngest of the whales seemed to have gotten separated from the others, become disoriented, and drowned under the ice. "The nation mourned," one newsman reported.

By October 22, more than two weeks after the whales had first been observed, fifty-seven holes had been cut, and Bonnet and Crossbeak had moved one and a half miles toward open water. But this seemed to be as far as they wanted to go. Nobody could figure out why. "Start thinking like a whale," the Inupiat elders advised their chain-saw crews. "We just didn't have the mentality," Brower said later, "to understand what the whales were trying to tell us."

It may have been a blessing in disguise. Clearly, something else needed to be done. The Coast Guard generally had two icebreakers available, but one was on a mission out of Alaska and the other was undergoing repairs in Seattle. Greenpeace convinced the State Department to negotiate with the Soviet Union, which also had icebreakers capable of breaking through an ice ridge grounded on the ocean floor, a mile long and a quarter mile wide. It was a delicate situation, in that a Soviet ship had never been allowed so close to Barrow. But with people in both countries engrossed in the televised saga of the gray whales, President Reagan gave the go-ahead and Premier Gorbachev didn't even have to blink. From somewhere in the Soviet Far East, the *Vladimir Arseniev* and the *Admiral Makarov* were under way.

About now the Eskimos realized that the whales were refusing to move any farther because, at a depth of only twelve feet, they considered the water too shallow. When the Eskimos shifted their chain-saw cuts into deeper waters, the whales became energized and eager to follow. They began moving into the new openings even before the saws had finished cutting. But polar bears had been seen in the area. Rescue workers armed themselves to fire warning shots in the event the bears threatened the whales. Volunteers began standing guard round the clock.

Meanwhile, oil scientists specializing in ice conditions conferred with Eskimo elders. They decided to drill through the ice where they could place a weighted line and offer depth soundings to help guide the Soviet vessels through the very shallow water. In a mission that became known as Operation Breakthrough, the Soviet ships arrived on October 25. They were flying two flags—the Hammer and Sickle, and the Stars and Stripes. For the first time ever an American aircraft was allowed to land on a Soviet vessel. The first thing Captain Sergei Reshetov reportedly said was: "How are the whales?"

They hoped the icebreakers would be able to carve out not simply a narrow channel but a huge opening along which the whales could be led. During the first night the icebreaker *Admiral Makarov* broke a path through the pressure ridge that was about twenty feet high and between two hundred and three hundred feet wide. The relatively thin ice in the area of the whales was then broken by the ice-strengthened freighter *Arseniev*. That afternoon Bonnet and Crossbeak left their last hole, dove, and swam into the carved-out channel. The Soviet ships backed out in front of them while TV helicopters circled overhead.

In Washington, President Reagan said: "It has been an inspiring endeavor. We thank and congratulate the crews of the two Soviet icebreakers who finally broke through to the whales. They were part of a remarkable team effort." Spoken in his last months in office, these may have been the kindest words this president ever had for the "Evil Empire."

The whales, however, weren't out of the woods yet. When the Eskimo chain-saw crews came to check out their progress before dawn on October 27, they found the channel refrozen and the pair having moved a mile and a half but gasping for air in a small opening. The Eskimos enlarged this hole, then cut a new series through a gap in the ice ridge while the icebreakers cleared one final path to within twenty feet of the whales. Soon Bonnet and Crossbeak had "escaped" into the Chukchi Sea.

It had taken thirteen days for the dozens of people involved in the rescue effort to cut through about eight hundred tons of ice and fashion more than a hundred breathing holes across a four-and-a-half-mile span. Arnold Brower, Jr., was the last human being to see the whales. The following morning he spotted them near the channel, spoke to them, and watched them dive in the direction of an open lead nearby. "Nothing like this will ever happen again in our lifetimes," he said afterward. "And probably not in the lifetimes of our children, or our children's children."

Some two hundred members of the press left for home. A weather satellite showed a series of ice-free leads nearly all the way to a wide-open area of the Chukchi Sea, about two hundred miles to the southwest, leading onward into the Bering. Conditions remained favorable for about a week, more than enough time for the whales to make their way through. But nobody ever saw them again.

That televised spectacle, which led the nightly news for days that fall of 1988, had marked my first awareness of gray whales. Like millions of other people, I had found myself enthralled and rooting for their rescue. So I might have been drawn to Point Barrow regardless, but I had another reason for visiting. I had a decision to make: whether to venture on to the native gray whale hunting villages along Russia's Chukotka Peninsula, far more remote even than Piltun. One of the Americans most familiar with Chukotka and its problems was Dr. Tom Albert, senior scientist in the Department of Wildlife Management for the North Slope Borough, headquartered in Barrow, some 350 miles north of the Arctic Circle.

George and I had first to fly from Sakhalin Island back to Anchorage, then catch another flight to the apex of North America. We struck up a conversation en route with a young sculptor who fashions dragons out of mammoth ivory and was on his way to do some prospecting in the northern Alaskan riverbeds. When we landed in Barrow it was nearly seventy degrees, as hot as it ever gets. The dust rising from the unpaved streets seemed as ubiquitous and impenetrable as the ice floes reaching right up to the July shoreline for as far as the eye could see. Yes, astoundingly, the broken ice showed little indication of melting.

If Nome evoked the Wild West, Barrow was an Oklahoma oil boomtown out of the 1920s with a frozen sea. We checked into the Top of the World Hotel and had dinner next door, at Pepe's North of the Border—eating probably the world's most expensive Mexican food at about as far away from Mexico as you could get. George and I drove a hundred-dollar-a-day rented pickup past yards cluttered with old snow machines and barking dogs and racks of caribou and seal meat, down streets where children raced around on four-wheelers in the midnight sun. We headed northeast into Barrow's outskirts, where a road sign warns of polar bears in the tundra ahead. I longed for pristine Piltun.

It remained for Tom Albert to give me a different perspective. "This is probably the only place in the United States, and one of the few places in the world," he was saying the next morning, "where you can go and see wild animals moving from horizon to horizon. Like the wildebeest in Africa, except here it's caribou, thousands or even tens of thousands of them at a time, passing right in front of you!"

We were sitting in his office inside a compound marked by a bowhead skull outside, where Albert and twelve others among the department's scientific staff share space with a community college. He explained he'd come here on a sabbatical from the University of Maryland in 1977 to conduct studies for the since-closed Naval Arctic Research Lab on how animals adapt to cold. He is a veterinarian who had started out practicing in a Chesapeake Bay farming and fishing community, where traditional and scientific knowledge combine readily. So Albert was simpático with the hunting-and-fishing mentality of the Eskimos, and with the bowhead whaling tradition that had been their lifeblood for centuries. When they'd established their borough government, after the U.S. Congress passed the Alaska Land Claims Settlement Act in 1971, the Eskimos had created the Wildlife Management Division, mainly to gather research that could justify their continuing to hunt bowheads.

"When I first came here, I don't think I even knew what the word *culture* meant," Albert said. He'd come to the definition through the Inupiat, and he'd ended up staying on to run the scientific program. Today its budget is about $2.7 million a year, three-quarters of the money devoted to bowhead studies. By 1914 commercial whalers had beaten the population down to about a thousand, but now it's up to 8,200 and increasing by some 3 percent annually. Enough for the Alaskan Eskimos to receive an annual subsistence quota from the International Whaling Commission.

About a third of the wildlife budget has customarily gone to looking at the offshore impacts of the petroleum industry: noise disturbance and the threat of oil spills. Oil basically created the North Slope Borough, a huge county that covers about one-seventh of Alaska and includes the two largest oil fields in North America. The one at Prudhoe Bay, about 200 miles east of Barrow, at its peak provided almost 20 percent of the oil consumed in the United States.

Long ago the Eskimos had believed the black liquid that oozed from their frozen tundra—*uqsruq*, they called it—to be the lethal poison of an angry spirit. In the late 1990s the North Slope Borough and the Alaska Eskimo Whaling Commission successfully sued the federal government to increase the quality of their impact assessment studies. It turned out that offshore seismic exploration definitely interfered with the bowhead migration, just as the Eskimos had claimed to little avail fifteen years earlier. When the oil company's air guns were booming, no whales would come within twelve miles.

Albert pointed to a picture on his wall of a gray whale, taken by his wife at Point Barrow some years back. When he arrived here, Albert said, "it was not uncommon for people to knock on my door and say, 'There's a whale out there—not a bowhead—and it's dying.' " Albert would go look, only to find gray whales that hadn't stranded but were merely feeding within thirty or so feet from shore. "At the time, they were so rare that many people here didn't even recognize them anymore. The older people did, but not the younger ones. But we've seen a steady increase in the last twenty years." Today, according to Albert, gray whales go around the point and at least a hundred miles farther into the Beaufort Sea. He figures that, as their numbers grow, they are repopulating their historic range.

He'd been one of the participants in the 1988 gray whale rescue and, in his aerial surveys, has seen quite a few grays washed up on the beaches. These are usually young, small whales—perhaps a dozen every summer—which, Albert says, "can tend to die from childhood diseases. And we think," he adds, "that a fair number of young polar bears who would normally starve their first winter away from their mothers are *not* dying, because they stumble upon one of these gray whale carcasses." I recall hearing about how, twenty years ago, when a limited but lucrative market existed for polar bear hides, the Eskimos of St. Lawrence Island would occasionally shoot small shore-feeding gray whales, haul them past the high tide line, and leave the carcasses for bait; one hunter is said to have killed twenty-two polar bears off a single gray one winter.

Today, however, Albert attributes some of the *increase* in the endangered polar bear population to this fact of nature. "So these gray whales are performing several important functions—not the least of which is, they're nice to look at. Yet another is also to allow for the physical survival of a bunch of people in Chukotka."

This was ultimately what I wanted to discuss with Tom Albert. He was the central liaison between the North Slope Borough and the Russian native villages across the Bering Strait. As Albert told the story, it traced back to discussions in the 1980s at the annual meetings in Barrow of the Alaska Eskimo Whaling Commission. "The conventional wisdom among Western scientists," Albert said, "was that the bowheads who winter in the Bering Sea came up the coast of Alaska, past Barrow, where we would count them, and in the summer they'd all be over in the Canadian Beaufort Sea. But that's not what the St. Lawrence Island Eskimos kept saying. They maintained the population count was inaccurate, because a number of bowheads went around their island and over into Russian waters off Chukotka for the summertime. But there wasn't any way for us to know, because we couldn't get over there."

In 1989, not long after the gray whale rescue, "the Ice Curtain began to melt a little bit," as Albert put it. An Alaskan physician organized a Chukotka expedition for about fifty American scientists. The Russians even sent a ship to Nome to pick them up. Tom Albert was among the group. Most of the trip's orientation was toward the larger city areas and institutions, but Albert wanted to find out about the whales. So did his two Eskimo companions, Ben Nageak—later mayor of the North Slope Borough—and Matthew Iya, a Native leader from Nome who could converse with the Yupik Eskimo people in Chukotka using their common language.

They all flew from the city of Provideniya north to the Chukotkan regional center of Lavrentiya, along the Bering seacoast. There the weather got bad, and they spent a week waiting and wandering around. Albert remembers, "We were the first foreigners many of these people had ever seen. We'd hand out postcards and trinkets to these kids who'd follow us everywhere. One of them spoke pretty decent English. He kept talking about his father, who he said was 'chairman of Cooperative Naukan.' I had no idea what that was, though it turned out to be a marine mammal hunting organization. Finally the boy asked what I did, and I told him I was a veterinarian. He got real excited, because that's what *he* wanted to be, too. When we left, I told the boy I'd write to him."

Albert went on to another Eskimo village called New Chaplino. As the helicopter circled in low, he saw a sixty-foot-long Soviet "catcher boat" anchored near the beach, having just towed a gray whale to shore. "I could hardly believe my eyes. There was the whale at this cutting-up station, and outside this awful-looking building were a couple hundred of what looked to me like fifty-gallon drums. The people told us they had oil in them, whale oil, that had not in recent years been taken away to

Moscow. And right up on top of the hill was the fox farm where they carted the whale meat. We were some of the first outside people to see one of these damn fox farms."

For years it had been rumored that the International Whaling Commission's annual quota of gray whales—generally between 169 and 179—allotted to the Soviets "for the needs of the native population" were really going to feed polar foxes. But the IWC membership feared to take action, lest the Soviets bolt the organization. Caged in numerous villages along the Chukotkan coast, some twenty thousand foxes were being raised annually for their furs. The skins were sent first by ship to Provideniya, then by plane to fur-fashioning factories in Irkutsk.

The fur trade had marked the ultimate subsumation of Chukotka's native cultures. For centuries, while their Chukchi brethren raised reindeer on the tundra, maritime Chukchi and Yupik Eskimos had hunted whales, walruses, and other marine mammals for sustenance. In the early 1960s Soviet authorities officially banned aboriginal whaling by the villages. Any private subsistence hunting became subject to heavy fines. Weapons and boats fell under sole control of the state-owned fox farms. Collectivization found many small villages shut down altogether, the native peoples being moved into larger ones such as New Chaplino.

"One of the reasons," Tom Albert had learned, "was they didn't want people out on these points of land where they might be somehow influenced by the imperialists only a few miles across the strait. Or where they might escape. Or form an avenue for an invasion. So they were crowded into apartment buildings for the first time and given jobs on the fox farms, where they could be good Soviet citizens."

The catcher boat, manned by a nonnative crew and operated by the Ministry of Fisheries, would deliver the gray whales to New Chaplino, where the meat and blubber would be made into a gruel by adding water and grain, then cooked and hand-carried to the ever-hungry foxes. Tom Albert couldn't wait to get out of there, but again weather locked him in. Through Matthew Iya's translations from the Yupik language, Albert was able to ascertain that bowheads did indeed pass regularly along the coast on their northward migration.

In 1991 members of the U.S. National Park Service went to Provideniya to see their Soviet counterparts. "I don't know how I got mixed up in that," Albert says, "but I got to go." On this trip he again encountered the boy who'd latched on to him in Lavrentiya, and who introduced Albert to his father, Mikhail Zelensky, who had an intriguing history. His parents had moved to the Far Eastern region when Stalin called upon dedicated young Communists to go proselytize among the natives. The father became Chukotka's first teacher, the mother chairperson of its first state farm. So Zelensky was a white Russian raised among the Chukchi, even able to speak their language. He was also something of a rebel, more sympathetic to supporting the natives' desire to cling to their culture than to indoctrinating them with Marxist dialectic. With the coming of perestroika, in 1987 the Naukan Native Cooperative was created as an effort to bring together Yupik and Chukchi peoples. It was the first enterprise in the Chukotka Autonomous District that was not state-owned. Its aim was to help bring back a traditional subsistence lifestyle. Mikhail Zelensky was so well thought of that he, a nonnative, was elected to run it.

Albert liked him instantly and felt similarly about Ludmilla Ainana, who had just formed the Yupik Eskimo Society of Chukotka. "These organizations began when the Soviet system was deteriorating markedly and there was getting to be a little bit more freedom and people wanted to get together," Albert said. "They'd never been able to do that before, unless they were stooges of the government. Everybody used to be afraid of their shadow, and with good reason, because over there your shadow might kill you."

If his borough could come up with some money, Albert asked the two native leaders, would they enlist some people to count bowhead whales in the region and write up reports? Zelensky and Ainana agreed instantly. Albert got an initial contract of nine thousand dollars to pay villagers a hundred dollars a month apiece for several months of spotting. In 1992 the borough paid for ten of them to come to Barrow with their results. Just as this unusual kind of "whale diplomacy" commenced, the Soviet Union fell apart.

"When the government collapsed, it basically said, "Hey, you guys want to be natives, go to it.' But they had nothing," Albert said. The fox farms had never made a profit. They'd been part of a "planned loss" economy, since bringing in the feed, equipment, and electric power had resulted in very high costs. Then, with the shift to free enterprise, government subsidies for the farms ended. So did centralized shipments of supplies. Prices for transportation and fuel soared. The fur traders had no money to pay for repairs to the whale catcher boat. "So the fox farms started going out of business, which is where they belong," Albert continued. "But now the people don't have the jobs anymore. And a lot of the old ways have been lost—how to hunt seals, make skin boats. They needed organizations like Ludmilla Ainana's to be able to get the word out, and like Mikhail Zelensky's hunting cooperative to go out and catch walruses and basically give them away to people who might otherwise starve."

In the years since, Tom Albert had made helping the native peoples of Chukotka a personal crusade. "These two organizations hardly had a pencil—no office, no phone, nothing—so we bought them each an apartment that could be used as a headquarters." He utilized funding from his own borough, as well as the U.S. Information Agency and more recently the National Park Service. Before long more than thirty people in ten villages were being paid to collect data on bowhead whales and other wildlife. Outboard motors, fishnets, binoculars, compasses, boots, and CB radios were funneled across the Bering Strait to the Chukchis and Yupiks. The only way to get the equipment through Russian customs was to declare it humanitarian aid. Even at that the bureaucracy remained a terrible hurdle. Everything needed to be approved by a committee in Moscow, which met only sporadically. Mikhail Zelensky flew there and sat on the doorstep until he got the okay.

The indigenous peoples wanted to resume their whaling tradition and requested weapons, "big pieces of equipment with a lot of political connotations," Albert realized. "Which might raise questions among certain conservationists like, are the Alaskan Eskimos arming the crazy natives over there in some giant conspiracy to kill all the world's whales?" He knew it wasn't like that and, after visiting Chukotka again in 1996, that the people's situation was becoming more desperate all the time. The IWC still had its annual aboriginal subsistence quota on gray whales in place; indeed, with the closing of most of the fox farms, it was no longer a sham. But, before the North Slope Borough would assist with weapons, Albert told them, they needed to establish their own regional counterpart to the Alaska Eskimo Whaling Commission, a by-the-book organization with a set of bylaws. Zelensky, Ainana, and Vladimir Etylin set up the Union of Marine Mammal Hunters of Chukotka, bringing together about one hundred men from thirteen villages.

"Before I came to Barrow, I was a member of every conservation organization known to man, and I still believe in all of it," Albert told me. "As a veterinarian, I've spent my whole adult life trying to help animals one way or another. One of the jobs I now have is to try to help some of them die, with as little problem as possible. So we've transferred some weapons to those people over there, to help them kill animals more humanely. If they don't eat gray whale or walrus, they're not going to eat."

In the past the whale hunting had been done with spears and rifles. "It takes longer than it should," Albert said, "but they had nothing else." So in 1997 the borough and the Alaska Eskimo Whaling Commission sent over twenty darting guns and one hundred projectiles for the Chukotkan native peoples to hunt gray whales.

"Chukotka is a very difficult place to get to, and to get around in," Albert continued. "But you can. You'll need an interpreter."

I told him that I'd applied for a visa, just in case, through an outfit in Anchorage that could arrange my "tour" for a substantial fee. Getting the visa itself required more than a month. It would be too late for George to accompany me; I'd have to go alone. The Russians still didn't make it easy for outsiders to visit the region.

"They don't want people to see what it's really like over there," Albert said. "Those folks have been abandoned, left to their own devices completely. I don't know how a lot of them make it through the winter. Well, they don't—unless they've managed to put away enough whale meat."

Seeing Tom Albert made up my mind. He said he'd send faxes to several key people in the native organizations to help smooth my way. "Before you leave," he added, "there's somebody I'd like you to meet."

We walked into the next room, where a young man was working at a computer. He'd been in the United States since March, having initially been sent under the auspices of the North Slope Borough to Louisiana State University for some additional biology training and to bone up on his English writing skills. "How do you do," he said, "my name is Gennady Zelensky." Yes, this was the boy who'd approached Tom Albert on the streets of Lavrentiya ten years earlier. He would soon return home, as a practicing veterinarian.

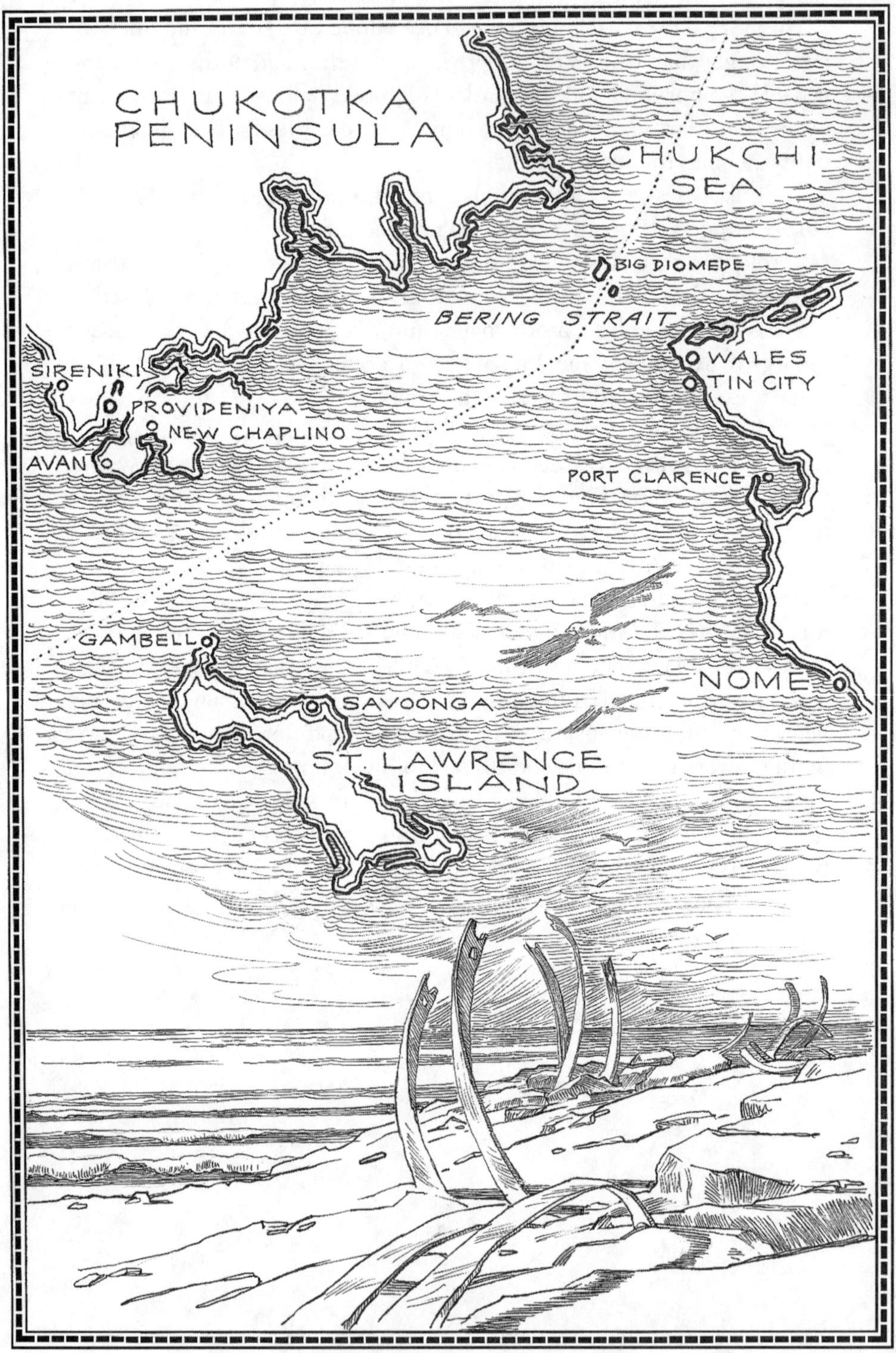
CHUKOTKA
PENINSULA
CHUKCHI
SEA
BIG DIOMEDE
BERING STRAIT
SIRENIKI
PROVIDENIYA
NEW CHAPLINO
AVAN
WALES
TIN CITY
PORT CLARENCE
GAMBELL
NOME
SAVOONGA
ST. LAWRENCE
ISLAND

Chapter 21

Catastrophe in Chukotka

There are many wonderful animals, but whales are best of all. As they pass by your skin boat, great and quiet, you immediately come to understand your place on the Earth, and you become warm inside. . . . It is very interesting to look at whales, but most importantly, it is useful and necessary in order to become a full-fledged person—a hunter.

—Nikolai Gal'gaugye, Sireniki, 1994

Northeastern Russia's Chukotka Peninsula is believed to have been part of ancient Beringia. Some of the flora of its eastern sector and that of western Alaska are so similar as to be considered proof that the Bering Land Bridge did once exist. Along the Chukotkan east coast in 1981, a new area of ancient aboriginal whaling was discovered—six large settlements containing as many as two thousand gray whale skulls. These skulls, and to a lesser extent other bones, had been used as primary materials to construct underground dwellings and meat caches. The skulls were measured and found to be predominantly of nursing calves and yearlings. Noticeably absent were skulls of adult gray whales.

During the late nineteenth century, as described in the 1984 scientific assessment *The Gray Whale*, there occurred "a period of severe population reduction of all principal cetaceans off Chukotka Peninsula as the result of Yankee commercial whaling and the vicious depletion of the California gray whale stock in the lagoons of Baja California. For the aborigines of the Chukotka Peninsula, this situation was realized as a severe reduction of aboriginal whaling, massive disasters and starvation, particularly between the 1870s and the 1890s."

The Chukchi people who still survive along this coastline tell a legend of the first whale, which has been recorded like this:

Old men say that on these shores, in days long gone by, there lived a young maiden. So beautiful was she that even the mighty sun, gazing at her in wonderment, remained always in the sky and the stars came out in daylight to behold her. Wherever she set foot lovely flowers bloomed and springs of clear water sprang from the ground.

The maiden often strolled by the side of the sea to watch the waves and listen to their whispering. When she was lulled to sleep by the rhythm of wind and waves, the sea animals came onshore to gaze at her with their round, unblinking eyes. One day an enormous whale swam by, and, seeing the crowd of animals on the beach, he swam closer. He was so charmed by the maiden's beauty that he forgot where he was and where he was bound. When the sun, exhausted, had sunk down onto the horizon to rest awhile, the whale swam inshore, touched his nose to the shoreline, and turned into a handsome youth. The youth took her hand and led

A Chukotkan shaman (courtesy Provideniya Museum)

her to a place on the tundra that was soft and covered with moss and flowers. Thereafter he remained onshore as a man and made the maiden his wife.

Soon it came time for the maiden to give birth, so the whale-man built a large *yaranga* and moved in with her.

Her babies were whales, and their father carried them out to a small lagoon where they lived. When they were hungry they came to the edge of the water and their mother fed them, but soon they outgrew the lagoon and asked to be sent to sea. Their mother was sad to see them go, but whales belong to the sea so what could she do? Her children swam away.

Soon the mother gave birth to more children, this time in human form. Still, the whale children did not forget their parents and often came to play close to shore.

Time passed and the parents aged. The father no longer went hunting, so the sons had to bring the food. Before their first hunt their father said: The sea is a good provider of the strong and brave. It is the home of your brothers the whales. Protect them . . .

Before long the father died and their mother became infirm. The sons had many children, and they needed more and more food. The descendants of the whales became the peoples known as the Chukchi and Eskimo, who hunt for animals in the sea.

There came a year when there were very few animals in their waters. The walruses had forgotten the sea roads to their village, and the hunters had to venture far out onto the icy sea. Some perished on the ice floes, while others drowned. Only the whales continued to frolic happily and noisily offshore.

One of the hunters said, "Why do we not kill the whales? Look, there are mountains of meat and blubber."

"Have you forgotten that they are brothers?" others objected.

"Brothers indeed!" scoffed the first. "They live in the water, not on the ground. An old wives' tale!" He turned on the others and began to prepare a big boat for the hunt. He took the strongest and most skilled of the rowers.

Capturing a whale proved easy. One of the whales swam right up to the boat, was harpooned and towed to shore. The hunter went to his mother's *yaranga* to boast of the rich gift he had given his people, but she lay dying. "You have killed your brother because he does not resemble you. What will you do tomorrow?" and she said no more, for her last breath had left her body.

•

Everything appeared to be arranged. Circumpolar Expeditions, run by a Russian living in Anchorage, would have accommodations and transportation waiting for me. My visa would permit travel from Provideniya to two outlying native villages, Sireniki and New Chaplino. Tom Albert sent some faxes from Barrow on my behalf, suggesting someone familiar with the villagers should accompany me, "because there are a lot of suspicions about who people are." I dropped George at the Anchorage airport, en route home to L.A., and he wished me well.

The day before my departure I visited with an Anchorage-based photographer who'd spent time in Sireniki. She had a few suggestions. If offered, I should accept and eat the traditional foods the same way the people do—but try to avoid raw walrus liver. Don't pet the dogs, because they muck around in human waste. She predicted an abundance of cockroaches, and she said she'd sometimes put a roll of duct tape around her sleeping bag to keep them away. She noted that fermented foods, particularly walrus, make you very tired. "My stomach has dropped" is a compliment to offer if they've fed you well. My stomach had definitely dropped by the close of our conversation. My last night in Anchorage, I feasted on a four-course dinner at a fancy restaurant. Where my next meal might be coming from, I had no idea.

The single-engine eight-seater, a Piper Navajo Chieftain, left from Nome shortly after noon. It's about a ninety-minute flight across the Bering Strait and slightly south to Provideniya. Our altitude tops out at close to ten thousand feet. I'm sitting in the cockpit next to Kevin, a burly guy who's been piloting in this region for more than twenty years. He asks if I'm bringing in a computer. How about a GPS or a cell phone? I answer no to all three questions. "All good answers," Kevin says. Otherwise he'd have to hold any of these items for me, or send them back home to Boston. How come? I ask. "Paranoid. The Russians don't know what's on those chips," Kevin says.

"All I care about," he adds, "is that the visibility's good." Mountains fifteen hundred feet high surround the entrance to Provideniya Bay, he explains. There's maybe a mile and a half between the mountains and, if you miss the approach, you've got to head straight for another mountain up ahead and make the turn and head back out again. "Takes nerves of steel to do that," Kevin says. He's talking to the Provideniya air traffic controller and scrutinizing a chart on the steering column. As we come

Aerial view of Provideniya (courtesy Provideniya Museum)

out of the clouds, I see snow on the mountaintops ringing the narrow bay. It's the beginning of August. "Right on course," Kevin says.

At the airport I'm met by the family with whom my stay has been prearranged. For several years Viktor Rekun and his wife, Lyuda, have been chaperoning Americans who visit the Chukotka region. The Rekuns' daughter, Yana, who attends a university in Anchorage and speaks fluent English, will be my translator. As we drive through a militarized sector and circle the bay into Provideniya, a first impression can be expressed by only one word: grim. The feeling is of a crumbling city. The paint has faded from dozens of tall concrete apartment buildings, many of which are abandoned because nobody can afford to heat them anymore. There is the original museum that burned along "Main Street." That falling-down building was once the music school. Along the pier where few ships dock anymore, tall empty cranes, hulks of rusting steel, and other piles of rubble are everywhere. The peeling statue of Lenin in what people still call their Red Square seems to stand guard over another junk heap just behind it. A coal-fired power plant belches smoke into the incessant drizzle. Since the early 1990s, I'm told, about half the population of Provideniya has left. Maybe 2,500 people remain.

My hosts live in the tallest and newest gray-stuccoed cement building in Provideniya. Five floors up a dank, grimy stairwell, their apartment is tidy and homey. Shoes are left in the entryway. I'd sleep on a comfortable fold-out couch in a wood-paneled living room containing a computer and a CD player, and would eat far more decently than I'd anticipated. But beyond my bedroom window, as the local museum director Vladimir Bychkov puts it, are "so many sad stories."

Bychkov is a friend of Tom Albert, who has provided Vladimir some funding for a museum education program. We meet in Bychkov's small office down the hill from where I'm staying. He speaks fluent English and looks, with his neatly trimmed goatee and dark, brooding eyes, like someone out of a Dostoevsky novel. "The general opinion of visitors here is, the federal government forgot about the people who live in our area," Bychkov says. "Those who live here have the same opinion. We don't know what's in the future, what we can do tomorrow. No heat. No electricity. Not enough food. Nobody knows how we can spend the winter here next year. Only God knows."

Then Bychkov changes the subject. "But let's go back to the history of this name, Provideniya. In 1848 two British vessels came here trying to find the lost expedition of Sir John Franklin. The *Plover* was under Commander Thomas Moore. It was September season in Chukotka when on a very stormy day Moore found the entrance to this very protected and very deep bay. He and his crew made the decision to spend the winter here. When it was over these very religious sailors believed only God helped people survive here. So Thomas Moore nominated this as Providence Bay. Many expeditions after this stopped in Provideniya."

Those expeditions had included that of the Western Union Telegraph. There is a story that, as late as 1960, the remains of a cabin built by the expedition still existed at the head of the bay. Port Providence was also then known as Plover Bay, after Moore's ship. "Bound for Plover Bay," Scammon headed his journal entry in the late summer of 1865. "September 20, weather squally. . . . All made St. Lawrence Island bearing east distance 3.4 miles." Colonel Bulkley's steamer was already docked in nearby Plover Bay, where Scammon soon arrived. Scammon's *Golden*

Gate took on some water, but the captain wasn't sure how long they'd need to remain there "on account of the strong winds." Bulkley's more powerful vessel was able to take off again for Alaskan waters. Scammon wrote to the colonel from Plover Bay on September 29:

> *Sir—The following day after your departure several canoes came alongside with Indians from different villages on the borders of this bay. They brought with them as articles of trade three reindeer, some half dozen walrus teeth, a few pieces of whale bone, and undressed deerskins. They were all treated kindly and a few ships biscuits given to each. We purchased the reindeer meat (three carcasses) for light plugs of American tobacco. . . . Four yards of gray flannel. Four butcher knives a few biscuit and a box of matches. For a few trifles, as a matter of accommodation to them I purchased three walrus tusks and whalebone and procured for Mr. Dall one pair deer horns. As it is contrary to the rules of the Company to trade with natives, except for supplies, I thought it well to report to you as above.*

In his journal Scammon added:

> We were very glad to get the deer meat and found it excellent. . . . Throughout our stay, they [the natives] appeared friendly and anxious to render us every assistance they could. . . . We were surprised to find several that could make themselves quite well understood in English or rather English and Kanaka. The three principal men that visited us gave their names as Mr. Knockum, Mr. Pizzlebo, and Mr. Enoch. Knockum was the most intelligent. . . . Pizzlebo informed he was a deer man, implying that he was one of the big men of his tribe and his father a great friend of Capt Moore of the *Plover*. . . . To my surprise Knockum and Enoch seemed to understand about the telegraph and remarked "spose make nope make talk Californy this place one time, all same time quick." They seemed anxious to have it pass through their immediate places of residence.

Scammon's young naturalist companion, William Healey Dall, wrote to Spencer Baird at the Smithsonian: "We obtained the under jaw and some of the baleen of a 'bowhead' at Plover Bay. The jaw is about 14 ft arch. I do not know if it will pay to send and shall not have it forwarded till I hear from you. We have done nothing else on whales this year. I have the head, cervical vertebrae, and four flippers of a leopard or 'hair' seal from Plover Bay, mad crating."

Colonel Bulkley had ordered Scammon to proceed on to the southern edge of the Kamchatkan peninsula and touch base with Major Serge Abasa, a Russian nobleman who'd been selected to oversee exploration and construction of the telegraph line across Siberia. Bulkley was concerned that Abasa's party might "get frozen in the Anadyr River, as winter seemed fast approaching. Already the lower hills were covered with snow, and ice was closing the upper part of the bay" in Provideniya. So, in early October, in what he described as a "heavy crass sea," Scammon set course for the harbor of Petropavlovsk, or Petropaulski as it was then known.

Having made port there successfully, early on, Scammon recounted, "a disturbance occurred among the crew on the berth deck, while I was on shore. . . . Some of the crew being in liquor were fighting among themselves, when ordered to desist they attacked the officers of the ship. The disturbance however was soon quelled and the men causing it confined in irons." Scammon had to prefer charges of "mutinous conduct," but "on promise of future good behavior" the four crewmen were released. By early November 1865, "the explorations for the season having terminated successfully," Scammon's ship was under way again for San Francisco through major storms.

As Colonel Bulkley wrote in his report of the year's activities, "The services of Captain Scammon have been of the greatest importance, not only as a thorough seaman, but particularly as an officer of the Government of the United States carrying our national flag."

The Esquimaux builds his insecure skin-tent on the most exposed place, so that the snow may blow away from it, and there it stands, his shelter and home through all the blasts of the long winter. I have seen no Esquimaux on the Asiatic side inhabiting underground winter houses as of old, the excavations and ruins remain, but the people are gone long since, and the present races occupy the ground with their deer skin habitations.

—Bulkley report, 1865

Still no roads lead to the Yupik Eskimo village of Sireniki. It's 9:00 A.M. on my first full day in Chukotka when, outside the Provideniya apartment building, the *vezdekhod* shows up. It's a tank by any other name, an all-terrain vehicle with steel caterpillar treads whose name is literally translated "going everywhere." My young Russian driver, Valiko, is wearing a black leather jacket and has a native sidekick named Slava and a cocker spaniel named Lady—yes, after *Lady and the Tramp*. "One of the best drivers, everybody knows him in all the villages of Chukotka," Igor Zagrebin, another of my guides, tells me. I will soon know what Igor is talking about when Valiko forges up and over the crest of a summer snowbank, which yet marks the spot where tundra meets mountain.

Igor is sprawled behind me in the back of the *vezdekhod*. He's in his forties, with curly red hair, a round pleasant face, and a walruslike mustache. Four of us are jammed into the front seats—me, Valiko, Slava, and my translator, Yana. Igor also speaks fluent English, first learned twenty years ago at the Moscow State Pedagogical Institute. Today he is manager of the Nature Ethnic Park Beringia for the Provideniya region. He has inspectors in all the villages who keep strict count of the hunting of marine mammals and collect data on local wildlife. He works in close liaison with Tom Albert and wears a cap inscribed ARCTIC SCIENCE NORTH SLOPE BOROUGH.

Early this past May, Igor says, it took him *eight days* to make this same sixty-mile trip from Provideniya to Sireniki. One entire day simply to journey the couple of miles across this river valley we're crossing. In a sudden heavy snowfall the supply truck Igor was in got stuck. The driver walked back to Provideniya, and a special Caterpillar tractor came to rescue them. Probably this is why, when I asked Igor how long our journey would take, he just shrugged his shoulders and replied, "Who knows?" This would turn out to be one of Igor's favorite expressions.

Provideniya's is the largest of a chain of fjordlike bays along this coastline, carved deep by melting glaciers during the last Ice Age. The glacial terrain that etches a circuitous path to Sireniki is an alpine series of rocks and ridges, mountain peaks sweeping into U-shaped valleys. Igor's description implanted itself in my memory: "From the mountaintops, the tongues of the glaciers go down on the valleys. Like a caterpillar, they hugged the ground and made such valleys."

As we wait for a barge to carry our *vezdekhod* across Provideniya Bay, I ask Igor whether the gray whales ever came in this close. "Sure. Last year, we saw a whale every day in the bay, all gray whales, a lot of them, and sometimes even two or three at a time. Because the Chukchi Sea was closed by the ice, and there was no place north for them to go, so all of them stayed here in the Bering Sea. They come close feeding. It's not bad feeding grounds, though it's rather deep and they try to find the shallow waters."

As Valiko backs down the barge after a twenty-minute trip up the bay, he hands me some earphones to help muffle the rest of the ride. Soon Igor spots a pair of sandhill cranes high-stepping across a wetland. He passes some binoculars and points excitedly at these tall, stately birds, whose elegant rust-colored plumes melt into the silver-gray of their expansive wings. All the way from Nebraska, as many as fifty thousand sandhill cranes are known to cross the Bering Strait in a few days.

"There's not a lot of wildlife here in tundra," Igor is saying. "In wintertime, sometimes ptarmigan and hares. No moose or caribou, only reindeer. Brown bears, sure, sometimes a lot, it depends on how much food they have. Polar bears, yes, sometimes they even come in fall time with the pack ice from the north. But the wildlife really is concentrated on the coast."

Alongside a river called Anuvam—which means "River of the Day"—our driver decides to take a break. Valiko opens the back of the *vezdekhod.* We all sit cross-legged in the large rear compartment, watching a steady rain that's begun to fall. Igor describes this weather as "very typical. This summer has been terrible, we have two days with sun and that's all." Valiko reaches into a little satchel, brings forth a bottle, and shakes its contents. Then he lights a match to a cork at the opening. "Spirits," he says with a smile, the first English I've heard pass his lips. The bottle works its way to Slava, Igor, then myself.

Whatever it is, it's strong medicine. One of the Russian toasts translates: "If you're going to run with the wolves, you have to be one of them."

Eventually Igor and I jump down and stand together in the rain beside the river. Sprays of rosy purple fireweed coat the floodplain as it meanders toward the surrounding mountains. Russians use the fireweed for tea, called Ivan's tea, Igor informs me. The native Chukchi who roam the tundra eat the young shoots raw, even the pith of the stems, because fireweed is a good source of vitamins A and C.

Igor is remembering what this region was like when he came here with his family in 1982, from Moscow. He'd taught geography for eleven years in New Chaplino. Under the Soviet system people willing to work in the hinterlands of Siberia received good salaries. Local stores were well-supplied with food and clothing.

"There was Russification for sure, and I was a part of this system," Igor says, "and I feel not very good about it now. Because we came here and tried to give our culture, our ideas and opinions, socialism—and the native people were left in-between. One foot in Russian culture, one foot in Eskimo and Chukchi. And then Russian culture, *kaput*!"—Igor slices a hand across his throat—"and a lot of people lost themselves. Just now it's a very serious situation in the native villages. Life is pushing them back to their traditional way of life. But many of them, they forget."

We hear a weapon fire in the near distance. We've been on the road for seven hours, and Igor estimates that we've probably got two more hours to go. Suddenly, as the *vezdekhod* lurches over a rise, a strange and unforgettable sight appears. The bleached lower jawbone of a gray whale forms a vertical arch, like a turret, where it leans up against what remains of a massive upraised head. A host of other whale bones lies scattered in the surrounding tundra. Another set looks like a whale resting upon a skin boat.

At my urgent request Valiko brings our vehicle to a grinding halt and I climb out. Igor says the carcasses were hauled onto the tundra over many years by the Soviets, with ropes attached to tractors. He remembers some of the village women weeping, because the bones should have been returned to the sea. In most Yupik legends animals are self-sacrificing. They give themselves to the people and, if properly treated, are reincarnated in the next generation. The women wept because the tundra was anathema to marine mammals; the gray whales were now "dead." The Soviets, however, had been impervious to this. They were concerned only about the foul smell—which also happened to attract Arctic foxes, for which the men of Sireniki set their winter traps. It was one of those paradoxes. It took as many as twenty years for all of the whale meat to disappear.

Now the bones are barren. They bear the look of an ancient monument. Our destination waits over the next rise.

Surrounded by mountains on all sides, Sireniki lies in a river valley, its dwellings spaced evenly along a terraced slope that leads down to the Bering Sea. One could easily traverse the perimeter of the dirt roads in less than an hour. The newest structures are blocklike, two-story buildings, fourteen apartments in each, their wooden frames covered over by a layer of concrete. They were raised during the 1980s. The oldest buildings, Igor points out, are a remaining handful of wooden, single-story homes at the edge of town, painted yellow and blue, "River Street, so-called." These date back to 1957.

Until then the people of Sireniki still lived in family *yarangas*, pole-supported, tentlike structures with roofs covered in walrus skin or canvas and walls fashioned from wood, metal, or the wreckage of ships. The Cold War, however, brought an intensified military presence to the eastern Arctic. With the soldiers came the perceived need for a more civilized—meaning collectivized—Chukotka. The nomadic Chukchi people were required to relinquish most of their family reindeer herds to the state cooperative farm system and were forcibly removed from the tundra. Seventeen communities of sea-hunting Yupik Eskimo peoples were uprooted and merged into larger settlements.

Sireniki, the oldest of these, was the only village that remained intact and in fact experienced considerable growth. Many Russians—teachers, engineers, technicians—came here as well, drawn by higher wages than they could earn in their own republics. At its peak Sireniki had a population of 711. In the summer of 1999 its census stands at 554. It's about evenly divided between Yupik and Chukchi, with most Russians having returned to whence they came since the collapse of the USSR.

As Igor and I walk past the border guard station, a young soldier marching back and forth behind the barbed wire orders us to a halt. He demands to see the footage I've been shooting in my video camera. He wants to make sure there's no film of his lonely encampment or the radar tower on the mountain beyond. I rewind a couple of minutes of tape, and the soldier cranes his neck over the viewfinder. Apparently satisfied, he nods and walks away.

Igor shakes his head and mutters, "Crazy country. Well, it was seventy years . . ."

Late afternoon, and it feels eerily quiet. Perhaps because I'm no longer bouncing along inside the *vezdekhod.* Then I realize: there's not a car or truck in sight. Anywhere. For a truck that customarily fills a village water tank at a nearby river, there's no fuel available. That's why the people are carrying buckets. There *is* electricity at the moment, since a bargeload of coal showed up a couple of days ago to fuel the old power plant at the top of the hill.

Looming above a grassy area about a hundred feet from the beach, supported on stilts, stands a series of buildings that are actually rows of empty cages. "This fox farm," Igor says, "was constructed on top of some dwellings, which dated to the first years after Christ." It arose early in the 1960s, not long after the first houses came to Sireniki. Here as many as four thousand polar foxes at a time were raised on gray whale and walrus before being slaughtered for their pelts. It closed at the end of 1997.

Not far away, above one of the widest areas of the beach, are the remains of a large rectangular structure, steel outer columns supporting the shell of a rooftop. "The ruins of another crazy idea," Igor says, "to construct here a factory which would utilize marine mammals. When they finished about half of it, they found out they would need to bring marine mammals here from *all* Chukotka to make this plant profitable. This is Soviet Union, very logical." He laughs and adds: "And then the ocean washed it out anyway." The storm, in the mid-1980s, had destroyed the factory a year after its completion.

"Look, gray whale! A blow!" Igor shouts. I quickly turn my head toward the sea, but too late. "Wait a little bit," he half-whispers.

We walk down onto the rocky, gray-cobbled beach. Not far from the high-tide line an escarpment is covered with strange protrusions. Igor says this is known as a cultural layer. Sod-house artifacts have been found here as old as the first century A.D., from the original of three villages. With every new storm this ridge reveals more ancient implements of bone, stone, and ivory.

Feeling every inch the archaeologist, I join Igor in examining objects from the sandy hillside. The white rib bone of a whale, once used to stir a pot. Pieces of pottery. A rectangle of ivory from a centuries-old walrus tusk, turned brown with age, with perfectly spaced round holes that appear to have been drilled along one side. "That was a runner for pulling a dogsled," Igor says. Next he holds up a piece of whale baleen that bears the look of old parchment.

Suddenly Igor cries out: "Blade!" He's cradling what I'd mistaken for a rock, a large and magnificently carved spear blade made out of slate. "This is my lucky day. This will go to the museum collection in Provideniya."

Twilight in Sireniki. The sun has emerged. A brilliant shaft lights up a craggy cliff, which slides at a steep right angle into the Bering Sea. Along the ledges, though not visible from where we stand, are massive bird colonies, especially auklets and murres. The people gather the eggs with nets, Igor says. The mountain itself is called Yakunit, or "Wing of the Bird."

Igor looks out to sea again and sighs. "No more blows." We hike back up the long hill to where Yana is preparing a wild mushroom and potato stew in the kitchen of our three-room "hotel," formerly the home of a Russian officer and the only square building in Sireniki. After dinner that night the soldier from the border guard station shows up again and wants to see my passport and visa. Upon reviewing it he admits to Igor that he's simply lonely. From the *vezdekhod* where he's camped outside, Valiko brings in the "spirits."

The Siberian Yupik still delineate their calendar year this way: December to January is the "month of ceremonies." January to February is "a lot of frost in the corridor." March to April is "birth of seals." May is "rivers open," and June is "summer woman" or "when the light starts to shine." July is "begin to collect eggs." August is "when the reindeer begins to cry" and "green shines golden." September to November, "small birds begin to jump in the water."

Some say the name Sireniki derives from the Yupik word for sun—*siqeneq*—and means "Valley of the Sun." But an old woman named Kawawa remembered otherwise, that Sireniki came from *siqhinaq*, Yupik for "antler": "When I was young, there was a mound of antlers on the hill not far from the village. People would go to those antlers to receive answers to their questions."

Pyotr Typykhkak (pronounced "Peter Tipikok") has no idea what the Russian-language translation for Sireniki would be. Sireniki is, after all, far older than Russia. What Pyotr, born in 1933, remembers are the edicts of the elders. No marine mammal bones could be discarded to the tundra; all remains had to be returned to the sea. They could not take boiled meat to eat at sea before the crested auklets arrived in the spring. Walrus skins prepared for covering the skin boats had to be painted after the kittiwakes came. There should be no killing of small dolphins or killer whales or small seabirds. During the migrations of the whale and the walrus, all entrances to dwellings had to be curtained, so no light could be seen that might frighten them away.

Igor, Yana, and I are sitting across a conference table from Pyotr, in the office of the village administrator. He is wearing a tight-fitting, weather-beaten gray sportcoat. He has short-cropped black hair and cataracts in his eyes. When he tells stories Pyotr gestures as though he is rowing the boat or tying the rope.

As a boy, Pyotr is saying, he couldn't sleep when the whales and walruses were migrating past. His mother told him it was important to "feed" the whale bones that stood in columns along the nearby hunting grounds at Imtuk, Antikok, and Singok. They were his ancestors, she said.

Pyotr's eyes were so bad that he never went beyond "third form" in school. He was eight when his father began teaching him the ways of the

The Sireniki whaler Pyotr Typykhkak
(Saunders McNeill)

hunter. Pyotr dressed for sea in fur pants and boots that his mother made from sealskin. A grass called *uokhkutak* grew along the coastline and was sometimes used as the inner sole for a hunter's boots. It helped keep one's feet dry and warm. The first outboard motors didn't arrive until 1948, so Pyotr started off as a rower. Every whaleboat had a mast and two sails, and used the strong currents for extra propulsion.

Traditionally the bowhead had been the hunters' whale of choice. When his people still lived in *yarangas*, Pyotr remembers they put the *mantak* and meat in enamelware and stored it in a warm place for two days. It became a little sour and very tasty. But it didn't work to use the meat of gray whale the same way. When they got a bowhead in the spring, they put small pieces into sealskin floats and then into the meat pits in layers of permafrost on the beach. The small intestine of the bowhead they used for sewing rain jackets, with overlapping strips so the water would run down the outside. These were good jackets, lightweight and waterproof. From the baleen the people made fishing lines and ladles for water.

But under the international laws made after the American whaling fleet put the bowhead into endangered status, the native people were no longer allowed to take them. They didn't take gray whales either—that was left to the Soviet catcher boat, the *Zviozdny.* The local people were confined to running the fox farm. When the meat got dropped off for the foxes, some also went to the people of Sireniki. "They got used to it," Pyotr says, and most now prefer the taste of gray whale over bowhead. But not Pyotr.

Autumn is when, Pyotr says, "sometimes a whale stops and starts to whistle. They show their tails and play. They often stand near the cliffs." He has seen bowheads and beluga whales swimming together, sometimes even playing. But never bowheads and gray whales. During both the spring and the fall migrations, the two species would often be observed simultaneously. Yet they would not approach one another. Curiously, birds and seagulls avoided the bowheads, too—whereas the wake of gray whales seemed always to be marked by the wings of birds. Pyotr says, "When a gray whale twirls near the coastline, tufted puffins like to swim and dive near it. They are not afraid of gray whales."

Sometimes the gray whales are a stone's throw from the beach. Pyotr watches them rub their bodies on the sand and gravel. They like to do this in freshwater, near the outlets of rivers. Especially the young ones enjoy playing in the rolling surf. The people didn't hunt these whales. The elders said, Never go after a young gray whale with its mother. That is something only the killer whales seem to be allowed by the gods.

Pyotr once witnessed the orcas trap a gray whale calf near Cape Uliakphen. He can still envision the pieces of *mantak* that appeared afterward at the water's surface. Killer whales had become more numerous around Sireniki in recent years. The elders called the bulls the foremen. Pyotr says: "There is a legend that when the surf is big at sea and it's very difficult to land, the orcas come and help push the skin boats to the beach with their bodies. The elders said you need to give something when you pass close to orcas at sea, like some tobacco."

The orcas savored the flippers and the tongue of their prey, the bowhead and the gray.

Traditionally, the flippers are a delicacy that go to the first harpooner.

When young Pyotr killed his first whale, he took small pieces of *mantak* from the flippers and threw them into the sea to express his gratitude.

Now approaching seventy, Pyotr continues to hunt. The hide of a walrus that Igor and I had seen on the beach was taken by Pyotr and his son. They use the skin boat—*baidarka*—in strong surf, because it's so lightweight. In winter they use the aluminum boat, because it can move more easily through the ice. The big wooden whaleboats are used only when the seas are very calm and the ice is floating in the early summer.

From one of the wooden boats, a gray whale had been taken early in June. Along with Pyotr, this afternoon Dmitrii Typykhkak and Nikolai Panauge would tell me the story.

There was no meat in the village at the time. Dmitrii, at thirty-eight Pyotr's middle son, looked out his window at a quiet sea and picked up the telephone to call his brothers. Mikhail, the eldest, is the motorist. Vladimir, the youngest, is the team leader. Dmitrii, called Dima by his family, is a shooter. He has one son, sixteen-year-old Andrey, who is on Sireniki's student hunting team and has taken a walrus.

On this day it was walrus the brothers planned to pursue. There were two hunting teams left in the village, with ten men on each. The Typykhkaks' team included Nikolai Panauge, twenty-seven. His father, Timofei, had been one of the peninsula's most renowned marine mammal hunters. Timofei disappeared, along with another son, a daughter and two other passengers, on September 8, 1995. They'd been going to set nets for salmon in an inlet west of Sireniki. Timofei's overturned skin boat was found in the water, with the front end severely damaged. That same day nine others died in yet another skin boat accident. Last seen at the entrance to Provideniya Bay, that boat's passengers had included a group of Russian and American scientists on their way to conduct anthropological studies in Sireniki.

The conjecture went on for months about what caused the near-simultaneous capsizings, which claimed the lives of fourteen people. The weather had been clear, the seas calm that day. Gray whales had been highly visible in the waters. Just about everyone in Sireniki suspected that both skin boats had accidentally struck—or been struck intentionally by—gray whales. The people talked about the twin incidents in the same whispered tones that they talked about the spaceship many believe they saw land in the village during the fall of 1990. According to eyewitnesses, the alien craft was sighted near the fox farm, and a large metallic-looking creature emerged. Upon seeing one of the farmworkers, the creature returned to his ship and flew away.

Today all the hunters would go out together. There would be two wooden whaleboats, each about twenty feet in length, each with a crew of five. Additionally, there would be three slightly smaller aluminum boats. All the boats were equipped with outboards, mostly Envinrudes donated by the Alaska Eskimo Whaling Commission.

They launched. After several hours no walruses could be seen on the ice floes. The hunters knew that during this season gray whales concentrate not far offshore, feeding near the now-abandoned Eskimo village of Singok, about nine miles northwest of Sireniki. The hunters decided to try for a midsized *antohuk*, as they called gray whales in Yupik.

Dima's whaleboat spotted it first. The aluminum boat, Nikolai's, sped on ahead. The first marksman sat in the bow. Nikolai, one of the two second marksmen, sat behind him. The captain was in the stern, one hand on the throttle of the motor. Moving through an open lead on the water, they homed in on the gray whale. It appeared to be about forty feet, bigger than they'd been looking for. Anatoly Kukilgin stood on the bow. The harpoon he held was about thirty inches long. Its sharp iron tip was attached to a long wooden shaft, a rope, and a buoy.

As the whale broke surface and spouted, Anatoly unleashed his weapon. It connected just below the neck. As the harpoon penetrated, the tip was designed to toggle, meaning that it twists and holds fast. Blood crested. Nikolai helped throw the buoy and the ropes into the water. The floats would keep the whale visible and, they hoped, close to the surface.

The whale made a run for the ice. The four boats spaced themselves at intervals where the ice met open water. The whale turned and dove. When it came up again the hunters decided to try another tactic to steer their prey away from the ice, where they could not follow. One of the men brought out a piece of baleen about a yard long. He leaned over the side and began slapping the water in a certain peculiar rhythm. It was the sound of orcas beating their flippers. The sound the orcas make just as they dive for an attack.

Tricked by one hunter imitating another, the gray whale remained within harpoon range. Anatoly threw another. A marksman on another boat hit the whale with a third. Now the whale was attached to three harpoons and three floats.

Then Dima switched from his wooden boat to an aluminum boat. Pyotr's son carried the darting gun. Sometimes it's called a push gun, because a pushing cartridge is placed in the barrel. The gun is designed to project a grenade that will explode inside the whale, killing the animal quickly. The gun is mounted on a long, heavy wooden shaft. A float is attached to the shaft to prevent the gun from sinking after a shot.

The hunters had no "bombs" for the darting gun, only self-made "bullets" the same size as the gun barrel that were sorts of pipe bombs. Still, if they hit the right place, the whale would die from the shock. You stand on a special footrest and aim for the heart. You must throw low and hard, skimming the surface of the water. You must throw with one hand or the projectile will go *too* low. You use your other hand for support.

Dima's first two bullets stuck in the whale's outer layer of blubber. Two more lodged near the heart. Only the fifth apparently penetrated the heart.

From the first harpoon to the final shot from the darting gun was about two hours.

Despite its massive size—well over thirty tons—this proved to be a very calm whale. The hunters knew well the gray whale's aggressive reputation. In 1996 two hunters from the village of Nunligran up north had died after one they'd wounded rushed their boat and overturned it.

Finally this whale did not come up again for air. It sank on the three floats. The hunters tried pulling the whale up with a long rope. They couldn't. They decided to use the boats to pull the whale closer to the beach, letting the ice push them. With the harpoon ropes they managed to get the whale into shallow water. They tied one end of the long rope to the tail and staked the other end on the shore.

All this took about an hour. Half of the whale was beached, half in the water. The hunters waited for high tide and tried pulling some more. No good. Then they waited for the low tide and started to butcher what they could. They took about half the meat and the *mantak*, and decided to return later for the rest. While they were cutting up the whale, the wind carried the ice floes toward the shore and closed the hunters in. They would need to wait the night.

By morning not much had changed. They loaded everything they had onto the boats. The whaleboats pushed off first through the ice, the aluminum boats following. It took a long time. It was evening of the second day before they got back to Sireniki.

The people know, when the hunters are out so long, that they may have taken whale. The people gathered their knives and their baskets and came down to the beach. About three hundred waited for the return. Men, women, and children. Now there would be meat in the village.

The flippers had already been divided among the hunters. These are considered the delicacy: tasty, rich in calories, high energy. Nikolai liked the flippers better than the *mantak*. Dima felt just the opposite.

Now the whale was on the beach. Pyotr Typykhkak greeted his sons, took a tape and measured the whale, and looked to see whether it was male or female. The people gathered around. They greeted the hunters with jubilation. Many placed their hands on the whale in a kind of blessing. Some of the women wept. Then they cut pieces of the whale to be taken home, to be eaten fresh or dried or placed in one of the four meat pits dug into the permafrost near the beach.

When the hunters went back out to claim the remainder of the whale the next day, they found only the rope. The ice had picked up the gray whale and carried it out to sea.

The meat from the June half whale had lasted only about a month. In June the village had also harvested fourteen walruses, in July eight more walruses and two bearded seals. During the first seven months of the year, altogether about twenty tons of meat. Now, once again, there was no meat in the village. And Pyotr, Dima, and Nikolai were the only hunters currently in Sireniki. The rest had been on St. Lawrence Island for two weeks, visiting American Yupik relatives, gathering hunting supplies, waiting for the weather to clear enough for them to make the sixty-five-mile journey by boat back to Sireniki.

Yet now was the time to prepare the walrus meat rolls for the winter cellars. "Everybody goes fishing and waits," Igor Zagrebin translates for Pyotr and the town administrator, Natalya Protopova. The people go to nearby Imtuk Lagoon and fish for Pacific cod, Arctic char, chum salmon, coho salmon, some flounder, even herring. By what the Russian government has divvied up to all the Chukotkan native villages under a quota shared with the United States through the International Whaling Commission, Sireniki is allowed a subsistence harvest of five gray whales and one bowhead for 1999. Igor continues, "If they do not harvest another whale and more walruses for the winter, it will be a very serious problem. Very serious."

Natalya is a short Yakut woman with a cryptic smile who came to the village as a teacher, then was elected the chair of the village soviet and later the head of village administration. She says, "The reason why hunters become very important is we have shortage of food, supplies from outside. We shall go to the store and see!"

Down the street the only store in Sireniki offers rice, buckwheat, sugar, a few packages of pasta, and TU-134 cigarettes. Ships' captains don't like to unload cargo here anymore because of the strong surf. There's little money to buy goods anyway. Pyotr does receive a small monthly government pension and distributes it among his sons. Natalya is considered a government employee but, since last December, has received no pay. She explains, "Everybody receives salary, but this salary is on paper only. A guy comes and each month writes down that I earned one thousand rubles, but the money is only in the book." The old Soviet state farm system has broken down into different associations—semiprivate, semimunicipal, semistate. All the offices, however, are in Provideniya. So is the only bank in the region, but there are no vehicles to get there and back. At the moment there is no doctor in Sireniki, and there is a shortage of medicine and vitamins. Asked if there is more crime in these tough times, Natalya replies, "Stealing food."

A year ago the Yupik marine mammal hunters would barter in the winter with the Chukchi reindeer herdsmen, trading spotted or ringed seal for reindeer meat. Now, though, the numbers of reindeer have declined drastically, from about two thousand in 1995 to fewer than five hundred in 1999. The herders, after the state left things strictly up to them, ate too many and paid no attention to the ratio of male to female reindeer. In the recent past, too, some Sireniki people had gone to Provideniya to barter fish, bird eggs, and mushrooms for items such as sugar, tobacco and tea. But in 1999 there had been few eggs to collect.

They were back to a situation basically unchanged since the days of Scammon. But when I ask Natalya if she'd prefer a return to the Soviet system, she gives a resigned shrug and says, "Yes, in that time, the stores were full of supplies for the year and the children had fruit and sweets. But now we have more freedom, to relearn our old ways, to negotiate with our relatives on St. Lawrence Island—who for years we were not even allowed to see." The irony is, those American relatives are now the ones being subsidized—by the U.S. government.

When, after two and a half days in Sireniki, my *vezdekhod* prepares to depart at noon, a large crowd of villagers has gathered. Igor asks if I would mind whether some of them came along with us to Provideniya. No, of course not, I tell him. But there is room for only about fifteen to crowd into the back of the vehicle. These are all women and children, selected largely on the basis of their need for medical or dental care. I notice that many are wearing American clothes, apparently gifts from other visitors. A Lakers jacket. A Bears cap. A Sitka, Alaska, sweatshirt. I've already given away most of the contents of the extra suitcase I brought for such purposes.

As the *vezdekhod* pulls away from Sireniki, I watch the silent crowd for whom there was no room walk back down the hill with their luggage. And I find myself uttering a silent prayer for the safe and quick return of the other hunters.

Over the winter of 1865–66, members of the Western Union Telegraph Expedition had remained ashore both in upper Siberia and in Alaska. Facing blizzards and temperatures of fifty below zero, they'd had a tough time laying out potential routes for the telegraph line. Still, Colonel Bulkley put a good face on things in his published report upon his return to San Francisco. Scammon was safely back there, too, and requested permission for a continued leave of absence from the U.S. Revenue Marine. According to William Dall, though, much of this was posturing. On April 26, 1866, Dall wrote his sponsor, Spencer Baird of the Smithsonian: "Capt. Scammon may resign because of Bulkley's humbugging and unbusinesslike & insulting way of doing business & with fussing over the Capt's powers and privileges without consulting him." Dall added that Scammon was the only one who really gave a damn about the scientific work.

Whatever Scammon may have confided to Dall, the captain went ahead with plans to return to Port Providence/Plover Bay, on a new ship that Bulkley purchased for the expedition. This was the *Nightingale*, a yachtlike clipper ship over 230 feet long, with the tallest of six masts being a "cloud-raker" 170 feet high. "Our vessel is . . . the fastest on the coast," Dall wrote excitedly to his father, and one with a "curious" history. It had been a merchant ship in the China tea trade and a slave ship, which Dall wrote was "sent to the Coast of Africa where she obtained a cargo of 900 slaves." After the Civil War broke out the Northern Navy had seized it and turned it into a blockade vessel to keep Confederate ships from leaving port. Scammon must have appreciated its much faster, much larger usage for a flagship.

On July 11, 1866, he cast anchor and wrote in his log: "On this trip I am accompanied by my family. Mrs. Scammon and child (Alexander Elfsberg Scammon). Chas K Scammon the latter a cadet in Land Service and assigned to duty and to report to Major Abasa, commanding the Eastern Siberian Division." His eldest son, Charles, was then sixteen,

Alexander three. Clipping along at about 120 miles a day, the *Nightingale* arrived at Plover Bay on August 14. "Five canoes came alongside. The Indians were not allowed on board, but were furnished with hard bread etc."

Bulkley's ship arrived three days later. Scammon and his wife did considerable entertaining aboard the *Nightingale* while all the materials from several expedition vessels were unloaded. On August 24 the first Western Union Telegraph pole on Russian soil was raised at Plover Bay. Scammon's flagship offered a twenty-one-gun salute. Now, it was believed, they were truly on their way toward success.

Scammon, September 16: "Several whales in the Bay. The boats of the brig *Victoria* started in pursuit—struck two but only secured one whale."

Dall, to his father on September 18: "It has been raining hard ever since we got here, and I have not been ashore yet. . . . We are still lingering in this dreary bay. . . . I have made many collections and have measured by barometer two or three of the highest peaks around the bay, naming one of them Mount Kennicott," after Major Robert Kennicott, Dall's mentor, who'd been dropped off to explore Alaska's Yukon territory almost a year earlier.

Also that day Dall wrote to Baird. He described the collections he'd made in the Chukotka region, "about 2,500 specimens. The best things are in the way of plants, mosses, and skull." These last included "a fine set of Tchookchee skulls—[obtained] at some personal risk." Dall would later use this series of crania of the wandering Chukchi to compare with skulls of Eskimos as well as Indian peoples from Alaska, Puget Sound, and California. He would thus become the first to confirm that the Chukchi of Russia and the Eskimos across the Bering Strait in America possessed a common origin. He wrote that they "closely resemble each other in their strongly marked Mongolian features" while differing considerably from Native Americans to the south.

Dall also wrote Baird about something else: "The expedition has been terribly mismanaged this season. We have been nearly six weeks in this Bay and nothing yet done for the Youcon [*sic*] division. . . . Col. B [Bulkley] has been completely under the control of a parcel of swindlers, sots and toadies whose advice he has followed to the detriment of the objects of the Company. They have been steadily at work since we left San Francisco to annoy, obstruct, and even oust all the honest hard workers in the service. On Scammon & the Sci. Corps they have centered a great deal of their venom. . . . I have been fighting them all the summer and have had a bitter and hard time of it. . . . I know Scammon looks out well for his own interest."

They were about to leave Plover Bay, however, for Fort St. Michael. There, Dall anticipated seeing Kennicott again and forming an alliance with him and Scammon that would mean "a grand fight is going to come off. . . . If Bob [Kennicott] stays, of course all is well, and I stay with him."

Dall had no way of knowing that things on the expedition were about to go from bad to worse.

> *From St. Lawrence Island, especially, frames of kyaks and umiaks are transported to Plover Bay and exchanged for tame-reindeer skins, walrus-ivory, and whale sinew and blubber. The distance traveled is about forty miles, occupying nearly twenty-four hours, and the voyage is never undertaken except under the most favorable circumstances and with all possible precautions.*
>
> —WILLIAM HEALEY DALL, *Contributions to North American Ethnology*, 1877

The evening before I left Sireniki, I'd been walking down the beach when I noticed two women standing together at the shoreline, looking far out to sea, as if maintaining a kind of vigil. "They come every twilight," Igor Zagrebin told me. "They are watching for their husbands' boats to return from St. Lawrence Island."

Not until we got back to Provideniya did I learn what had happened four days earlier. Almost eighty people, predominantly from Sireniki and New Chaplino, had been staying with friends and relatives on Alaska's St. Lawrence. Over the past eleven years this had become a kind of summer ritual. They were there on special ninety-day invitations from a church in Gambell, because the visiting villagers weren't allowed to have international passports or travel visas. Now many of the women were anxious to return home and start collecting wild onions, mushrooms, and berries for the hard winter ahead. The men included a number of marine mammal hunters, also concerned about the soon-to-be-worsening weather. They'd all been given clothes, outboard motors, gasoline, even cash by their comparatively wealthy American Eskimo kin. At five in the morning they gathered on the Gambell shoreline beside their boats.

The skies looked potentially threatening, and there was considerable debate about whether to attempt the crossing. New Chaplino is at the tip of a peninsula almost due north of Gambell. But the Russian authorities required them to pass through customs at the port of Provideniya, tucked on the other side of a peninsula to the southwest. That was a voyage of some sixty-five miles. Knowing how rough the surf around Sireniki might be after that, those villagers opted to stay put. Some from New Chaplino, though, were worried that if they didn't get home on their government's due date, they might face restrictions on their future movements.

So, while about forty natives remained behind at Gambell, thirty-seven others on fourteen boats set off early that Tuesday morning. Two of the craft were wooden whaleboats between twenty and twenty-six feet long; the rest were aluminum skiffs, between sixteen and twenty feet long. All contained fifty- to sixty-horsepower outboards. All were loaded down with gear. In a crossing they anticipated would take about six hours, they planned to stay within sight of each other. At the last minute an American photographer named Heidi Bradner—there doing a magazine piece about cultural links between Alaskan and Russian Eskimos—decided to go along.

Russian authorities continued to forbid these citizens from carrying radios or electronic navigation devices, so they had to make their way by compass or hand signal communication. About halfway across, the conditions took a severe turn for the worse. Strong winds came up. A thick fog descended, limiting visibility to about two hundred feet. The seas began to rise—first to eight feet, then ten, then twelve. An aluminum skiff carrying a father and son from New Chaplino got its rope stuck in the propeller. This brought the boat to a rapid halt, with water pouring over the sides. They were swamped. That was the last seen of them.

By that first nightfall two of the skiffs had arrived in Provideniya. Six more made it late that night. One of those held Heidi Bradner, after a harrowing fifteen-hour journey. Now she happened to be recuperating in the apartment building where I was staying. Days later she still shook as she recounted to me the story. At the recommendation of two members of a British film crew, she'd chosen to travel with a skilled captain of one of the wooden whaleboats. She'd seen two nearby skiffs take on water and start to disintegrate, with all four of their passengers taken aboard her boat. A man in trouble on another skiff tossed his outboard motor and gas tank onto the whaler before also jumping aboard. This made Bradner's boat top-heavy, "and in basically a split second we all threw everything overboard. I lost my cameras. They lost things they really needed to survive this winter."

She still wasn't sure how the nine of them had made it. Another boat had also arrived after a day and a half at sea. Two people were now confirmed dead. Six others, apparently aboard three boats, were still missing. Five of them were from New Chaplino, the last from nearby Yanrakynnot. Their wives and children had established tent encampments here in the harbor. A search remained ongoing. The U.S. Coast Guard had sent two C-130 aircraft and a helicopter over two thousand square miles of American waters. Russian authorities, searching their own side of the Bering Sea with one helicopter and one boat, refused the Coast Guard's offer to expand its effort into their territory. Three days in, the Coast Guard suspended its search, believing the missing boats—if they hadn't capsized—had likely drifted into Russian waters. "Our hands are tied," said a Coast Guard lieutenant.

That was where things seemed to stand when, on a Sunday afternoon five days after the disappearances, I was having lunch with my hosts in Provideniya and a desperate voice crackled over a CB radio that Viktor had on in the apartment. I recognized the British accent instantly. This morning I'd hiked down to where the *Achiever* was docked. A sixty-three-foot sailing ship that its filmmaker captain had "bought cheap in New Zealand," it was being used by Clive Lonsdale and Robin Smith to shoot a documentary about the Bering Sea's marine mammals for Survival Anglia, an independent TV station based in Norwich, England. We'd talked over coffee about their visit in 1994 to one of the fox farms in Lorino, and their having been able to film a gray whale hunt. Today they'd planned to take a cruise around the bay, because orcas had been spotted and they thought they might be able to witness an attack on a gray whale. Instead, it now seemed they were trying to help rescue two of the missing boats.

Another voice, this time speaking Russian, came over the radio. "That's Afenasi," Viktor said. Afenasi is a bilingual native of Provideniya who was serving as a local guide for the British filmmakers. I couldn't understand the exchange, but Viktor was shaking his head in dismay. "The Russian border guards are ordering them to return to port in Provideniya," Yana translated. "They are saying they need permission to save these people!"

Later I would be able to piece together what was going on. The day before, the Coast Guard had received a fax from Moscow granting it permission to continue the search across Russia's maritime boundary. A C-130 was about nineteen miles away when the metal hulls of the two skiffs suddenly blipped across its radar screen. As the U.S. plane set out in that direction, the pilot began trying to radio any vessel that might be close enough to help. The Russian border guards heard the transmissions, but nobody understood English and the pilot spoke no Russian. About forty-five minutes went by. Then Robin Smith, on the deck of the *Achiever*, heard an American voice over his radio. "I thought, that's strange, and listened—nothing. I flipped through the channels awhile, until again this scrambled message came over." Smith made radio contact and agreed to set course in the direction of the missing skiffs. Then the border guards, recognizing the Britishers, asked to speak to Afenasi—and instructed him first to return to Provideniya to receive proper authorization.

"When they made us come back," Smith continued, "that took us an hour, maybe two. Then we set back out again. We were maybe an hour from the two skiffs, still talking to the aircraft, which had them in sight." The C-130 had initially dropped the boats a radio and watched the Eskimos paddle toward it with their hands. Food, water, parkas, and other supplies were also dropped. "They had frostbite on their hands," said Smith, "and they'd also been drinking salt water. One of the men was already suffering from terribly swollen joints in his knees." Low on fuel, the C-130 gave way to another aircraft and a helicopter, which volunteered to try to hoist the four men aboard. But when told a rescue boat was on its way, the Eskimos chose not to abandon their skiffs.

Meantime, a Russian patrol boat had entered radio communication with Afenasi. "It turned out they were actually much closer than we, maybe six miles away." So, in five-foot seas, the patrol boat managed to complete the rescue.

"There was still confusion over whether another boat was missing," Smith went on. "The Coast Guard really wanted us to confirm or deny this, which we couldn't do. Then the border guards came on the radio and told Afenasi, 'Oh, we know there's only two boats.' So the Coast Guard, thinking it was all over, headed off out of radio range. At which point the border guards came back and said, 'There's a third boat missing.'

"It just happened that, when we were going back in again, another border guard station from Sireniki came on the radio and said a freighter had seen one of the boats offshore, and if we were in the area could we go and look? So we headed over that way and spotted the third one drifting offshore. The native people from Sireniki were already on their way, moving in to pick them up.

"After which, the Provideniya border guards call again and want us to bring those guys back to *them*! Well, they could barely stand. They were in no condition to be taken on another bloody sea voyage. We said no, we'd only bring them if they wanted to come. I think in the end the authorities said they could stay in Sireniki."

I walked down to the Provideniya beach that night, where the vigil was continuing. Quite a few people were camped in tents alongside a bonfire and their boats. Kids skipped stones in the water. A fellow gazed through some binoculars out to sea. When the patrol boat finally arrived late that night, the four men who'd been without sustenance or shelter for five days were brought to the border guard station to give statements while their families waited outside. Then the one who could hardly walk was allowed to enter the hospital.

The governor of the Chukotka Autonomous Okrug [District] responded to the crisis by decreeing a monthlong ban on any boat travel between the villages under his domain and St. Lawrence Island.

New Chaplino remained in mourning when I arrived there in Igor's jeep, about fifteen miles east of Provideniya by dirt road. In 1958 the Soviets had moved the Eskimos here by boat over a three-year period from their ancient settlement at Old Chaplino. All of the elder hunters died during the transition—"yes, of broken hearts," Igor said. Today, one of the hunters planned to keep looking for the bodies of the father and son lost on the crossing from St. Lawrence Island.

If anything, New Chaplino felt even emptier than Sireniki. In the "municipal store," I made a quick inventory of largely bare shelves: brushes, feather dusters, soap dishes, toilet paper, shoes, a few clothes, including a couple of fur hats. The promised supply ship hadn't docked in months. The exchange rate at this time was about twenty-five rubles to the dollar. Behind the counter I spotted a decent-looking radio selling for five rubles. "They bought this radio ten years ago," Igor explained, "and they do not raise the prices."

New Chaplino had resumed subsistence hunting of gray whales in 1998. They took seven, including some in October, so the meat could be frozen to last the winter. In 1999 they had landed one more in July. As we walked down the beach, approaching the remains of the whale, Igor cried out in amazement: "Look, they cut everything! Even the head. Usually people don't bother about the heads."

On our way out, the empty cages of two now-closed fox farms rattled in the wind. The jeep radio was playing, incongruously: "The hills are alive, with the sound of music."

Feeling trapped in the sadness of it all, I couldn't think of much to say as we headed back to Provideniya. Then, high on a mountainside, I noticed the letters about two-thirds of the way up to a thirteen-hundred-foot summit. Big, block-like, white letters that appeared to have been painted there. "No," Igor said, "those letters are all made from rocks and stones, each at least fifty feet high."

What do they say? I asked. " 'Glory to the Soviet soldiers,' and below that, 'Glory to the great October.' The military climbed up there and placed the stones. On another mountain over there, they have changed the message to 'Glory to the Russian Border Guards.' "

I tried to imagine how they clambered up there and clung to the mountain in fashioning their rock art. How long did it take? I asked. "Who knows? But what else were soldiers to do here? It keeps them busy." Maybe this was their monument to Communism, their Stonehenge? Igor laughed heartily and said, "Crazy job."

For my last stop in Chukotka, a bus with the biggest tires I've ever seen will conduct us up the coast to the ancient settlement of Avan. It had been abandoned in 1942, because the Soviet Navy was afraid their shells might fall on the village. I'm going there with Igor, Yana, and an American couple with two teenage sons who've come from Nome to Chukotka on a brief "adventure tour." The adventure begins when the bus straddles the shoreline, up to the axles in water. Igor says the lake is usually a little farther out than this. This journey continues at a snail's pace for well over a mile. Somehow, while the American mother clings to her seat on the other side of the bus, we avoid tipping over into the water.

Avan was once considered an ideal location by the Yupiks. Surrounded by mountains, a narrow spit dividing a freshwater lagoon from the sea served as a barrier between two aquatic environments rich in living organisms: harvestable fish, birds, and marine mammals. Nobody knows how long Avan was inhabited. Today its tall, windblown lyme grass enshrines subterranean dwellings along a meadow reaching to the sea. Those dwellings are all made of whale bones, leaning obelisklike out of the earth, whale-bone foundation poles lacking the walls and roofs. Between the jaws, Igor says, they would place turf, then cover it with walrus skin.

"All the scientists think this one was a men's club," Igor continues, as we step over a whale-bone archway, being careful not to fall in. "Because usually when the men prepared to go hunting, they would spend several weeks here in some rituals, avoiding women and sex."

Chukotka's "Whale-bone Alley" (courtesy Provideniya Museum)

Above the whale-bone meadow is a steep cliffside of nothing but black rock. Yet it is terraced and possible to climb. For centuries this cliffside had served as Avan's cemetery. As we make our way up, Igor points to a large square coffin, which was once lined with fur. "This was the grave of a rich man," he says. "Maybe a shaman or a leader of the tribe. Sometimes they would be placed with their knees up." The wind howls as Igor adds, "All of the artifacts were looted long ago."

Our ascent continues. At the summit there seem to be hundreds of rock coffins spaced at neat intervals, open now to the elements. Igor and I walk to the edge of the cliff and sit down, meditating on the Bering Sea. "There are many amphipods offshore here," he says.

The spout of a gray whale frames the horizon. I can't tell whether it is moving toward Sireniki or New Chaplino. I'm thinking of how, in the legends of Chukotka, the union between human and whale is utterly vital. It is also the source of enmity between villages: those who would kill the whale, those who would not. The end of one such legend—about a meeting of a woman with a whale, the birth and upbringing of a little whale, and its death—is this: "In the fall the Nunagmits expected the little whale, but he did not arrive. Then they knew for sure that their neighbors had killed the whale. They decided to take revenge. They lay in wait for the strongest, bravest man—there was one sea-hunting from a kayak—and they began to pursue him. They caught up with him when he cast a mooring line at a cliff and was starting to climb up on it. They shot arrows at him and killed him. Since that time the neighbors were at war with each other. . . . And they say that in Nunak there is a ditch where they had kept the little whale."

I describe this legend to Igor Zagrebin and ask him what he thinks. "Who knows?" he says, as the mist of the gray whale fades into the Bering Sea.

Chapter 22

NORTHERN CODA: END OF THE EXPEDITION

The attempt which was made by the Western Union Telegraph Company, in 1865–66 and '67, to build an overland line to Europe via Alaska, Bering Strait, and Siberia, was in some respects the most remarkable undertaking of the nineteenth century. Bold in its conception, and important in the ends at which it aimed, it attracted at one time the attention of the whole civilized world, and was regarded as the greatest telegraphic enterprise which had ever engaged American capital.

—GEORGE KENNAN, *Tent Life in Siberia*

AFTER an uneventful five-day trip across the Bering Sea from the Provideniya region, Captain Scammon's flagship *Nightingale* anchored near Egg Island in Alaska's Norton Sound on September 24, 1866. The weather had already turned cold and wretched. Scammon was ill when he fired two guns to notify Colonel Bulkley's *Wright*, already in the harbor, of his arrival. An official party, including Bulkley, came to confer with the captain the next morning.

They brought the news that Major Robert Kennicott, the telegraph expedition's scientific director, had died there the previous spring, at the age of only thirty-two. On May 13 he'd gone to look for trails and check on a hoped-for break in the ice when he apparently suffered a heart attack. The *Nightingale* would need to transport his body back to San Francisco. William Healey Dall would be appointed to take Kennicott's place.

Young Dall was devastated. Upon hearing the news he penned a letter to his mother: "I write in great sorrow. The man who trusted me, loved me, believed in me and whom I loved better than anybody else in the great world is dead! . . . If I had been with him, it might not have been." Dall described the cause as "heart disease added to by ingratitude and worry," and vowed: "I shall carry out his work in Natural History to completion, if God gives me strength and health."

Dall also wrote a circular letter sending word to all American scientific societies and then to Scammon's brother, Jonathan Young Scammon, president of the Board of Trustees of the Chicago Academy of Sciences: "It has been a heavy blow to me. . . . I frankly confess I would not choose to stay in this dreary country, already white with snow, but I believe it to be my duty, and for the best interests of the Academy and the purposes of science." He added as a P.S.: "Captain Scammon has stood by us nobly." Jonathan Scammon's ensuing tribute described Kennicott as "a scientific soldier who died upon the field of battle."

In another letter, dated September 29, 1866, Dall elaborated on what he believed had happened to his mentor. "He was murdered . . . by slow torture of the mind. By ungrateful subordinates, by an egotistic and selfish commander [Colonel Bulkley], by anxiety to fulfill his commands, while those that gave them were lining their pockets in San Francisco. I am so nervous from rage and grief that I can hardly write. And it must all be kept down, for I am surrounded by the same pack of sycophants and thieves. The honest men among the officers of upper grades can be counted on the fingers. Captain Scammon is the only man here that I can rely on and confide in. He lies prostrate on a bed of sickness, and disgusted with the conduction of the whole enterprise, will leave it on the return to San Francisco."

Then a furious storm struck, carrying gale-force winds and sleet. Several times on the night of September 30, the *Nightingale* ran aground in the shallow bays around Fort St. Michael. The ship's water casks were damaged, and there wasn't enough drinking water to make it back to San Francisco. Scammon would need to return for water to Plover Bay. The next day, however, found him too sick to be on deck. He lay confined in his cabin, attended by his wife. Dall came to say good-bye. With a first mate in command, the *Nightingale* left again for Chukotka.

While Scammon's ship battled adverse winds on the crossing, his former vessel—the *Golden Gate*—met with disaster on October 4 at the mouth of the Anadyr River in Siberia. Before it could finish unloading supplies for the forty-seven men who were to work there constructing the telegraph line, the anchored ship found itself trapped and crushed by moving ice. The men were stranded on the shoreline in already subzero temperatures. Facing a long winter, they began by building a house out of the telegraph poles.

A year later the *San Francisco Daily Bulletin* would report on how they made it through:

> They found the Chukchee natives, on the coast of whose territory they were stationed, very kind to them. They assisted the men in various ways, and traded with them to some extent. . . . The party . . . were so fortunate soon after landing as to kill 150 deer. The meat and skins were preserved, and the latter used for clothing by the men. Two or three weeks afterward sixty more deer were killed, and geese were procurable in great quantities. Many were shot, and some killed with clubs. Fish were also plentiful, so the party had no lack of fresh meat—a circumstance which enabled them to escape the dreadful scourge of scurvy. During the winter the cold was intense and the snow very deep, but the party had a good house, plenty of fuel, and the men were warmly clad. Miraculously, only one failed to survive.

Meantime, it had taken eleven days for Scammon's ship to reach Plover Bay, from where, apparently still unable to obtain enough water, it sailed on again for Petropavlovsk on the Kamchatka peninsula. Not until November 13 did the *Nightingale* embark from there for San Francisco. When they arrived three weeks later, it was to learn that, on the fifth and last attempt aboard the *Great Eastern*, a cable across the Atlantic had been completed the previous July. Two years of work by the Western Union Telegraph Expedition were apparently for naught.

December 11, 1866, a letter from the secretary to the chief of marine, Harbor of San Francisco, to the Honorable J. Y. Scammon in Chicago: "Sir: Your brother, Captain C. M. Scammon, Chief of Marine, W. U. Telegraph Expedition, arrived here in command of the flag-ship, on the 8th instant, from Petropaulski, Kamschatka. He has been very ill on board ship for the last three months, and being much reduced is unable, at present, to write you, and requested me to do so. His physician, however, thinks that with proper care he will recover."

January 14, 1867, a letter from Scammon to Mrs. Caroline H. Dall, Boston: "At the request of your son, Lieut. W. H. Dall of the W. U. T. Co., I improve the earliest opportunity—after recovering from a long sickness—to write concerning him. We left Mr. Dall at Michaeloffski the 1st of Oct. last, in excellent health and spirits. . . . During my intimate

Captain Scammon (at center with arms folded) with his fleet officers, aboard the Nightingale *in 1866, Mrs. Scammon at far right* (courtesy Bancroft Library, University of California, Berkeley)

associations with him for the last eighteen months, I became much attached to him, and interested in his welfare. As a scientific gentleman he is considered of great promise, and has been eminently successful in collecting." Scammon went on to describe the illness that had confined him to his cabin for three months as "chronic gastritis."

Even before the expedition's demise the captain had apparently decided to part company with Colonel Bulkley. In mid-March 1867 Scammon took over command of the revenue cutter *Joe Lane.* Three days later he received a letter from a senior captain in the Revenue Marine concerning "serious charges having been preferred against you by Col. Bulkley and the Russian Telegraph Expedition such as Drunkenness, incapacity to command, Smuggling, peculations, etc. which reflect seriously upon your character." It was suggested that Scammon demand an official court of inquiry. Several of his officers drafted a letter describing him as a peerless commander and sailor. That seems to have been where the matter ended, although Bulkley also accused Dall of stealing money and equipment from the expedition and wouldn't authorize payment of his salary. That would take two years, and many letters and affidavits, for Dall to receive. At that time Bulkley dropped out of the picture; he eventually died broke in Guatemala.

On March 30, 1867, a treaty between Russia and the United States ceded the territory of Alaska for $7.2 million. The following day the Western Union Russian American Telegraph Company announced the close of its $3 million enterprise. It didn't end right away in Siberia, however. By the time those expedition members got the news, they'd cut nearly twenty thousand poles, and they kept putting them up across the region for another seven months.

When I was traveling on Sakhalin Island, the last thing I'd expected was to encounter someone familiar with the telegraph expedition and its historic import. Mikhail Vysokov, who has headed the Sakhalin Center for Documentation of Contemporary History since 1991, proved the surprise. The Russian historian had file cards with more bibliographical references to the telegraph expedition than I'd seen anywhere else. As we sat with an interpreter in his book-lined study in the capital of Yuzhno Sakhalinsk, I asked Vysokov whether he believed the expedition had proven instrumental in bringing about the Russian sale of Alaska to the United States.

"Undoubtedly," the Russian scholar replied. "The expedition's scientific work going on in Russia's Alaskan holdings led to more American understanding about the territory. It pushed interest in America toward purchasing Alaska, and interest in Russia toward selling it. I think it's one of the major reasons why Russia lost Alaska."

Vysokov had examined much of the diplomatic correspondence between the Russian government and the telegraph company. "Western Union was planning to pay back its huge investments by having a monopoly on all cable traffic between Europe and the U.S.—until the success of the Atlantic cable killed it. The Russian government was very upset by the refusal of the company to complete the telegraph regardless. Because they themselves had decided to meet Western Union halfway, and to build a line from Irkutsk to the mouth of the Amur River. They'd already invested quite a bit of money in the project."

Even though it ended up unsuccessful, in Vysokov's view the Western Union Russian-American project was extremely important precisely because it *did* push the Russian government into sinking substantial funding into reliable communications. Not only with all of Siberia but on into Japan and China. In 1871 a Danish company was able to build a telegraph cable from Vladivostok to Nagasaki, opening the first such link between Europe and Japan. At the time the British were great enemies of Russia and so hastened their own plans to lay a cable from India that would connect to Hong Kong and Japan.

Vysokov said: "In terms of creating a world system where everyone is connected by telegraphic lines, the Western Union Expedition's construction in Siberia really accelerated this. What happened from the fifties through the seventies of the nineteenth century is equivalent to what we're seeing with the Internet today, that kind of a revolution. One side effect, as far as the Russian government is concerned, was that they felt they'd been burned by the capitalists. Western Union created a certain impression that one could not trust private American companies."

Then the historian added: "Moreover, trying to govern their 'colony' in Alaska was costing the Russians a lot of money. They wanted to sell it. And I think if Alaska was still part of Russia today, it would be one of the most forgotten and backwater parts of the entire Russian Federation. Similar to what you see in Chukotka, unfortunately."

What ended in failure for the enterprise as a whole marked the beginning of success for several of its individual members. No fewer than four lengthy, influential books came out in the expedition's immediate aftermath. The first was written by the expedition's staff artist, Frederick Whymper, published in 1869 as *Travel and Adventure in the Territory of Alaska.* He'd generally traveled on Scammon's ship and had journeyed with Dall in the Yukon during the winter of 1866–67.

The second was authored by George Kennan, like Dall only nineteen when he joined the expedition. He returned to the United States after his five-thousand-mile journey across Siberia, began lecturing about his experiences, and wrote *Tent Life in Siberia,* published in 1871. Kennan's book was translated immediately into Russian and, according to Vysokov, "was so popular that it went through three editions, one after another. It really informed the opinion of the Russians about this huge territory, of which they knew very little in the seventies and eighties of the nineteenth century." Kennan himself would return to Russia several more times, becoming an authority on the country. Many years later a distant cousin would become America's ambassador to the USSR during the Cold War. "I feel that I was in some strange way destined to carry forward as best I could the work of my distinguished and respected namesake," wrote George Frost Kennan in his own autobiography.

Dall, who learned of the termination of the enterprise in the spring of 1867, decided to stay on in Alaska and continue his scientific explorations. As he wrote in a letter that summer:

> Sometimes in this far-off coast, I have been camped in the open air on the white snow, and waking when all was still and watching the bright Northern Lights play over the sky, and the tall pines bend and rustle with a passing breeze, it has seemed to me as if God was very near. . . . I have travelled in winter with the thermometer down from 8 to 40 below zero, over 400 miles. When the spring came I took up my paddle and started in an open canoe with one white companion from Fort Youkon, 650 miles up stream of one of the largest and most rapid rivers in America. I did it in 26 days, and I was the first American to do it. I turned back and came down the same distance in 6 days *with* the current and then pushed on 700 miles more down to the sea, in ten days more. I obtained 4,550 specimens of birds, plants, insects, &c. I learned enough of two Indian dialects and the Russian language to get along with any of them.

Dall remained mostly in Alaska until the late summer of 1868. He traveled back and forth in the Yukon Territory between St. Michael and Nulato, taking copious notes on everything he saw. Then he published, in 1870, an illustrated volume of more than six hundred pages, *Alaska and its Resources*, the standard authority on the territory for years. Dall went on to become an eminent scientist with the Smithsonian Institution and for the U.S. government.

As for Scammon, he never published anything directly about his experiences with the telegraph expedition. But his time as its flagship commander—and the resulting relationships with Dall and the Smithsonian—marked the nascency of his new life as a writer and naturalist. And the gray whales he'd pursued so ruthlessly in their Baja nursing grounds, and watched the native peoples pursue in their northern feeding grounds, were about to assume center stage again for him in an entirely different way.

PART THREE

Chapter 23

Christopher Reeve and the Gray Whale

The moaning of the whale is a stirring of profound thoughts.

—Victor B. Scheffer, former director, U.S. Fish and Wildlife Service

For my last night in Chukotka, my Russian hosts had prepared a feast including red caviar and a veal roast, washed down by numerous toasts of a cognaclike mountain berry vodka. Even the *vezdekhod* driver, Valiko, came up to say good-bye. "If he wanted," Viktor said, "he could drive that thing up thirty flights of stairs!" While their eldest daughter, my translator, Yana, returned to finish college in Anchorage, Viktor, his wife, Lyuda, and their five-year-old planned to leave Provideniya soon for the Russian interior. He had been in the wholesale food business but saw little opportunity here. We watched TV—now almost indistinguishable from the American variety, with numerous product commercials, talk shows, even MTV—and news that President Yeltsin had named a new prime minister for the fourth time in thirteen months. This one, Vladimir Putin, was "an old Soviet KGB guy," my hosts informed me, shaking their heads.

The next morning I stopped by the Provideniya museum's little gift shop and purchased a strange sculpture of a Chukchi shaman carved from a reindeer antler. It would remind me of what I'd witnessed in Sireniki. Igor Zagrebin came to the airport to see me off in the rain and fog. After a couple-hour delay the eight-seater plane was airborne for the short ride across the Bering Strait to Nome—and across the date line into tomorrow evening. I called Jim Stimpfle upon arrival, completing a full circle of my journey north with a quick dinner together before flying to Anchorage that night. As I sat waiting in the Cheers bar during my next layover, I felt that America had never seemed quite so opulent and garish. For the next twenty-four hours I continued straight through Seattle, St. Louis, and home to Boston. I'd been away from my wife for six weeks, gone altogether for two months. Summer was nearly over.

For gray whales as well it was a difficult period of adjustment. I learned that, in July, the increase in their strandings had been designated an "unusual mortality event" by a working group of the National Marine Fisheries Service. Five scientists—including Oregon's Bruce Mate and two I'd met in Baja, Jorge Urbán and Paco Ollervides—were starting a study of the "potential causes and implications." I placed a call to the research team's leader, Dr. Burney Le Boeuf of the Department of Biology and Institute of Marine Sciences at the University of California, Santa Cruz. He was most interested in what I'd observed in Alaska—grays showing up in Ketchikan and off the Kenai Peinsula, grays staying into the summer around Sitka and Kodiak Island. Later these sightings would appear in the report, which documented the numerous other anomalies of 1999 as well: the three-week delay in the southbound migration in January and the whales' more southerly distribution in Mexico (counts 30 percent lower in San Ignacio Lagoon than the previous year at the same time that numbers in Magdalena Bay soared from 68 to as many as 400, while many more grays were being seen all the way to the southern tip of Baja).

The last phenomenon was attributed to lower-than-usual water temperatures in the lagoons owing to La Niña conditions, which Jorge Urbán had spoken about at San Ignacio in March. Following two years of the strongest El Niño pattern ever recorded, La Niña results in the opposite cooling effect. This was why the ice had taken longer to recede in the Arctic, delaying the gray whales' return north in 1999. The cold, productive waters associated with La Niña had also resulted in an abundance of krill, on which I'd seen the gray whale surface-feeding off Monterey—and others had glimpsed, more prevalently and for more prolonged periods than ever before, off central and Southern California. The scientists would come to a "hypothesis that the whales were undernourished and took advantage of opportunities to feed on unusual prey species in several areas."

Their overall "general hypothesis" would be "that insufficient body reserves and malnutrition" were a "major cause" behind a final count between late December 1998 and September 15, 1999, of 274 stranded gray whales, twice as high as any previous year, dating back to the first time such counts were made, in 1985. Also unique was that adults and immature whales—not newborn calves—predominated in the strandings. There were 120 dead whales found in Mexico (71 of these in proximity to the lagoons), 43 in California, 31 in Oregon and Washington, 8 in British Columbia, and 72 in Alaska. "Since the whales lose between 11 and 29 percent of their body weight between the southward and northward migrations (Rice and Wolman, 1971), the data suggest that the whales were more likely to exhaust their limited energy reserves on the return to the foraging grounds," the study would say.

Which all led back to what was happening on those foraging grounds, where higher-than-normal mean sea surface temperatures exceeded 6 degrees Centigrade for nine of the ten years between 1990 and 1999. Amphipods, the gray whale's favored food, are temperature-dependent. "Warm temperatures result in low age at reproduction, small adult size, small brood size, and reduced life span," according to scientists. "Benthic sampling studies southwest of St. Lawrence Island and the Chirikov Basin indicate that a significant decline in productivity and a change in the dominant benthic fauna occurred in the 1990s. These data reveal more than a 50 percent decrease in benthic biomass from summer 1990 to 1993–94, a decrease that continued through to 1998 and 1999."

At the same time the gray whales remained a target of Washington's Makah Tribe—and, it seemed, of the Japanese. On October 1, four months after they had unsuccessfully sought to thwart the tribal canoe on the day before the whale was killed, seven whaling protesters were formally charged in federal court in Tacoma. The Coast Guard continued to hold *The Bulletproof*, one of the Sea Defense Alliance's Zodiacs. *The Seattle Times* reported: "Yesterday also marked the official opening of the new hunting season, and two tribal whaling canoes have been out on the water for paddling practice nearly every night."

In mid-November the Sea Shepherds' Paul Watson met with Washington Governor Gary Locke, calling his attention to a state law giving gubernatorial authority to intervene in federal or tribal matters that were contrary to the state's interest in conserving species it had classified as "protected"—which the gray whale is. "I have always been against the hunt," Governor Locke told the activists and said he would consider such intervention. He would also ask his secretary of state and attorney general to look into whether the Makah might be illegally going after "resident," rather than migratory, whales.

As it happened, the Makah did not hunt again during the gray's southbound migration. However, at an annual strategy session of Greenpeace whaling and fisheries campaigners in Gloucester, Massachusetts, I learned of the latest strategy of Japanese whaling interests. It focused on a forthcoming meeting of the Convention on International Trade in Endangered Species (CITES). There Japan's Fisheries Agency would propose to downgrade the level of protection given three whale populations—the South Antarctic and northwestern Pacific minke whale, and the Eastern Pacific gray whale. Their argument was that these populations were now abundant enough that their products could be traded internationally. The "down-listing" by CITES would be the crucial first step toward pushing the International Whaling Commission (IWC) to lift its moratorium on commercial whaling.

"You start to trade one species, it's so difficult to differentiate between others without DNA analysis. The illegal sale of whale meat would escalate," as Juan Carlos Cantú of Greenpeace Mexico put it. Gerry Leape, at the time the organization's U.S. whaling campaigner, told me that Japan's commissioner to the IWC "has said publicly he feels it's his obligation to bring down the price of whale meat in his country, so he will encourage export from any nation that wants to trade. Japan would love to see the gray whale down-listed by CITES."

The Russian Federation had made it clear they supported this effort, because of the tremendous currency that a reopened whale meat market with Japan would generate. Over the summer, at a meeting in the Russian Far East, the chief of the Federation's fisheries agency stated that the commercial moratorium was costing his country $50 million a year in lost profits. Suddenly a Russian company launched a hunt for beluga whales in the Sea of Okhotsk. Like other small whales and dolphins, belugas are not protected under the rules of the IWC. Now thirteen metric tons—the first-ever commercial trade of beluga meat—were en route to Japan. The International Fund for Animal Welfare (IFAW) dispatched a film crew to investigate and located the importing company in Hokkaido. The fund, with offices in fifteen countries, began contacting government representatives and exposing the incident in the press. Russian authorities then ended the hunt, stating it was not "sustainable" and was antithetical to Russia's commitment to environmental protection. Still, this had clearly been a test balloon in advance of the coming spring's CITES meeting in Nairobi—just how *much* resistance would there be? "To us," as Greenpeace's Leape described, "whales are facing the most serious threat since before the global moratorium took effect in 1986."

•

Then there was the ongoing battle over the proposed salt factory at San Ignacio Lagoon. Things were heating up. At the end of September a press conference was called in Los Angeles. The Campaign to Save Laguna San Ignacio—a joint effort of IFAW, the Natural Resources Defense Council (NRDC), and a coalition of Mexican environmental organizations that now numbered fifty—was announcing a "Mitsubishi Don't Buy It" campaign in California.

The eminent whale scientist Roger Payne attended and pointed out a Mexican federal prosecutor's recent finding of 298 violations of the country's environmental laws at the current saltworks alongside Scammon's Lagoon, despite the ESSA company's claims of a forty-year record of protecting the Baja region. "None of these violations was punished or fined, and most were unremediated," Payne said. "They include unsafe storage of large quantities of fuel, illegal dumping of PCBs, and the dumping of batteries directly into the bay. That doesn't sound to me like a particularly good environmental record."

Pierce Brosnan, the movies' latest James Bond, described visiting the lagoon with his son. "This part of the world is sheer magic—to see the trust and love of these great creatures as they brought their calves to us to stroke and touch." Brosnan noted that he was now a grandfather and would be more than "deeply saddened" but "really *pissed off*" if the salt project went forward. "It's unnecessary, it's shameful neglect."

Later I would learn the intriguing origins of the groups' boycott of Mitsubishi in California. During the spring of 1998 a representative of IFAW in Tokyo had been approached by a Japanese senator. Someone high up in the Mitsubishi Corporation wanted to meet with the animal welfare group. This company official opposed the saltworks project but refused to talk on the phone, write, or e-mail. It would have to be a discreetly private, one-on-one, meeting. So, after the November 1998 UNESCO World Heritage Commission meeting in Kyoto, where the lagoon's status as a previously designated World Heritage Site was discussed, Jared Blumenfeld, IFAW's director of habitat for animals, traveled to Tokyo.

The rendezvous took place at the Imperial Hotel. The official began by explaining his belief that the saltworks controversy was hurting his company's image. Mitsubishi was so huge and diversified, whether they extracted salt from another lagoon didn't even round out a decimal point in their budget. Besides, the man believed Mitsubishi needed to be "a green company." And he didn't think the New York office—the "clearinghouse" for information on the saltworks—was giving a clear enough picture to Tokyo of what was going on. So he was willing to offer some advice.

Blumenfeld asked about the 750,000 letters that the NRDC/IFAW–led campaign had generated. The man shook his head and replied that they made absolutely no difference. How many would it take? Blumenfeld persisted. Two million? Three million? The man replied that even *20 million* letters wouldn't alter Mitsubishi's perspective. The powers that be would continue to see these as coming from ill-informed people overseas, being led along by duplicitous environmental groups, who if they knew the real truth would be on Mitsubishi's side. What about the full-page ad, Blumenfeld wondered, the one they'd placed in *Yomiuri*, the world's biggest-circulation newspaper, comparing the saltworks expansion to tearing down the ancient cedars of Yakashima? Again, the man shook his head. The ad had certainly reached the target Japanese audience—but it had probably infuriated them even more about American meddling in their affairs.

Well, what *would* make a difference? Blumenfeld asked in exasperation. The Mitsubishi official replied that if the UNESCO World Heritage Commission declared San Ignacio Lagoon an "in danger" site, the company would probably pull out. That was because UNESCO, the scientific and cultural arm of the United Nations, had played an important role in Japan after the Second World War and become a critical element in the country's psyche.

So that was one front on which IFAW and the NRDC would continue to push. In August 1999, UNESCO sent a team of four international and three Mexican government experts to the lagoon. Their report was provided to the Zedillo administration in late October. Although press statements emanating from both Mitsubishi and Mexico alleged that the report gave the project a "green light," another interpretation seemed just as valid. The UNESCO review stated: "The associated installations (industrial area, infrastructure, and pier) would involve major changes to the land inside and near the World Heritage site. At least a part of the area would be transformed into urban and industrial land. . . . This would constitute a substantial and significant change from the current conditions of the site."

Yet clearly a new strategy was needed. What about trying to affect the monolithic corporation's bottom line? "What Mitsubishi has against them," Blumenfeld reasoned, "is that nearly all of their entities have the same name." California, where people had a close relationship with the coastal gray whales and were already sensitized to the saltworks issue, also happened to be Mitsubishi's biggest marketplace in the United States. Using the Internet, IFAW made available a list by zip code of Mitsubishi Motors dealerships and branches of Union Bank of California, a majority-owned subsidiary of Bank of Tokyo–Mitsubishi. The notion was, people from those areas could write and make their anti-saltworks views known. When folks from "the neighborhood" were, for example, threatening to boycott Mitsubishi in favor of the GM or Chrysler dealerships down the road, the echoes were likely to reverberate all the way to Tokyo.

And this was only the beginning. The fastest-growing sector in the financial portfolio, from asset managers to mutual funds, is socially screened moneys, now worth something on the order of $4 trillion. Late in October new full-page ads appeared in *The New York Times* and *The Wall Street Journal.* The latter paper had, two months earlier, run a news story headlined ECOLOGY GROUPS USE GRAY WHALE TO RE-ENGINEER MEXICAN PROJECT. Now the eye of the whale loomed above an advertisement headed, "When these money managers make a killing in the stock market, it's not at the expense of an entire species." No fewer than fifteen mutual funds and asset managers of $13 billion were listed "calling on Mitsubishi Corp. to abandon their Laguna San Ignacio salt factory plans" or "they won't be purchasing Mitsubishi stock and . . . will be calling on other investors to do the same."

By early November the movement had spread. A mutual fund worth $54 billion was sitting down with Mitsubishi to discuss its opposition to the expansion plan. The city councils of Los Angeles, San Francisco, Oakland, Berkeley, and Sacramento had all voted unanimously to call upon Mitsubishi to cease and desist, announcing they were investigating their respective contracts with Mitsubishi companies as the first step toward potential divestment. As the gray whale commenced its latest southbound migration toward California, the momentum it had generated was astounding.

Around this same time Kathryn Fuller, the president and CEO of the World Wildlife Fund–U.S. (WWF) flew from Washington, D.C., to Mexico City for a meeting with President Zedillo. The World Wildlife Fund is the largest nonprofit conservation organization in the world, operating in more than one hundred countries, with about 5 million members and a $350 million annual budget. Its Mexico branch office alone had more than forty employees, including Roger Payne's daughter, Holly, who for months had been helping prepare a study of economic alternatives to the San Ignacio salt factory. The WWF's three-hundred-page report, based on research commissioned from independent economists and scientists, concluded that both ecotourism and sustainable fisheries could provide as much, if not more, long-term economic development to Baja California Sur as the saltworks. Fuller was arriving for a press conference on the study's release; Mexico's environment minister, Julia Carabias, had arranged Fuller's visit with Zedillo to discuss WWF's conservation agenda for his country and several items the organization hoped his administration might address during his final year in office.

The president, it turned out, was, like Fuller herself, a passionate scuba diver. "We spent quite a bit of time," she would recall, "talking about his diving among these deepwater aggregations of sharks and rays, from these tiny islands off the west coast of Mexico. He told of taking his kids out diving with whale sharks on the Caribbean side. It was clear that this man cared deeply about the marine environment."

For a while their talk shifted to matters like Mexico's monarch butterfly sanctuaries. Fuller had brought along then-and-now satellite imagery demonstrating how much—even in supposedly protected areas—logging had encroached upon the butterflies' mountain habitats in only eight years. Finally, she broached the subject of San Ignacio Lagoon. She discussed the WWF study, how the jobs created from ecotourism and sustainable fisheries would be permanent, with the revenues staying in the communities, where they belonged. "Zedillo listened thoughtfully and asked a lot of probing questions," Fuller remembered. The president wondered why they couldn't have all *three* kinds of development, provided of course that the salt extraction facility was carefully managed. The WWF executive responded that she opposed it as a matter of principle, because the factory would be built inside a biosphere reserve. The decision was fundamental regarding what type of development would be permissible in a protected area, not only in Mexico but worldwide. If a salt factory was okay, this set a precedent for other extractive industries, such as minerals or oil and gas.

As the hour-long meeting came to a close, Zedillo was noncommittal. But he expressed interest in reading the WWF report.

The next morning Fuller was called to a meeting with Herminio Blanco, Mexico's commerce minister, who also chaired the ESSA salt company's board. He wanted to assure her that the work on the new environmental impact assessment was "comprehensive and responsible." Fuller replied she had no reason to question that, but it simply wasn't going to change her opinion. At the press conference that afternoon, Fuller was accompanied by a longtime friend of Zedillo's, Julio Gutiérrez, retired chairman of a large electrical supply manufacturer and a central figure in a $50 million trust fund that supported conservation programs around Mexico. He said he agreed that, as a matter of principle, a salt factory should not be built in a biosphere reserve.

A strong argument along these same lines soon arrived on Zedillo's desk. It was a letter handwritten in Spanish by the Netherlands' Prince Bernhard, dated November 24, 1999. Recalling that he had learned about the proposed project five years earlier from Homero Aridjis, the prince expressed his surprise that it remained viable. He wrote: "Mr. President, it seems to me that to build a massive saltworks in San Ignacio Lagoon goes against the very idea of what a biosphere reserve is. . . . No short-term economic benefit could historically justify the destruction of the very nursery where the gray whale mates and gives birth in Mexico. Future generations of Mexicans and of human beings will thank you for taking a decision in defense of this fabulous animal, which is part of the mythology of the oceans."

It's about a two-hour drive from Boston to IFAW's headquarters in Yarmouthport, Massachusetts, on Cape Cod. This seemed an unusual location for an organization with a membership of 2 million, a budget around $70 million, and a global staff of more than two hundred people. The fund had ended up in Yarmouthport during the early 1970s, when the organization's Canadian founder, Brian Davies, had his charitable status revoked there because of the "political nature" of his opposition to the country's seal hunts. Also a pilot, Davies had wanted to move somewhere close enough to fly to the hunt, and an ex-wife of his lived on the Cape. Hence, Yarmouthport.

The IFAW offices are in a sprawling complex off Route 6A. I visited in November to meet with Jared Blumenfeld and President Fred O'Regan, to get their take on how the lagoon battle was progressing. O'Regan, a stocky fellow of Irish descent, had come to IFAW in 1997 after serving as a regional director of the Peace Corps in Washington, D.C. Blumenfeld, originally from England, is in his early thirties, the son of two prominent residents of Cambridge, the author-philosopher

Yorick and the sculptor Helaine. A lawyer, Blumenfeld left the NRDC for IFAW in 1995, when he met Homero Aridjis at an IWC meeting and ended up getting the saltworks issue high on the agendas of both the U.S.-based organizations. "Without Homero and Betty," he said, "the salt factory would have been built a long time ago."

Blumenfeld had come to believe that this was the most important environmental issue facing the planet. "If you let a company the size of Mitsubishi build a huge industrial plant in the middle of one of the world's most protected areas, you can do it anywhere." O'Regan viewed this as a test case in much the same way. He pointed to "a vacuum that's been created in the whole environmental regulatory regime worldwide. The strength, commitment, and resources of individuals—and even nations—has waned up against the massive power of corporations, and the tremendous growth of the free trade movement calling for less regulation. The line in the sand is thicker. Somebody has to play the role that governments are not, and show these corporations that we're watching and are not going away."

Both men felt that, although their previous top-down effort hadn't worked, the bottom-up boycott was having more of an effect. It had moved from preaching to the converted to reaching out for an entirely new base, from a conventional "yell-at-the-bad-guys" approach to strategizing how to get at the corporate decision makers. This was no longer an environmental campaign but a political one. Blumenfeld was optimistic: "I think we'll win this." O'Regan was far from certain: "The odds are very much against us. They've got the money, the internal political contacts. This could have some dire consequences. But we hope to make this a self-perpetuating machine, where, at the very least, we'll be gnawing at their ankles. Even if they build the saltworks, we won't stop. We'll be out there next lying in front of the tractors."

If this seemed an audacious statement from the CEO of a well-heeled conservation outfit scarcely in the same category as the Sea Shepherds, O'Regan's next words offered an explanation. He began talking of his yearly trips to San Ignacio Lagoon, the bird life, looking out across "those magnificent flats with the mountains in the background, like something out of a Clint Eastwood movie"—and, of course, the gray whales. "They've been involved in some interesting social history—and here they go again. They're our best buddies in this campaign, a magnificent megacreature that people can relate to. If you could get everybody just to go to that lagoon, this would be over! They wouldn't buy a Mitsubishi paper clip."

Later that month a column cowritten by Homero Aridjis and Serge Dedina—the initial "whistle-blowers" in the saltworks saga—appeared on the editorial page of the *Los Angeles Times.* Datelined Mexico City, it began: "The most consequential political campaigns for the future of Mexico are not those being orchestrated by the dinosaurs of the long-dominant Institutional Revolutionary Party, or PRI, or by members of the opposition parties. The campaign with the greatest implication for democracy in Mexico is the one being waged over the gray whale." The authors went on to say that the battle over San Ignacio Lagoon "will determine whether Mexico evolves into a nation that permits private citizens and nonaligned organizations to participate in national policy and local economic development debates." Presciently, they concluded: "The fishing communities and environmental organizations involved in the gray whale campaign are the real key to a more open and pluralistic political system in Mexico. A real and vibrant democracy will only occur in Mexico when independent groups are allowed to emerge from the shadows of the PRI."

At the same time this column was published, Homero and Betty Aridjis came to New York City to visit their daughter, Eva. My wife, Alice, and I took a train from Boston to meet them for dinner. Homero read us his new poem, "The Eye of the Whale," which he had written after the previous winter's visit to the lagoon. This was where—as Homero had it—in the beginning God had "created the great whales." We talked of how the Bible has, of course, always had its geographical Genesis in the Middle East, where, as far as we know, there weren't any such whales. Homero's eyes twinkled, and he chuckled as he said, "I am decentralizing Creation."

Later that night Homero would describe the environmental struggle as being akin to "a religious movement, because you must have local missionaries and you must convert people." Over a glass of red wine the poet-activist admitted that he was *cansado*—tired—because the ruling powers in Mexico try to wear you down and make you feel utterly alone, as if nothing you do has any real meaning. He held on to my sleeve for a moment, looked deep into my eyes, and said that Betty would not allow this. "She says to me, 'Homero, the butterflies are waiting, the whales are waiting.' "

And so they were. Not long after the Aridjises departed for Mexico again, Alice and I made plans to move later in the winter to Baja, where I would immerse myself in seeking to tell the gray whale's story. There was, however, another man on the East Coast who had spent intimate time with the whales and with whom I hoped to speak. Upon returning from my first visit to the lagoon, I'd watched a PBS Home Video of a 1995 episode from the wildlife documentary series *In the Wild.* It was an hour-long program, "Gray Whales with Christopher Reeve." The narrator, the actor's friend Stephen Collins, revealed that this was the last film Reeve made, shortly before the horseback-riding accident in May 1995 that left him paralyzed. The star of *Superman* had journeyed to both the feeding grounds in the Bering Sea and the nursery area at San Ignacio Lagoon. Reeve had accompanied some Inuits of St. Lawrence Island on a hunting trip. In Baja his lagoon guide had been none other than Francisco (Pachico) Mayoral.

There was little doubt that it had been an awe-inspiring experience for Christopher Reeve. I had no idea if he would be interested in reflecting on it, or whether that period in his life might remain too painful to discuss. I knew that he and Bobby Kennedy, Jr., lived in the same general area of New York and were close friends. When I spoke to Bobby that fall, he said he was going over for dinner soon and would raise the subject with Reeve. Bobby called to tell me, "Chris says he'd be happy to talk with you."

Driving toward Reeve's house through the woodlands of Westchester County, I mulled over a number of questions. How did he see the subsistence hunting issue? Had he been as profoundly affected by "meeting" the gray whales as so many others had, including myself? If so, what did this mean to him today, in light of all he's gone through? As I turned down a bucolic road and up a hill into the circular driveway of a large Victorian home with a gray-shingled roof, I parked beside a small basketball court and wondered too how I would react to seeing him in his current condition. Nervously I rang the bell and was ushered by a staff aide into an outer room.

In a few minutes I walk down a long, narrow hallway to where Christopher Reeve waits in his study. His wheelchair faces a picture window, overlooking a lovely pond surrounded by stately old oaks as well as maple and birch trees. Although I know something of what to expect, having watched him in televised appearances and interviews since the accident, I am nonetheless taken aback. His feet, wearing casual running shoes, are strapped onto the wheelchair. His hands are similarly held down along black leather rests. His head is kept in place by an attached support resembling a boxer's headgear. A seat belt is wrapped around his midsection. He's hooked up to a metal ventilator box that's kept on a shelf attached behind the chair, with a translucent tube leading to the tracheal incision just below his Adam's apple. This is how he breathes, and is able to speak. A first sight of this athletic man, six foot four and so utterly immobilized, is wrenching no matter how much one might have mentally prepared to witness it.

Then Reeve smiles, a warm smile that immediately puts me more at ease. He asks that I take a seat across from him. As I briefly explain my book project, he seems to enjoy hearing about our parallel travels among the gray whales. "You've answered all my questions," he says. He's wearing a gray turtleneck sweater and white chino slacks. His penetrating blue eyes fall upon my tape recorder. "You might want to put that on my knee," he says quietly. "You'll probably hear me better that way."

A potted palm graces one side of the study, alongside a tray table with several copies of his best-selling book, *Still Me*, and another book that his wife, Dana, has recently written. On the walls are framed posters from many of his stage and screen credits: *In the Gloaming*, *Rear Window*, *The Aspern Papers*. I notice that any vestige of the *Superman* films is absent from his surroundings.

How had he ended up doing a film about gray whales? I ask. "What happened was, the producer of the program *In the Wild* asked me, what animal or part of the world would I like to go to and make a documentary about? My first thought was to do something on wild horses. Because I'm an avid horseman, you know. I thought it would have been fairly easy to do, since I just wanted to follow the herd of wild mustangs out in the West. But they rejected that. So my next choice was the gray whale, because their long migration really fascinated me. The fact that they would go so far to breed—at least a five-thousand-mile swim from one habitat to another, with all the dangers they face along the way, like being attacked by orcas off of Monterey. Besides, Baja and Alaska sounded like a great trip. Quite simply, that's how it came about."

His voice is relatively high-pitched, and it's not easy for Reeve to speak. He must pause midsentence to suck a quick breath through the ventilator. Still, this is not distracting; rather, it's a kind of rhythm to which he's long since adjusted.

He'd arrived on St. Lawrence Island in July 1994, flying a small plane from Nome himself. "I've been a pilot for twenty years," Reeve continues. "It's a hundred-mile-long island, and they've only settled one little corner of it. The interior is totally undeveloped, which fascinated me, to see such a beautiful spot that had never been used for crops. There's not a single tree on the island. So nearly all the houses are built out of driftwood off the beach. Some material gets flown in. They do have a few buildings with concrete foundations, and tin roofs. Really their whole culture revolves around salmon. And walrus. And the gray whale."

It wasn't one of those quick in-and-out shoots. Reeve had lived in a cabin on St. Lawrence for two weeks. "Our problem was, for the first week we literally did not sight a whale. And we knew we had to stay there until we saw at least one." He passed the time reading, doing some fishing with the Inuits. He said, "I particularly liked helping a little bit to build a whaling boat. The old-timers there still know how to do it. One walrus hide sort of split in half can make a boat. The women do all the stitching. I assisted in stretching the skin onto the hull."

Accompanied by two native hunters, Reeve finally saw a gray whale from the skin boat he'd helped to make. He was told the islanders were allowed a quota of two grays per year at that time. He found their quest legitimate, because, he said, "I think whatever is essential for survival is okay. It's the senseless killing of animals for sport that I completely object to." On this day, however, there was no attempt at taking a whale. "On St. Lawrence, we didn't really get very close."

The following February, at San Ignacio Lagoon, getting close would no longer pose a problem. For Reeve the two-day trip by boat from San Diego and the week he spent camped alongside the lagoon is a time he has never stopped cherishing.

"We made contact quite quickly with the whales," he remembers. "With just a bit of patience, we waited and they came over to investigate. They would use the bottom of our dinghy to scratch barnacles off their backs, and one time we got lifted completely out of the water. That was the only time I was a little worried. But they're so benign, such gentle giants.

"We befriended a mother and a calf. The process would start with when they approached the dinghy, and we'd spend some time petting. I was surprised by the texture of the skin. It's very soft, like chamois. And then making eye contact. The eye is so big, and you sense an intelligence in it, that these are creatures who have knowledge and understanding that we're only beginning to appreciate.

"These same two whales kept coming back to visit us. What surprised me most of anything was the specific relationship. I'd thought it would just be random contact. I was amazed, I thought the mother would be more protective of her calf. But the pair of them seemed to take to us, until we got to the point where I could go in and swim with them—

which we had to get special permission from the Mexican government to do. I started by hanging over the side and sticking my head underwater with a snorkel and touching them. This gradually led to going in with full scuba gear. I'd sit on the edge of the dinghy and slip into the water. So that there was no big splash, no commotion, just gently gliding in and joining them.

"My biggest worry then was the little guy. Because he didn't swim all that well, and I was afraid of getting hit by his fluke or by the tail. The visibility in the lagoon is very poor. I mean, you probably only see about twenty feet. It's very shallow in some parts too, only between fifteen and thirty feet deep. I remember I'd be down and constantly turning around and looking in all directions, knowing that at any minute this huge creature could come cruising by from any direction."

But did he feel the whales were aware of where he was? "Absolutely. I'm convinced they made a conscious effort not to hurt me. I think they absolutely knew that they could easily crush me, that I was a very fragile little creature, and respected me as such. That's what I mean by how benign they are. This was to me the most powerful part of the experience, realizing their incredible sensitivity. One time I was under and between the mother and the calf, thinking—I could be a sandwich here. But they gave me room. We swam right along together, and they would simply keep pace. It was wonderful."

Reeve pauses for a moment, as if basking in the memory, then continues: "I felt it was like gradually getting acquainted with another person. First there's social contact. Then you strike up a conversation. Then you find you have a common interest, and then you start doing things together."

I ask if he's ever had any similar experience with another wild creature. "No, not at all. That by far exceeds any experience I've had, with animals in any context. You know, as someone who was an experienced rider, I'm used to bonding with a horse. Maybe that helped me in bonding with these whales in the lagoon. The fact that I was very relaxed may have helped."

He adds: "But even Pachico said this was a unique circumstance, something special with a mother and a calf."

I ask Reeve to elaborate on his relationship with Pachico. "He was a very calm presence. Somebody who's totally at home on the water, and who knew where the whales were. When we went to his home to have lunch with his family, my first reaction walking up the beach was, I felt sorry that somebody had to live in such cramped quarters. But once we were inside, and I was trying to make tortillas with his wife—and doing a terrible job of it—it was such a happy household. They had so little, so little space and obviously not much money. Yet they seemed more content than people in more civilized society. It just reminds you that happiness is something you can find in any setting; it doesn't depend on material things. I really appreciated seeing up close an old man who'd obviously had such a fulfilling life—and still was."

I respond that it feels, in many ways, as if this trip to the lagoon had a profound effect upon him as a human being. "It did. The only other experience I've had like this was when I went to Kenya in 1993 to scout locations for a movie I was going to direct. We went way up north, to the remotest areas, where the Samburu people live. The only other foreigners they'd come in contact with were Italian missionaries and occasional doctors. I remember coming to these villages, and they thought I was a doctor. Women would bring their babies up to me, trying to tell me what was wrong with them and asking for my help. This was a contact as unusual as the contact with the gray whales but actually more intimidating, because here were fellow human beings with whom I could not communicate and who were very much in need. There was a sense of desperation that made me feel uncomfortable. My heart went out to them because all I could say was, 'I'm not a doctor, there's nothing I can do.' And how was I going to tell them I was looking for a suitable film location? But they were very hospitable and invited us to join them for a meal.

"So this was the other major alien encounter in my life," Reeve says, and pauses for breath. There's a glass of grape juice on a tray table with a straw, and for a while no one has entered the room to offer it to him. I ask if he's thirsty. He is, and I walk over to hold it while he drinks.

The silence in the room is palpable as Reeve continues: "I'm very grateful, particularly now that I'm sort of sidelined, that I have these experiences to look back on. The fact that I was injured at age forty-two was a blessing really. I've flown my own plane twice across the Atlantic. I've flown the space shuttle simulator. I've swum with the gray whales. I've interacted with the remote Samburu people. I've been a film star. A political activist. And fathered three children. I can look back and say, Well I never was a couch potato. I always used to say that a harsh punishment would be compulsory sunbathing. There is no way I could lie on a towel on the beach. I would have to be windsurfing or bodysurfing or diving. But lying there and cooking like a strip of bacon—no! Couldn't do it. Never."

I ask about his relationship with Bobby Kennedy, Jr. "We're good friends." Had they ever talked about the gray whales? "Yes we have, actually. In fact, Bobby was over here one time and we were discussing the salt factory Mitsubishi and Mexico had proposed to build at San Ignacio Lagoon. My son Will, who is seven, was listening in. All of a sudden he walked over to the cupboard and got out a jar of salt. He said, 'You should take this to those people and tell them we don't need any more salt! We already have enough!' Isn't that wonderful?"

We laugh, but Reeve's mood changes as he expands upon his views concerning the saltworks expansion plans. "We should do *everything possible* to stop it!" he says emphatically. I envision him pounding his fist on the chair—if he could. "Because it's not like the gray whales have an alternative where to go. This lagoon has been the place for millions of years. It's totally instinctive. You can't put up a sign saying 'Closed for Saltworks,' and direct the whales to move further down the coast. There are no more lagoons, and it's an absolutely unique environment and breeding ground. We don't have the right to mess with that. Not for any reason at all."

When he describes his time among the gray whales as "very much" a spiritual one, I ask if he might go so far as to say that it gave him strength when the accident happened a few months later. "I list it along with some other experiences I've had," Reeve says. He recalls flying solo across the Atlantic at 27,000 feet—"hardly Lindbergh or anything, because I had the radar and loran and high-frequency radio, all the necessary equipment. Nevertheless it's an awe-inspiring, humbling, and spiritual experience to be just a little dot up in the clouds above this vast ocean."

He speaks, too, of paying visits to terminally ill children for the Make-A-Wish and Starlight Children's Foundations. "When I first started doing that, I was very apprehensive walking into those homes. I thought it would be a very sad and painful experience, that I wouldn't know what to say. But it always ended up the opposite. Here was a ten-year-old kid who was going to die in a couple of weeks, with such joy in his eyes because Superman had come to visit him. I played a character who is an archetype, really a piece of mythology. I happened to be the custodian of that part for a ten-year period. I could be wearing blue jeans and a T-shirt, it didn't matter because the children knew I'd played that character. And the character is more important than the actor. It was such a powerful, healing force to talk with them about who the character is, what he represents. I always emphasized the fact that while he was powerful, the most important attribute was that he was a friend."

There is another long pause before Reeve continues: "I would say that all these sorts of things—which have to do with overcoming fear—have helped me in my situation."

Yet there must have been times of anger and bitterness, of wondering why this situation happened to him.

"Sure, you think, What did I do to deserve this? I compare it to the aftermath of a hurricane. The next day you come back to your house and it's in pieces. So you mourn. It's necessary to experience that. Your beloved furniture and precious memorabilia were blown away. But once you've taken the time to acknowledge that those things are gone, the challenge is to use all your resources to move forward. Like the hurricane victim, how long are you going to stand and cry about it? Your job is to find a purpose afterwards."

So Christopher Reeve had indeed moved on and discovered a new purpose. He had become the leading spokesman for people with spinal cord injuries, working tirelessly to raise money for research and to focus attention on the needs of the disabled. In 1999 he had formed the Christopher Reeve Paralysis Foundation to help fund the search for effective treatments and a cure.

"Of course, when you make that decision to move forward," he continues, "it's in private, through conscious willpower but with the help of family and friends. It's not something you can face alone."

You must realize, more than ever before, how much those relationships mean, I say.

"Right, exactly. Those are the benefits that come out of tragedy. A tragedy in a neighborhood brings everybody together, where previously they may not have had much contact with one another. What would be nice is if we all could have that contact, that closeness and caring, without a disaster to make it happen. People ask what's my wish for the millennium. I say, do we move in that direction? Do we realize that we're 6 billion people living in a village? We have the technology to instantly communicate with each other, so let's build on that. Let's understand that we're really one big extended family. It's an era where we have the means to achieve incredible things—but we have to be more selfless than before."

Dana walks in, accompanied by a pale Labrador retriever named Chamois. Named, I think to myself, after the term Christopher Reeve used to describe the skin of a gray whale. We've been talking for almost two hours, and he's clearly tired. It's time to take my leave.

"So this is where I hang out," he says as I stand beside him and we look out at the pond together. "I wrote my book here, dictated it into a transcriber. I would just let it go without worrying about structure. Then the next day I'd see it on paper, review the pages, and spend about two hours rewriting. We were doing about seven pages a day. We worked nonstop from May to October, then turned it in."

I tell him I'll send him a copy of Homero Aridjis's new poem about the gray whales. "That would be nice," he says.

Driving back down the hill on a misty autumn day, I feel still inside, thinking about a "gentle giant" named Christopher Reeve and drawn back to my first morning on San Ignacio Lagoon.

Chapter 24

Scientific Puzzles in San Diego

Humanists are the shamans of the intellectual tribe, wise men who interpret knowledge and transmit the folklore, rituals, and sacred texts. Scientists are the scouts and hunters. No one rewards a scientist for what he knows. Nobel Prizes and other trophies are bestowed for the new facts and theories he brings home to the tribe.

—Edward O. Wilson

Late in January 2000, I flew to San Diego, where the gray whales make their final southbound turn into Mexican waters from the United States. The grays are believed to have utilized San Diego Bay and probably the adjacent Mission Bay as calving and breeding areas in the era before whaling and then heavy ship traffic. "In an earlier day," according to the late whale expert Raymond Gilmore, "the ladies of San Diego dared not be rowed across the bay to their favorite picnic site on North Island, because of the prevalence of gray whales spouting in the harbor."

The late Carl Hubbs, head of the Scripps Institution of Oceanography, once went so far as to say that the grays were largely responsible for the city's founding, since in its early years whaling was the principal local industry. During the early 1870s weekend outings brought substantial numbers of people to the rendering station on Ballast Point, where they would picnic at the Old Spanish Lighthouse and watch six men in a dory fight a thrashing gray whale near the kelp beds.

In a prescient look at what San Diego would become, Scammon wrote in the *Overland Monthly* in March 1872:

> A great outlet . . . for the animal products of Alta California was subsequently found at the Bay of San Diego. There the Jesuits es-

> tablished a Mission in 1769. At this point, or in this latitude, it may be said, is the beginning of the prolific lands—that reach northward to the forest and mineral regions—whose resources are being rapidly developed as the tide of immigration flows in, revolutionizing the whole system of enterprise and industry. The valleys now produce abundant crops, and the boundless pastures are covered with bleating flocks, that yield their snowy fleeces at semi-annual gatherings. Instead of the tedious carriage over mountain-roads, or along the slimy sea-shore, to the *pueblos*, every indentation of headland in the least protected has been transformed into an *embarcadero*; and at many exposed points, where the heavy ocean-swell forbids the use of surf-boats, wharves have been run out, or slips are suspended from the high, shelving cliffs beyond the beach-waves, so that the produce of forest and field finds easy transport to the great commercial market, San Francisco.

Eventually, San Diego would also become a center for studies of the marine environment. First there was Scripps, from which in the winter of 1946–47 Carl Hubbs and his wife initiated an annual census as the gray whales migrated past San Diego toward the Baja lagoons. Up to this point the grays were thought to be all but extinct. Soon their passage was becoming an event. On December 31, 1956, the *San Diego Union* trumpeted in a front-page story: 10,000 CROWD POINT LOMA TO WATCH WHALES. 42 GRAYS AS SEASON GOT UNDERWAY.

This marked the origin of the whale-watching phenomenon. In 1955, according to Gilmore, "a wild man named Chuck Chamberland hung out a shingle on the municipal sport fishing pier offering to take people whale watching for one dollar." When Chamberland quit, Gilmore himself—already the first biologist to perform detailed studies on the gray whale's migratory patterns—took over the tour boat operation. His first year, 1959, 330 passengers came out with him. Other skippers soon joined the parade, and within three years the number went past 4,000. Gilmore wrote in 1961: "As the gray whale continues to increase, more and more people see it from shore or from boats, and it works itself more and more completely into the recreational and aesthetic values of the great urban community that is Southern California—in a manner that no other whale has done elsewhere in the world." At seventy-seven, Gilmore was still leading whale-watch cruises when, on New Year's Eve of 1984, he suffered a heart attack on the boat ramp and never revived.

The scientist Raymond Gilmore with a gray whale skull (Steven Swartz)

Today the industry Gilmore pioneered has mushroomed into a billion-dollar-a-year enterprise in at least eighty-seven countries. The most recent study showed that 4.3 million Americans went whale watching in 1998, almost half of the worldwide total. Along the California coast that year, a minimum of 1,012,000 people paid at least fifteen dollars each to do land-based gray whale watching. Additionally, more than sixty-five operators use 140-plus boats for whale-watch tours in twenty different communities in California.

The world's first marine theme park, SeaWorld, is based in San Diego. It was here that Gigi (in 1970) and then J.J. (in 1997) became the only whales ever maintained in captivity, juvenile grays observed by thousands in their tanks, and from whom scientists learned much before their release off Point Loma to rejoin the migration.

I've arrived at Point Loma before sunrise. Located ten miles west of San Diego's city center, this sector today hosts a top-secret military base. I pick up my security pass at the Space and Naval Warfare Systems Center, SPAWAR for short. Then I follow down California Highway 209 a van whose passengers know the code to the locked gate. "We're going to a point that's nicknamed Stiff Cliff, because it's right next to the [Rosecrans National] Cemetery," Lisa Schwartz informs me upon arrival. She is into her third season working on a master's thesis for San Diego State University about gray whale migration patterns. "It's a bluff with an extremely good view, 125 meters up. So we can see out miles and miles, and get some really accurate theodolite data from that elevation. It looks like today is going to be very clear."

One goal is to see whether the whales change their ways in the presence of certain boats or greater numbers of craft. One weekend Lisa counted sixty vessels, particularly sailboats, in a half hour off Point Loma. They've also observed as many as twenty whale-watching boats around a single gray. The whales seem most disturbed, tending to break their breathing rhythms, when approached quickly by boats.

Another of this project's aims is to determine whether the whales are affected by a pipeline that extends four miles into their migration corridor, through which all the treated sewage from San Diego passes into the bay. "We've tended to lose track of whales just as they were crossing over the pipeline," Lisa says. "We didn't know if they did a deep dive, went offshore or inshore. Acoustical analysis says the pipeline doesn't make any noise, but this is something that needs to be thought about at least, whether it may need to be designed differently."

Lisa's team must all know the observation codes. For example, F means first surface blow. FS is a surfacing with no blow, B is simply a blow, U for fluke-up dive, D for fluke-down dive. There are also BC for bubble cloud, BR for breach, HR for head rise, US for unidentified large splash, and UB for unidentified behavior. The longest they've been able to track a particular whale, Lisa says, was two hours; generally, they hope to follow one for thirty minutes.

Another vehicle pulls up. Dr. Jim Sumich emerges. He's one of the committee members for Lisa's thesis and comes out to Point Loma on a regular basis. He's been studying gray whales for thirty years, his own doctoral dissertation possessing the rather elaborate title "Latitudinal

Distribution, Calf Growth and Metabolism, and Reproductive Energetics of Gray Whales, *Eschrichtius robustus*." Today a marine biology professor at Grossmont College, Sumich is right up there among the leading experts on this marine mammal. He's funded his ongoing research with royalties from a marine biology textbook, the first of its kind, published in 1976 and now in its seventh edition.

Lisa calls out: "Was that a no-blow rise? Did you guys see that? To the left of the kelp patch there."

Sumich says, "When they're snorkeling through, unless you're staring right at them, you'd never see them." Snorkeling is what's defined in the computer as a no-blow rise, meaning the whales are basically doing their exhalations underwater. "So when they come up, they only inhale," Lisa adds. "They look just like a wave and are extremely hard to see. This might be a type of evasive maneuver, maybe to get away from boats."

Many gray whales have altered their migration along Southern California in recent years for perhaps the same reason. During the late 1970s, on a ten-hour count, between 350 and 400 might be observed going by Point Loma. "If we get a tenth of that now," Sumich says, "it's a pretty good day—even though the population is quite a bit larger. In the last couple of decades, there's been a dramatic shift of animals offshore." Aerial surveys in the late 1980s had first detected about half the migrating grays going west of San Clemente Island about seventy miles out, another 40 percent traveling west of Catalina Island, and only the remaining 10 percent coming down through this mainland corridor.

Sumich thinks it likely that increased boat traffic closer to shore is the reason. "Sadly enough, the one single activity we can identify that's grown the most during the eighties is whale watching. As fewer and fewer animals are available, you get more and more of a concentration on the remaining ones. So a lot of times out here you'll see one little whale come through with two or three big commercial whale-watching boats and maybe eight or ten private vessels on it."

This shift to offshore movement has raised more mysteries about how the grays know where they're going. As we sit down along the hillside, Sumich explains it like this: "Before, the simplest explanation was that when you go south along the coast, you keep the sound of the surf in your left ear. If you can't hear it, you veer left. If it gets really loud, you veer right and just keep going. But now, when we see the animals making these long open ocean jumps from the northern Channel Islands to San Clemente over very deep water, or from San Clemente to the Coronados, they've got to be using other kinds of cues. And I don't have a clue what they are.

"What we do find out here is that San Clemente and Catalina are islands that are tilted in opposite directions. On Catalina, the west side is an almost vertical cliff face that drops off into really deep water. On San Clemente, the west side is a gently sloping, shallow-water, kelp-bed kind of environment. In spite of those differences, about 99.9 percent of the grays going by those two islands pass both on the west side. So this may be another little rule of coastal migrating. If you run into a point, you stay on the right side of it. Because if it's an island, it doesn't make any difference. But if it's a peninsula, it's a dead end."

The same phenomenon occurs farther south, where for example the gray whales could easily pass the deep channel of Baja's Cedros Island on the east but always do so on the west. In the 1950s Raymond Gilmore had been the first scientist to hypothesize that this might be traced to a "cultural memory." Before sea levels rose at the end of the Pleistocene period, Catalina and San Clemente weren't islands at all but connected to the mainland by land bridges. So in those days only to the west would migrating gray whales avoid getting bottlenecked.

It's almost mid-morning when Jim Sumich and I head off to find a coffee shop. I'm curious to know how he became so intrigued by the private lives of gray whales—especially since he'd never even seen one growing up in the hinterlands of Oregon. "Like a lot of people who ended up working on whales, especially in the seventies and eighties," Sumich explains, "we all started on cockroaches or fruit flies or, in my case, sea urchins." Those last had been the topic of his master's thesis at Oregon State, after which he'd landed a job teaching biology at Grossmont College and found himself "casting about for something a little more fun."

In 1970 Sumich got invited to a wine-and-cheese reception at the San Diego Natural History Museum for a discussion of the first organized tours to Scammon's Lagoon to look at gray whales. It turned out to be, he says, "one of the most interesting evenings of my life." He recalls the museum's education director beckoning everyone present to "get aboard *now*," because within five or ten years these animals would probably be extinct. Jim raised his hand. If that was so, how could they be promoting seventeen trips without any idea how the whales would respond to boats?

Afterward an older gentleman wearing khakis and a battered Panama hat approached him. "He started pounding my chest with his finger saying something like, 'Listen, you young whippersnapper, until you know something about gray whales, you need to keep your mouth *shut.*' This was Raymond Gilmore. I thought about what he said for a few seconds and realized I agreed with him. So I decided to get more informed about gray whales. I came up with more ridiculous and seemingly inappropriate interpretations. And then there was Gigi. The first gray whale I ever saw—and she was in a tank."

Gigi was the large calf captured in Scammon's Lagoon and hoisted onto a vessel that brought her to SeaWorld early in 1971. She was studied in captivity for a year, then released back into the wild. Sumich was fascinated by her. He came twice a week the entire time she was there. And, he thought, with Gigi so readily accessible, there were some basic questions about gray whale biology that marine scientists ought to be asking.

At this same time Dale Rice and Allen Wolman published their landmark *The Life History and Ecology of the Gray Whale.* The book was based upon having examined 317 gray whales landed under a research permit at a California whaling station since 1959. Reading it, Sumich noticed some glaring inconsistencies and omissions. The scientists had found that northbound whales were significantly thinner, relative to length, than southbound ones when only their girth was measured—despite the blubber thickness remaining about the same. The obvious question—how does blubber thickness correspond to fat content?—was simple enough to test for but hadn't been. Nor had any tissue samples been put away for future genetic analysis.

In eleven years of northbound sampling, Rice and Wolman had never observed any mothers with calves. Since their boats were looking a hundred miles offshore, the presumption was that the pairs must be even farther from the coastline. To Sumich that seemed like a lot to ask of a newborn. Since the sampling stopped each year around March 25, wasn't it an equally valid notion that the mother-calf pairs were making the journey *later,* when nobody along the coast was paying systematic attention? This basic tenet of gray whale behavior would be confirmed in 1980, when Michael Poole, the first researcher based at Piedras Blancas Point lighthouse, near the Hearst Castle in San Simeon, verified the nearshore route taken by females with calves *two months after* the other grays.

What Sumich found most difficult to accept had to do with the grays' feeding and then fasting for all those months. "Pregnant females are the ones to cease feeding and leave first, and then return last to resume feeding. Well, they don't get grotesquely fat like chipmunks do, and they're not like hibernating mammals, which may lose 50 percent of their body weight. During this whole period of weight loss, they're active—swimming, giving birth, nursing. My basic question was, How do they do this? And if they do it, what's the minimum size an adult animal can be and still pull it off?"

To answer this meant first to pin down what kinds of energy demands newborn calves make on their mothers. "They're feeding on Mom, who's not feeding. As far as the energy unit is concerned, it's almost one word: mom-calf. But we didn't have a very good idea of how fast calves grow, how rapidly they add material, how much energy they burn." Sumich's start toward his Ph.D. happened to coincide with the initial reports of "friendly" gray whale behavior at the San Ignacio Lagoon. "Very naïvely I thought, aha! I can go down there and do some measurements and tests like they should have done with Gigi. These would be kind of like noncaptive captive animals!"

In 1978, when he and a partner drove down the Baja Peninsula from San Diego, they found getting to San Ignacio Lagoon a formidable task. "It took two and a half days to go from the town to the lagoon, the road was so bad," Jim recalls. "Everything was washed out, in what was probably the wettest winter on record there. I still have the map we drew on the way. 'Right turn by the hubcap leaning against the rock,' that sort of thing."

They'd arrived in January and pitched their tents where the Kuyima camp is now. Sumich started venturing out in an eleven-foot inflatable with a five-horsepower engine. Gray whale mothers were out there with their new calves all right—but they wanted absolutely nothing to do with him. As soon as he got close, they'd head in the other direction. "We tried everything. It was terribly frustrating. We were there about a month, and absolutely nothing. We were literally breaking camp and packing up in total defeat—when it started raining. It rained four and a half inches that night, and we were stuck at the lagoon. It was going to be another ten days before we could get out."

Sumich was learning about what Mary Lou Jones and Steven Swartz came to call "lagoon time"—no matter how much you planned, the elements dictated your days. Jones and Swartz were then in their second season studying the grays, ensconced down the way at Rocky Point. Sumich already knew Swartz, having taken his classes to visit Swartz's education department at SeaWorld. When SeaWorld wouldn't grant him a leave of absence, Swartz had quit and teamed with Jones to set up camp alongside the lagoon. Now the couple came to Sumich's rescue. "We were pretty much out of supplies, until they came through with bread and tortillas, peanut butter, and beer and coffee. This was also at a time when the whales were moving down the lagoon, so we moved our little sampling down to Rocky Point."

And then it happened. "We learned the first thing people now know—that friendly calf activity happens *late* in the winter season, rather than early, when their moms are more protective. In those last ten days we knocked 'em dead! And we got some very interesting results that made me understand I could return the next year and really put it together."

His efforts to unravel the riddles of a calf's metabolic rate required considerable cooperation. First a stopwatch would be used to measure the length of time one held its breath. Then the calf would need to come within arm's reach, so that a small meteorological balloon held over the blowhole might capture the next breath. Over a five-year period in the lagoon, Sumich got between fifty-five and sixty exhaled air samples from calves. He could then get a handle on how much oxygen is extracted out of each breath, and how the volume of oxygen consumption changes as a calf grows. Which all relates to how much energy it's getting from its mother.

The first friendly calf Sumich worked with, he remembers, had "a mom with a big scar right below her dorsal hump on both sides. It looked like possibly even from a harpoon; it was eighteen inches in diameter and really deep and filled with cyamids. So you could recognize her from a half mile away. She brought her calf in, and it was all over us. Saved my Ph.D., some of the best data we've ever had from *any* calf. Mom was really shy, always stayed under our boat just out of fingertip range.

"The next year, she wasn't there. The year after that, she came back with another calf. This time the calf was friendly and *she* was friendly. The next year, she was by herself and friendly. Then the last year I had contact with her, she was back with another calf. Every other year, right on schedule."

Except that time, everything changed. "We'd just shut down for lunch, too far from shore to go in, so I turned the motor off. About twenty seconds later we heard this loud noise"—Sumich whacks the table—"and there's this big whale a quarter mile away with a calf, steaming toward us and putting up a bow wave like this! I'll bet she was going at least ten knots, and she came up about fifty meters in front of us. She did exactly the same type of chuffing blow I've heard with captive dolphins, when their training or work situation isn't going the way they want. It's more like a frustration blow. She did this, then dove. Next thing we knew she came up vertically under us, had our whole boat five feet out of the water. Knocked us all to the floorboard, and we all got hurt, though just minor bruises. Three times she did this. I'm lying there reaching up, trying to get the outboard started so we can get the hell out of there. She followed us all the way back, until we finally escaped in over a shallow mudflat where she couldn't go."

Sumich shakes his head and continues, "Now I don't know what that means. I don't know if she followed us home to kill us, or like a great big puppy. It's the same whale. And maybe the same behavior. It may *all* be the way a thirty-ton animal plays. It was extraordinary, and it stayed with me. I get less and less comfortable trying to attribute some particular thing to this friendly behavior."

A couple of years ago Sumich had a beautifully different, but equally baffling, encounter. "It was March, so the calf was six weeks old, out there with Mom about fifty yards away. We were stopped, the motor on idle, swinging around back and forth and hoping they would approach. All of a sudden, Mom swam under her calf and picked it up. Literally the calf was draped over her rostrum, almost completely out of the water. Mom swam the calf right to the edge of our boat and just held it there like, 'I've got a pretty baby, don't I?' We were all standing up petting the pair of them. I'm trying to remember what happened after that, and I don't.

"Sometimes you look in that eye and you can see all sorts of things going on," the marine scientist continues, remembering a time on the lagoon with a group of high school biology students. The wind was blowing, roughing up the sea. They were all watching a calf cavorting. Sumich was alone in the bow when its mother surfaced with a spy-hop right in front of him. "I reached out my hand. The boat was bobbing and bouncing, and she and I were just dancing. I don't know how she moved that body to stay with the boat. She never let me touch her, but that animal was completely in tune with my fingertips, and I don't know what else."

Who could explain the dance, or the touch of a whale? Over a ten-day period this group of students had petted numerous mothers and calves. Jim figures one of the young women must still hold the San Ignacio Lagoon record for having kissed sixteen whales in a single day! But as the last voyage arrived one girl had never seemed to be in the right place at the right time. "She was in tears, convinced that there must be something wrong with her. We'd been out all day, and I was about two-thirds of the way back, when this whale surfaced real close. I slowed down. It came over to the boat and put its nose right in front of that kid. She reached out and touched it, then planted a kiss. Everybody else went crazy."

After a brief silence Sumich adds: "She said she would never forget that, and I think she meant it. We still get a couple cards from her every year, and she always mentions it. She's heading off to medical school now."

How then does he account for such phenomena? "Everybody has their own take on this, and I guess mine is rather different. I think it's a learned behavior, passed down from mothers to calves. And it's as natural as teaching tourists that this is an okay thing to do. Friendly encounters are not gray whale behaviors, they are gray whale–human interactions. We all know it started in the mid-1970s, when finally somebody was naïve enough or brave enough to just sit there and let it happen when this huge animal is steaming toward you. It must have been so damned hard for Pachico, because all the other local fishermen at first thought we were absolutely loco to let the animals approach us like that. When they went to check their lobster pots, they'd always come down the mudflats and skirt the channels. They wanted *nothing* to do with these whales.

"And now look at them! In twenty years there's been more of a change of behavior in fishermen and the way they relate to whales than the other way around. That's probably true of most of us. And you have to wonder what would have happened, before the whales made the connection between boats and harpoons, if one of Scammon's skiffs had just gone out there, shipped their oars, and put their hands out."

Jim Sumich has puzzled long and hard about why gray whales travel such a distance to give birth in the Baja lagoons. "They don't need to go there for the water temperature or energy conservation," he notes, "because normal healthy calves are well insulated to handle water that's ten degrees Celsius colder than they get in San Ignacio." While whales who have come to mate turn around and head back north relatively quickly, the mothers and calves delay into early April. Newborns breathe three to four times as often as their mothers. By the time they leave the lagoons, however, Sumich has observed, "their diving, breath hold, and breathing patterns are very similar to adults'."

So they grow up fast. And Sumich thinks it may have to do not so much with the long migration ahead as with "buying a little predator-free time in Mexico, to get ready for the contest they'll face once they get past San Diego or Point Concepción." Specifically, he's talking about the grays' primary nemesis of recent years: killer whales.

Sumich has always found it remarkable that killer whales have been reported outside the lagoons—but never inside. "If a gray whale can do it, a killer whale that can flip itself up on the beach and get a seal can sure as hell go into San Ignacio if it wants to. So why don't they?" Maybe, I suggest, there are some fundamental rules in the whale world. He mulls this over momentarily, then says: "It may be a very simple thing—that most of the lagoon is so shallow, an orca's echolocation probably doesn't work very well. They may be just too vulnerable. There are a lot of big gray whales in there. If killer whales went in kinda blind and feeling their way around, they could be in trouble. Regardless of how they look, they're pretty fragile animals. One fluke slap across the face from a big whale could put them out of action permanently."

So for gray whales, "coming to the lagoons may be an evolved behavior to avoid predation at the most sensitive time in their life cycle, during mating and with very young calves."

Even though most female gray whales are nine or ten when they have their first calves, they don't reach full physical maturity for another thirty years. During that period their body weight and length about doubles. We don't know with any certainty their life span, or when they might reach a post-calf-bearing stage. There are no growth records in a baleen whale, except in the ovaries of females. The Rice-Wolman studies in the 1960s noted ovarian scar tissue forming where the egg erupted during pregnancy, and remaining perpetually. The two females found to have the most such tissue represented twenty-seven and twenty-five pregnancies. If they had their first calves at ten, and reproduced on average every other year, those whales would then have been in their mid-sixties. At the time they were dissected and studied, both were pregnant. Since we are no longer killing whales to obtain such data, that's still as much as we know. The same types of observations aren't possible with whales that wash ashore dead because of rapid decomposition.

Jim Sumich's painstaking work on the energy demands of gray whale calves has led him to conclude that the large adult females could fast without much trouble through their long round-trip journey. But this is not necessarily so for the ones bearing their first calves. "They don't have the same fat storage capacity, so they're probably going to run out of gas trying to do a six-month loop," Jim says. "I'd bet most of these probably need to do at least supplemental feeding on the way back north."

He supposes, too, that many of the younger grays don't bother to make the migration at all. In the late 1970s Sumich first proved to doubting colleagues that some gray whales do spend their summers off the Oregon coast. He'd taken a group of oceanographers less than a mile offshore, where they witnessed four grays feeding over the next ten miles. "My guess is that once they're weaned—and most of them will get weaned in the north, at around seven months old—for the next couple of years or maybe three to four, a lot of juveniles are wanderers. We find them in Baja's Gulf of California in the summertime, and off Oregon and Washington and Alaska. We know they're solitary foragers, and are simply everywhere within the whole geographic range of gray whales—at *any time* of year. In fact, this may be an important part of how they learn the migration. They go out and fiddle around through it, and never go into really deep water. If they do that, they're going to be fine and they'll probably find food. Since they're smaller and not reproductive, and don't need the dense food resources the adults do, they can do this all year long."

Sumich is well aware, though, that he can do no more than make educated guesses—about this and many other matters surrounding gray whales. On the one hand, they're more accessible in the wild than any other species, and both of the only whales ever maintained in captivity have been young grays. He was able to test out certain of his lagoon hypotheses on J.J., the baby gray rescued off the beach and sequestered at SeaWorld in 1997–98. "J.J. was extremely cooperative in swimming around right beside us," rather than in midpool. That first day Sumich got more exhaled air samples from J.J. than all of his lagoon calf measurements combined. What he learned about J.J.'s oxygen intake and lung volume measurements bore out his assumptions from the wild calves. He also found that the breathing pattern at birth is basically the same as when the whales are a year old, which Sumich believes is more a product of their blood chemistry than anything else.

On the other hand, as Sumich points out, "we're plagued by turbid water almost everywhere in the total life cycle of the gray whales"—unlike humpbacks, which can often be clearly seen even several hundred feet underwater, such as in Hawaii. Take the acoustics work, for example, that one of Sumich's graduate students is currently doing at San Ignacio. They both believe that gray whale calves most likely possess a vocal repertoire at birth, although it's not known where anatomically their sounds come from. Indeed, a whale may go hours or even sometimes days "without apparently saying anything." And most of the time the whales aren't at the surface and so can't be seen, even if they're only ten or twenty feet away.

Sheyna Wisdom *has* detected what she believes are two more previously unknown gray whale sounds. One of these, recorded twenty-six times through an underwater hydrophone, immediately preceded a calf's breaching on sixteen of those occasions. "Several times Sheyna heard it and said, 'Okay, there'll be a breach in five seconds'—and boom!" Sumich recounts. "Sort of like the calf was saying, 'Watch this!' But we still don't know if the sound is Mom telling the calf to breach, or the calf announcing this to Mom, or has nothing to do with either." In the other instance a certain sound finds calves that are swimming away from their mothers immediately turning around and coming back. Again, though, Sumich says it's unknown whether the sound emanates from the calf or might be "Mom saying, 'Get your ass over here!' Because there are a lot of whales around, and we never know for sure how far away the sound is coming from."

Then there's mating behavior, of which Sumich says, "Again, none of us truly knows anything—because, invariably, the only thing we ever see is when it's *not* happening! I mean, face it, a male who's waving his Pink Floyd around in the air so we can see it from a boat and get great photographs isn't getting the job done!" His first couple of years in the lagoon, Sumich started counting whales in what he surmised were courting groups, then realized that there was no way he could be sure of the numbers.

"Just like the assumption we make all the time that, if you see a calf and there's a big adult in its company, that's its mother. We've never tested that—it could be Dad, or an uncle, who knows? I guess I'm just getting more and more comfortable with my ignorance as time goes by. But I've been doing this for a long time—and it doesn't get old."

I'm riding in Jim Sumich's car to Torrey Pines Beach. We're going to look for gray whales—in fact, the remains of a pair of dead ones that washed up within a couple of hundred yards of each other last winter. Behind us in a pickup are two of his marine biology students. This afternoon's goal is to pinpoint where the high tides from recent storms have deposited what's left of the whales. Eventually, some of the bones will be reassembled into a skeleton for display in the new science lab building at Grossmont College.

At the base of a long hill, the inshore waters are lined with surfers catching sizable waves on a hazy January day. We turn right and drive for about a mile down the shore, where Sumich points out: "You're going to start seeing a lot of 'wildlife,' because at this point it becomes an informally designated, clothing-optional beach." Sure enough, a few shorebirds and a plethora of naked people soon dominate the seascape. A mile farther we park near a couple of hundred-foot-high cliffs sprinkled with ice plant vegetation. "We should be just about on top of our whales," Sumich says.

Even though it's been a year since the whales died, the smell remains strong. Soon we've stumbled upon a flipper buried in the sand. A buzzard, one of the students notes, "is gnawing on some of our friend." Sumich sets up a theodolite to get a precise measurement of distance from the beach, rather than stepping it off, and expresses the need for a reference point. "If we painted the rock, would that help?" the other student asks.

Later they'll need to gather and soak the bones for a couple of days before steam-cleaning them. For today it's enough to know that the whales are still in the same basic vicinity. Isn't it unusual, I ask, to have two whales die in the same place? "Absolutely," says Sumich. "I don't know what it is about the currents that happened to bring these two in right here."

These were not small whales. One had been twenty-six feet long, the other thirty-one feet. And strandings in this size range were not common in the past. "Since the 1930s," Sumich explains, "the history of stranded dead grays has been that almost all the mortality is of young whales. First of all, postnatal calves a week or less old. If they survive their first week, most seem to make it through their first year. Then there's another big mortality peak in yearlings toward the end of that initial winter. Probably a lot of it is food-related, after they're weaned. By the age of fourteen or fifteen months, most calf age classes will have experienced about a 60 percent mortality. That's not atypical for big, long-lived mammals. But from the data we gathered a decade ago, it looked like if you could survive to your second birthday, you had it made. This situation is now changing."

Jim Sumich wasn't yet ready to sound an alarm about the 274 dead gray whales counted in 1999, the majority full-grown adults. But data being gathered by a colleague at the National Marine Fisheries Service (NMFS) is another indicator that something may be very much awry.

Wayne Perryman works out of NMFS's Southwest Fisheries Science Center in the San Diego suburb of La Jolla, one floor below his boss, Bob Brownell, and two floors above the office of Dave Weller. The Scripps pier is right down the hill, and gray whales are often seen in the bay below. Among other duties Wayne Perryman runs what's called the photogrammetry program.

We're standing in a lab down the hall from his office, where Perryman is excitedly showing me the cutting-edge cameras he uses for his aerial photography. They've all come from the Army or the Navy. "When the military builds a new airplane," explains Perryman, a onetime naval officer, "the last thing they design is the pod that the camera goes in. So when they go to a new fleet, say from F-4s to F-14s, they have to make new cameras because the old ones won't fit anymore. What I've learned to do is be standing there with my hand out. Because the military doesn't want these cameras thrown away, I mean they cost a hundred thousand dollars each. So they've been great about giving us photographic equipment we could never afford—tens of millions of dollars' worth."

Perryman holds up a camera with an extremely long lens, whose three parts—lens cone, shutter deck, and motor drive system—he's disassembled and put back together in seeking the best possible resolution. "What's unique about this is what's called forward image motion compensation. That means if you tell the camera how high you are and how fast you're going, it'll move the film the same speed your image at the ground is traveling. So every time you take a picture, it's as if you're standing still. The military needed that so their reconnaissance planes wouldn't get shot out of the sky. Also they could scream along at tremendous speeds and still collect high-resolution images. Well, our speeds are certainly less, but these cameras allow us to fly at about a hundred knots and six hundred feet and get those same type images of migrating whales."

He films from a small twin-engine Italian Partenavia, an observer aircraft with a big Plexiglas nose. "Tomorrow I hope to go out and photograph maybe sixty or seventy southbound-moving gray whales and be able to tell something about their condition, just based on measurements from photographs. No one's ever done this before. People have looked at sizes of whales, of course. It's not a big deal to take an aerial picture of a whale, and if you know your altitude you can measure how long it is. But here we're trying to take the next step, and not only ask how large the animals are but say something about their comparative shape. Gray whales naturally change shape at the different times of their lives, with a near-term pregnant female looking very different from another whale."

Perryman brings out a three-paneled photograph. One frame depicts a pregnant gray, another shows a female that has given birth to a new calf, and the third shows a mother with a three-month-old. Each female expresses a vastly varying shape. Then he displays more photos: a big, boxy mother moving south, a normal-sized one going north. Again, the differences are not subtle.

"Gray whales are interesting because, unlike humpbacks, the females fast during the time they're lactating," he continues. "Building a fetus is not that big a deal. But feeding that new calf—while it grows from four

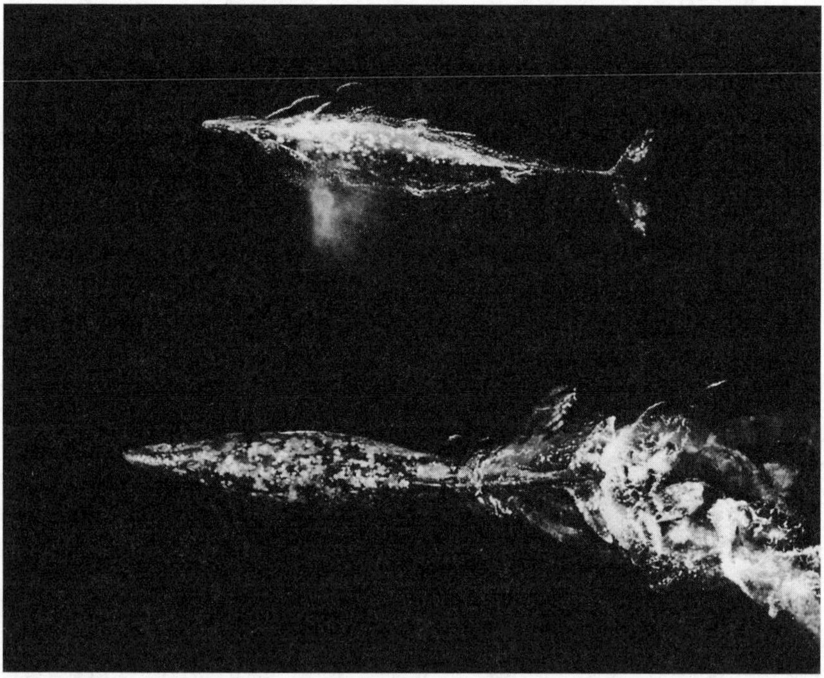

Pregnant and nonpregnant gray whales heading south (Wayne Perryman)

and a half meters up to maybe eight meters before they get to the Bering Sea in May or June—that's physiologically a tremendous demand. Unless the mother has stored up enough fat the previous feeding season to support that calf, it's not gonna work. So my primary interest is looking at the condition of females, because if anything's going to happen to this population, that's probably where we'll see the significance."

Perryman has already begun to notice a distinct change. In comparing the relationship between width and length for southbound grays photographed in 1997 and 1998 with that for those photographed in 1999, he's found that the latter are considerably thinner.

Additionally, for six consecutive years Perryman has conducted surveys of gray whale calves moving north with their mothers. Aerial and shore-based observations have shown that over 95 percent of these will pass within 1,300 feet of Piedras Blancas Point, near San Simeon. In 1997 the direct calf count achieved a high of 501. In 1999 it fell to a low of 141. That, Perryman says, represented a 60 percent drop from the previous year.

In a short paper titled "Eastern Pacific Gray Whale Calf Production in 1999 Is the Lowest in Six Years," Perryman and two colleagues wrote:

> After correcting annual counts for animals passing during "off-effort" observation periods, for animals passing far offshore, and for calves not spotted by primary observers, we produced estimates of 1,000, 601, 1,141, 1,439, 1,316 and 400 calves for 1994 through 1999, respectively. Calf production indices calculated from these estimates (calf estimate/population size) are 4.5 percent (1994), 2.6 percent (1995), 5.1 percent (1996), 6.5 percent (1997), 5.1 percent (1998), and 1.6 percent (1999). We suggest that this year's calf production index, the lowest of the six years, reflects diminished reproductive success corresponding to a reduction in the condition of the population. Condition in this context refers to the nutritional state or relative "fatness" of the animals, which may be critical to an individual animal's fitness for migration, pregnancy, and/or lactation.

What does Perryman suspect might be causing this? For one thing, he describes what's being called a major regime shift in the Bering Sea in the last couple of decades. Scientists believe there has been a drop in the flow of carbon that enables benthic organisms to flourish. This is a probable reason cited by Dave Rugh in Seattle for why the southbound migration has been averaging about a week later than observed before 1980—implying that the grays need to stay longer in Arctic waters to get enough to eat. In addition to this longer-term trend, Perryman notes that both of the years that saw low calf counts (1995 and 1999) were preceded by a shortened feeding season for gray whales. After especially cold Arctic winters, the Bering ice sheets receded very slowly. When the grays had arrived in the spring of 1998, ice had still covered their feeding grounds.

"My educated guess," Perryman says, "is that calf production is related to the length of the feeding season. If a pregnant female can't get enough to eat, she may abort spontaneously or reabsorb that fetus. That wouldn't be unheard of for mammals. Whether it's human beings or gray whales, if a female is in poor condition, I suspect the probability of carrying a fetus to term is diminished." If, however, the whales were losing their calves in the lagoons or failing to lactate enough to support them, "we'd be finding scads of gray whale calves on the beaches. And we haven't. That's why I believe they are losing them up north, early in the pregnancy rather than late."

Perryman differs with those scientists who attribute all this to gray whales' reaching their so-called carrying capacity, or population limit based on available food supply. He says there's simply no real data about how many gray whales were able to support themselves in the years before whaling twice jeopardized their survival. "All we have is current information, which shows that their reproduction is fluctuating more than we would have guessed."

As the whale biologist talks, I think back to portions of the Bering Sea being ice-covered when I was there in the summer of '99 because of La Niña conditions, which would have marked a second consecutive year of diminished feeding time for gray whales. "Absolutely," Perryman responds, when I bring this up. "My prediction is, this will be another low calf year in 2000. And I suspect we will have more strandings again as well."

He suggests I join him at Piedras Blancas Point in the spring, when the next calf count is being conducted. "I'll be there," I reply.

And how were their cousins, the western grays, faring off Sakhalin Island? My scientist host at Piltun Lagoon, Dave Weller, is also putting me up on my visit to San Diego. From his tales of what had transpired at Piltun since my visit there in July, it had been an auspicious period.

First, in late September 1999, during a severe storm, there had been an oil spill. The underground pipeline flowing from the Molikpaq to the

Floating Storage Unit had separated at a coupling—and dumped about 1.5 tons of oil into the Sea of Okhotsk. "But we didn't see any physical evidence on the beaches—no oiled animals, birds, or anything," Dave says. "We knew something was up only because we could see they'd moved the tankers to the other side of the Molikpaq, and were flying a helicopter consistently around the rig itself. Since they couldn't get the oil to the storage unit anymore, they shut down and stayed that way up to the time we left, according to our records."

The scientific team conducted their last photo-ID survey on October 13 and gave themselves a weeklong window of time to depart by helicopter in order to make their connecting flight back to the United States. "Sure enough, come the fourteenth, we packed up all our gear and equipment and were ready to leave—when we got socked in by high winds, rain, and miniblizzard snowstorms. It was too cold and wet even to go outside the cabin. We had nothing to do because we'd packed all the computers and data."

The days passed. "We played cards. People slept a lot. Read the same magazine over and over. Endless hours of guitar playing. We were running out of patience—and provisions. After seven days, the *last* day we could still make our flight connection, we finally hit a beautiful sunny day and the helicopter was able to land. About ten o'clock in the morning, we loaded everything in and off we went."

When they landed in Moscow, all their 1999 gray whale DNA samples were confiscated by customs. Eventually these were returned to a friend's custody, but the samples had yet to reach Dave. Undaunted, he was working on a paper about a population estimate based on three years of photo-ID work. Having started with 69 whales going into the season, they'd ended up with 88 in the catalog. That, of course, is far fewer than the 250 "officially" designated, and, Dave says, "I don't think it's going to grow significantly." The oil companies were still questioning whether Piltun's might be a subgroup of a larger gray whale population somewhere else, but Dave doubts it. Based upon his team's work, for the first time the western gray whale will be listed not as simply endangered but as critically endangered in the Red Book of the International Union for the Conservation of Nature.

And it appeared that funding would be forthcoming for another four-month season at Piltun, where new components are to include tracking the whales by radio (for local movements) and by satellite (hoping to find out where the winter breeding area is). "That is, if we're lucky," Dave cautions. "We could put ten satellite tags on the whales and track them to the end of Sakhalin Island and lose the signal. But if the tags do stay attached for several months and transmit proper signals, we might be able to follow them down to Korea, the South China Sea, or wherever it is they go."

As for the moment, houseguests raiding the refrigerator have been somewhat chagrined to discover, right between the V8 juice and the applesauce, a rather murky jar of six-month-old western gray whale fecal matter. "It's valuable stuff, and we haven't decided yet where to have an invertebrate biologist take a look at it," Dave Weller explains.

About two-thirds of the way back to Los Angeles from San Diego, on a cliff some 125 feet above the Palos Verdes Peninsula, the American Cetacean Society's Gray Whale Census and Behavior Project has been ongoing since 1979. Come here to the Point Vicente coastline on any day from December 1 to at least May 15, sunrise to sunset, rain or shine, and you're sure to find at least four people scanning the horizon. Their official binoculars are Fujinon 7 × 50 with built-in reticle and compass. Additionally, they use spotting scopes to confirm and detail sightings. During the 1998–99 migration, seventy-three volunteers contributed 9,280.5 effort hours. One of these volunteers stayed here above Long Point for 166 days, recording not only surfacings and breaches but any other species in the vicinity as well as hourly weather data over Pyramid Rock and Whale Rock.

I approach Joyce Daniels and ask why she spends as many as seven hundred hours in a season here. "First of all, could you stand over there so I can watch?" she strongly suggests. Then, eyes trained through the binoculars, she says simply, "I love whales." She's been to the Baja lagoons on a number of occasions. She's gone to see the humpbacks in Hawaii. This year she'll be heading for the Dominican Republic to snorkel with humpbacks.

It's a human phenomenon every bit as amazing as friendly grays approaching small boats. In fact, less than ten years ago the whale counters of this peninsula had their own resident "friendly"—a young whale they named Johnny Paycheck, who hung around from Veterans Day into April doing shallow surfacings like a Baja gray and going from boat to boat.

I've stopped here because the data gathered by this citizen's effort are considered vital by the scientific community. The official gray whale abundance estimate of, in 1999, more than 26,600 is based upon Dave

Rugh's scientific analysis of observer data from Granite Canyon, near Carmel, California, on the southbound migration. However, Rugh considers the Piedras Blancas Point station equally important in obtaining the northbound count on calves, along with this one at Point Vicente because here both the north- and southbound migrations are chronicled.

Alisa Schulman-Janiger, a marine science teacher at nearby San Pedro High School, is the society's project director at Point Vicente. "We've found that whales close to shore don't fluke as much as the ones farther out," she tells me. "Offshore you see them raise their tails longer, make more surfacings, and tending to travel in bigger groups." They count more whales moving north than south, because they're nearer the coast then. The mothers, Alisa says, seem to favor nooks and crannies of this cove as "good places to nurse the kids."

They've seen it all from Long Point: gray whales entangled in fishing nets and hit by boats. Gray whales, especially recently, coming up with mud-smeared faces from feeding. Gray whales in the company of dolphins, staying outside the occasional red tides that drift in. Gray whales changing their migration path once oil rigs arrived. "But if a rig isn't active anymore, we still see the whales using it apparently as a reference point for navigation."

In her report on the 1999 work, Alisa had written: "This season our calf count dropped dramatically: only 15 southbound newborn gray whales were tallied (2.2 percent of southbound migrants) from 28 December to 2 February, with no more than 2 calves per day. In contrast, last season we counted a record 106 southbound calves (8.6 percent of southbound migrants), up to 10 calves per day."

This one has been a long, gray January day without much activity. I'm about to depart when Joyce exclaims: "Blow! Two blows! Just on the other side of Pyramid!" The other three women at cliffside are all eyes.

Alisa aims the spotting scope. Turning to me, she says, "Very seldom do I get a whale that blows more than once in ten seconds. If I see five blows in six seconds, I'd feel very confident we've got five different whales."

Six blows in ten seconds. This is quite a pod. I find it impossible to leave. Forty-five minutes have passed by the time I next glance at my watch.

Finally, as dusk closes in and the whales have journeyed beyond range of our sight in the direction of Baja California, Alisa Schulman-Janiger looks at me once again and says, "Now you've seen what keeps us going."

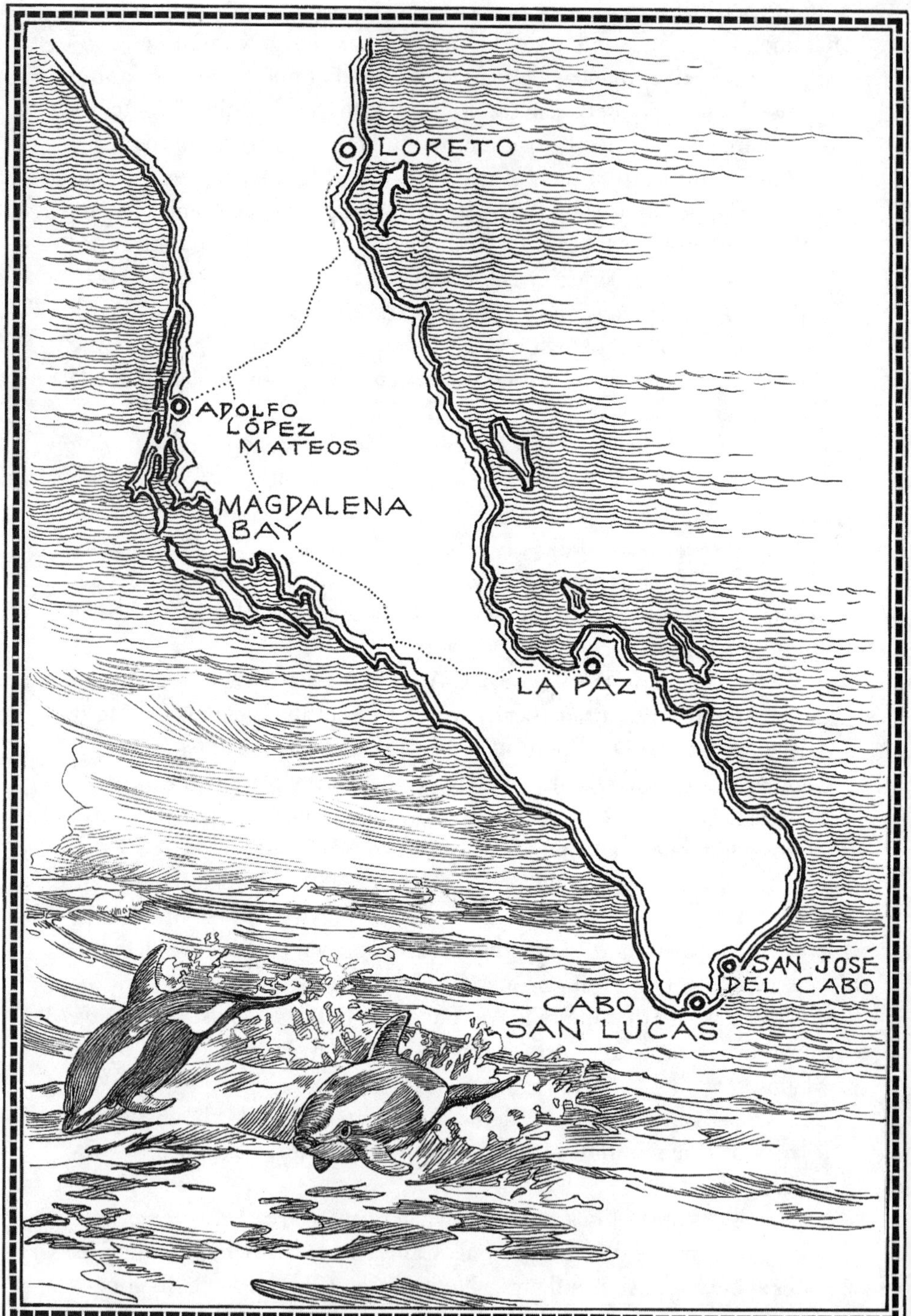
LORETO
ADOLFO LÓPEZ MATEOS
MAGDALENA BAY
LA PAZ
SAN JOSÉ DEL CABO
CABO SAN LUCAS

Chapter 25

MYSTERIOUS EVOLUTIONS

> *Our ship, with its great iron heart beating on through calm and storm, is a burly, noble spectacle. But think of the hearts of these whales beating warm against the sea, through darkness and night. On and on for centuries.*
>
> —JOHN MUIR, letter home from Alaska, 1879

BETWEEN his continuing sojourns in Alaska during the late 1860s, William Healey Dall settled temporarily in San Francisco. While writing up his Alaskan findings and cataloging collections from his explorations in the Yukon, Dall agreed to serve as a volunteer curator of zoology for the California Academy of Sciences. Founded in 1853, the academy was the only such scientific society on the West Coast of North America. Its mission, as outlined by the five doctors, school superintendent, and realtor who established it, was "to bring together persons with collecting urge and the curiosity to know." They hoped to show that San Francisco could be every bit as scientifically sophisticated as cities with similar societies in New York, Philadelphia, and Chicago—the last having Jonathan Young Scammon as its board chair. By 1867 the California Academy of Sciences' gatherings were being patterned after those of the Royal Society of London and Philadelphia's Academy of Natural Sciences. Some two dozen resident members assembled biweekly; the elected corresponding and honorary members included Charles Darwin and Thomas Henry Huxley.

The first record Scammon kept of attending one of these meetings is dated September 2, 1867. There was a presentation of various specimens, including the jawbone of an extinct mammal. Scammon listened to lectures about California's redwoods, volcanoes, and using trigonometry to determine the height of Mount Hood and Mount Shasta. He heard a lively discussion about earthquake tremors. Afterward Dall introduced Scammon to some other members. And the captain quickly became a regular attendee.

Scammon had saved considerable information from his whaling days—notes and sketches, distribution data, and measurements taken by both himself and other seamen. In January 1868 the *Proceedings* of the California academy reported that Scammon was preparing a book about whales. The academy also published a list of cetaceans, which was described as having been taken from Scammon's notes. By the middle of that year, Scammon had become a corresponding member of the academy and finished an article detailing his longtime observations of the gray whale. He showed it to Dall, who suggested he send it along to Spencer Baird at the Smithsonian.

At the same time interest in the gray whale was being evinced by Dr. Edward Drinker Cope. He was a precocious paleontologist-zoologist a few years older than Dall, at Haverford College and the Academy of Natural Sciences of Philadelphia. In 1866, in some limestone beds of southern New Jersey, Cope had discovered the second significant American dinosaur skeleton, named by him *Laelaps aquilunguis*, or "eagle-clawed terrible leaper." That June, Cope had also written Dall, asking about some notes on cetaceans that Dall had sent the Smithsonian, "especially descriptions made from two specimens of the 'Gray Whale.' " These were apparently the ones Dall had gathered in Monterey, just before the Western Union Telegraph Expedition. Cope wrote that he was working up a monograph on cetaceans at the Smithsonian's behest, "to which the Gray Whale of California also pertains. It has not been properly defined hitherto, and thus notes are an important part in accomplishing this desirable object." He added that he would be "giving of course all credit and acknowledgment where due." Subsequently, he wrote Dall asking him to try to obtain cetacean bones and measurements from Alaska.

In the September–October 1868 *Proceedings* of the Philadelphia academy, Cope published "On Agaphelus, a Genus of Toothed Cetacea." This was a new genus he had named to include both the extinct Atlantic Ocean "scrag whale" (*Agaphelus gibbosus*) and the Pacific gray whale (*Rhachianectes*). Scammon was not mentioned in Cope's study. Then, early in 1869, Cope learned about the article Scammon had prepared about the gray whale and requested a copy from the Smithsonian, maintaining to Baird that he'd use it only for background information. Baird passed it along, in turn seeking a kind of "peer review" on its authenticity. Cope quickly wrote back that Scammon's study appeared to be an important contribution to zoology. He wanted to edit and publish it immediately, although he couldn't afford to print the "figures"—drawings by Scammon.

Apparently, Scammon had no idea about any of this. Cope first read the captain's manuscript aloud at a Philadelphia academy meeting late in April, then published it in the academy's *Proceedings* in September. Scammon's was preceded by a forty-page systematic synopsis about cetaceans written by Cope—largely based upon material gathered by Scammon concerning such characteristics as anatomy, bone structure, and fetal development of the gray whale. Scammon's article—printed without his knowledge—was ten pages long, titled "On the Cetaceans of the Western Coast of North America." It had been edited and interpreted extensively by Cope. Buried in Cope's own systematic synopsis was a page and a half of straight description of gray whales, largely in the same words that would appear later in Scammon's book. There were glaring mistakes in some of the published figures, which Cope would later claim were the printer's fault. He even misconstrued the captain's service with the U.S. Revenue Marine, making him part of simply the Marines.

Scammon was justifiably furious. He wrote an outraged letter to Baird, who responded on November 2 decrying Cope's "absurd blunder." Baird expressed indignation that Cope hadn't given proper credit or even correct reference to Scammon, saying he'd "have it out with him" when the fellow returned from a trip south. He promised to reclaim Scammon's drawings. Then Baird set out to assuage the captain's feelings, telling him this wouldn't affect "in the slightest" the "well merited admiration in which you will be held by all students of so important a branch of zoology as that of the cetaceans." There would be, said Baird, ample opportunity "hereafter of putting you in your proper light before the scientific community." Baird suggested that Scammon not rush to publish anything more but wait and gather additional observations. The Smithsonian wished for more specimens of gray whale skulls, cervical vertebrae, and scapula, "if you can do no more in the way of securing the [full] skeleton," for future comparisons.

Baird continued: "The whole subject of the Natural History of the Pacific Cetaceans is especially your own, and you are the right person to continue it," with full support of the Smithsonian. He enthused over Scammon's drawings, suggesting that his eventual volume be among the report of the Geological Survey of California, "where it can be superbly illustrated." Publishing in California rather than three thousand miles away, Baird indicated, would allow Scammon more control. The Smithsonian's secretary concluded with a request for Scammon's first legitimately published article, "Fur Seals," in the November 1869 issue of San Francisco's *Overland Monthly.*

After this Cope did not treat Scammon so cavalierly. During the 1860s the captain had collected three "Blackfish"—known today as pilot whales—off the coast of Baja California. In designating these *Globiocephalus scammonii,* Cope wrote late in 1869: "The present species is named in honor of Capt. Scammon, who has furnished us with a mass of information on the subject of the Marine Mammalia, and an amount of novelty in connection with it seldom equalled in the history of zoology."

Scammon wrote back to Baird "with many thanks for your constant kindness," noting that there existed "unusual opportunities just now for getting up and down the coast as far south as 'Scammon's Lagoon,' and probably if all could be arranged, I could bring a good deal to pass."

Indeed he would, and had. In an era when zoologists were yet concentrating on discovery and classification of species, Scammon was years ahead in focusing on animal behavior. The gray whale had first been *mentioned* as such scientifically only in 1861. Scammon's article in 1869, however much Cope tried to hijack credit, was the first detailed scientific description of the gray whale. And it would be recognized a generation later by Frederick W. True, a Smithsonian zoologist, as "the basis of our knowledge of the cetaceans of the west coast of North America. . . . The proportion of original matter is seldom equalled in zoological writings." Scammon was speaking an entirely new language, one deduced from observations of animals in the wild.

By contrast, Cope had never seen a live whale. He was a student of dry bones and teeth. In his 1869 overview the Philadelphian retracted his theory published the previous year that gray whales represented a separate genus from Atlantic scrag whales. Having examined additional material of the Pacific gray whale, Cope stated, he now recognized its distinctness from the Atlantic specimen he'd assigned to *Agaphelus gibbosus*; it should simply be of the genus *Rhachianectes.* In fact, Cope's first correlation was really the only right one he made. The Atlantic scrag whale (his *Agaphelus*) and the Pacific gray (his *Rhachianectes*) *were* related. Over the ensuing years Cope's continuing conjectures on this gray matter would prove both voluminous and muddled.

In 1872 Cope proposed the name *Eschrichtius davidsonii* for the first fossil cetacean species to be described from the American West Coast.

(This is now considered a rorqual of the family Balaenopteridae). The genus *Eschrichtius* was a decision that Cope had already liberally used for a diverse array of fossil mysticetes from the eastern United States. A century later, when Remington Kellogg—considered the father of modern whale paleontology—sought to sort out Cope's taxonomy of these fossil whales, he would write: "It is now certain that not one of the mandibular types of the fossil mysticetes referred to as *Eschrichtius* by Cope exhibits even a remote resemblance to the mandible of *Eschrichtius robustus.*"

Today, it's the gray whale alone, of both the Atlantic and the Pacific, which is known as the earliest available, correctly formed name *Eschrictius robustus*—not *Rhacianectes glaucus,* as Cope had ended up calling it.

Also in 1872 Scammon had published a short note describing a *living* rorqual—the North Pacific minke whale (*Balaenoptera acutorostrata*)—that is a modern relative of Cope's fossil *Eschrichtius davidsonii.* In what we might see as belated "poetic justice," a 1986 reevaluation by a San Diego paleontologist appeared in the *Journal of the Society for Marine Mammalogy.* "It is here proposed that this new name *Balaenoptera acutorostrata scammonii* new subspecies, [be] after Captain C. M. Scammon, the original describer," wrote Thomas A. Deméré.

The irony of all this is that, without Cope's early attempt to undermine Scammon, the captain might not otherwise have been as motivated to pursue what became his pioneering work, *Marine Mammals of the North-Western Coast of North America.*

Cope became best known in his later years for his discoveries of dinosaur bones in the American West—and for a legendary feud with a rival paleontologist from Yale named Othniel Marsh. Only three of the twenty-six dinosaur genera named by Cope are still classified as he named them. Eventually he would either resign or be forced out of the Philadelphia Academy of Natural Sciences. Financially ruined in his later years, Cope would sell his house and move in with his museum fossil collection. He spent his last days in 1897 on a cot, surrounded by piles of bones.

In his will Cope directed "that after my funeral my body shall be presented to the Anthropometric Society and that an autopsy shall be performed on it. My brain shall be preserved in their collection of brains, and my skeleton shall be prepared and preserved in their collection in a locked case or drawer . . . open to the inspection of students of anthropology."

Cope's skeleton eventually wound up at Philadelphia's Wistar Institute, a repository for anatomical specimens. In 1993 the dinosaur paleontologist Bob Bakker wrote a description of Cope as the type specimen for our species, *Homo sapiens.*

Edward Drinker Cope may have been an all-too-human prototype, but the gray whale seems to have had the last laugh. The gray's would-be classifiers have been confused ever since. Here's the way the vertebrate paleontologists Samuel A. McLeod and Lawrence G. Barnes have summarized the saga:

> Until 1937 the gray whale was known to science under the name *Rhachianectes glaucus.* In that year two zoologists discovered that some subfossil bones (bones that are younger than the Pleistocene Epoch or "Ice Age") from Europe, called *Eschrichtius robustus,* belonged to the same species as the living gray whale from the Pacific. Because the subfossil bones were [first] described in 1861 whereas the living gray whale was not described until 1868, the name [that was] applied to the subfossil bones has priority. Unfortunately, the zoologists called the species *Eschrichtius gibbosus,* believing the gray whale to be the same species as the "scrag whale" described by a New England whaler, Paul Dudley, in 1725. But the subfossil bones are preserved in a museum for scientists to reexamine, whereas there are no specimens of Dudley's "scrag whale," only a written account. Most zoologists, therefore, acknowledge that the "scrag whale" is not identifiable without a specimen, and this is why the gray whale, including subfossil remains from an extinct Atlantic population, is currently known scientifically as *Eschrichtius robustus.*

This, however, constitutes merely a taxonomist's conundrum or, as the case may be, a comedy of errors that dates back to Cope. Meanwhile, the origin of this particular species remains shrouded in a whole other series of riddles. Dr. Lawrence Barnes set about outlining this to me one afternoon at the Los Angeles County Museum of Natural History, where he is curator of vertebrate paleontology.

Larry Barnes is the keeper of the one—and only—true fossil gray whale skeleton. It was discovered in 1970 on the Palos Verdes Peninsula in San Pedro, not far from where the American Cetacean Society conducts its contemporary gray whale census. The West Oil refinery was grading a slope for a new road when its tractor ran over something unusual. It had clipped through the lumbar region of a fossilized skeleton, leaving some vertebrae sticking out of the sediment. The natural history museum was alerted, and a volunteer named Paul Kirkland set out to examine the situation.

"He started digging and, about fifteen or twenty feet into the embankment, there was the skull," recalls Barnes, who was himself a student at the University of California, Berkeley, at the time. "The skeleton was nearly complete. The lower jaws were absent, but the skull and midpart of the skeleton, all the ribs and the flippers, were preserved. During a Pleistocene interglacial period, when the polar ice caps retreated and much of the Los Angeles basin was inundated by the sea, the Palos Verdes Peninsula was an island. This skeleton had been deposited on the

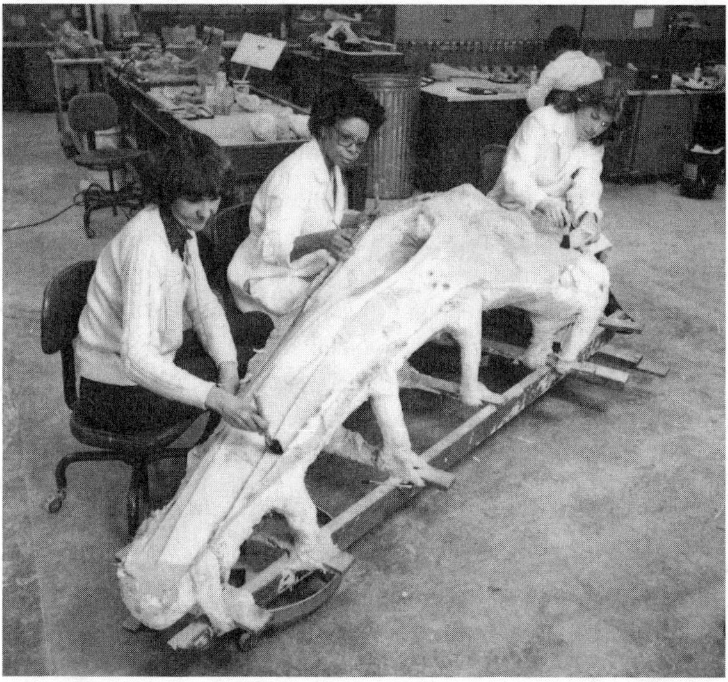

Workers at the Natural History Museum of Los Angeles County (left to right), Diana Weir, Pat Tyus, and Hilary Hicks, cleaning a gray whale fossil found in 1970 (courtesy Lawrence G. Barnes)

landward side of 'Palos Verdes Island,' in very shallow water, and was found belly-up. Most whales in fact die that way. They start to decompose, the gases are trapped in the throat and belly, and when they sink, they plummet and land upside down on the bottom of the sea."

The skeleton appeared to have settled onto the finely bedded, fine-grained sand relatively quickly. Numerous cuts on the bone of the whale's snout are evidence of either a shark attack or a postmortem scavenger. A single tooth of a great white shark lay on the carcass. There was also a fish skeleton on its shoulder blade, and many clamshells lay around the whale.

In 1971 numerous museum staff and volunteers gathered at the site. They encased the bones in plaster and loaded them onto a truck for transport to the museum. Preliminary study of the skull and other features found these to be very similar to today's gray whale. The strata in which it was found suggest that the specimen was between 50,000 and 120,000 years old. Compared with most other fossil whales, this one is quite recent. The only other such record of a gray whale in the North Pacific was some ear bones found on Alaska's North Slope and considered to be at least 50,000 years old.

Barnes says: "So literally we no have evidence of gray whale evolution before about 100,000 years ago, no fossil record beyond the middle of the Pleistocene—even though their shallow-water habitat should prove more favorable to their preservation as fossils. The anatomical differences between grays and other baleen whales are so significant that they must have had a separate lineage for several millions of years. But nothing discovered from back that far looks like their direct ancestors."

Barnes contemplates for a minute and then adds: "So they're an enigma. Where did they come from? Do you remember in *Star Trek IV*, how they went and got the humpback whale from Earth of the past and brought it into the future? Maybe that's how the gray whales got here." He grins, then spreads his hands wide and says, "Well, we'll find the fossil ancestor of the gray whale sometime. Just when we say we haven't got the ancestors, we always seem to find *something*."

When Barnes was in school he recalls paleontologists talking about "primitive" versus "advanced" characteristics. The current jargon for the latter is "derived." Barnes says, "So if I throw out the word 'derived' to you, that means a highly evolved organism. That's what gray whales are. Marine biologists who refer to them as 'primitive' are making a misinterpretation. Bone-wise, morphologically, they are *very* specialized whales, designed for unique living areas and feeding behaviors, anatomically quite bizarre. If anything, fin whales and blue whales and humpbacks are much more primitive."

Barnes elaborates on some of the grays' many "derived" characteristics: an arched snout, heavy and deep jaws, the lack of one phalanx. "They are 'primitive' inasmuch as they have big, round, flexible front flippers—but so do sperm whales. They're also 'primitive' in that they have a low number of baleen plates. The contradiction is the fact that those same plates are very thick, which is a highly evolved characteristic. So I'd say on a scale of one to ten, gray whales are at eight or nine as far as being highly changed from the ancient whales. You'll see some *really* primitive whales in our lab, and then you'll start to understand why I say these things."

I'd read that the terrestrial progenitors of whales are believed to have been bear-sized, long-snouted animals called mesonychids, followed by a half-land, half-water creature known as *Ambulocetus*, literally, a "walking whale" that invaded brackish estuaries nearly 50 million years ago. Genetic research by Japanese scientists and others has recently implicated hippopotamuses as the closest living relatives of whales. (Many other types of now-extinct animals that evolved from mesonychids are actually closer to whales, but among the few surviving derivatives of mesonychids, Barnes says, the hippos show the most similar genetic signals.) Jim Sumich, who's also an expert in evolutionary biology, had spoken of how similar-looking he'd observed hippo and gray whale calves to be. He also related "sitting on the edge of a lake in South Africa's Kruger National Park listening to the hippos blow—and it was so much like sitting in San Ignacio Lagoon listening to gray whales."

My lesson in the origin and evolution of whales is about to begin in Barnes's behind-closed-doors laboratory on the top floor of the natural history museum. He explains quickly that there are two living suborders of whales: Mysticeti (baleen-bearing whales like the gray and the bowhead, and their primitive toothed ancestors) and Odontoceti (echolocating toothed whales, dolphins, and porpoises), which together make up seventy-five to seventy-seven living species in thirteen or fourteen families. In one of his dozens of cabinet drawers, Barnes is looking for a member of a third suborder—the long-extinct Archaeoceti.

I'm looking down at a most unusual skull. "This goes back to the last part of the Eocene epoch," Barnes explains, his voice beginning to rise with excitement. "It was found in 1969 by a college student, in a gully in eroded former cotton fields of Mississippi. It's about 35 to 38 million years old." He picks it up gingerly by a long, slender intertemporal region in front of where the brain was, noting that "modern" whales possess no such handle. "I'm going to tilt it up, and you're going to see a mouthful of teeth. Notice the very serrated back molars. This isn't a very old animal, these teeth aren't fully erupted yet." He points out the palate, the internal nostril openings, the eye, and a groove I can lay my thumb into and feel where the optic nerve leading to the brain used to be.

"If you want to pet it, that's one of the oldest whales you'll ever get a chance to see," Barnes says. I give the head a quick couple of strokes, and Barnes returns the Archaeocete to its resting place.

He describes the highly rotatable, flexible neck as about a foot long and hippolike in structure. "When we did a reconstruction of this species in 1978, my colleagues branded me ridiculous for giving it a hind leg," Barnes says. "Well, now we know they had not only hind legs but kneecaps, an ankle joint, all the way down to toes."

So they could walk on land? I ask. "I think they could have gotten out on the sand and pushed around, almost as good as a crocodile. They didn't have the fixed rigid elbow a modern cetacean has, but a rotational elbow. There was apparently a macroevolutionary event at this point. Something happened within a very short period of time." Such as? "Lack of walking. Spending more time at sea. It didn't take millions of years—probably a few hundred thousand, which is just a blink of the eye in comparison to the totality of geologic time."

As we walk past the rows of cabinets and sit down at a long table, Barnes continues: "The Archaeocetes were around for about 15 million years. By virtue of the fact that all modern whales descended from them, they're still with us—just like the birds apparently represent a continuation of the dinosaurs."

Barnes brings out another cranial specimen. "The animal in front of you is *Aetiocetus cotylalveus.* It's about 24 million years old, slightly before the Oligocene-Miocene boundary, found on the coast near Newport, Oregon, first named in 1966. This is the Rosetta stone for the Mysticetes, the very first *toothed* baleen whale discovered and described, the type species of the genus, and the holotype for the species, meaning the single specimen on which the species name is based." It actually belongs to the Smithsonian, and Barnes had brought it here in his lap on an airplane. He notes certain similarities with the Archaeocetes—the long nasal bones, long space between the brain and the nose area, similar position of the eye. It can still be picked up by its handle, the elongate intertemporal area, and on its palate I can see the tooth sockets. "Three incisors, one canine, four premolars, and three molars—just like your garden-variety shrew, with different types of teeth from front to rear, just like your dog and cat," Barnes says.

But this surprisingly crocodilian face has a flat snout, like that of a baleen whale, rather than a tubular one. "The ear, the structure of the braincase, that's all baleen whale, too. There's nothing about this animal that says it couldn't someday become a gray whale—or a fin whale, a humpback, right whale, or bowhead. Any of those can come from *Aetiocetus*—and I think did." The next specimen Barnes brings down from a shelf is about 15 million years old—with a larger and more triangular braincase, what he calls a "surfboard snout," and no longer any evidence of teeth. Pure baleen whale.

I'm puzzling over why, evolutionarily, these types of whales didn't need their teeth anymore. "Efficiency in food catching," Barnes explains. "Moving from an ambush predator—which like an Archaeocete, or a crocodile, would have gone after individual large-sized fishes—to an animal which then can engulf large amounts of water and thus take in an incredible amount of small food items like oil-rich fish and krill. That allows you to have a bulk feeder, which is harvesting the sea and, like a blue whale, growing to be a hundred feet long. Most of the early whale ancestors were relatively small, about twelve to thirty feet long. The dolphins never got big because they weren't bulk feeders and had to spend a lot of time finding one or two fish. Whereas a rorqual simply says, 'There's a school, I'll take the whole thing!' "

Except for minke whales, which are only about twice the size of the typical ancestral whales, as a rule baleen whales grew larger through time and bigger than their toothed whale counterparts. In the next room, where models of the *T. rex* skulls were created for Steven Spielberg's *Jurassic Park*, Barnes moves on to display a cetothere from the Middle Miocene epoch. "Another icon," he says fondly, "among the most complete baleen whale skulls known. Laureano, can we look at those nice mandibles there?" Barnes's young Argentinian colleague Laureano Clavero has been reconstructing this relatively complete whale skull and jaws for a couple of months.

"The next part of the story is in the basement," Barnes announces. The three of us take an elevator down to the museum's main floor, where a sixty-five-foot-long fin whale skeleton—an animal harpooned off

Monterey in the 1920s—is hanging from the ceiling. Then we continue into the bowels of the cavernous building. Barnes puts a key into a storeroom door. Inside are dozens of shelves lined with crania of all sorts.

First Barnes wants to show me another "icon," the next step in whale evolution. "A strange cetothere, looking even more like a modern rorqual. This is a very famous whale; scientists and students travel thousands of miles to come here and see this." Then comes the coup de grâce, as Barnes lifts up a plastic tarp to unveil, encased in plaster, "the world's only fossil gray whale skull." He's moving a finger rapidly from spot to spot. "Skinny arched snout, *giant* nostril opening, those really short and wide nasal bones typical of gray whales. Weird skull. Bizarre."

He shakes his head. Laureano asks why the nose is curved down. "You always ask why," Barnes replies, "I always ask what. All cetotheres have a surfboardlike snout, all gray whales an arched snout. Why? Adaptation. Why? Gray whales turn sideways and feed on their sides along the bottom. They're not just skimming along the surface, maybe that's it. Yet the right whales which *do* feed at the surface in the horizontal position also have an arched snout. But for them, it allows really long baleen to collect a lot of food and water at one time."

Barnes gives the whale skull another long, admiring look. "It weighs three hundred to four hundred pounds, and takes about four people to move it. That's the way we like to find them, all in one piece. In fact, a good 80 percent of our specimens are not jigsaw puzzles. A lot of people think we put pieces together, but we don't like to if it's not necessary." When the gray whale fossil was brought in, various volunteers had to remove the sandy matrix from the skull. Although the sand grains weren't cemented onto the bone, and could be scraped or brushed away, the bones were extremely fragile and tended to crumble. Great caution was taken during the cleaning, which took more than two years. Barnes adds, "I've never had the nerve to flip this one over. It's too fragile, like chalk. So it still has its original sealed plaster jacket on."

Eventually we take a walk around the bone warehouse, one of several maintained by the L.A. County museum, whose collection is second only to the Smithsonian's. Barnes shows me a sea cow from Europe, part of a walrus from Orange County, more whale bones. What's the oldest piece? He goes right to a *Tyrannosaurus rex* skull between 65 and 70 million years old, and says I can go ahead and put my hand in its mouth if I'd like.

As we make the long trek back upstairs, I wonder if Barnes will hazard a guess on when gray whales—as we now know them—actually made their planetary appearance. "I personally feel theirs is an old family," he says. "The rorquals seem to be their closest morphologic relatives, and rorquals are present in the late Miocene time, about 12 million years ago. Logically, you'd expect there to have been gray whales back then, but we sure can't find them."

Barnes says fossilized baleen has been found—"I think approximately seven occurrences in the world, of which I have one in the lab. And none of these, even the old ones, is at *all* like the gray whale." He's no longer even so sure, having published a paper in 1984 equating the fossil I've just seen to a "modern" gray whale, that this is true. "I now think it may be some extinct related species. There are enough differences, we might revisit that issue."

In Barnes's view there's simply no way around it: gray whales fall into the mysticete category only because they filter their food with baleen plates. "In our parlance, we say they're monotypic, they are utterly unique—one species, one genus, one family. *Eschrichtius robustus.* They're our mystery whale."

This "mystery whale" was first known to science in 1861 on the basis of subfossil remains from Europe—even though Wilhelm Lilljeborg, the first to examine the bones found on the coast along Sweden's Gulf of Bothnia, mistakenly believed them to be from a previously unknown species of finner whale. *Balaenoptera robusta,* he labeled it. A Norse zoologist colleague, Daniel Eschricht, also examined the subfossil and took exception to Lilljeborg's conclusion. Clearly this didn't belong among the species of *Balaena* but should be in a different genus, maybe even in a different family. Eschricht then wrote to John Edward Gray, keeper of the Zoology Department at the British Museum of Natural History. Coincidentally, a whale's cervical vertebra had also washed ashore in England in 1861 and been sent to Gray. He immediately associated it with the species in Lilljeborg's hands but agreed with Eschricht that this was no rorqual. So Gray proposed *Eschrichtius* as a subgenus, named in honor of the gentleman who'd alerted him.

Scammon is said to have heard British whalers referring to the California gray as a scrag whale a few years earlier, sometime during the mid-1850s while at Magdalena Bay (although he does not mention this among its "most prominent" nicknames in his book). In a history of Nantucket written in 1835, Obediah Macy noted that in the colony's early years—before 1672—a whale known as "scragg" had sometimes entered the harbor and been pursued and killed by settlers. This was said to have marked the start of Nantucket whaling.

In 1725 Paul Dudley, a naturalist and later chief justice of Massachusetts, wrote an essay about the various whale species frequenting the New England shoreline, which was published in the *Philosophical Transactions of the Royal Society of London.* "The Scrag whale is near a-kin to the Finback," Dudley said, "but instead of a Fin on his Back, the Ridge of the After-part of his Back is scragged with a half Dozen Knobs or Knuckles; he is nearest the right Whale in Figure and for Quantity of Oil; his Bone is white, but won't split." This description could apply to only one whale: the gray. As we've seen, it was cited by Cope in 1868, in making his first (and only correct) correlation between the Atlantic and Pacific grays.

As far as whaling historians can tell, Scammon was the first American to refer to the Britishers' "scrag whale" as a gray whale. Whether he was aware of the Atlantic connection before Cope published in 1868, we don't know. One scholar believes that it wasn't the whale's color that made Scammon go for the gray designation, rather that the captain's own research had led him to the British Museum's John Gray. Hence, Gray's whale. If this be true, then Scammon was again way out in front of everybody else. "Historically," as Larry Barnes says, "it took a long time to realize that Scammon's gray whale—then called *Rhacianectes*—was in fact the same animal found in Sweden and called *Eschrichtius.*" Etymologically, we ended up with *Eschrichtius robustus* for all—the former word being initially a token of Gray's gratitude (before Cope applied it wrongly to numerous other species), and the latter harking back to Lilljeborg's misidentified *robusta.* Before the zoological dust ultimately settled on *Eschrichtius robustus,* things had gotten so mixed up that the Smithsonian's table on cetaceans from the late nineteenth century listed three distinct gray whale categories: *Rhacianectes glaucus* (occurring only in the Pacific), the scrag whale *Agaphelus gibbosus* (only in the western Atlantic), and *Balaenoptera robusta* subfossil *Eschrichtius* (on both sides of the Atlantic but not the Pacific). This type of confusion was not infrequent a hundred years ago, when travel between museums, and communication between scientists, was more costly and time-consuming than now.

Those early taxonomists, though, had no patent on confusion when it came to gray whales. Barnes tells of being in Washington, D.C., sometime in the late 1970s or early 1980s. "I was walking the Smithsonian racks where they had all the big whales stored, with James Mead, their curator of modern marine mammals. I looked underneath one rack and said, 'Oh, nice gray whale braincase.' Jim turned on his heel, came back, and said, 'That's not a gray whale, that's a minke whale.' I said, 'I know gray whales, we have them in our museum.' Jim said, 'Oh yeah?' So we got down on our hands and knees and started crawling underneath these racks. Before we were done, we'd found others, and Jim ultimately located more from other East Coast locales that turned out to be gray whales."

Mead went on in 1984 to coauthor with the cetologist Edward Mitchell a paper, titled "Atlantic Gray Whales." Besides the "scrag whale" account of 1725, he cited the "*sandlooegja*" of Iceland and the "*otta sotta*" of English-Basque origin as still earlier references to these gray whales. Mead and Mitchell also accounted for seven European and nine U.S. eastern seaboard subfossil specimens, starting with a New Jersey jawbone found around 1855 that Cope had thought was a minke whale. Amateur divers off Myrtle Beach, South Carolina, had made the next U.S. find in 1959, although a storm buried what they couldn't gather on the first try. Then had come the oldest Atlantic specimen, the partial skull of a juvenile gray whale from the Chesapeake Bay that was radiocarbon-dated as being more than ten thousand years old. The most recent specimen was traced to A.D. 1675.

In Europe the first gray whale bones had actually been excavated in 1829 from the Happy Union Tin Stream Works in Cornwall, England—but they hadn't been described until 1872. Today the oldest European samples date from about 8,300 years ago to A.D. 1610. Barnes was among the first to postulate that, just as there are in the Pacific, there might have been two stocks of Atlantic gray whales—one on the west and one on the east side. This would not be incompatible with the "cosmopolitan distribution" of most of the world's big whales, which maintain separate east and west populations in the large ocean basins.

His notion was borne out in March 2000, with publication of a doctoral dissertation by Ole Lindquist of the Universities of St. Andrews and Stirling in Scotland. Following up leads in various European archives, particularly in Iceland, Lindquist had amassed considerable new historical evidence about Atlantic gray whales, his sources dating as far back as a thousand years. He based his positive identifications on numerous descriptions of the gray's type of baleen, large tongue, "knobs" on its lower back, unique bottom-feeding habits, and propensity to rest in coastal shallows.

Lindquist also pinpointed who ate the grays' meat and rendered their blubber into oil. He traced the hunting of grays to the coastal inhabitants of four regions: around the southern North Sea and English Channel from prehistoric times to at least the high Middle Ages; in Iceland from about A.D. 900 to 1730; in New England by European settlers for almost a hundred years starting in the mid-seventeenth century, and possibly earlier by the Indians; and by the Basques during the same period.

From analysis of his data Lindquist concluded that an Eastern North Atlantic population of grays had wintered and calved in the Mediterranean waters off northwest Africa, then taken a migratory route to Iceland off Galicia and western Ireland; a portion of this population had visited the English Channel and southern North Sea during the summer, with some straying into the Baltic. A Western North Atlantic population had wintered and calved along the coasts of Florida and South Carolina, also migrating to Iceland, with some perhaps stopping off in the Bay of Fundy. The two stocks had converged and mixed in Icelandic waters.

Lindquist came up with records indicating that the Atlantic gray whale could be every bit as ferocious as its Pacific relatives. A fellow interviewed in the Old English *Aelfric's Colloquy* (1055) said he much preferred going after fish to hunting grays (*hran*), "which with one blow can sink or kill not only me but in addition my companions." From Denmark, 1657: "It is difficult to kill and dies slowly. . . . If one comes upon it in the sand, one cannot get near it because it throws up the surrounding sand and moves vigorously in an extraordinary way. But once the force of the waves has driven it into the shallows [or narrow parts of bays] and it has been run through in several places by spears, it lies dead."

Lindquist concluded that Icelandic peasant fishermen caught both adults and juveniles with spears and lances, possibly after entangling them in tail ropes. It is intriguing that no mention of the grays appeared in a compendium of whale lore in Norway during the mid-thirteenth century. Known as the *Royal Mirror*—an educational work aimed at young Norwegian princes—this was an otherwise exhaustive look at the North Atlantic's cetaceans, including detailed descriptions of the bowhead, white whale, narwhal, and walrus. Also recorded here were imaginary beings, such as "evil whales." Based on some later legends depicting the gray as a whale that could never be approached from the front because of its "magical eyesight," Lindquist hypothesized that its omission from the *Royal Mirror* relates back to strict taboos concerning the gray when Iceland was first settled, between A.D. 870 and 930.

Lindquist confined himself primarily to European accounts. But early American sources contain reports about unidentified nearshore whales that may well have been grays. In 1620 the *Mayflower* pilgrims saw "large whales" that "came daily alongside and played about the ship." In 1635 Richard Mather described "multitudes of great whales . . . spewing up water in the air like the smoke of a chimney . . . making the sea about them white and hoary." In 1705 a missionary reported whales off New Brunswick and Nova Scotia "in such abundance" and "so close to the land they could be harpooned from the rocks." Delaware Bay and Long Island Sound contained large numbers of "scrags." By the early 1700s seven small whaling factories had sprung up along Long Island's Southampton and East Hampton beaches. They produced more than six thousand barrels of presumably gray whale oil in 1707; given that about thirty-six barrels could be obtained from an average-sized gray, this meant that over two hundred were taken that year in one locale alone.

In the lore of New England Native Americans, Maushop was a giant being who often fed his people by catching whales and depositing them onshore. According to legend, Maushop created Nantucket. When the natives gave up their lands to the English colonists, they retained the rights to whales that drifted ashore. There is no evidence that Indians went to sea pursuing whales before the settlers came. But it is recorded that, in 1726, at least thirty vessels from Nantucket went whaling along the eastern seaboard. Their twenty-foot boats were the same length as the Native American canoes, with each manned by an English captain and five Indians. The probability is that their quarry included some of the last of the Western Atlantic gray whales.

On both sides of the ocean, only a generation into the eighteenth century, the Atlantic grays appear to have been the first whales rendered extinct by human beings.

"The question is," Larry Barnes was saying, "should we transplant gray whales back to the Atlantic? Should we do *Star Trek IV*?" He laughed, then added: "I think we ought to. And it's doable. I've sometimes thought about a big barge with an open bottom. Just take some grays from Mexico down through the Panama Canal, then back up the other side. It would take a few million dollars. But from a paleontologic-biologic standpoint, I'd like to see balance restored to the world, as far as animals are concerned."

Barnes wasn't the first to raise this possibility with me. When I had met at San Ignacio Lagoon the year before with Roger Payne of the Ocean Alliance, Payne said he'd been thinking about the same thing for years. He envisioned catching a small group of grays in the lagoons, airlifting them across to the Atlantic side, and emplacing satellite tags to keep track of where they went. Payne's idea, too, had to do with restoring balance. "They could have a profound effect on the productivity of the North Atlantic, which has been lost," Payne said. "Somebody calculated that the amount of mud put into suspension every year by the existing population of gray whales in the North Pacific is twice what the Mississippi River suspends. Well, think about this in terms of ocean topography, and things like shellfish beds, in the North Atlantic. It's certainly worth finding out what gray whales might do for this situation."

In his thesis Ole Lindquist speculated that the North Atlantic and North Pacific grays might once have mingled. But climatically that would have been at least nine hundred thousand years ago—when a warm interglacial period found the polar ice cap melted enough for a connection through the Arctic Ocean. Others, however, suggest that the only way to have periods of continental glaciation is for the Arctic to be ice-free, as a source of moisture for everything piling up on land. This could mean that a link between the Atlantic and Pacific gray whales is more recent than generally supposed. Barnes wonders, too, whether the Pacific grays might have been pushed a couple of thousand miles south during the Last Glacial Maximum, when all the shallow seafloor areas they utilize in the Bering, Chukchi, and Okhotsk Seas were either completely icebound or drained of water. If so, might they then have bonded through the Panamian Seaway? That would have been only about eighteen thousand years ago.

When ice melted back, sea levels rose, and the Bering Land Bridge disappeared at the close of the last Ice Age, the Pacific grays probably received renewed access to those amphipod-rich feeding grounds. Most likely, marine scientists such as Jim Sumich believe, their population had been much smaller during the preceding Ice Age. When their production pattern had to shift south, they would also have had to forage in deeper water. Plankton and fish-feeding whales cruising the open ocean don't face the same problem. Sumich thinks grays would have been likely to suffer during glacial periods and thrive during interglacial periods, such as the one we're in now.

As coastal, bottom-feeding denizens, however, the grays are also considered the most flexible of whales—precisely because of the varying conditions they've had to adapt to. Go back about fifteen thousand years, and not only was their favorite Chirikof Basin feeding area in the Bering Sea inaccessible but the calving lagoons in Baja were dry beach. Perhaps there was then enough rainfall in Baja or Southern California for river channels to fashion lagoons farther offshore that have now been buried in sediment. Or perhaps gray whale calves that have been born in recent years in Santa Monica Bay or at the entrance to San Diego Bay were simply the offspring of whales revisiting their species' onetime embayments of choice.

Who knows how many times the grays have been excluded from their customary breeding haunts and dining areas? Between ten and a dozen Ice Ages over the past million and a half or so years have been well-documented. One marine scientist I spoke with figures that, when all those soft-bottom crustaceans were covered over, the grays probably fed at the ice edge on the vast epontic zooplankton that live on the underside of the ice, hence initiating their side-sucking behavior, which persists to this day. Maybe they're food generalists who capitalize on that unique niche of benthic amphipods whenever, geologically speaking, they can.

In all likelihood, then, their flexibility is the biggest reason behind the California gray whales' amazingly rapid return from near eradication at the hands of man.

At our first meeting Larry Barnes told me about a remote spot in southern Baja California on La Paz Bay, north of the capital. Because of erosion—and excavations over the past decade and a half by a Mexican mining company—Barnes said, "some Oligocene Age marine sediments, about 25 to 28 million years old, are exposed in rocks above the coastline there, amid a series of repetitious phosphate beds in the siltstone and sandstone." Phosphoretic rocks in some parts of the world are famous for their fossil content, and among these have been found the oldest whales in all Mexico. "These include the earliest of the echolocating toothed whales or dolphinlike animals, and the earliest of the baleen whales, which still had teeth. We've discovered about fifteen species that are new to science, none of which are yet named. Every trip there we seem to collect one or two more."

Early in February 2000, at Barnes's invitation, I met up with him in La Paz for a "field trip." A photographer for *National Geographic* and a filmmaker from Los Angeles and Mexico City would accompany us, along with the German geologist Tobias Schwennicke and the Mexican paleontologist Gerardo González-Barba. "Tobias loves rocks, Gerardo loves sharks, and I love whales. We make a good team," Barnes summed up.

We're driving around tabletop mesas and mauve-colored cliffs with clean, interbedded layers of volcanic ash. Barnes says these were laid down in an ocean environment about 30 to 15 million years ago. The cliffs are topped by a thick bed of water-laid, pink volcanic ash, which is quarried for masonry and visible in geometric patterns along the walls, streets, and sidewalks of La Paz. A paved road carries the many workers who travel back and forth to La Paz by bus every day. They do strip mining and open-pit mining to dig out phosphates for use in agricultural fertilizers. These are offloaded onto a freighter that docks regularly in the harbor before heading across the Sea of Cortés to Manzanillo on mainland Mexico.

It's a long, slow, winding drive up canyons and into valleys. Our two vehicles finally come to a halt halfway down a cliff that overlooks the sea. In this area, amid the sagebrush and cardon cactus, last November the scientists and students discovered seven whales in only two days: members of the primitive toothed mysticete family Aetiocetidae and contemporaneous primitive baleen-bearing mysticetes of the family Cetotheriidae. They've named this spot Mesa Tesoro, "Treasure Mesa." They rarely need to do any pick-and-shovel work. They simply gather the pieces on the surface, already weathered out by nature. Walking up a slope covered in mudstone and shale, about fifty feet from the car, Barnes leans over and picks up a piece containing a whale's rib bone. "You need to be able to read the rock," he explains, pointing out a distinctive white pattern. Aeons ago, once a whale was encased in the rock, water in the sediment gravitated toward it and formed these mineral-laden nodules around it. That's the giveaway for cetacean fossils within. Between the beach below and the apex of these hills, there are at least seventy phosphetic beds covered with such nodule-bearing rocks. "The ones in this horizon," Barnes continues, "might have been exact contemporaries, could even have seen each other. But three feet of difference in the rock might represent five thousand years."

Schwennicke, our group's geologist, explains that these marine sediments were laid down when the Baja California Peninsula was part of the Mexican mainland, between 23.7 and 30 million years ago. We're currently standing at about 100 feet above sea level. Water depth here had varied several times, from very shallow to a maximum of about 650 feet. The phosphate beds are largely in a middle range of fine-grained sediments, their fossils having been deposited by strong currents during "storm events" such as hurricanes.

Barnes adds that this particular stretch of rock, which continues northward for approximately eighty miles, is all about the same age. "In the entire terrain, if you keep walking you're going to find a whale about every hundred yards in any direction either vertically or laterally. That's typical of the abundance of marine mammal bones in originally laid ocean sediments throughout the world." We're witnessing here the same stage of evolution as can be found along Washington's Olympic Peninsula, or in New Zealand and portions of Latin America. "You find me the rocks, I'll find you the whales in them," Barnes says unequivocally.

We're standing beside a substantial whale skull a little farther up the hillside. "This is the real keeper," Barnes says proudly. It's recognizably exposed and heavy enough to take at least two people to carry it. "We assume the rib I showed you to be part of the same individual." I ask if he has any idea how big this whale once was. "This one's not hard to figure out," Barnes says, fingering the occipital region. "Let's see, about twenty-four inches across his zygomatic arches here. The skull would have been about five and a half feet long. On these animals the skull is about 20 percent of the body length. So it's going to be twenty-five feet long, plus or minus. We've not collected this kind of cranium here before—that variety, size, width, style."

At the end of a long day the scientists and students carry the rock-enclosed skull in a sling to the back of the station wagon. Barnes can't seem to help but show me the 28-million-year-old nostrils and ears. "This is a really good specimen," he says, "the kind we can certainly identify and publish on."

To everyone's surprise, when we got back to La Paz a Mexican colleague had arranged a whale watch for our group for the next day. We would need to drive north for a while and then west across the Baja Peninsula about four hours to reach the upper sector of Magdalena Bay and the little Pacific town of Puerto López Mateos. I had yet to go there, and, even though it was early in the breeding season, gray whales ought to be there in some abundance. Caught up in the excitement of the moment, I didn't remember what Paco Ollervides had told me the year before, at the Magdalena Bay port of San Carlos. "Here the whale watching is chaotic," he said. "But it's even worse in López Mateos. Because it's very narrow up there, just one channel. Boats, whales, everything is happening within fifty feet of width and three miles long. So it's like an alley. Here, at least, the whales have places to go."

On the drive up Barnes talked again of how frustrating it was never to have found a gray whale older than around a hundred thousand years. "Of course, sometimes the ancestors of modern groups of whales are masquerading in some other family, and our analysis simply hasn't recognized them yet." He went on about how different today's Baja environment is from when the earliest whales lived here, around 30 million years ago. "The water was warmer, and the coast of Mexico was a broad shelf. Probably there were no lagoons or mangrove swamps and things like that. But it's interesting when you consider that the very first whales had come from the land—remember our Archaeocete?—and probably would return to freshwater sources to drink. Now today we have the gray whales returning to these shallow waters to have their babies. It's almost like they still have this tie to being near the land, like their ancestors did millions of years ago."

We're passing a portion of Highway 1 that Barnes says was once a series of Ice Age lake beds. Pleistocene bison and camels had walked around the shores of these lakes. At another locality, in Pliocene sediments, about 3 to 5 million years old, cetacean ear bones include those of belugas—today strictly Arctic whales—and of a long-snouted dolphin, a pontoporiid, whose closest living relative now exists only in Brazil, Uruguay, and Argentina. It is amazing to think of the past diversity of strange marine animals that once lived along Mexico's shores.

When we turn off the main highway, about twenty miles from Puerto López Mateos, the cornfields give way to flat desert land again. Entering town the pavement turns abruptly into sandy streets. An image of the gray whale is painted on the school. More gray whale caricatures decorate many of the restaurants and storefronts as we follow other *ballena*-bearing road signs toward the pier. At the end of the road a large parking lot opens onto a compound of wooden stalls selling whale T-shirts and whale key chains and ironwood whale sculptures. There's a bandstand with a semi-life-sized color depiction of a whale swimming above a Coca-Cola sign and over large letters reading: "*VII Festival Internacional de la Ballena Gris 2000.*" Whale-watching *pangas* line the dock that begins just beyond two kiosks dispensing life jackets.

The town was founded in the 1950s, when the Mexican government built the fish cannery that's a backdrop for all this. During the late 1970s and early 1980s, impoverished immigrants from the Mexican mainland flooded in to pursue the abundant fish and shellfish. Eventually the cannery was privatized and the bay's resources went into decline. In 1995 the processing plant laid off almost all of its three hundred permanent workers. Many of the local fishermen and former cannery employees went into the whale-watching business. By the close of the 1990s, almost twelve thousand whale watchers were paying annual visits to Magdalena Bay.

A white bus pulls up with a full load of elderly Americans. Boat captains eagerly await them, at twenty dollars a head for a two-hour run. Our 1:00 P.M. trip is all arranged. It takes about ten minutes for the filmmaker with us, who's shooting a documentary about scientific collaboration between Mexico and America, to fasten a tripod for his video camera with ropes onto the bow of our *panga*. I watch some kind of vaporous effluent rising from the cannery's smokestack as we motor off down the narrow waterway that leads to the Pacific. Past contoured white sand dunes that sweep in voluptuous shapes down to the sea. Past dense mangrove thickets, with egrets and cormorants and great blue herons along the shoreline. No whales in sight, but our driver, Vicente, assures us we'll soon be among them.

"Could you ask him where we're going?" I say to Barnes.

"He says, 'To the nursery.' "

I should have known what was in store when I heard that. At San Ignacio Lagoon the sector where the gray whales are born and spend their first weeks is off-limits to whale watching. Not here. Of course, at first it is heart-stirring to witness what I was about to: mothers swimming beside newborns that can't be more than a week or two old. They aren't even gray yet, but still pink. Yesterday, Vicente tells us proudly, his tour party even saw a whale give birth. We must have been in the vicinity of more than ten mother-calf pairs during the hour we spent in the nursery. For the most part these whales are still skittish, and understandably so.

We start heading back. The filmmaker is sitting cross-legged next to his mounted video camera on the bow, waiting for perhaps that one great shot. I've got my own video camera hanging around my neck, just in case. Sure enough, there's another mother with her calf, just the two of them, just ahead. We close in. Slowly at first. Then we're on them. Looking through our viewfinders. They start to run. So do we. It's subtle how the chase begins.

Suddenly, from behind, I hear the sound of more motors. Two additional small boats, laden with tourists, cameras ready, are racing toward us. One boat takes up a position to the left of the whales. We're on their right. The other boat hems them in from behind. Looking through a lens is like wearing blinders. You don't really see what's happening around you. But for some reason I can't shoot. I lower my camera. I'm appalled at what's transpiring. The channel really is little more than an alleyway—no more than fifty feet wide—and we have the two whales surrounded. Corraled. No place to go but straight ahead.

In their claustrophobia the whales are swimming faster and faster.

The baby is staying on the surface and desperately trying to keep up with its mother. It appears to be gasping for breath. My companion has stopped filming and looks mortified, too.

"Tell the driver to turn away from them!" I shout.

Our *panga* veers off. The whales are able to change course away from the other two boats. The other captains follow our lead and leave them alone as well.

I'm staring at the camera tripod on the bow. It looks like a harpoon gun.

Are we still their predators?

Chapter 26

Scammon's Legacy

—Where once the waters were alive with different varieties of marine animals . . . now only are seen a few stragglers making their periodical migrations. The sea-beaches of island and coast, once the herding-places of these amphibious animals, whose peltries were highly prized among the enlightened classes of both Europe and America, are now deserted; except at the most inaccessible points, there are but few found, and their wild and watchful habits plainly tell that the species is nearly annihilated.

Charles Melville Scammon, 1869

With the Western Union Telegraph Expedition at an end, early in 1867 Scammon had resumed his duties with the Revenue Marine aboard a graceful sailing schooner, the *Joe Lane.* The captain's assignment was to patrol the coast from Washington's Puget Sound to the southern tip of Lower California. The latter point, Cabo San Lucas, was his first port of call around the beginning of February. There Scammon was to meet a fellow San Franciscan who would prove as seminal an influence upon his future as William Healey Dall.

J. Ross Browne had been traveling the Baja Peninsula for over a month. The previous fall some American businessmen representing the Lower California Colonization Land Company had approached him about putting together an expedition of California scientists. The entrepreneurs wanted to find out whether establishing a land grant settlement in Baja might be worth doing. Browne marshaled some recruits and left San Francisco by ship in late December. He was en route home again, after a rather harrowing experience at Magdalena Bay, when he and Scammon ran into each other in Cabo.

Scammon was quite familiar with Magdalena Bay, where he'd first hunted gray whales in the mid-1850s. Undoubtedly this would have been a major topic of his first dinner with Browne, who'd exhausted his water supply on an arduous nine-day mule trek across the desert and been unable to find any freshwater at Magdalena Bay either. At their darkest moment Browne and his companions had seen the masts of two whaling vessels anchored several miles out. Several frantic hours were spent trying to signal them before one of the masters rowed ashore to see what was afoot. Browne obtained some vitally needed casks, as well as a whaleboat loaned him to survey the bay. He determined that "Magdalena Bay affords no suitable location for the nucleus of a colony of civilized people," as he would write in his report.

Still, Browne wasn't willing to rule out the possibility of some such "suitable location" along Baja's western coast. Here was where Scammon's knowledge came in. For several days the two men sailed together aboard the U.S. steamer *Suwanee*, doing more coastal surveillance. Browne and Scammon then headed home on their separate vessels—but not before Browne had enlisted the captain to write a report detailing all he knew about the region. They made plans to meet again soon in San Francisco.

J. Ross Browne was far more than simply an adventurer-explorer. "There can be no question that Browne was one of the most widely traveled, observant, and versatile men of his time," the literary historian Duncan Emrich wrote in 1950. "It is unfortunate that his ability as a writer and the influence he exerted on other writers of his time has not been more widely recognized."

Four years older than Scammon, Browne had spent seventeen months cruising the Indian Ocean on a whaling ship. The daily journal he kept became the basis of a book titled *Etchings of a Whaling Cruise*, published in 1846. More than five hundred pages long, the book included "History of the Whale Fishery"—which Scammon would draw upon at length for his section "The American Whale Fishery" in *Marine Mammals.* Browne included his own sketches and illustrations, something Scammon would also do.

Browne intended his *Etchings*, above all, as an exposé of the abuses seamen suffered at the hands of unscrupulous whaling captains. Herman Melville, who'd had his own whaling experiences but had yet to write *Moby-Dick*, reviewed Browne's book for the *Literary World.* Melville compared it with Richard Henry Dana's *Two Years Before the Mast*, published six years earlier. He called Browne's work an antidote to romantic ideas that "unquestionably presents a faithful picture of the life led by the twenty thousand seamen employed in the seven hundred whaling vessels which now pursue their game under the American flag."

It appears that Melville saw more than this in Browne's book. One Melville biographer, Charles Roberts Anderson, has detected considerable parallels between *Etchings* and *Moby-Dick*, the latter published five years afterward, in 1851. Browne's ship, actually the *Bruce* but renamed the *Styx* for publication purposes, was mastered by a "Captain A." Anderson describes Captain Ahab as "distinctly reminiscent of Capt. A in more than the provocative initial," and continues: "It is worthy of note that the route Melville ascribes to the *Pequod* is strikingly similar to that taken by Browne in the *Styx;* and if Melville ever took the shadowy voyage described in *Moby-Dick* no record of it survives save in the vicarious one he may have found in *Etchings of a Whaling Cruise.*" Anderson adds that in Browne's description of "trying out" a captured whale, "Melville found inspiration for one of his most brilliant passages. . . . Certainly no one, not even Browne himself, could carp at an appropriation handled in such a masterly manner."

Nor were Scammon and Melville the only contemporaries indebted to Browne. Another was Mark Twain. The two men were well-acquainted when both lived in the San Francisco area. Based upon his travels to the Near East, Browne wrote a humorous book published in 1853, *Yusef; or The Journey of the Frangi.* As the poet Joaquin Miller later noted, "It is clear to the most casual reader that if there had been no *Yusef* there would have been no *Innocents Abroad.*" Browne spent most of that decade working along the frontier at customhouses and Indian reservations with the nascent U.S. Treasury Department, a tour of duty that produced two more books. According to the literary historian Emrich, "his *Peep at Washoe* and *Washoe Revisited* were clear forerunners of Mark Twain's *Roughing It.*" Browne said as much in a letter written in 1870, recounting having run into Twain in London: "I believe he is writing a book over here. He made plenty of money on his other books—some of it on mine."

During the 1860s Browne had set about educating himself concerning mineral resources, and he was the first to recommend that the U.S. establish a Bureau of Mines. That was how he ended up making the resource-seeking trip on which he met Scammon. Both were strong-willed men known for their scrupulous honesty who had gone from whaling backgrounds into government service. What encouragement as a naturalist Scammon received from Dall, he seems to have received from Browne to become a writer. Soon after his return to San Francisco, Scammon was quick to provide the requested report on Lower California. Browne made this part of his overall findings submitted to the land company, and Scammon's work was summarized in several outlets: the daily *San Francisco Bulletin, Merchant's Magazine*, and the *Commercial Review.* Browne incorporated the complete *Report of Captain C. M. Scammon . . .* into his 1869 book, *A Sketch of the Settlement and Exploration of Lower California*, and later into his massive *Resources of the Pacific Slope.*

Most important for Scammon, though, was Browne's introducing him into San Francisco's burgeoning literary community. The first issue of the *Overland Monthly* appeared in July 1868. Browne, after a falling-out with his East Coast publisher, immediately began contributing to the magazine, whose regulars soon included Twain, Joaquin Miller, and Bret Harte, the last also its literary editor. Browne shepherded Scammon over to introduce him to Harte and to the publisher, Anton Roman. While Browne departed for a brief and disastrous stint as U.S. minister plenipotentiary to China, Scammon's first article ("Fur Seals") for *Overland Monthly* led the November 1869 issue. The writers weren't bylined, but the last issue of each year contained a catalog of who'd written what; until then, the authors' identities were a popular guessing game among readers. Such Scammon series as *Pacific Sea-Coast Views* did much to educate those readers about horizons near and far.

Browne, having turned his attention to California development schemes for several years, died suddenly on December 8, 1875, at only fifty-four. He had lived just long enough to see the 1874 publication of Scammon's *Marine Mammals*, in which the captain credited and quoted from Browne's whaling history for many pages. Although we have no record of what Browne and Scammon may have discussed about whaling, Browne's disdain for the industry—and feeling for whales—colors his legacy.

In one of his lectures on whaling, Browne described his experience like this: "There is a murderous appearance in the blood-stained decks, and the huge masses of flesh and blubber lying here and there, and a barbarous wildness in the looks of the men, heightened by smoke, dirt, and the fierce glare of the fires, which can only be appropriately compared to Dante's pictures of the infernal regions. . . . Hour after hour, and day after day, have I turned that horrible grindstone, to keep the harpoons, spades and boarding knives sharp—turned it till my hands were turned into a blister, and every muscle of my body done . . . till I turned every remnant of romance out of my head."

He told of a shipmate's son "sitting in the forecastle with the tears streaming from his eyes and the whale gurry from his head—for here we have a touch of natural feeling described from hard experience."

Browne went on to extol the whale as "this Royal potentate of the great kingdom of ocean—this sublime sovereign of the everlasting waters—whose majesty and power have been sung by the prophets of old, and whose inestimable services in the civilization of the world have been chronicled by the rarest and brightest intellects of ancient and modern times."

The first evidence I have been able to find of a plea for the conservation of whales appeared in a little book published in London in 1808. Its author was listed only as Mrs. Linzorn. Titled *The Balaenic Games; or, The Whale's Jubilee*, it's an all-verse tale about a feast held by the whales for all the other animals. Near the conclusion are these words:

> Henceforward your sovereign shall rule the great deep,
> And without molestation, his dignity keep;
> No more with his blood shall the billows be dy'd,
> Nor his consort, his children, be torn from his side;
> No more shall he struggle with unequal foes,
> But till Nature's last summons, in plenty repose:
> Harpoons, our great dread, shall to plough-shares be chang'd,
> And man from the traffic of blubber estrang'd.

The prescient Mrs. Linzorn proceeded to prophesy a new form of lighting that would replace the need for whale oil.

But the electric light was still more than half a century away. In Scammon's day, as he would write of the aftermath of discovering the gray whale lagoons, "The try-works were incessantly kept going—with the exception of a day, now and then, when it became necessary to 'cool down,' in order to stow away the oil and clear the decks—until the last cask was filled. Nor did we stop then; for one side of the after-cabin was turned into a bread-locker, and the empty bread-casks filled with oil; and the mincing-tubs were fitted with heads, and filled, as well as the coolers and deck-pots; and, last of all, the try-pots were cooled, and filled as full of oil as it was thought they could hold without slopping over in a rough sea."

Yet one wonders when pangs of conscience may have begun to work on Scammon. Could it have been in 1866, when he pasted a newspaper article into his scrapbook about a whaling captain who, "in a fit of mental aberration jumped overboard at sea . . . [and] was buried at Scammon's Lagoon"? There was evidence of a change of heart by the time of Scammon's 1869 report, when he looked back upon a Baja region "where once the waters were alive" and now but "a few stragglers" were seen.

That same year Scammon kept up a steady correspondence with his naturalist friend Dall, about to publish his epic book *Alaska and Its Resources.* Scammon wrote Dall that he was "delighted to hear that you have got through with your M.S. and that your book will be out by new years. Don't fail to send me a copy as soon as it is possible. I am itching to see it for I am confident it will be good to say the least." Toward the end of the letter, Scammon added: "I have been amusing myself since we came to San Francisco in writing a few magazine articles, which seem to find a ready market."

In late December 1869, Scammon forwarded to the Smithsonian's Spencer Baird a manuscript "on 'Leopard Seal,' with drawing, thinking you might possibly like to look at it or perhaps have it published with the Sea-Otter." Clearly, something new was brewing for Scammon, during a time when his latest Revenue Marine command had "the finest sort of officers that it has ever been my grand fortune to fall in with" and when he was seeking help from these officers "in collecting &c. when it will in no wise interfere with their official duties."

During the spring of 1870, Scammon received a letter from his eighty-five-year-old father, Eliakim, whom he had rarely seen since going west as a young man: "We received the 'Overland Monthlies' you sent us. Your communications surprised and pleased me very much. I did not know I had a literary don in California!" (Eliakim would die that November.)

May 15, 1870: "Dear Dall—I drop you a line this morning at the suggestion of my friend, Prof. Whitney [J. D. Whitney, president of the California Academy of Sciences], to inquire of you about your publisher, or if you can recommend any to me in case I should decide to publish a book of abt. 350 pages. . . . Prof. W with many others of my literary friends advise me to put my papers together."

The following autumn, while serving as captain of two ships at once, at the Strait of Juan de Fuca Scammon collected the type specimen of the Davidson piked whale, *Balaenoptera davidsonii.* He would forward this to the Smithsonian. An 1871 report to the Secretary of the Treasury extolled Scammon and several others in the Revenue Marine for employing "their leisure hours in prosecuting researches into the natural and physical history of the coast."

In 1871 Scammon initiated discussions with the new publisher of the *Overland Monthly*, John H. Carmany, about publishing his proposed book. In a letter to Dall, Scammon bemoaned that "our Bret Harte" had left the magazine. (Lured by a large salary and demand for his rising literary reputation, Harte had gone to New York.) Scammon also wrote Dall that, "as I can only work at it [his manuscript] in leisure hours it will be some months yet before I shall have it ready."

Scammon's letters reveal that he was suffering from an ongoing illness, a factor also delaying his labors on the book. In May 1872 the *San Diego Union*'s advance notice described his "exquisite lithographs . . . as faultless in execution and as true to life as it is possible for the lithographer's stone to produce them." Another review, of his *Overland Monthly* contributions, noted that Scammon "has the merit of never writing except when he has something to say."

The scrapbooks Scammon kept during this period are filled with lore of the times: newspaper articles about Livingstone's African expedition in search of Stanley, the saga of Mexico's Maximilian and Carlota, interviews with a number of Brigham Young's wives. From the *Hawaiian Gazette* of October 24, 1871: "The very latest from the Arctic fleet! 33 ships lost, 1200 shipwrecked seamen. The whaling fleet cruising in the Arctic." Another clipping, headed "A Modern Crusoe," told of a fellow marooned in the Antarctic who survived upon biscuits and pelican flesh.

Scammon saved, too, an 1872 story on what was becoming of the whaling fleet.

> The sad mishaps to whaling vessels in the past year, have struck a severe blow at the trade, which had been on the decline for some time previously. Fifteen years ago New Bedford alone sent out three hundred and twenty-nine vessels; now that number comprises the whole of the American vessels engaged in whaling. But one vessel was added to the trade in 1871, while seventy-two were lost to the business. Of these, twenty-six were abandoned in the ice, six wrecked, three condemned, two sold, thirty employed in other business, and three broken up. Fortunately for the interests of commerce and domestic economy, whale oil is not an indispensable article, and whalebone [baleen] has to a large extent been

> supplanted by the bamboo, rattan, willow and steel, in light fabrics of general utility.

Also in Scammon's scrapbook were what can only be called incidents both magical and long-forgotten. One such newspaper story told of the Smithsonian's Professor Baird's coming to see "a wonderful fish" caught near Eastport, Maine, and being "as yet unable to place it in the known lists of the animal kingdom." It was over thirty feet long and seemed to be part beast, part fish: "one enormous dorsal fin, two side belly fins, and a broad, shark-like tail," along with "two huge legs, terminating in web feet," a small-teethed mouth and a series of gills, "which overlap each other like the flounces once the style in ladies' dresses." The skin was dark and tough, like that of an elephant or a rhino, and its "immense body . . . was estimated to have weighed when captured about eleven tons."

Beyond Scammon's scrapbook, one searches archives in vain for mention of the "wonderful fish."

In September 1872, briefly but fortuitously, the most famous scientist in America entered Scammon's life. Louis Agassiz had trained and influenced a generation of zoologists and paleontologists, including William Dall. Originally from Switzerland, Agassiz had come to Boston twenty-five years earlier, at the age of forty. He'd become a professor at Harvard, where he founded the Museum of Comparative Zoology. He'd urged creation of a National Academy of Sciences, become a founding member in 1863, and that same year been appointed a regent of the Smithsonian. He dined with Thoreau at Emerson's home. His friend Longfellow delighted in looking through his laboratory microscope. Henry Adams wrote in his *Education* that Agassiz's lectures on "the Glacial Period and Paleontology . . . had more influence on his curiosity than the rest of the college instruction altogether."

Considered the father of glaciology (Agassiz formulated the theory that a great Ice Age once gripped the earth), the professor was simultaneously at odds with Charles Darwin. Agassiz came from a European background in *Naturphilosophie*, trained to seek metaphysical interconnections within the world of living things. Thus did he reject the notions of evolution in *Origin of Species* (1859), despite Darwin's having previously found evidence in Agassiz's own work, which, Darwin wrote, "accords well with the theory of natural selection." Rather, Agassiz defined a species as "a thought of God" and believed that "classification, rightly understood, means simply the creative plans of God as expressed in organic forms."

This understanding was more in line with the teachings of Emanuel Swedenborg, the eighteenth-century mystic who wrote that "the natural world, and all that it contains, exists and subsists from the spiritual world, and both from the divine." Scammon and his older brothers were all followers of Swedenborg's thought. Whether or not Scammon discussed metaphysical realms with Agassiz, we do not know. But there is no doubt that the captain was deeply impressed by Agassiz, for whom marine biology (especially fossil fishes) was closest to the heart.

Their one meeting took place when Agassiz stopped in San Francisco on his way home from a trip around Cape Horn. Then in his mid-sixties, Agassiz was ailing. A large crowd gathered at Pacific Hall to hear his lecture "The Natural History of the Animal Kingdom." Scammon was among Agassiz's hosts. After viewing the captain's drawings to be included in his *Marine Mammals*, Agassiz wrote Scammon an enthusiastic letter, which concluded: "Your practical knowledge of these animals and the faithfulness and excellence of the representations will make the work standard, and it will give me the greatest pleasure to do everything in my power to obtain subscribers for you in the Atlantic States and in Europe." Later the *San Diego Union* reported that Professor Agassiz had "declared that it was the first time he ever saw the whale properly exhibited on paper."

Agassiz also enlisted Scammon to do some collecting for him. As Dall reported in a letter to Baird early in 1873, "Agassiz has written Scammon that he may draw to the extent of $1500.00 for expenses in collecting cetaceans &c. for the M.C. [Museum of Comparative] Zoology. . . . I suggested to S. [Scammon] that he had better take advantage of Agassiz' help to get the larger skeletons & skulls for which the S.I. [Smithsonian Institution] cannot afford to pay, and to send the smaller and less expensive things as far as possible to the [Smithsonian] National Museum."

Agassiz, however, would not live to see publication of Scammon's book. He died in Boston on December 14, 1873. Scammon attended a memorial gathering at the California Academy of Sciences. A copy of the earlier program—"Welcome to Agassiz San Francisco September 1872"—he taped into his scrapbook. Scammon's book would be "dedicated to the memory of LOUIS AGASSIZ. As a humble tribute from the author."

•

Agassiz's former star student Dall had been engaged in completing a "Catalogue of Cetaceans" for Scammon's opus. The captain wrote Dall that he did not wish the catalog "to come out in the Academy or to be published anywhere till the book comes out . . . in its proper place giving you full credit for what you do in the matter."

Dall to Baird, January 12, 1873:

> I am poorly, not ill but feel played out. . . . I have been working very hard for Scammon neglecting my own work and spending strength which I could ill spare, but it is not an agreeable task, and had I realized how little he appreciates the necessity and value of exact accuracy, and how absurdly suspicious he would be when he gets a dyspeptic fit, I never would have undertaken the job. However, for the sake of making the most of his contribution for science, and rendering the great amount of material which he has brought together, available, I shall persevere as the worst of the work is over. The Revenue Service is a dog-eat-dog institution, as you are doubtless aware, and when Scammon gets a little bilious, as he has once or twice since I took hold of the matter, he begins to suspect not only me but everybody else of wishing to curb his thunder, and has not always kept his foolishness to himself. At other times he has entertained me with cock-and-bull stories of somebody unknown, who had taken all the books on whales out of the public libraries and had been heard of in search of somebody to draw figures for him.
>
> Notwithstanding his eccentricities, which I do not wish to have mentioned, Scammon is doing a work, which supplemented by thorough identifications, cannot but be very valuable and creditable to him. He mentioned the other day a fact, that if he had the authority from the Department he could obtain almost without expense, skulls of various species of whales from Monterey and elsewhere and bring them to San Francisco on the cutter. He suggested that if a favorable opportunity should offer, you might represent this to the proper officers, as a request from the S.I. [Smithsonian] stating that he felt unwilling to attempt anything of the kind for fear he should be regarded as neglecting or exceeding his proper duties, for the promotion of science and his own in-

vestigations on whales. He sent a very valuable lot of things via P.U. S.S. Co. the other day including the complete skeleton of a dolphin and a skin of a sea elephant.

Dall to Baird, January 26, 1873: "Scammon has returned from a cruise in better condition of mind & body and we are now working as smoothly as could be desired."

By now, Scammon had taken command of the propeller-driven *Oliver Walcott*, the first steam vessel ever built on the Pacific Coast. His latest exploits included rescuing a ship disabled from scurvy, providing vegetables to the crew, and towing them into port. A third son, Lawrence, had been born in 1872, although Scammon's firstborn, Charles—who had detested his father's forcing him to become a cadet with the Western Union Telegraph Expedition—had gone off into the hinterlands of Alaska.

At the close of 1873, *The Nation* described the forthcoming publication of Scammon's "large quarto," which contained new and valuable information "concerning the habits of whales, of which but little is generally known, as well as of their food, their diseases, etc."

But something had gone awry between Scammon and Dall. From Dall to Baird, January 9, 1874: "With regard to Scammon I consider that he has treated me in a very ungentlemanly way, and I have not had any intercourse with him since I returned." Dall to Baird, January 18, 1874: "At all events I will have nothing further to do with him, unless I have an unmistakeable apology for the insult to which he subjected me last spring. His book will be out in March . . . and the whole will be much more extensive than he first contemplated."

That is all we know of the rift between the two men. It coincided not only with publication of *Marine Mammals of the North-Western Coast of North America*, which contained Dall's rigorous and systematic catalog, but with lingering ill health that forced Scammon to heed his physician's advice and take an indefinite leave of absence from the Revenue Marine. The captain's papers contain this telling sentence: "I became an invalid from long and continuous sea life." In a horse-drawn carriage filled with furniture, Scammon, his wife, and two of their sons moved more than seventy-five miles away from San Francisco not long after the book came out. Scammon had purchased a small apple farm in Northern California's Sonoma County, in the rural hamlet of Sebastopol. He was about to turn fifty.

•

In the preface to his book, Scammon wrote: "The chief object in this work is to give as correct figures of the different species of marine mammals . . . as could be obtained from a careful study of them from life, and numerous measurements after death. . . . It is hardly necessary to say, that any person taking up the study of marine mammals, and especially the Cetaceans, enters a difficult field of research, since the opportunities for observing the habits of these animals under favorable conditions are but rare and brief. My own experience has proved that observation for months, and even years, may be required before a single fact in regard to their habits can be obtained."

The book, more than 350 pages long, bore a gold imprint of a sea lion on its cover and was bound in green, red-brown, or black cloth depending upon the desire of the customer. In separate chapters it discussed ten whale and fourteen dolphin species, along with six pinnipeds and one sea otter. It included some of the first descriptions ever published of elephant seals and other species. Scammon wrote of once seeing a herd of rare Arctic banded seals—later called ribbon seals—on a beach at Point Reyes, California. Later marine biologists usually discounted this as an erroneous observation, until, in 1962, a live ribbon seal—the only other record of one in the Pacific United States—was lassoed by a cowboy on a beach near Morro Bay. About one-third of Scammon's book was devoted to whaling history and its techniques. It included thirty-five natural history drawings, twenty-seven plates of marine mammals, six scenes of whaling sites, and numerous detailed drawings of the industry's equipment—most of which were signed by Scammon.

Gray whales, as we have seen, were his featured attraction, the subject of his frontispiece drawing of San Ignacio Lagoon, his first chapter, and much more throughout. And his account of what happened when he first encountered the grays inside the lagoon that came to bear his name is strikingly reminiscent of that of Melville and the sperm whale. Compare this passage from *Moby-Dick*, 1851: "From the ship's bows, nearly all the seamen now hung inactive; hammers, bits of plank, lances, and harpoons, mechanically retained in their hands, just as they had darted from their various employments; all their enchanted eyes intent upon the whale, which from side to side strangely vibrating his predestinating head, sent a broad band of overspreading semicircular foam before him as he rushed. Retribution, swift vengeance, eternal malice were in his whole aspect, and spite of all that mortal man could do, the solid white buttress of his forehead smote the ship's starboard bow, till men and timbers reeled."

From Scammon's *Marine Mammals*, 1874:

> Early the next morning, the boats were again in eager pursuit; but before the animal was struck, it gave a dash with its flukes, staving the boat into fragments, and sending the crew in all directions. One man had his leg broken, another had an arm fractured, and three others were more or less injured—the officer of the boat being the only one who escaped unharmed. The relief boat, while rescuing the wounded men, was also staved by a passing whale, leaving only one boat afloat. The tender being near at hand, however, a boat from that vessel rendered assistance, and all returned to the brig. When the first boat arrived with her freight of crippled passengers, it could only be compared to a floating ambulance crowded with men—the uninjured supporting the helpless. As soon as they reached the vessel, those who were maimed were placed on mattresses upon the quarter-deck, while others hobbled to their quarters in the forecastle. The next boat brought with it the remains of the two others, which were complete wrecks. Every attention was given to the wounded men, their broken limbs were set, cuts and bruises were carefully dressed, and all the injured were made as comfortable as our situation would permit; but the vessel, for several days, was a contracted and crowded hospital. During this time no whaling was attempted, as nearly half of the crew were unfit for duty, and a large portion of the rest were demoralized by fright. After several days of rest, however, two boat's crews were selected, and the pursuit was renewed. The men, on leaving the vessel, took to the oars apparently with as much spirit as ever; but on nearing a whale to be harpooned, they all jumped overboard, leaving no one in the boat, except the boat-header and the boat-steerer. On one occasion, a bulky deserter from the U.S. Army, who had boasted of his daring exploits in the Florida War, made a headlong plunge, as he supposed, into the water; but he landed on the flukes of the whale, fortunately receiving no injury, as the animal settled gently under water, thereby ridding itself of the human parasite.

Describing the mouth of San Ignacio Lagoon, the captain wrote this tender passage about the gray whales he had so avidly pursued:

> It was at the low stage of the tide, and the shoal places were plainly marked by the constantly foaming breakers. To our surprise we saw many of the whales going through the surf where the depth of water was barely sufficient to float them. We could discern in many places, by the white sand that came to the surface, that they must be near or touching the bottom. One in particular, lay for a half-hour in the breakers, playing, as seals often do in a heavy surf; turning from side to side with half-extended fins, and moved apparently by the heavy ground-swell which was breaking; at times making a playful spring with its bending flukes, throwing its body clear of the water, coming down with a heavy splash, then making two or three spouts, and again settling under water; perhaps the next moment its head would appear, and with the heavy swell the animal would roll over in a listless manner, to all appearance enjoying the sport intensely. We passed close to this sportive animal, and had only thirteen feet of water.

About no other species of whale did Scammon wax so eloquent. He used the California gray, too, to exemplify how whalers were disturbing the balance of nature to the extent that entire species might not survive. Indeed, the book described populations of sea otters in Washington and Oregon that would later be exterminated.

However well-received by reviewers and the scientific community, *Marine Mammals* fared poorly with the general public. Out of a printing of some one thousand copies, only about half were sold. Despite the handsomeness of the publication, the ten-dollar asking price was undoubtedly too steep for its time. In the spring of 1877 the publisher, Carmany, wrote to Scammon: "All attempts to sell your book have failed me, even with the copyright included, and of course I have the 500 copies in sheets on hand, which I will gladly hand over to any person—*copyright included*—who will liquidate the balance of indebtedness due on publication account."

Marine Mammals continued in stock at the printer's office, where Scammon would buy copies for two dollars apiece. He would send one gratis to any scientist who asked. In 1906, however many remained were destroyed in the San Francisco earthquake and fire. In 1969, inspired by the growing interest in whale watching and especially gray whales, a facsimile edition was published. Today an original 1874 edition of Scammon's book sells for more than two thousand dollars.

—Among the noticeably beautiful residence properties in Analy Township is that of the above named gentleman [Scammon]. It is situated one mile north of Sebastopol, at which point he is the owner of thirty-five acres of rich and productive land. Captain Scammon purchased this land in 1874 and commenced its improvement the same year, building a beautiful and convenient cottage residence and suitable out buildings. His residence is finely located upon high ground, which is approached by a beautiful drive way, and his grounds are highly improved, shade trees, flowers, etc., surrounding his home. The view from his study window is one of surprising beauty, overlooking as it does the Laguna with its placid waters and the beautiful meadows on the lower plateau. Captain Scammon is devoting his lands to general farming. In fruit culture he confines himself to a family orchard and vineyard, in which he has some of the most valuable and improved varieties of fruit and table grapes grown in Sonoma county. All his stocks are of the best. The cattle are improved by the famous Holstein and Jersey breeds, and the horses are improved by thoroughbred stock.

—*History of Sonoma County*, 1888

The farm where Scammon spent his twilight years was reminiscent of his boyhood home in Maine. Joining him as a next-door neighbor in 1876 was another former whaler, Jared Poole, originally from Martha's Vineyard and married to Scammon's wife's sister. He'd been one of the first to equip a whaling vessel for the Arctic and was the man who had first led Scammon to the San Ignacio Lagoon. According to Scammon's descendants, the two old captains did not get along. Scammon's wife, Susan, enjoyed sitting on the porch with her "spy-glass," checking up on the Pooles.

Despite going into semiretirement, Scammon scarcely lived the life of a gentleman farmer. Indeed, even in failing health he kept up with all his former interests and acquaintances. By 1880 he felt well enough to apply for active service once more in the Revenue Marine, heading for Florida and a side-wheel steamer. There he studied the habits of the manatee, which would be the subject of his last published article. Like Scammon, the *Overland Monthly* couldn't be kept dormant for too long. After ceasing publication during a business slump in 1875, it bounced back for a time in the late 1880s and took Scammon's piece on the "Sea Cow."

After contracting malaria in Florida, Scammon was forced to return to California. By the spring of 1882, however, he was back on duty in Galveston, Texas, charged with a steamer that cruised from Mobile, Alabama, to the Rio Grande. Contemplating a new book focusing on the marine mammals of the Gulf of Mexico, Scammon began sending Spencer Baird sketches of several new varieties of porpoises. "It is my desire," he wrote, "that the Mottled Porpoise, should be a *Bairdii* of the Atlantic. For it would be a hard case, if you cannot have even a Porpoise, in the Atlantic, as well as the Pacific." Scammon was referring to Dall, in 1873, having named a new species of dolphin *Delphinus bairdii.* Scammon is also said to have renewed ties with Dall, although it is not recorded whether they met again or merely corresponded.

The captain began writing poems—"The Wreck," "The Rescue," "On Watch," "Ocean Warn," "Sea Life"—during the 1880s. Not until 1894, at the age of seventy-one, did he accept a half-pay retirement from the Revenue Marine.

Scammon's eldest son, Charles, helped out on the apple farm for a while before heading off to the Yukon during the Klondike gold rush. He became a gambler and died in a shack in Alaska. Alexander, the second son, bought another farm near Sebastopol and at one point worked on horticultural projects with Luther Burbank. The youngest boy, Lawrence, designed electrical fixtures and was a part-time artist. A series of etchings he did of California's Spanish missions were well known in their time.

Through the auspices of Scammon's only biographer, Lyndall Baker Landauer, I managed to track down the family's descendants in suburban Sacramento. Mildred Scammon Decker, the captain's granddaughter, was then ninety-one. In December 1998 we met in the living room of the condominium where Mildred's son, Tom Decker, resided with his wife, Edy. The couple's grown son, Dave, had picked up his grandmother at her home nearby. Along a stairway leading to the Deckers' second floor were a framed series of black-and-white photographs. One of these depicted a handsome, bearded man with fair skin and short, straight hair. He was wearing a uniform and appeared to be very proud of the fact.

"That's the captain," Mildred said, pausing a moment before continuing: "I'm planning to be scattered out to sea when I go. I used to think that when I died, I'd see my grandfather—and George Washington. They were both characters, weren't they?"

She laughed, then glanced at a list of notes about Captain Scammon that she'd jotted down for our meeting. She didn't really need it. Mildred was a sprightly woman with childhood memories that, as she put it, "still flow like a moving picture." In her mind's eye the two-story white frame house to which Captain Scammon and his wife, Susan, had moved remained more vivid than her parents' residence next door.

She brought out a photograph of the old Scammon house, which still stands on East Oakland's Grande Vista Avenue, formerly known as Orange Avenue. "There's the pine tree," she said, "that Grandpa planted in the front yard in 1898, after they moved down there from the farm."

The move had been prompted by Scammon's loss of active-duty pay and failure to make the Sebastopol farm a moneymaker. When the last of his surviving brothers died and left Scammon five hundred dollars in his will, the family returned to urban life. In the foothills east of San Francisco Bay, an area called Fruitvale at the turn of the century, the Scammon residence was not as grand as some of the Victorian mansions that also lined the street but spacious enough to house visiting relatives from Maine or Chicago. Horse-drawn ice wagons passed by regularly. One well-to-do family owned a two-cylinder automobile. The elder Scammons had a relatively new invention of their own: a telephone.

Mildred and her older sister, Dot, enjoyed the elegant dinners at their grandparents' table. It was made of solid walnut, surrounded by handsome leather chairs with a sparkling decanter of wine on a sideboard. Clearly, Grandma ruled the household. She often shouted orders at the captain, who'd become almost deaf. Mildred's older brother poked fun at Grandpa for removing his false teeth and gumming his food.

Scammon himself rarely spoke, yet Mildred was mesmerized by him. She loved running her hands over the furniture he'd fashioned in a basement wood shop. He kept his tools in a handmade chest, which the little girl was told he'd had on shipboard. With these tools the captain had assembled a large rectangular wardrobe, lined with wonderful-smelling aromatic woods. There was an oak tongue-in-groove piece, with a secret drawer in back, which Mildred discovered.

From a ladder in the upstairs bathroom, Mildred and Dot could climb through a hole in the ceiling into the attic. It was filled with books—she would remember a collection of Emanuel Swedenborg's works—and the musty smell of old papers. On the floor was a bearskin rug, constantly shedding. The sisters invented stories about how Grandpa shot the bear, maybe in Alaska, or perhaps faraway Siberia.

If you knew how to shine a light down into a small covered closet, you could glimpse the captain's navy blue uniform: the epaulets on the collar, the double sets of brass buttons. His sword rested in a corner—an American eagle perched at the top, balance scales carved into the steel scabbard alongside an engraved barefoot, pipe-smoking, rakish-looking figure. On one wall was a letter commending Scammon's service in the U.S. Revenue Marine, signed by President Lincoln.

Sometimes, Mildred would stand in the doorway of the captain's study. He would be sitting behind a tall desk. Surrounding him were stacks of correspondence, handsomely bound notebooks, and a series of scrapbooks filled with old newspaper clippings. He seemed to save everything about nautical matters, including storms and shipwrecks. Years later Mildred would come upon this piece from 1894 in the *San Francisco Chronicle*: "Horrible tale of shipwreck off the barren coast of Alaska and of gnawing famine." The article went on to tell of the fifteen survivors of the whaling bark *James Allen*. "Nine of these men had resorted to a diet of human flesh when rescued from the jaws of starvation. They had devoured the corpse of one of their mates, and were starting in on another when the United States Revenue Cutter *Bear* arrived on the scene."

Mildred recalled that Scammon wrote constantly, always dressed in a black alpaca coat and a skullcap that helped ward off the chill of the coal-heated house. His blue eyes were penetrating when they chose to acknowledge her presence. He was not a tall man, perhaps five foot eight. Yet an aura about him made the captain feel huge.

Once in a while an old friend would drop by and drive him down to the estuary a few miles to the south. A long, wide channel separated the towns of Oakland and Alameda, leading out into the bay and eventually to the Golden Gate Bridge. Here Charles Melville Scammon had once embarked for Mexico's Lower California, as well as South America, Alaska, Russia, and China.

On May 2, 1911, first his wife, Susan, and then Scammon died within twenty-four hours of each other. They were both in their mid-eighties and had been married for sixty-three years. The funerals took place in their family parlor. Every Memorial Day Mildred's father, Lawrence, would take her to the grassy knoll under a small palm tree at Evergreen Cemetery. Her grandparents were buried side by side in unmarked graves. The family could not afford headstones.

William Healey Dall would write the obituary that appeared in *Science Magazine* on June 9, 1911. Of the Western Union Telegraph Expedition, Dall recalled, "To his intelligent and kindly cooperation the scientific corps of that expedition owed much of their success. Captain Scammon early became interested in the natural history of the marine mammals of the Pacific Coast, and in those days before the invention of photographic dry plates, spared no trouble in gathering measurements, drawings and other data bearing on the cetacea." Dall concluded that Scammon's *Marine Mammals* "forms the most important contribution to the life history of these animals ever published, and will remain a worthy monument to his memory."

Dall died in 1927, having described more than 3,500 species of fossil and living mollusks, published more than 1,600 papers, and authored numerous books on subjects ranging from birds to Alaska and including a biography of Baird. His own obituary in *Science* called him "one of the last pillars from the fast disappearing class of systematic naturalists."

During the years between the passings of Scammon and Dall, Mildred had come to learn considerably more about her grandfather. Her curiosity was aroused when she came upon his *Marine Mammals* on a shelf in her aunt's knitting room. Grandpa wrote that book, Mildred thought, and began leafing through the pages. One sketch particularly caught her eye. Captioned "California Grays Among the Ice," it depicted a host of whales, their massive bodies craning skyward along the tumultous surface of an ice-packed sea.

Uncle Alex was more than pleased to tell Mildred the stories. He spoke of Father, as he always called him, with the utmost respect. Yes, Scammon had been considered among the very best of the old whaling captains. He'd even discovered the hidden lagoons halfway down the coast of Lower California.

"Later on, he felt terrible about that," his granddaughter continued. "He hated whaling really. He didn't like killing the mother whales out there with their babies. That's why he went into the Coast Guard [Revenue Marine]. That's what my uncle said."

> *The large bays and lagoons, where these animals once congregated, brought forth and nurtured their young, are already nearly deserted. The mammoth bones of the California gray lie bleaching on the shores of those silvery waters, and are scattered along the broken coasts, from Siberia to the Gulf of California; and ere long it may be questioned whether this mammal will not be numbered among the extinct species of the Pacific.*
>
> —SCAMMON, 1874

Scammon's *Marine Mammals* was published the same year that a bill to "impose a penalty on every man, red, white or black, who might wantonly kill buffaloes" was passed by both houses of Congress before being pocket-vetoed by President Grant. It was also the same year that Alvan Southworth, secretary of the American Geographical Society, traveled several thousand miles up the Nile into Central Africa. "Ivory traders, wars and civilization" were eliminating the elephant, Southworth warned, "and it is thought that a few years more will suffice to extinguish the last vestige of the African colossi."

However, among those who chased the whales, not many ever expressed regrets. Or, if they did, their remarks were cloaked in the glamour of it all. Here is England's W. F. Scoresby, writing as early as 1820: "There is something extremely painful in the destruction of a whale, when thus evincing a degree of affectionate regard for its offspring, that would do honour to the superior intelligence of human beings; yet the object of the adventure, the value of the prize, the joy of the capture, cannot be sacrificed to feelings of compassion."

One whaling captain, M. E. Bowles of the ship *Jane*, did envision what was coming. He published this account in July 1845, in the weekly *Polynesian* newspaper of Hawaii: "The idea sometimes advanced by captains of ships, who ought to know better, that 'there are now as many whales as ever there were, that they had only been driven from particular grounds, &c.,' is preposterous in the extreme. . . . That the poor whale is doomed to utter extermination, or at least, so near to it that too few will remain to tempt the cupidity of man, I have not a doubt."

Another account, *Prentice Mulford's Story: Life by Land and Sea*, describes the author's days as a cook aboard the schooner *Henry* in 1857. Since this ship hailed out of San Francisco and worked the coastline of Magdalena Bay, it's likely that the author and Scammon were acquainted. Of the gray whales, Mulford wrote, "They knew the doors to these lagoons leading out into the ocean as well as men know the doors of their houses." And what he saw of the gray whales' destruction clearly pained him in retrospect: "It is a mighty death, a wonderful escape of vitality and power, affection, and intelligence, too, and all from the mere pin's prick of an implement in the hands of yon meddlesome, cruel, audacious, greedy, unfeeling pigmies. Spouting blood, bleeding its huge life away, shivering in great convulsions, means only for us forty barrels more of grease, and a couple of hundred pounds of bone to manufacture death-dealing, rib-compressing, liver-squeezing corsets from. And all the while the calf lingers by the dying mother's side, wondering what it is all about."

Where does Scammon ultimately fit into this picture? In one of his scrapbooks, the captain had taped a four-page newspaper. It was called *The Lower Californian*, published by the "Exploring & Colonizing Expedition of the Lower California Company at the (projected) city of Cortez, Magdalena Bay." A front-page "Notice to Whalers" dated October 10, 1870, began: "The excessive & indiscriminate killing of whales during the last 10 years in these waters, destroying both cows and calves, is injuring materially the fishery. The Lower California Co. has therefore determined without delay to enforce its rights under its franchises and notice is hereby given to all whom it may concern that from this date all taking of whales is prohibited in, or upon the bays, harbors and coasts of lower California, both of the Pacific Ocean and of the Gulf of California between the 24th and 31st degrees of latitude, without first obtaining from the said company a duly certified license to prosecute the fishery which in such case will be subject to the supervision of an officer of the company."

Did such scenes as his frontispiece for *Marine Mammals*—the gray whales being devastated by whaleboats—continue to haunt Scammon? With whaling fleets that spanned the globe by the mid-nineteenth century, Herman Melville had wondered "whether Leviathan can long endure so wide a chase, and so remorseless a havoc." Scammon used the same word—"havoc"—to describe the whaling he'd been involved in, which he wrote resulted finally in nothing but "decaying carcasses . . . bleaching bones strewed along the shores . . . [and] havoc."

There is no evidence that Melville and Scammon, born six years apart, were aware of each other. Nor do Scammon's descendants know why Melville was chosen for his middle name. Both were whalers who became writers, and both left us some of the most vivid descriptions of whaling that exist. Their curious kinship is a conundrum, like the love-hate relationship between man and whale.

Ultimately we do not know if Scammon felt, in some way, like Melville's protagonist: "In his heart, Ahab had some glimpse of this, namely: all my means are sane, my motive and my object mad." We do know that Scammon retained a fascination that could be described as compassionate for the gray whale, whose destruction he, more than anyone else, made happen. He described the grays lying for hours on the lagoons' sandy bottoms in a few feet of water, until the rising tide floated them. He described their quickness, their "unusual sagacity," their fondness for their young. He described the hunting of the "poor creatures"

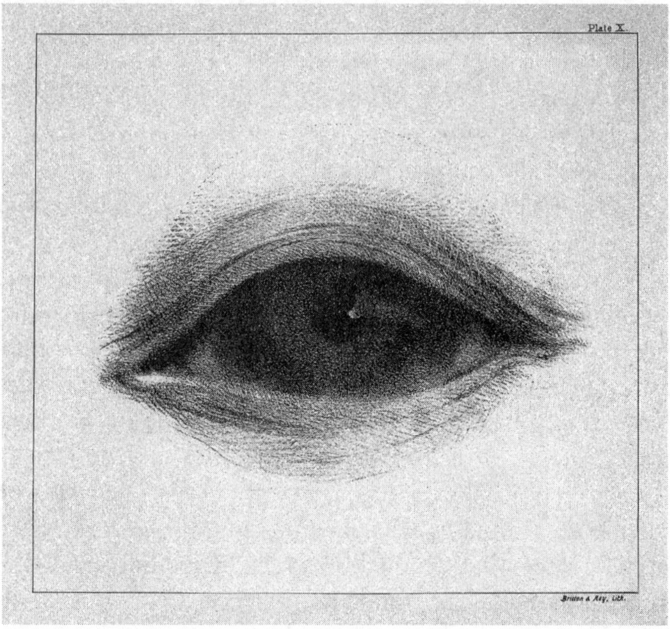

Scammon drawing, Eye of the Whale

by northerly Indian tribes, "launching their instruments of torture, and like hounds worrying the last life-blood from their vitals."

Scammon never explained just why he got out of whaling, whether it was his disdain for the business or simply the fact that profits weren't the same. In retrospect, had it become for him a conservation issue, as some of his entries indicate? Was he in a sense like Prometheus bound? Or did he have an eye-to-eye experience with the gray whales—which turned them, in his mind, from sources of oil and blubber to maternal, communicating, nurturing animals who mysteriously altered the course of his life? Did Scammon possibly undergo the type of conversion that so many people experience today in visiting the gray whales at the San Ignacio Lagoon?

Given Scammon's measured words—he was a man of the Victorian era—we will never be able to answer these questions definitively. It may be said, though, that the great hunters always, in a sense, "fall in love" with their prey. Especially if this is *dangerous* prey—whether elephants, lions, or gray whales. A life-or-death situation creates an intimate involvement between hunter and hunted, and cannot help but have a profound impact upon the individual. Consider, for example, Theodore Roosevelt, who after starting out as a major predator became our first great conservationist. By the late 1880s, as his biographer Edmund Morris puts it, Roosevelt had come to realize "the true plight of the native American quadrupeds, fleeing ever westward, in ever smaller numbers, from men like himself." With several other well-born sportsmen, he had formed the Boone and Crockett Club, looking to end the relentless and wasteful slaughter of big-game animals. Later, of course, Roosevelt was responsible for creating America's system of national parks.

During the twentieth century other hunters would undergo similar conversions. Here is Aldo Leopold, a passionate hunter who went on to found the Wilderness Society, writing in 1909 of what happened to him on a mountain in New Mexico: "We reached the old wolf in time to watch a fierce green fire dying in her eyes. I realized then, and have known ever since, that there was something new to me in those eyes—something known only to her and to the mountain. I was young then, and full of trigger-itch; I thought that because fewer wolves meant more deer, no wolves would mean a hunters' paradise. But after seeing the green fire die, I sensed that neither the wolf nor the mountain agreed with such a view."

Here is Ernest Hemingway, writing of an antelope hunt in 1951: "It's a trophy, men, if you glassed them right and took the biggest buck and didn't shoot a doe mistaking ears for horns. It is probably shot through both shoulders too and it is still living and will try to get up, looking at you, as you come with the knife. From the eye you can tell that the buck is thinking, 'What the hell did I do to deserve this?'. . . . The author of this article, after taking a long time to make up his mind, and admitting his guilt on all counts, believes that it is a sin to kill any non-dangerous game animal except for meat."

Harold J. Coolidge, a longtime American big-game seeker who was part of the last high-volume hunt in the name of science, would become president of the International Union for the Conservation of Nature and help establish the World Commission on Protected Areas. Britain's Sir Peter Scott, son of the legendary Antarctic explorer and another renowned hunter, would be among the founders of the World Wildlife Fund in 1961 and initiator of the Species Survival Commission to investigate what was happening to the great whales. Today in England there even exists a group of hunters turned preservationists that calls itself the Penitent Butcher's Society.

So Scammon was well ahead of his time—not only as oceanographer and naturalist, but as conservationist. More than a century later his descendants would look back upon Scammon's life and times. "He set out to do one thing and, in the course of the hunting, he saw the depletion," his great-great-grandson Dave Decker believes. "He saw the big picture, the future of what his own industry was doing. I think that's pretty phenomenal, for a man of the sea to have that kind of vision, far beyond his peers and his era."

Tom Decker, Scammon's great-grandson, added: "He turned his life around for the benefit of nature. If the captain was alive today, I think he'd be out there fighting to save the ocean environment."

Ironically, Tom Decker had taken Scammon's legacy a full circle. As a fifteen-year-old he had worked a summer at one of the last whaling stations to operate in the western United States. It was not long after World War II, at Fields Landing near Eureka, California. The station primarily rendered humpback, fin, and sperm whales. Tom assisted his father, who worked the boiler. The teenager used to take a big garden rake to the ivory sperm whale teeth at the bottom of the boiler, wash them off, and bring them in a bucket into the boss's office for counting. He'd been there the day a cable broke hauling in a seventy-foot-long finback and decapitated the foreman. He'd be long gone by the time he heard about the boss committing suicide. As the whaling historian Campbell Grant put it, "Captain Scammon saw the beginning of whaling in California; his great-grandson saw the end."

While working at Fields Landing, Decker was befriended by a visiting marine biologist, Raymond Gilmore, who would go on to become the father of whale watching and the leading expert on gray whales. Tom loaned Gilmore a copy of his great-grandfather's book. Gilmore ended up being invited home for dinner, where Tom's mother, Mildred, mentioned, "Oh, you'd probably be really interested in seeing my granddaddy's sea chest." After viewing Scammon's logbooks, charts, drawings, and more, Gilmore immediately alerted the University of California. "There's a gold mine here," he told the Bancroft librarian. And that was how the captain's history came to see the light of day.

One day in the early 1950s, Mildred and Tom Decker received an invitation to visit Gilmore at the San Diego Museum of Natural History. Gilmore and the museum's curator were about to embark on a cruise down to Baja. As their boat pulled out of the slip, Gilmore waved goodbye and shouted out: "Here we've got the granddaughter of Captain Scammon and we're going down to San Ignacio and Scammon's Lagoon! You ought to be onboard here with us." Mildred shook her head. "Maybe next time!" she remembered shouting back.

"You know," she said as she told the story, "I think I'd like to visit those lagoons someday."

Chapter 27

VICTORY AT SAN IGNACIO LAGOON

The great, intelligent eyes stared back into his; was it pure imagination, or did an almost human sense of fun also lurk in their depths? Why were these graceful sea-beasts so fond of man, to whom they owed so little. It made one feel that the human race was worth something after all, if it could inspire such unselfish devotion.

—ARTHUR C. CLARKE, *The Deep Range*

FROM the dead of a Boston winter, my wife and I flew to join the gray whales again in Baja California, settling into the home my friends and I had built above the Sea of Cortés. We had quite a reunion planned for 2000 at San Ignacio Lagoon. George Peper, my photographer companion in Alaska, would drive down with my friend David Gude and their wives from Los Angeles. Richie Guerin and his wife, Jessie Benton, daughter of the late American artist Thomas Hart Benton, would drive up in a new motor home from southern Baja, accompanied by her son and grandson. We'd meet halfway. I'd been in touch with Maldo Fischer at the lagoon, who indicated it would be fine for all of us to camp out again on the premises of his family's *palapa*.

Homero and Betty Aridjis were to be at the lagoon as well, part of what was becoming an annual pilgrimage to the whales while planning the next year's strategies for the saltworks war. A new three-thousand-page environmental impact assessment on the proposed expansion was rumored to have been completed. However, it was anticipated that President Zedillo would hold off releasing the results and announcing his decision until after the presidential election in July. In Mexico City, though, Homero was keeping the pressure on. He had arranged through a friend, the city's director of municipal services, for posters to be placed at dozens of bus stops. Each one showed the regal tail of a gray whale rising above the lagoon waters. Emblazoned above it was a question: "*¿Patrimonio de la humanidad o de una empresa?*" Do the whales belong to humanity, or to a corporation? One such larger-than-life image loomed from a high-rise above the traffic-clogged passage onto Mexico City's Fifth Avenue, the Paseo de la Reforma. It cast a long metaphorical shadow on the presidential palace.

I set out a few days ahead of everyone else with a neighbor, Dennis O'Brien, and his three dogs in Dennis's modified four-wheel-drive Dodge Ram truck. We stopped the first night at Magdalena Bay, where I wanted to catch up on the latest acoustical work by Paco Ollervides. Paco and my Irish friend from Idaho hit it off immediately. Dennis told of being out on the Sea of Cortés with his wife, watching two gray whales "do something I've never heard anybody else talk about. They were just lying on top of the water. One would sound, then come up and hit the other one in its midsection with a big thump, lifting it right out of the water."

Paco smiled and nodded. "Yes, like playing a game—as Roger Payne has talked about with right whales, picking logs up onto their shoulder."

Dennis continued, "We sat there for two hours with the engine off, in awe that about twenty whales were all around us."

Paco's sound studies on gray whales were progressing well. He already had more recordings than last year. He'd detected no new sounds beyond the ten previously identified, but he had begun to notice distinct behavioral changes in the presence of whale-watching boats: more tail slaps with direct approaches, more breaches with indirect approaches.

Farther north, at Puerto López Mateos, recent changes in the tides had made the entrance too shallow for the whales to go in. So more grays were concentrating around San Carlos, along with tourists who canceled out at the other location. I was pleased to learn this about the site of the terrible chase I'd been part of a month earlier. Nature still seemed, at least, to be on the side of the gray whales. Paco added, however, that there was pressure by the boat captains to open a mangrove area north of Devil's Curve to whale watching. "Of course, we recommend not to do that, because the place where they give birth is in those channels."

By this eighth week of the season, nine whales—one more than last year at the same time—had stranded themselves in the bay. Only last week a particularly painful stranding had occurred. Paco explained: "The bathymetry is very different here this year. There are new sandbars, and even experienced boat drivers have run aground. We had a live stranding in one of these places. I gave the whale five hours to live, and he lasted five days. We kept water on him during the day, but we couldn't rescue him. This was a thirty-five- to forty-ton adult, and the water was too shallow for a big boat to get through. We needed a medium-sized boat, and the Mexican authorities have one at the dock. But they say, well, the key is in La Paz. The harbormaster here has a patrol boat, but he says it's only for safety of the dock. I tell them, this is an emergency. But they have different interests, different attitudes.

"So in the morning, we would arrive and the seagulls were pecking at the whale's back and he was bleeding. Tourists were crying. This is the same emotion you feel when you get to touch a whale, but the flip side, when you are seeing it suffer and you can do nothing. It didn't want to die. Finally, you see the eye closing."

None of us said anything for a while, until I asked Paco if he'd heard any rumors about the outcome on the saltworks expansion. "No," he said, "nothing official yet."

By the following evening, Dennis and I arrived in the ESSA corporation's hometown of Guerrero Negro. We had dinner with Shari Bondy, who'd come here in 1988, fallen in love with the whales (and one of the Scammon's Lagoon whale-watching captains), and taken up residence to run tourist cruises herself. Bondy described a phenomenon that she began observing in the mid-1990s. "First I noticed nursing groups forming, of all the whales born that same day or over a couple of days, usually about six mother-calf pairs. And almost every group had with them what looked to be a crusty old whale, like an elder." Each of these whales was barnacle-covered, the many layers of scar tissue causing it to appear almost white. Each was also small, thin, even emaciated. When she would throttle down her engine about six hundred feet from a nursing group and begin an approach, invariably this "elder" whale would break away from the mother-calf pairs and come toward her boat. It would swim very slowly. The elder would make a couple of tight circles around and under the *panga*, just out of reach. It wouldn't spy-hop but would instead do what Bondy refers to as a "head-lift," bringing one eye out of the water as if to check out her boat. Then the whale would return to its nursing group.

"I would bet money these sentry whales are the matriarchs," Bondy said. "Maybe they're the baby-sitters, too. We have no idea whether these nursing groups might be related—and perhaps this is great-grandma." She's observed elders to be the first whales to enter the lagoon, as early as November. During the 1999 wintering season she saw something new. It was ten elders clustered together, rather than with the mothers and calves. "The next day, one of these died. And then they dispersed. It made you wonder if they hadn't been paying their last respects."

At around nine the next morning, I placed a call home to see how my friends' trip plans were progressing. Richie answered. "I knew it was you!" he exclaimed, almost before I could say hello. "Five minutes ago I had the radio on, and they started broadcasting part of a speech President Zedillo gave yesterday afternoon in Mexico City. I thought I heard him say, '*No salinera.*' I couldn't believe it! I asked our cook if that's what he heard, too. He said, '*Sí, sí. No salinera.*' "

My mind raced to interpret this news. "You mean," I stammered, "they've canceled the saltworks project?"

"That's what Zedillo said!" Richie exclaimed again.

The announcement by the forty-eight-year-old Zedillo, in his final year of a six-year term in office, had been given at a conference on biodiversity policy at the presidential residence. Almost as soon as Zedillo stood to speak, he began talking about the wonders of San Ignacio Lagoon. "This is a place that has had minimum interference by humans—one of the few places like that left on the planet," he said. It was an echo of Homero Aridjis's—and other environmentalists'—rhetoric. The president did lash out at those he asserted "have used this project to seek notoriety and even, I have to say, to profit financially and politically," clearly referring to the expansion's opponents. He also insisted that the saltworks would not have harmed the whales or other wildlife. These things notwithstanding, Zedillo continued, the saltworks would change the landscape "in a place unique in the world both for the species that inhabit it and for its natural beauty, which is also a value we should preserve. . . . I have made the decision to instruct representatives of the Mexican government . . . to propose a permanent halt to the project." He added: "Fortunately, Mitsubishi is backing up the Mexican government's decision."

The timing couldn't have been more auspicious. Later this afternoon activists including Homero and Bobby Kennedy were to arrive at the lagoon for a strategy session that would become a victory celebration. If we left Guerrero Negro immediately, we'd probably get there around the same time. Perhaps they'd know something about the ultimate reasons Zedillo had suddenly given up on a $180 million investment that would have earned between $85 million and $100 million in estimated

Celebrating at San Ignacio Lagoon (left to right), Joel Reynolds, Bobby Kennedy, Jr., Homero Aridjis, and Jean-Michel Cousteau (George Peper)

annual export revenues. We drove by the image of the gray whale's tail on the front gate of the salt factory and sped toward San Ignacio.

In the oasis town I dropped in at the Kuyima camp office to ask José de Jesús Varela Galván about Zedillo's decision. "It's a surprise for us," he said. "We thought the decision would wait until the next president is elected. But, for us, of course it is good, we did not like this project."

This year the NRDC/IFAW gathering at the lagoon was being held not at the American-owned Baja Expeditions campground but rather at Maldo Fischer's new Campo Cortés. It seemed fitting that a Mexican—and such a man as Maldo—would be able to host the group at this, probably its last, annual pilgrimage to the whales. When Dennis and I pulled up, we learned that Homero and Betty had arrived and were already out on the water. A van from the little lagoon airport carrying Bobby Kennedy, Jared Blumenfeld of IFAW, and others followed immediately in our tracks. After stowing his gear in a tent, Bobby headed off alone to explore the seashore.

I saw Homero and Betty wandering in from the beach, their pant legs rolled up. Jacob Scherr, the NRDC's director of international programs in Washington, embraced the poet and said, "Such a long fight, Homero. It's wonderful. Wonderful." Homero smiled and replied, "Well, the president was a little bitter at the environmentalists."

There was all kinds of speculation on how the decision came about. It turned out that Zedillo had come to the lagoon the previous weekend, accompanied by Environment Minister Julia Carabias, ESSA Director Juan Bremer, and others. "But according to the people at Kuyima," said Mark Spalding, a professor of international relations at the University of California, San Diego, "Zedillo gave no impression to anyone what his thoughts were." The long-awaited environmental impact assessment (EIA) had been submitted to Mexican authorities the day *before* the president's speech, allegedly accompanied by a building permit request from ESSA. Mitsubishi had a press conference scheduled for immediately *after* the president's speech, supposedly to announce the EIA's findings. "I think they may have gotten blindsided," Spalding guessed. Scherr argued that, on the contrary, he thought Mitsubishi had been looking for a way out for a while.

Blumenfeld, who like the others had learned of the cancellation via a phone call from the media, had immediately dialed Mitsubishi International Corporation headquarters in New York. The company's press spokesman, Steve Wechselblatt, had gone completely silent when he popped the fateful question. "Are you there?" Blumenfeld asked. "I said we just heard that the project has been canceled, is that true?" Wechselblatt said, "I can't comment. I'll have to get back to you." When he called five minutes later, it was to say, "Jared, you're right. You won."

That night under the big tent, while the tide of an approaching new moon rose to crescendos against the lagoon shoreline, Blumenfeld inaugurated the festivities by dousing everyone with champagne. Then, one by one, people stood to express how they felt. "I've been in a condition of total and utter disbelief now for over twenty-four hours," Jacob Scherr began. "I don't think I will ever forget the precise moment I heard." He'd been standing on the tarmac at La Guardia Airport when *The New York Times*'s Mexico City correspondent rang his cell phone. Boarding the plane, Scherr turned to the passenger next to him and said, "This is the best day of my life." Now he revealed, "I had to restrain myself from wanting to scream and shout and hug everybody."

He looked across the table, through the flickering kerosene lamplight. "The reason I am here, and NRDC is here," Scherr continued, "is because of Homero and Betty, who called asking whether we'd be willing to join with the Grupo de los Cien in this fight. Little did I realize that we would be sitting here together almost five years later, celebrating one of the most important environmental victories of our generation."

Mexico's TV soap opera star Angélica Aragón, here with her young daughter, warned: "Be very vigilant. It is not out of conviction that our government has taken this stand."

Bobby Kennedy said, "I knew we were gonna win this." He went on to speak of bringing his children to the lagoon when they get a little older.

Joel Reynolds, director of the NRDC's Marine Mammal Protection Project out of Los Angeles, said: "I think if we turned all the lights off in here, the whole tent would glow anyway. This is a coalition that's crossed state, regional, and international boundaries, has taken on the largest corporation in the world—and that corporation has given up. That is an amazing achievement. . . . Yet if we could not save Laguna San Ignacio, our environmental future wouldn't be worth a damn."

Betty Aridjis concurred, saying that to have proceeded with the saltworks "would have been the thin edge of the wedge which would let them get away with *anything* in the future."

Homero was shaking when he rose. "I was prepared to come this year dressed for battle," he said. When the *Chicago Tribune* correspondent called his home in Mexico City, saying, "I have good news for you," Homero was sure the man was mistaken. "When I learned it was true, I felt disarmed, as one who is running and suddenly has to stop. Then Betty said to me, 'We shall go to the lagoon to watch the whales in a different way—free and enjoying their right to exist as an animal species that human beings have to respect.' "

Homero inhaled deeply, and for an instant I thought I heard the exultant sighing of a gray whale. He went on, "Globalization sometimes has its good side. In this case, that an isolated lagoon in Baja California Sur, known by very few people, became a symbol to defend the environment. For us, this has been a victory of life—against corporations, against governments, against greed."

Jean-Michel Cousteau, president of Ocean Futures out of Santa Barbara, California, said: "My faith in humans has gone up a notch, and it was pretty low." He recalled sitting down with President Zedillo not long ago, a visit that lasted nearly an hour. "I can tell you that there is a human being who happens to be a diver, and was torn about this issue. The indication I felt was, if he had a chance to *not* see this project go forward, he would take it."

Jorge Peón, the local custodian of the airport, spoke of when the people of the lagoon came to have schools and electricity, that would truly be cause to celebrate. Bobby Kennedy immediately went over and shook his hand.

In a cynical age the gray whales had sounded a triumphant call for decency and compassion.

The morning whale watch departs at 9:30. Homero is wearing his monarch butterfly cap. Also on our *panga* are Jacob Scherr and IFAW's board chair, Anne Fitzgerald. Pancho Mayoral, the third of Pachico's sons, is our guide. On a flat-calm day, Pancho says he thinks the whales have good moods and bad moods. He's barely finished the sentence when one spy-hops right behind the engine. Anne says she feels like singing a lullaby. Pancho says he'd once tried "La Cucaracha," but it hadn't seemed to work very well. Whale-spouts are all around us.

Homero is writing in a little notebook. As pelicans pass overhead, Betty says, "They look like a flying string quartet." Homero says, "I'm quoting you." A breach. Anne says, "We're surrounded." Homero says this morning is "quiet, serene, full of whales." Pancho says, "In this weather you get beautiful rainbows when they spout." As whales pass under our boat, the water turns an extraordinary aquamarine. A mother nudges her baby in our direction. I find myself able to pick out individual whales by their markings, for the first time. Silently, I name them.

For a full ten minutes Homero is scribbling furiously. When he finishes he says this is not a poem, "it is a bio-message," though he cannot describe what it is about. Another *panga* pulls alongside us, and a reporter from Mexico City's TV Azteca reaches out to Homero with a microphone. At the close of the interview, I hear him recalling the first time he reached out to a whale here in 1997. "It is a magic touch, a communication of another kind, beyond species, where you are speaking in some supersensory way. You can feel them. And you feel that they can feel you. How people become transformed because they touch a whale is inexpressible. You can't explain why."

Not far away a whale has opened its cavernous mouth and is allowing everyone to stroke its baleen. I see Bobby Kennedy lean far over the rail. He appears to have placed his entire head inside the jaws of the whale.

That afternoon the visitors boarded a chartered plane and flew across the lagoon, out into the Pacific, and above the mangroves, to the little fishing village of Punta Abreojos. Two weeks earlier the community had declared, after putting the issue to a vote, that it stood united in opposition to the salt plant. ESSA had promised that the factory would bring jobs, electricity, and paved roads. But the town's size and character would have changed forever; a pier for large cargo ships and fuel tankers, constructed in the midst of productive lobster and abalone beds, would have been only the start. Courageously, the one thousand residents elected to preserve what they have. Now the local fishing cooperative, which had led the fight against the expansion, had invited the activists to a meeting. But nobody had any idea what really awaited them.

As the plane touched down on a narrow airstrip, Homero looked out the window. His jaw dropped. The runway was surrounded by a caravan of at least a hundred cars and trucks. People honked their horns and waved banners out their windows. Dozens more people were standing together and applauding. Immaculately dressed schoolchildren sang a song of greeting and held banners aloft. "My God," Homero heard IFAW's President Fred O'Regan marvel, "the whole town came out!" One of the signs—*Patrimonio de la Humanidad*—was almost a replica of what Homero had inscribed on the bus stop posters in Mexico City. Tears came to the poet's eyes. Betty put her arm around him.

Antonio Zúñiga, president of the fishing cooperative, stepped forward to greet them. "This is the best news we could have hoped for," he said. The guests were ushered into waiting vehicles and driven to the nearby school. Folding white chairs were assembled around the outdoor basketball court, with the Pacific Ocean as the backdrop. On the other side of a pillar from Bobby Kennedy sat the Aridjises. On display in front of them was a large posterboard where the children had placed drawings of the lagoon alongside photographs of gray whales. About forty children lined up and began to chant and sing some more. They wagged their fingers as they cried out, "*¡No salinera! No polución!*" Their next words translated: "For a better life and world! Take care of our world for a better future!"

The guests were asked to take turns at a standing microphone in midcourt. Besides the TV star Angélica Aragón, the warmest welcome was reserved for two men. Bobby Kennedy had been here before, developed relationships with a number of members of the fishing cooperative, and gone out with them in their *pangas.* Homero Aridjis's 1995 appearance on a half-hour segment of Mexico City's *60 Minutes* program had been shown in the high school here, inspiring the townspeople with the firmness of his opposition to the salt factory.

Bobby began his brief speech in Spanish, talking about coming together to celebrate a great victory after five years of war. Then he shifted to English. "Today we drove through your beautiful community," he said, "and we understand why you have something here that you wanted to preserve. And you had the courage to go into battle against your own government, and the largest corporation in the world."

Homero seemed overwhelmed by the occasion and thanked the people in but a few words.

After this ceremony the guests were taken on a tour of the processing plant that adjoins the fishing cooperative's offices. At long wooden tables women wearing masks and rubber gloves were shucking huge clams and abalone. Bobby walked through with Aragón, sampling the delectable fresh fruit of the sea. The irony, as Jacob Scherr would point out later, was that while globalization threatened to destroy the ambience of this town and the lagoon, another result of globalization has kept it alive. A nine-ounce can of Punta Abreojos's abalone wholesales for $40; an entire case is worth $1,500. In 1999 the town earned about $5 million in abalone revenues. And much of this came from shipments of abalone to Japan, where the rainbow-colored shellfish is considered a delicacy. One visiting Japanese journalist said he'd seen more abalone here on a single table than probably in his entire life.

The guests were escorted a few miles outside town. There, on the very site where ESSA had planned to construct its pier, stood a large canopy. Beneath it, on wooden tables covered with white linen cloths, lay a beautifully prepared feast of barbecued oysters, fresh lobster, and fish. A man presented a huge platter to Homero. This was, the poet would say later, one of the most moving days of his life.

As the group reboarded the plane late that afternoon, many townspeople gathered once again on the runway. "*¡Adiós!*" they shouted, "*¡Vaya con Dios!*"

That night, as Fred O'Regan brought out another case of champagne in the tent, no one could stop talking about the reception in Punta Abreojos. "When you saw the conviction, the gratitude on the people's faces," said Patricia Martínez of Baja's Pro Esteros organization, "you knew that all the writing and shouting and fighting had been worth it." Before departing the lagoon, some of the group would meet again with the town fathers of Punta Abreojos, looking to define their future needs. "We're going to continue to help these communities," said O'Regan, "not just step away and leave them hanging." The plan was for several experts to spend some months in Punta Abreojos and the settlements directly bordering the lagoon, to help the people develop better-managed fisheries and shellfish aquaculture. There was even mention of assisting the locals in producing some "whale-safe" or "organic" table salt from a small portion of the natural salt flat.

As the latest round of champagne passed down the table, Homero spoke of how, since the beginning, the Mexican government had sought to minimize the importance of the gray whale. They had emphasized the steps being taken to ensure no impact upon the landscape, or the more obscure species of fish and wildlife. All through the campaign Homero had remained relentless and admittedly single-minded. "The gray whale," he said, "is a universal symbol."

And so, in the end, it seemed to be for President Zedillo. Having heard that—only five days before making his decision—Zedillo had been here at the lagoon, I set out to learn all I could about what had transpired. My search led me first to the Kuyima camp, whose whale-watch guides had taken the president and his entourage out onto the lagoon, and eventually to Kuyima's founder, José de Jesús Varela Galván. Overcoming some initial reluctance to talk, I managed to piece together the story.

Zedillo's interest in paying a personal visit to San Ignacio Lagoon had initially been sparked by watching a videotape sent to him by a Mexico City TV station. It showed the lagoon's "friendly" whales approaching boatloads of tourists. In early March 1999 the president had come for the first time on a quick outing with his family, landing by helicopter along the shoreline. During their hour on the lagoon, one whale had approached two of the president's four children, who touched it. But none would come near Zedillo or his wife, who told their *panga* captain, Víctor Ramírez, that she wasn't feeling well. There were rumors that Zedillo finally went overboard in his diving gear to try to observe the whales, but these could not be substantiated.

Afterward Zedillo was told about a group of schoolchildren from Punta Abreojos who had also come to the lagoon that morning. They had traveled five hours by car, bringing telescopes and binoculars to observe the black brant geese. When the helicopters came the birds had flown away. The children had returned home without observing any. The president was said to have been considerably disturbed by this news.

Not long thereafter Zedillo's teenage son received a letter from another adolescent, the son of Patricia Martínez. The letter said something like, "Please tell your father we need to preserve this beautiful lagoon, for the future, for your children and mine."

When the president returned to the lagoon on Saturday, February 26, 2000, he was accompanied by a contingent of about twenty-four people. Two nights before this the president had flown into Guerrero Negro from Mexico City. He and his family had gone whale watching at Scammon's Lagoon on Friday, then met with ESSA Director Juan Bremer, who came along on Saturday morning as Zedillo made a stop in the Vizcaíno Desert to check out the biosphere reserve's program to preserve the endangered peninsular pronghorns. After that he and the others had flown in two helicopters into Kuyima.

José had greeted the president. His brother Carlos, who supervises the camp's operations, had learned of the visit only the day before but arranged for all points of entry to Kuyima to be sealed off. Four *pangas* were at the ready. Ramírez had left this camp for the one down the way at Baja Discovery, so Luis Murillo, born at the lagoon and a boat captain for Kuyima these last six years, would guide the president.

The visiting dignitaries divided up. One *panga* would carry ESSA's Bremer, the Vizcaíno Biosphere Reserve's director, Víctor Sánchez, and two of the president's sons. His third boy would accompany Environment Minister Julia Carabias. As well, Zedillo had brought along three prominent Mexican writers—Héctor Aguilar Camín, Angeles Mastretta, and Federico Reyes-Heroles. The latter gentleman, a historian, was invited to join the president's *panga*, along with Zedillo's wife, Nilda Patricia, and the couple's young daughter. Kuyima's José also rode with the president, who immediately perched himself on the bow, a camera draped around his neck.

They had not been out as long as minute, José estimated, when a lone whale could be seen steaming toward them. It was not a large gray but a calf. Luis cut back on the throttle. Zedillo raised his camera as his wife and daughter placed their hands in the water and began to splash. And there it came, gliding up along the starboard side. As Nilda leaned the young whale lifted its head. The wife of the president of Mexico had kissed a whale! Zedillo's shutter clicked away. He was ecstatic. His wife was beside herself. "That," as one of my friends would say later, "was a *very* political whale."

Another "friendly" gray approached the president's boat during the hour they cruised out around Rocky Point, and everyone aboard the *panga*—with the exception of Zedillo, who was quite content to chronicle the moment in pictures—ended up touching a whale. Afterward, all gathered under the Kuyima *palapa* for a festive lunch of fresh fish (corbina), oysters, lobster, a meat dish called *machaca*, and the customary rice and beans. Julia Carabias, who had previously been to Guerrero Negro and Magdalena Bay but never to San Ignacio Lagoon, hadn't touched a whale but witnessed many "friendlies" in the vicinity and said she'd "never seen anything like this." ESSA's Bremer also expressed amazement, noting that he'd seen many backs of whales while out on Scammon's Lagoon but never any so close as today.

It was President Zedillo, though, who was consumed with questions. How is it possible that the gray whales are so friendly here? he wanted to know. How old are the whales? When do they arrive here and when do they leave? How do the local people live during the whales' off season, when ecotourism in the lagoon shuts down? What are the principal fisheries on the lagoon?

Zedillo did not bring up the saltworks project. Nobody else mentioned it either.

"The president said he didn't need to touch a whale," said José de Jesús Varela Galván. "For him, this trip was perfect. I think he and his family have a very special day."

Late afternoon. The sun has begun to fade, and I am looking across the lagoon to where the earth bends on the horizon beyond the mesas. On the shoreline at Campo Cortés, Jean-Michel Cousteau is saying how surprised he's been to see so many single whales still around; usually, the males have left by now. He is planning to title a documentary tracing the gray whales' migration *Migratory Hurdle*, for the long obstacle course they must run. We talk about having met more than a year ago at the Makah reservation in Neah Bay, during the course of Cousteau's unsuccessful attempt to convince the tribe to conduct a symbolic hunt. "A culture that does not evolve," he says sadly, "does not survive."

Homero sits down on the ground beside us. He says that he's sorry he wasn't on Jean-Michel's boat, because Cousteau is always the one who touches the whales. "Just lucky," Cousteau says. "No," Homero says, beaming at him, "it's magic."

About this time a car pulls into the camp bearing George Peper and David Gude. They've already dropped off their belongings at Maldo's *palapa*, where they received directions to find us. We drive off together for Kuyima. Our other friends, bringing the motor home up from Cabo, ought to be here by sometime tomorrow. I've arranged for two *pangas* for us—one to be piloted by Maldo, the other by Pachico. I can't imagine a better combination.

The next morning at Campo Cortés, the men and women who had fought so long and hard for the whales assembled for a group photograph before their departure. We said our good-byes—to Homero and Serge Dedina, the two who'd ignited the saltworks tinderbox; to Joel Reynolds, Jacob Scherr, Fred O'Regan, and Jared Blumenfeld, who'd continued to stoke the fire; to Bobby Kennedy, who walked in the shoes of the fishermen; and more. As the two buses pulled away for the airport, I felt my eyes mist over and wondered when our paths would cross again.

My other friends had arrived just in time to bid a hasty farewell of their own. Except for us the camp was suddenly deserted. And, ominously, the sky was growing dark to the south and a wind was increasing by the minute. "I thought you told us it never rains here," Jessie Benton turned to me and said. She'd been in the camper for two days, with her son Anthony—like his grandfather, an artist and muralist—and ten-year-old grandson, Atticus. They were a little beat up from traveling three hours on the miserable road from the mission town of San Ignacio but very excited about going out among the whales. And the wind was howling, the sky turning onyx.

It begins to pour. Later we will learn it has not rained like this in at least four years, maybe six. As we drive past the airfield runway, the small plane supposed to take Homero, Bobby, and the others back to San Diego is sitting stranded. The prospective passengers remain in the buses. I'm grateful the storm struck before they could take off. Less than a mile farther on, two carloads of us reach Maldo's *palapa*, dash inside, and huddle there. The cane-branched roof starts to leak in several places, and we locate a couple of buckets. The wind rattles the roof to such a degree it feels as if it will blow away. Outside the salt flats are turning to butter.

If the rain keeps up my friends' long-dreamed-of trip will be for naught. They'll need to be back in Cabo in time for Atticus to fly home to school in Kansas; he can't afford to miss more than a week. I think of Jim Sumich's story of his first trip to the lagoon in 1977, the sudden downpour that kept him marooned for ten days beyond his planned departure—and gave him the opportunity, at last, to meet the whales up close. Had the rain not fallen Sumich's life might have taken a far different turn. I look over at David Gude, playing cards at a table with Anthony and Atticus. And I remember, too, the night two years ago when our motor home nearly tumbled into the ravine before we'd encountered and been rescued by John Spencer. Had this mishap not occurred—leading us to John and then to Pachico—my own life, too, might have taken a different turn.

The lagoon—and the gray whales—have their ways and means.

At sunset there is a break in the weather. Atticus is playing on the mudflats, a hundred yards out at low tide, a small, lonely figure silhouetted against the stormy sky. A few hours ago the water was lapping at our feet. Now it is so far away, and in the distance the whales are blowing. Atticus is scrounging around, discovering things, clams, crabs, shells, octopus. He will come back to the *palapa* full of stories, specimens, and astonishment. For the moment it has ceased raining. The water is a deep turquoise, and a thin yellow sun is shooting through the long, black clouds. It is strikingly cold; we can see our breath. It looks as if more heavy rain is coming from the south. Soon we will gather in the *palapa* and eat raw oysters and fresh clams, and drink a little red wine. And none of us will mention the weather.

At intervals all night long I toss and turn in my sleeping bag on the shell-covered floor as the rain keeps falling. Then the morning dawns clear, and the lagoon is utterly calm. We wait for Pachico to arrive in his *panga.* At 9:15 A.M. his skiff—the now-familiar *Suzy Q*—pulls in to the shoreline on a high tide. Maldo brings his *panga* around to meet Pachico. Not long after, twelve of us are skimming rapidly over the glassy, shallow water.

Soon, as far as the eye can see, there are whales—blowing, breaching, rolling. The horizon is filled with smoke, from their blowholes. We can see Maldo's boat being approached. A whale makes the rounds of each of our friends' supplicant hands. Pachico's eyes seem to blaze as he scans the horizon. Approaching Rocky Point, he slows the boat to idle. There, right alongside us, a mother gray and her baby surface. The baby gently lifts its head. Jessie and her grandson lean over the edge and pet the velvety skin. The whale bows its head as they stroke the nose and above the eye. It's looking up at them, then raises its head effortlessly higher to look at everyone else. The mother is just underneath, quietly lifting her baby to be admired. Slowly, the pair sinks from sight.

There are two tour boats anchored on the lagoon, each around eighty-five feet long. The *Spirit of Discovery* and the *Royal Polaris* have woven their way through the narrow channel after four-day voyages from San Diego, 440 nautical miles to the north. Each vessel has several aluminum skiffs stowed on deck, to be lowered for its twenty-five or so whale watchers to clamber down a ship's ladder and aboard. A number of their *pangas* are in the water now, circling near the ships.

I recall that Bruce Mate, the Oregon State University marine scientist who pioneered satellite tracking of whales—and the first biologist to take a stand on citing concerns about the saltworks in 1995—is the naturalist onboard the *Royal Polaris*. Only later, reviewing my files, will I realize that this ship was the very *first* to experience "friendly" approaches to its skiffs, on February 16, 1976. A week after that the scientist Raymond Gilmore came to the lagoon on another vessel and wrote of a "friendly" gray: "As it swam around the larger boat it nosed under the outflow of water from the cooling system of the generators."

Now, in an instance of déjà vu, that is precisely what we are witnessing as Pachico motors toward the hull of the larger vessel. Except there is not one whale but *four pairs* of mothers with their calves playing in the outflow and rubbing against the big boat. Three other camera-laden *pangas* are in the same vicinity. Maldo is keeping his at a respectful distance, so the whales won't feel hemmed in.

The current is surprisingly strong in here. "It's as though the whales are riding it," Jessie says. Now the little ones on the surface begin moving from boat to boat, reaching up to be petted. Atticus leans, spreads his arms, and circles a calf's head with a hug. A huge mother rises perpendicular to gaze at the two of them, calf and boy. The high-set nostrils, the meditative, elephantine eye, "like the eye of an unfathomably old, unjudging god, prior to the faintest dream of being human," a friend once wrote after coming here. Is now and ever shall be . . . Jessie touches her as she goes by. Mother and baby pass under our boat and scratch themselves against it. So delicately that, while we can feel the vibration, the boat doesn't move. Anthony is petting the whales on the other side.

And it goes on. Time and again the whales return to our boat and to the others. Sometimes great volumes of air are exhaled underwater, rising in visible vertical columns until the air breaks at the surface in wide rings of foam and bubbles.

> *. . . for there is no splendor greater than the gray*
> *when the light turns it to silver.*
> *Its bottomless breath*
> *is an exhalation.*
>
> —HOMERO ARIDJIS

At one point a *panga* closes against the *Spirit of Discovery*, lessening the space between them, and a mother whale slowly comes up from underneath her little one and, lifting her head out of the water, very gently pushes the boat aside, making sure there is plenty of room for us all and no one will get hurt.

This species of whale manifests the greatest affection for its young, and seeks the sheltered estuaries lying under a tropical sun, as if to warm its offspring into activity and promote comfort, until grown to the size nature demands for its first northern visit.

—CHARLES MELVILLE SCAMMON

We are inundated in a plethora of gray whales. It is absolute joy. Stunningly beautiful . . . and totally natural.

The whales—they are my family.

—FRANCISCO "PACHICO" MAYORAL

A leviathan, wild and untamed in its element, allowing us the privilege of the deepest kind of communication and trust. People are shouting, "Oh, my God!" and the phrase is one not of epithet but of faith.

He is king over all the children of pride.

—JOB 41:25

Long after time has stopped . . . long after the heart is full . . . Pachico turns wordlessly and aims his finely tuned engine toward home. A sudden surge indents the water in a deep double swirl, like the design of yin and yang.

And the whales came out
to catch a glimpse of God
between the dancing furrows of the waters.
And God was seen through the eye of a whale.

—HOMERO ARIDJIS

Epilogue

AN UNCERTAIN FUTURE

Nature is loved by what is best in us.... And the beauty of nature must always seem unreal and mocking until the landscape has human figures that are as good as itself.

— RALPH WALDO EMERSON, "Nature"

IN the months that followed I set about to determine what other factors lay behind President Zedillo's decision to cancel the saltworks project. It proved to be a detective story with plenty of clues on many fronts. I stopped in La Paz on the drive back south from San Ignacio Lagoon and met with Jorge Urbán. The university professor had been in charge of reviewing potential effects on gray whales for the new environmental impact assessment (EIA). His report had been in the ESSA company's hands for several weeks.

While Urbán had concluded that no likely problems—in terms of reduced salinity, changes in water temperature, or noise—would affect the grays inside the lagoon, he did foresee a potential impact during building of a channel where salt water would be sent to pumps and then onto the salt flats. He'd recommended that no construction take place during the whales' winter season. Urbán also envisioned the pier at Punta Abreojos as posing a hurdle during the migration, since the whales travel closer to shore than the pier's two-mile extension into the Pacific. "Also the big ships make noise, and would be a source of potential oil contamination and physical injuries to the whales from direct impacts," he said. He'd recommended that ship movements be controlled—allowed to go no faster than three knots after entering the bay, turning off their engines and being pulled by tugboats inside two miles.

These things, of course, could be accommodated by the planners. Yet Julia Carabias, the environment minister, told the *Los Angeles Times* that what ultimately persuaded Zedillo was the effect of flooding tens of thousands of acres of desert just inland from the lagoon, which would have altered the landscape forever. Mitsubishi's Executive Vice President James Brumm echoed this at a news conference in Mexico City immediately after the president's announcement, saying: "What we would have done is flood salt flats. It wouldn't look natural. We came to appreciate a number of arguments by people that this is an area that should be left as is for ecotourism."

Later, when I met at IFAW headquarters with principal organizers of Mexico's fifty-group environmental coalition opposing the project, the attorney Alberto Szekely said the government had felt "betrayed" by the EIA because of its objectivity. I discussed this with Homero Aridjis, who said he "heard from sources close to the government that there were many factors." One of these, according to Homero, was financial. In December 1999, Zedillo had asked the ESSA corporation for an economic feasibility study of the expansion and was allegedly shocked to learn that none existed. Then the president discovered that the lion's share of the profits would go to Mitsubishi—"and he finally understood what we had been saying since 1995," said Homero, "that the new saltworks would provide only two hundred jobs, of which a mere fifty would be for the local population."

Homero also pointed out that around the same time Mitsubishi Corporation had been awarded a $278 million contract to build two new thermoelectric power stations on mainland Mexico. The poet speculated this "could be a consolation contract given by the Mexican government to its Japanese partners."

In mid-February 2000, two weeks before Zedillo's announcement, Minister of Commerce Herminio Blanco had flown to Tokyo for a meeting with the Japanese foreign minister and the Mitsubishi hierarchy. There are differing accounts of what transpired. The Japanese TV journalist Teddy Jimbo says, "My sources tell me that Mitsubishi, including its very top executives, was shocked to learn from Blanco about the president's decision but reluctantly accepted it." Homero heard an opposite story from his Mexican sources, that "Blanco was told by the Mitsubishi people that they wanted *out* of the project. Mitsubishi retreated because they were feeling the effects of the boycott in the United States. They have many different economic interests, and for them the marine salt was very little proportionately and it was not worth risking the prestige of the corporation any longer."

Mitsubishi International's James Brumm, who oversaw the saltworks controversy from his New York office and sits on the board of Mitsubishi Corporation in Tokyo, was careful to steer a middle course. "Mitsubishi and Mexico were proceeding on parallel tracks," he told me. "As partners we had independently come to the conclusion we were not going to move forward, and it was a matter of who talked to whom first. Blanco's trip to Japan presented the opportunity."

So we may never know just who blinked first, or why. This much, however, is certain: by 2000 the issue had captured the attention of millions, and it was not about to go away. The "Save Baja Whales" Web site mounted by IFAW had received about 2 million visitors over a four-month period. In California the NRDC had prodded the Coastal Commission and forty-three cities to pass resolutions against the project. Polls taken in Mexico revealed public opposition there to be at around 69 percent. High-powered Washington, D.C., legal and PR consultants hired by Mitsubishi seemed unable to stem the tide. Brumm conceded that the environmental groups had waged "a very effective campaign," and that "obviously our strategy was not."

Another full-page ad against the project—arranged by Homero and including nine Nobel Prize–winning authors—was scheduled to appear soon in *Reforma* and *The New York Times.* Instead, IFAW took out an ad in the *Times* on March 31 headlined, TO MEXICAN PRESIDENT ERNESTO ZEDILLO AND MITSUBISHI CORPORATION: THANK YOU FOR SAVING LAGUNA SAN IGNACIO! Then IFAW President Fred O'Regan kept a second promise he'd made to Brumm should ESSA bow out: he took the Mitsubishi executive out for an expensive lunch. Brumm would recall that, when Jared Blumenfeld asked Brumm whether he thought the project might ever be revived, "I think I said something like, 'God forbid!' " At IFAW's invitation he then brought his wife and daughter on a whale-watching trip off Cape Cod. It was, Brumm said, "just magnificent."

In Mexico, on July 2, three months to the day after Zedillo's decision, opposition leader Vicente Fox was elected to the presidency, ending seven decades of domination by the Institutional Revolutionary Party, or PRI. Andrés Rozental, another leader of the anti-saltworks coalition, said afterward that all the Mexican people signing petitions against the project represented "the first indication of a strong, pent-up desire by civil society to participate in decisions like San Ignacio, and the first indication of what later became a groundswell of the antiestablishment vote to elect Fox." Particularly striking was that three of Fox's closest advisers—campaign manager Héctor Elizondo, foreign policy adviser Adolfo Aguilar Zinser, and Senate Environmental Committee chair Luis H. Alvarez—had all been among the politicians who came to meet with the local people, and the whales, at San Ignacio Lagoon in 1999.

Ultimately, how much had the whales themselves influenced outgoing President Zedillo? "Oh, hugely," IFAW's Blumenfeld believes. "Whatever anyone says about all the technicalities, it was the whales." Or, as Homero Aridjis put it, "For me, the nursery of the whales was always sacred. This is not just an ecological victory but a spiritual one."

The gray whales received a reprieve in another arena in 2000. First, on April 20, as a Makah tribal canoe closed in on a gray whale not far from where they'd killed their first one a year before, a young female anti-whaling activist on a Jet Ski raced toward the crew. The noise and movement of Erin Abbott's craft caused the whale to dive out of danger from a harpoon throw. A Coast Guard boat proceeded to run right over Abbott, who spent the next four days recuperating in a hospital with a broken shoulder blade and fractured ribs. For violating the government's exclusionary zone, she faced charges that could bring jail time and a fine of up to $250,000. But she had saved the whale and, on June 9, a federal appeals court in Seattle rejected the federal environmental assessment that had allowed the Makah to proceed with their hunt. The U.S. government's review, concluded the judges in a 2 to 1 ruling, was "slanted" in favor of the hunt; the National Marine Fisheries Service was mandated to look again at the potential impact on resident gray whales and other environmental consequences before the Makah could resume the hunt. "My whales have at least a stay of execution," the tribal elder Alberta Thompson said.

Yet later that summer Japan sent its whaling fleet into the North Pacific to resume hunting sperm whales and Bryde's whales—two still-endangered species that were nearly wiped out before the 1986 commercial moratorium—under the guise of "scientific research." Already, at a meeting of the Convention on International Trade in Endangered Species (CITES) held in April in Nairobi, Japan had tried—and failed in a vote of member nations—to downgrade the level of protection given to the gray whale and two other species. A month later genetic scientists in New Zealand announced that gray whale meat had been sold commercially in Japan in 1999, falsely labeled as minke whale (140 of which, under the International Whaling Commission's rules, Japan is allowed to harvest annually in the Southern Ocean for stock research purposes). The University of Auckland scientists were unable to determine whether the gray whale meat came from the critically endangered Western Pacific population, the ones being studied by Dave Weller's team at Piltun Lagoon.

Then, at the IWC's annual meeting in July, Japan and Norway—flanked by seven of their patron states and Denmark—succeeded in blocking an effort by other nations, including the United States, to adopt a Southern Pacific Sanctuary for whales. A fisheries minister from Dominica resigned after claiming that Japan bought his country's vote. Later that month the Japanese government gave the green light to its whaling fleet. In August the United States joined fourteen other countries in a diplomatic protest to the Japanese. Then, on September 13, President Clinton directed that Japan be denied access to allotments for fishing in U.S. waters and ordered cabinet members to examine other options, including possible trade sanctions. By then, however, the damage had been done. Japan "harvested" forty-three Bryde's and five sperm whales (as well as forty minkes) before returning to port. They claimed to be studying fish consumption by whales—according to Japan's Institute for Cetacean Research, cetaceans consume three to five times the amount of marine resources that are harvested for human consumption. The U.S. commerce secretary, Norman Mineta, saw it differently, stating that Japan was "paving the way for outright resumption of commercial whaling."

The looming question is whether the U.S. response has been compromised or at least muted, because of having allowed the Makah to hunt gray whales again, precisely what Japan began pushing for in the early 1990s. One of the Makah's chief supporters, Eugene Lapointe of the IWMC–World Conservation Trust, phrased it like this in a late September press release: "President Clinton's threatened trade sanctions against Japan . . . reflect nothing more than made-for-media campaign hype prompted by his political party's desire to keep control of the White House." Meantime, the Vancouver Island–based World Council of Whalers was gearing up to hold its Third General Assembly in New Zealand.

If a large-scale resumption of whaling is one dark cloud on the horizon, the impact of global climate change is certainly another. Toward the end of April, I had joined the National Marine Fisheries Service scientist Wayne Perryman at Piedras Blancas Point in San Simeon, California. He was conducting a seventh annual count of northbound migrating gray whale females with their calves, supervising a professional team of four observers. From mid-March into June, as long as whales keep passing, they stand watch in paired shifts, with high-powered binoculars, over the course of twelve-hour days. Thermal imaging sensors on loan from the Navy are mounted on tripods and used to record data on the whales' nighttime movements. Aerial censuses are periodically conducted as well, to make sure that mother-calf pairs aren't migrating beyond range of the observers' sight. They almost never do. This is an ideal spot to see gray whales, especially mothers and their recently born, most of which will pass very close to the shore, in the lee along a bed of kelp.

On a point directly below the Hearst Castle, this is one of the West Coast's most scenic locations. I'm leaning back in an elbow of a rocky promontory overlooking the sea, hunkered down alongside Perryman and against a twenty-five-mile-an-hour wind from the northwest. About forty feet below, huge waves send a jet spray of seawater that falls just short of drenching us. In the spring huge flocks of Pacific loons settle in to feed in these waters, a million or more passing through in the course of their migration. Peregrine falcons nest near a group of sea lions, and an elephant seal colony occupies a beach at the base of a small hill. Behind me the meadows bloom with purple lupines, yellow mustard, and bright lavender iceplants.

Perryman explains that the gray whale mothers usually keep their calves on the inside, between themselves and the shore, apparently as a protective measure. "Almost 90 percent go by as a single unit," he says. "You can tell they're calves just by their size and the way they swim. At this point they're still really transitioning out of dog-paddling. So the heads keep floating up out of the water. The cue that they're coming, though, is the mom—because we can easily see her blow three miles away."

Then Perryman adds, wistfully: "But it's not as much fun to do this in the bad years." Nineteen ninety-nine had been one such, and this year looked even worse in terms of the numbers. "On a normal year at this time, you'd expect to have seen a hundred cow-calf pairs. We've seen twenty-one. You can stand out here all day and see zip."

Perryman hunches his shoulders against the wind, and continues: "I've never seen whales that look as thin as some of the ones this year. We had a juvenile going by yesterday that looked just terrible. The whale looked pasty, washed out. It hung around, roaming back and forth, wasn't swimming very well. With a few, I'm afraid their skeleton structure is showing through their blubber layer."

All of this, unfortunately, was bearing out the belief of Perryman and other marine scientists that something was very wrong on the gray whales' feeding grounds in the Bering Sea. This seemed to be affecting not only their body condition but their reproductive capabilities. When I spoke to Perryman again after the migration, his total calf count for the season had reached only 96—the lowest figure since he started the survey and well below "the reasonable range to expect of somewhere between 350 and 450."

At the same time the count of stranded gray whales—either found beached or floating dead—had by November reached another record number. It stood at 360, well above 1999's previous known high of 274. The majority of these whales had been detected in Mexico, with 207 found dead either inside or near the breeding lagoons, an 83 percent increase over the year before. The remainder of the tentative count on the northbound migration route was California, 57; Alaska, 56; Washington, 23; Canada, 15; Oregon, 2.

"If the whales were starving in 1999, they were starving in 2000 as well," according to the draft of the report for *Cetacean Research Management* by Le Boeuf, Mate, Urbán and other marine scientists. The report went on to hypothesize that gray whales' low body reserves and inanition was caused, in part, by "the depressing effect of increasing water temperature over the last decade on amphipod biomass."

The probable chain of events with the grays' predominant food source had to do, first, with the warmer Arctic temperatures that dominated the decade and brought about a decrease in amphipod production. Then, secondarily, came the La Niña cooling trend of 1997–98, which resulted both of the following summers in late-melting ice that prevented the grays from reaching the feeding areas early enough and long enough. Additionally, new studies are indicating that, in years when ice is slow to recede, not enough food is available for the bottom-dwelling amphipods. One hopeful sign is that in 2000 an ice breakup happened early—"so these animals going back with their tanks dead-empty are at least going to be able to get to the feeding grounds," as Perryman puts it.

In the Arctic the general trend of rising temperatures has caused sea-ice thickness to decline by more than 40 percent since 1958; it's been estimated that the Arctic's year-round ice pack could completely disappear within another fifty years. That would spell disaster for ice-dependent wildlife such as walrus, bearded seals, and polar bears. It's quite possible, too, that the Northwest Passage could open up, exposing cetaceans to increased ship traffic. The prognosis for the gray whales' summer feeding habitat is, overall, not good. A recent study for the World Wildlife

Fund/Beringia Conservation Program summarized the situation: "Scientists studying global warming believe Arctic ecosystems and their wildlife will be far more vulnerable to climate changes than those at the lower latitudes. Temperate and tropical animals, fish, and plants may be able to shift their geographic ranges northward to stay within comfortable climatic ranges. But for temperature-sensitive wildlife living near the poles even a modest amount of warming leaves no options. For the 'organisms of the tundra and polar seas,' writes biologist Edward O. Wilson, 'the North and South poles are the end of the line. All the species of the high latitudes, reindeer moss to polar bears, risk extinction.' "

What might gray whales be trying to "say" to us? Here is Melville in Moby-Dick, describing having come upon a whale nursery: "The young of these whales seem looking up toward us, but not at us. . . . Floating on their sides, the mothers also seemed quietly eyeing us.—Some of the subtlest secrets of the seas seemed divulged to us in this enchanted pond." So it is at San Ignacio Lagoon. Never is there a moment's fear that the whales will do us harm. Their huge size makes them more sensitive, not less. On the surface, externally, they are mountains of mottled gray, leathery skin and barnacles. Beneath the surface, internally, they are as big as the imagination.

Imagine . . . What if lions in the jungle suddenly allowed you to pet them? What if elephants suddenly slept at your feet? The Mexicans say the gray whales are "tame." Yet they are not domesticated. We did not "break" them as we might a horse. They tamed *themselves*—to come to *us*, their time-honored enemy, in the place where they give birth. And, mysteriously, it feels as if this is how it should be, how it *used* to be. The commonality is primordial. We are molded of the same clay. *Eschrichtius robustus. Homo sapiens.*

We are thus, in a phrase, biblically bound. Beyond Genesis and the Book of Job, we are told in Matthew 12:40: "For as Jonas was three days and three nights in the whale's belly, so shall the Son of Man be three days and three nights in the heart of the earth." And D. H. Lawrence, in a chapter on *Moby-Dick* that concludes his *Studies in Classic American Literature*, informs: "In the first centuries, Jesus was Cetus, the Whale. And the Christians were the little fishes."

Our Western culture is, of course, not alone in the symbolic meaning we ascribe to the leviathan—"bigness, largeness, the mass that moves upon the seas," as the Tse-shat peoples of Vancouver described the gray whale. The whale sustained and took care of the ancient peoples. The shamans knew. In trance states a trained shaman could go to the whales and communicate with them. Western humanity does not understand the nature of "a god" in the same fashion. Except, it seems, in our subconscious. *Moby-Dick* again, as the hunt approaches, as Captain Ahab wrestles with the demons of his inner being:

" 'What is it, what nameless, inscrutable, unearthly thing is it; what cozening, hidden lord and master, and cruel, remorseless emperor commands me; that against all natural lovings and longings, I so keep pushing, and crowding, and jamming myself on all the time; recklessly making me ready to do what in my own proper, natural heart, I durst not so much as dare? Is Ahab, Ahab?' "

Lawrence writes: "What then is Moby Dick? He is the deepest blood-being of the white race; he is our deepest blood-nature. And he is hunted, hunted, hunted by the maniacal fanatacism of our white mental consciousness. We want to hunt him down. To subject him to our will."

In more recent times it is our psychoanalysts who have addressed this dilemma we have fashioned for ourselves, especially since the advent of industry and high technology. "Man feels himself isolated in the cosmos," wrote Carl Jung. "He is no longer involved in nature and has lost his emotional participation in natural events, which hitherto had symbolic meaning for him. Thunder is no longer the voice of a god, nor is lightning his avenging missile. No river contains a spirit, no tree means a man's life, no snake is the embodiment of wisdom, and no mountain still harbours a great demon. Neither do things speak to him nor can he speak to things, like stones, springs, plants, and animals."

James Hillman, in an essay titled "Animal Kingdom," elaborates: "The reading of living form, the self-expressive metaphors that animals represent, is what is meant by the legends that saints and shamans understand the language of animals, not in the literal speech of words as much as psychically."

Which brings us, once more, to the gray whale, the whale that, despite our history in seeking to destroy it, wants to live closest to us. If they are being forgiving toward us, the implications are enormous. This is surely, in part, why they touch us so deeply. Like gray whales in their lagoons, human beings too must seek solace, a centering focal point, a place to go that remains relatively untouched and pure. A place to remind ourselves of our *basic* nature, not surrounded by all we have built. So do we commune with the whales at San Ignacio.

The fight to protect this "sacred nursery," a fight that captured the attention of so many—whether they'd had direct contact with these whales or not—represents something beyond environmental awareness or fervor. What is hurting them is hurting us. As the oceans go, so go we. Can we survive global warming? Noise pollution? The wanton carelessness about our habitats? Can we pretend to endure anything that the whales cannot? Can we come to grips with the suicidal tendency to destroy what sustains us? Is this what the gray whales are reaching out to communicate?

The answers to these questions are as yet unseen, hidden, perhaps entwined in our unconscious, in the great mystery of our relationship with these most majestic of nature's creatures. It may only be a mystery because we don't yet have the senses to perceive it, though we bump into it occasionally in the dark. And we glimpse it, at the rippling edge of life, bursting startlingly above the surface—in the eye of a whale.

Notes

Prologue: "Whales of Passage"

Background on lagoon region: "Conservation and Development in the Gray Whale Lagoons of Baja California Sur, Mexico," thesis by Serge Dedina and Emily Young (U.S. Department of Commerce NTIS PB96-113154, 1995). Basic information on the biology and migratory habits, and on native peoples' relationships with gray whales: Reviewed by marine scientists after being derived from numerous sources, including *Gray Whales*, by David G. Gordon and Alan Baldridge (Monterey Bay Aquarium, 1991); *Gray Whales*, by Jim Darling (WorldLife Library/Voyageur Press, 1999); *The Gray Whale*, by Steven Swartz et al. (Academic Press, 1984); *Gray Whales: Wandering Giants*, by Robert H. Busch (Orca Book Publishers, 1998); *Guardians of the Whales*, by Bruce Obee and Graeme Ellis (Alaska Northwest Books, 1992); *Men and Whales*, by Richard Ellis (Alfred A. Knopf, 1991); *The Golden Bough*, by Sir James George Frazer (Macmillan, 1922); *A Pod of Gray Whales*, by François Gohier (Blake Books, San Luis Obispo, Calif., 1988), and *The World of the California Gray Whale*, by Tom Miller (Baja Trail Publications, Santa Ana, Calif., 1975).

Version of the Makah Tribe's Thunderbird myth: *On the Northwest*, by Robert Lloyd Webb (University of British Columbia Press, 1988). Passages cited from Charles Melville Scammon: 1969 facsimile edition of *The Marine Mammals of the North-Western Coast of North America* (John H. Carmany & Co., San Francisco, 1874). "Gray whale's ferocity . . .": Newspaper article, Scammon's scrapbooks, among his collected papers in the Bancroft Library, University of California, Berkeley.

Chapter 1: Close Encounter at San Ignacio Lagoon

Portions of my description of whale watching on the lagoon first appeared in my contribution to *Whale Watching*, ed. by Nicky Leach (Discovery Communications, 1999). Dr. Bruce Mate reviewed the description of a gray whale birth in San Ignacio Lagoon, drawn from sources including "Observation of a Gray Whale Birth," by James G. Mills and James E. Mills, *Bulletin of the Southern California Academy of Sciences*, Dec. 1979. Background on Baja California came from numerous sources, including *Baja California*, 4th ed., by Wayne Bernhardson (Lonely Planet, 1998); *Almost an Island*, by Bruce Berger (University of Arizona Press, 1998); *Sierra, Sea and Desert*, ed. by Patricio Robles Gil, text by Bruce Berger (Exportadora de Sal, 1998); *Saving the Gray Whale*, by Serge Dedina (University of Arizona Press, 2000); *Resources of the Pacific Slope*, by J. Ross Browne (D. Appleton, 1869), *Baja California Plant Field Guide*, by Norman C. Roberts (Natural History Publishing, La Jolla, Calif., 1989), and *The Cave Paintings of Baja California*, by Harry W. Crosby (Sunbelt Publications, 1997).

Sources on Scammon and San Ignacio Lagoon: "Report of Captain C. M. Scammon, of the U.S. Revenue Service, on the West Coast of Lower California," in

J. Ross Browne's *Sketch of the Settlement and Exploration of Lower California* (1867); *Men and Whales at Scammon's Lagoon*, by David A. Henderson (Dawson's Book Shop, Los Angeles, 1972); "Ballenas Lagoon," by Scammon, *Alta California*, Mar. 31, 1865; "Interesting Geographical Fact: Large Saltwater Lake or Bay in Lower California," article from mid-1860s in Scammon scrapbooks.

Hubbs-Flynn expedition: "Nursery of the Gray Whales," by Lewis Wayne Walker, *Natural History*, June 1949. White expedition: *Hunting the Desert Whale*, by Erle Stanley Gardner (William Morrow, 1960). First "friendly whales": "Getting Close (Very) to Great Gray Whales," by Jack Goodman, *New York Times*, Feb. 2, 1975; "Gray Whales Losing Fear of Man, Evidence Indicates," by V. I. Murphy, *San Diego Union*, July 19, 1976; "The Friendly Whales of Laguna San Ignacio," by Raymond M. Gilmore, *Terra*, 15 (1): 24–28, 1976; unpublished accounts among Gilmore papers, University of California, San Diego.

Lagoon research and background: "Demography and Phenology of Gray Whales and Evaluation of Whale-Watching Activities in Laguna San Ignacio, Baja California Sur, Mexico," by Mary Lou Jones and Steven L. Swartz in *The Gray Whale;* "At Play in Baja's San Ignacio Lagoon," by Steven L. Swartz and Mary Lou Jones, *National Geographic*, June 1987; *So Remorseless a Havoc*, by Robert McNally (Little, Brown, 1981); "Into San Ignacio Lagoon," by Curt Cureton, Point Lobos Natural History Association newsletter, Spring 1993; "In the Presence of Whales," by Annie Gottlieb, *Quest*, July–Aug. 1977; "Gray Whales on the Winter Grounds in Baja California," by Dale W. Rice, Allen A. Wolman, and David E. Withrow, *Report of the International Whaling Commission* 31, 1981; "Mothers and Calves," by Steven L. Swartz and Mary Lou Jones, *Oceans*, Mar. 1984.

Chapter 2: The Whaler Who Became a Naturalist

Lagoon whaling and figures: *Marine Mammals* [*MM*] by Scammon; Henderson. Scammon background: *Beyond the Lagoon*, by Lyndall Baker Landauer (Associates of J. Porter Shaw Library, San Francisco, 1986); " 'The scene of slaughter was exceedingly picturesque,' " by Wesley Marx, *American Heritage*, June 1969; "Captain Scammon of Scammon Lagoon," by Campbell Grant, *Pacific Discovery*, Sept.–Oct. 1969; "Introduction to the Dover Edition" (*MM*), by Victor B. Scheffer (Dover, 1968); Scammon book and papers: "Charles Scammon: Whaler Turned Naturalist," by G. Kenneth Mallory, Jr., *Oceans*, July 1977; "Destruction of the Devilfish," by Joan C. Watkins, *Old West*, Summer 1969; *Salted Tories*, by Lloyd C. M. Hare (Marine Historical Association of Mystic, Conn., 1960). Brothers' background: Scammon papers, Bancroft Library. Quotations on Scammon: Author's interviews with Bruce Mate, Alan Baldridge, and Pieter Folkens.

Chapter 3: The Poet and the Saltworks War

Early salt extraction history: Scammon, Henderson. Background on Exportadora de Sal: Author's interviews with Francisco Guzmán Lazo, Mexico City, June 14, 1998, and Julio Peralta, Guerrero Negro, Mar. 1, 1999; "The Proposed Salt Works in Laguna San Ignacio," report by Mark J. Spalding, Dec. 8, 1996; "The Gray Zone," report by Katherine Angelo Hanly/Investigative Network, 1997.

Background on Mitsubishi: "Does Japan Own Baja California?" *ECO*, June 28, 1996; "Mitsubishi: The Diamonds Lose Their Sparkle," *Economist*, May 9, 1998; *The Corporate Planet*, by Joshua Karliner (Sierra Club Books, 1997); "Rousing a Sleeping

Industrial Giant," by David E. Sanger, *New York Times*, May 20, 1990; "Corporate Goliaths," by Charles Gray, *Multinational Monitor*, June 1999; *Mines, People and Land: A Global Battleground*, by Roger Moody (Minewatch, London, 1992); author's interview with Teddy Jimbo, San Ignacio Lagoon, Mar. 2000. Mitsubishi and whaling: "The Gray Zone." Salinas trip to Japan: Author's interview with Andrés Rozental, San Ignacio Lagoon, Mar. 1999.

Background on Homero Aridjis: Author's interviews, Mexico City, June 13, 1998, and New York, Aug. 9–10, 2000; interview with Pete Hamill, July 22, 1998; biography prepared by Betty Ferber de Aridjis; "A Declaration by 100 Intellectuals and Artists Against Contamination in Mexico City," [Mexico City] *News*, Mar. 1, 1985; "The Image of Our Future," by Homero Aridjis, *NPQ*, Spring 1989; "The Mexican Wasteland," interview with Aridjis, *Newsweek*, Oct. 30, 1989; "Face to Face with Torquemada," review of *1492* by Allen Josephs, *New York Times Book Review*, June 16, 1991; "Group of 100: In Defense of Whales," advertisement, *New York Times*, May 10, 1994; "For PEN's New President, Latin America Is Not on the Margins," by Sam Dillon, *New York Times*, Sept. 24, 1997; "Homero Aridjis," *Los Angeles Times* interview by Sergio Muñoz, Dec. 7, 1997. Some of the material by the author first appeared in his article "Poetry in Motion," *Amicus Journal*, Fall 1998.

Background on Japan and whaling: *The Blue Whale*, by George L. Small (Columbia University Press, 1971); "Why the Japanese Are So Stubborn About Whaling," by Kathy Glass and Kirsten Englund, *Oceanus*, Spring 1989; "Friends for Whales in Japan," by Karl Schoenberger, *Los Angeles Times*, Jan. 4, 1991; "Whaling Rejected at IWC Meeting," by David Phillips, *Earth Island Journal*, Summer 1993. IWC history: "International Management of Whales and Whaling . . . ," by Ray Gambell, *Arctic*, June 1993.

Saltworks expansion: Author's interviews with Aridjises; Serge Dedina, San Ignacio Lagoon, Mar. 2000; Bruce Mate, Newport, Oreg., Apr. 27, 2000; Steven Swartz, Boston, Oct. 27, 2000. Translations of Aridjis columns in *Reforma*, Feb. 15, 1995; Mar. 10, 1996; Aridjis op-ed, *Los Angeles Times*, Mar. 12, 1995; ad in *New York Times*, May 10, 1995; "The Friends of the Whales Fight a Saltworks," by Paul Sherman, *New York Times*, Apr. 27, 1995; "Ecologists Fear Baja Salt Mine May Threaten Gray Whales," by Juanita Darling, *Los Angeles Times*, Mar. 6, 1995; "Close Encounters," by Dwight Holing, *Amicus Journal*, Summer 1997; "A Baja Dilemma," three-part series by James Bruggers and Marie E. Camposeco, *Contra Costa Times*, Mar. 1997; "The Gray Zone."

Chapter 4: The Makah Tribe: Hunting the Gray Whale

Author's interviews, Neah Bay, with Charles Claplanhoo, Nov. 3, 1998, and Apr. 30, 2000; Gary Ray, Apr. 30, 2000; John McCarty, Nov. 3, 1998.

Scammon quotes: "About the Shores of Puget Sound," *Overland Monthly* [*OM*], Sept. 1871; *MM;* scrapbooks. Makah history: "Makah Cultural and Research Center," *Museum Exhibit Leaflet*, 1995; "Subsistence and Survival: The Makah Indian Reservation, 1855–1933," by Cary C. Collins, *Pacific Northwest Quarterly*, Fall 1996; "Ozette: A Makah Village in 1491," by Maria Parker Pascua, *National Geographic*, Oct. 1991; *On the Northwest; The Indians of Cape Flattery*, by James G. Swan (Smithsonian Institution, June 1868); *Nootka and Quileute Music*, by Frances Densmore, (U.S. Government Printing Office [GPO], 1939); *The Gray Whale*, chapter on aboriginal whaling; *The Northern and Central Nootkan Tribes*, by Philip Drucker (U.S.

GPO, 1951); "Makah," by Ann M. Renker and Erna Gunther, in *Handbook of North American Indians*, vol. 7, *Northwest Coast* (Smithsonian Institution, 1990); *Contributions to North American Ethnology*, vol. 1 (U.S. GPO, 1877); *The Whaling Equipment of the Makah Indians*, by T. T. Waterman (1920; rept., University of Washington Press, 1967); *Indian Days at Neah Bay*, by James G. McCurdy (Superior Publishing, Seattle, 1961).

Makah hunt/IWC/Japan/U.S. agencies: *Memorandum in Support of Plaintiffs' Motion for Summary Judgment . . . , Jack Metcalf et al., Plaintiffs*, v. *William Daley et al., Defendants*, U.S. District Court for the District of Columbia, May 21, 1998; *Order Granting Defendants' Motion for Summary Judgment*, Judge Franklin D. Burgess, Sept. 21, 1998; *Brief in Opposition to U.S. Government Support at the 49th IWC for the Makah Tribe Proposal to Hunt Gray Whales*, filed Sept. 10, 1997; Verbatim Record(s) of International Whaling Commission 48th and 49th Annual Meetings, Aberdeen, UK, June 24–28, 1996, and Monaco, Oct. 20–24, 1997; U.S. Delegation News Release, "Whaling Commission Approves Combined Russian-Makah Gray Whale Quota," Oct. 23, 1997; internal documents of U.S. Department of Interior/National Marine Fisheries Service obtained under FOIA in course of legal action, dated Apr. 3, 1995; Apr. 27, 1995; May 30, 1996; Aug. 27, 1997; author's interview with Robert Brownell, La Jolla, Calif., Jan. 17, 2000; "The Makah Indian Tribe and Whaling: A Fact Sheet Issued by the Makah Whaling Commission," July 21, 1998; "Makah Whaling: Aboriginal Subsistence or a Stepping Stone to Undermining the Commercial Whaling Moratorium?" by Leesteffy Jenkins and Cara Romanzo, *Colorado Journal of International Environmental Law and Policy*, Winter 1998; *PAWS News*, Fall 1998; *Sea Shepherd Log*, Winter 1998–99; "The Great American Whale Hunt," by Richard Blow, *Mother Jones*, Sept.–Oct. 1998; numerous articles, *Seattle Times*, 1995–98 (archived at seattletimes.com/search, "Makah Tribe").

Chapter 5: A Tribal Elder and the Gray Whales

Parts of this chapter appeared in the author's articles "The Monkeywrenchers," *Amicus Journal*, Fall 1987, and "Tribal Tradition and the Spirit of Trust," *Amicus Journal*, Spring 1999. Author's interviews: Paul Watson, Nov. 1, 1998; Alberta Thompson, November 2, 3, 1998. Scammon quotes on Makah: *MM.* Makah legends: see Chapter 4 on history. Scammon quotations on Tatoosh: *OM*, Sept. 1871. Alberta Thompson background: "Some Makahs Oppose Hunt," by Lynda V. Mapes, *Seattle Times*, Oct. 30, 1998; "Alberta Thompson," by Brenda Peterson, *New Age*, July–Aug. 1998; "Honoree Has Love of Whales," by Keith Thorpe, *Peninsula Daily News*, Mar. 17, 1997. Hunt controversy: "Protesters Shadow a Tribe's Pursuit of Whales and Past," by Sam Howe Verhovek, *New York Times*, Oct. 2, 1998; "Permission Granted to Kill a Whale. Now What?" by Robert Sullivan, *New York Times Magazine*, Aug. 9, 1998; *Seattle Times* archive.

Chapter 6: Return to La Laguna

Author's interview with Robert Kennedy, Jr., New York, May 5, 1998. "The Art of Being JFK Jr.," by Eric Pooley, *Time*, July 26, 1999. *The Riverkeepers*, by John Cronin and Robert F. Kennedy, Jr. (Scribner, 1997).

Sea turtle deaths: "Mitsubishi/ESSA Causes Sea Turtle Deaths," by Nathan LaBudde, *Earth Island Journal*, Spring 1998; "Poisoning a Sanctuary for the Sake of Salt," by Robert F. Kennedy, Jr., and Joel Reynolds, *Los Angeles Times*, Aug. 5, 1998.

Aridjis death threats: "PEN Leader in Mexico Tells of Threats," by Julia Preston, *New York Times*, Aug. 29, 1998; "Protective Poet Finds Himself in Need of Bodyguards," by Mary Beth Sheridan, *Los Angeles Times*, Sept. 9, 1998. The meeting with residents of San Ignacio Lagoon occurred on Feb. 23, 1999.

Gray whale mating behavior: Author's interviews with Bruce Mate, Steven Swartz, Marilyn Dahlheim, Seattle, Apr. 25, 2000; Mary Lou Jones and Swartz, *National Geographic*, June 1987; "Potential for Sperm Competition in Baleen Whales," by Robert L. Brownell, Jr., and Katherine Ralls, *Report of the International Whaling Commission*, special issue 8, 1986; "Sperm Competition in Grey Whales," letter by Brownell and Ralls, *Nature* 336, Nov. 10, 1988; "Migrant Gray Whales with Calves and Sexual Behavior of Gray Whales in the Monterey Area of Central California, 1967–73," by Alan Baldridge, *Fishery Bulletin* 72 (2), Apr. 1974; "Many Observe Courtship of Pacific Gray Whales," AP, *Riverside Enterprise*, Mar. 20, 1955; "Reproductive Behavior of the Gray Whale *Eschrichtius robustus*, in Baja California," by William F. Samaras, *Bulletin of the Southern California Academy of Sciences* 73 (2), Aug. 1974; *The California Gray Whale . . .*, by Roy Chapman Andrews, Memoirs of the American Museum of Natural History, New Series, vol. 1, pt. 5, Monographs of Pacific Cetacea, no. 1 (Mar. 1914).

CHAPTER 7: JOURNEY TO THE PILLARS OF SALT

Author's interviews in Punta Abreojos: Feb. 26–27, 1999; in Guerrero Negro, Feb. 28–Mar. 1, 1999. Scammon entry into lagoon: Derived from *MM;* Henderson. White quotation: "Hunting the Heartbeat of a Whale," by Paul Dudley White, M.D., with Samuel W. Matthews, *National Geographic*, July 1956. Saltworks background: "San Ignacio Saltworks: Salt and Whales in Baja California," *Semarnap Working Papers*, 1997; Fact sheets, Questions and Answers, Exportadora de Sal; Spalding report; *Sierra, Sea, and Desert;* Dedina and Young thesis; fact sheets of IFAW Campaign to Save Laguna San Ignacio; "The Center for Marine Conservation's Efforts to Protect Gray Whales of Baja California: San Ignacio Lagoon Project," report; Bruce Mate letter to colleagues, Jan. 26, 1995; NRDC fact sheets. Gray whale birth figures by lagoon: "Gray Whales on the Winter Grounds in Baja," by D. W. Rice et al., *Report of the International Whaling Commission*, 1981. Deaths at Scammon's Lagoon: "L.A. Man Dies When Whale's Tail Hits Boat," by D. Anthony Darden, *Los Angeles Times*, Feb. 27, 1983; "Whale-Watching Accident Claims Second Life," by Bob Williams, *Los Angeles Times*, Mar. 3, 1983. Scammon quotation: *Journal aboard the Bark Ocean Bird on a Whaling Voyage to Scammon's Lagoon, winter of 1858–1859*, by Charles Melville Scammon, ed. and annotated by David A. Henderson (Dawson's Book Shop, Los Angeles, 1970). Aridjis quotes: "Save the Whales? Mine the Salt? Mexican Standoff," by Julia Preston, *New York Times*, Mar. 10, 1999; "Whales for Salt," by Homero Aridjis, *Reforma*, Mar. 7, 1999.

CHAPTER 8: WHALE WATCHERS: THE SCIENTIST AND THE ARTIST

Oil exploration at lagoons: "Bid to Head Off Drilling at Whale Lagoon Begins," by Robert A. Jones, *Los Angeles Times*, Dec. 29, 1976; "A Park for Whales . . . and Tourists . . . and Oil?" by Patrick Heffernan, *Environmental Policy and Law* 3 (1977). Author's interview, Roger Payne, San Ignacio Lagoon, Mar. 6, 1999. "An Unacceptable Risk," ad in *New York Times*, July 15, 1999. Payne quotations: *Among Whales*, by Roger Payne (Delta Books, 1995). Author's interview, Francisco (Gerardo) Hernández Zamora, Mar. 3, 1999. Jorge Urbán visit to lagoon, Mar. 6, 1999. Gilmore quotation: "Calving of the California Grays," by Raymond M. Gilmore and Gifford Ewing, *Pacific Discovery*, May–June 1954. Lagoon 1998 count: "The 'El Niño' 1997–98 Effect on Gray Whales That Visited Laguna San Ignacio, B.C.S., Mexico," by Urbán et al., paper presented at Somemma conference, Apr. 18–22, 1999. Cave paintings: *The Hidden Heart of Baja*, by Erle Stanley Gardner (Morrow, 1962); "A Legendary Treasure Left by a Long Lost Tribe," by Gardner, *Life*, 53(3), 1962.

Chapter 9: Sound Check: Echoes from Magdalena Bay

Magdalena Bay background: Dedina, *Saving the Gray Whale;* Scammon's *Sketch* . . . and *MM;* "Present Condition of the California Gray Whale Fishery," by Charles H. Townsend, *U.S. Fish Commission Bulletin* 6 (1886); "A History of California Shore Whaling," by Edwin C. Starks, *California Fish and Game Commission* No. 6 (State Printing Office, San Francisco, 1922). Author's interviews with Greg Brennan, Ed Brennan, San Carlos, Mar. 7, 1999. Author's interview, Francisco (Paco) Ollervides, San Carlos, Mar. 7–8, 1999. Sound basics: *Sounding the Depths: Supertankers, Sonar and the Rise of Undersea Noise*, principal author Michael Jasny (Natural Resources Defense Council, Mar. 1999); "The Sound and the Fury," by Dwight Holing, *Amicus Journal*, Fall 1994. Scammon quotations: *MM.*

Sound impacts: Dahlheim interview; author's interview with Peter Tyack, Woods Hole, Mass., Nov. 9, 1999. E-mail to author from Lindy Weilgart, Mar. 27, 2000. "Quick Look: Playback of Low Frequency Sound to Gray Whales Migrating Past the Central California Coast—January 1998," report by Peter L. Tyack and Christopher W. Clark, June 23, 1998; "Bio-Acoustics of the Gray Whale (*Eschrichtius robustus*)," dissertation by Marilyn Elayne Dahlheim (University of British Columbia, Oct. 1987); "Noise Called Serious Threat to Sea Life," by Deborah Schoch, *Los Angeles Times*, June 27, 1999; "Man's Roar Ripples Through Whales' World," by Tim Friend, *USA Today*, July 6, 1999; "Navy Sonar Under Fire," by Martha Bellisle (AP), *Anchorage Daily News*, July 30, 1999; "Navy Plans Ocean Assault," by Nathan LaBudde, *Earth Island Journal*, Summer 1999; "Dead Whales Underscore the Threat of Human Noise to Ocean World," by Wendy Williams, *Boston Globe*, June 27, 2000; "U.S. Navy Cancels Tests Believed to Harm Marine Mammals," Environmental News Service, May 29, 2000. Acoustic thermometry project: "Global Thermometer Imperiled by Dispute," by Malcolm W. Browne, *New York Times*, Oct. 27, 1998.

Whale sounds: "Sea Sleuths, Underwater Wire Tap Fail in Efforts to Make Whales Talk," by Sid Fleischman, *San Diego Daily Journal*, Feb. 17, 1950; *Arctic Alaska and Siberia*, by H. L. Aldrich (Rand McNally, 1889); "The Gray Whale," by Raymond M. Gilmore, *Oceans*, Jan. 1969; "Listening to Gray Whales," by Edwin J. C. Sobey, *Sea Frontiers/Sea Secrets*, Jan.–Feb. 1986; "Sounds Produced by the Gray Whale, *Eschrichtius robustus*," by James F. Fish, James L. Sumich, and George L. Lingle, *Marine Fisheries Review* 36 (4), Apr. 1974; "Sound Activity of the California Gray Whale," by Robert L. Eberhardt and William E. Evans, *Journal of the Audio Engineering Society* 10 (4), Oct. 1962; "The Controversial Production of Sound by the California Gray Whale," by Paul V. Asa-Dorian and Paul J. Perkins, study for Office of Naval Research, Mar. 1967; "Sound Production by the Gray Whale and Ambient Noise Levels in Laguna San Ignacio, Baja California Sur, Mexico," by Marilyn E. Dahlheim, H. Dean Fisher, and James D. Schempp, in *The Gray Whale;* Dahlheim thesis and interview; Jones and Swartz report in *The Gray Whale;* "Gray Whales, *Eschrichtius robustus;* in Laguna San Ignacio, Baja California, Mexico," by Steven L. Swartz and William C. Cummings, report for Marine Mammal Commission, Feb. 1978 (first mention of whales and outboard engine); *Gray Whales: Wandering Giants* (Aristotle quotation); *Guardians of the Whales* (Swartz on female calling calf); "Behavioral Changes in Gray Whale Population in Ojo de Liebre," by Shari Bondy, e-mailed to author, May 1999; Tyack interview; Andrews study; *The Charged Border*, by Jim Nollman (Henry Holt, 1999); author's interview with Nollman, Boston, Nov. 19, 1999. Payne quote: *Among Whales.*

CHAPTER 10: ORCAS AND GRAYS ALONG THE SHORES OF MONTEREY

California shore whaling: Scammon, *MM;* "Present Condition of the California Gray Whale Fishery;" "Shore Whaling for Gray Whales Along the Coast of the Californias," by Hazel Sayers, in *The Gray Whale*; "A History of California Shore Whaling"; *Men and Whales* ("The California Gray Whale Fishery"); "Whalers from the Golden Gate: A History of the San Francisco Whaling Industry, 1822–1908," master's thesis by Richard William Crawford (San Diego State University, Spring 1981); "Offshore Whaling in the Bay of Monterey," by Edward Berwick, *Cosmopolitan*, 1900; "The History of the Whaling Industry in California," paper by Robert Weinstein, date unknown; " 'The Whalers Cabin' and 'The Whaling Station Museum,' " brochure of Point Lobos State Reserve, Carmel, Calif.; "Rugged Point Lobos Is Crown Jewel of California Coastal Parks," by Eric Pryne, *Seattle Times*, May 14, 2000.

Author's interviews in Monterey with Steph Dutton and Heidi Tiura, Apr. 24–26, 1999; Richard Ternullo, Apr. 25, 1999; Alan Baldridge, Apr. 24, 1999. Dutton and Tiura background: "Paddling in the Path of Giants," by Paul McHugh, *Paddler*, July–Aug. 1998; "Snorkel Breathing," by Steph Dutton, *Sea Kayaker*, Apr. 1996.

Orcas and gray whales: Scammon, *MM;* Andrews, 1914; "An Essay on the Natural History of Whales," by Paul Dudley, *Philosophical Transactions of the Royal Society* (London) 33, 1725; author's interview with Wayne Perryman ("Old Tom"); Nollman quote: author's interview; "Gray Whales, *Eschrichtius robustus*, Avoid the Underwater Sounds of Killer Whales, *Orcinus orca*," by William C. Cummings and Paul O. Thompson, *Fishery Bulletin* 69 (3), 1971; "Screaming Whales Studied," by Bob Corbett, *San Diego Evening Tribune*, Mar. 13, 1970; "Killer Whales Attack and Eat a Gray Whale," by Alan Baldridge, *Journal of Mammalogy* 53 (4), 1972; "Killer Whales (*Orcinus orca*) Chasing Gray Whales (*Eschrichtius robustus*) in the Northern Bering Sea," by Donald K. Ljungblad and Sue E. Moore, *Arctic* 36 (4), Dec. 1983; "Attack on Gray Whales (*Eschrichtius robustus*) in Monterey Bay, California, by Killer Whales (*Orcinus orca*) Previously Identified in Glacier Bay, Alaska," by P. Dawn Goley and Janice M. Straley, *Canadian Journal of Zoology* 72, 1994; "Observations of Killer Whale (*Orcinus orca*) Predation in the Northeastern Chukchi and Western Beaufort Seas," by John Craighead George and Robert Suydam, *Marine Mammal Science* 14, Apr. 1998; "Scientists Report Rare Attack by Killer Whales on Sperm Whales" AP, *New York Times*, Nov. 9, 1997; "A Killer Whale–Gray Whale Encounter," by G. Victor Morejohn, *Journal of Mammalogy* 49 (2) May 1968; "An Observation Regarding Gray Whales and Killer Whales," by Bryan R. Burrage, *Transactions of the Kansas Academy of Science* 67 (3), 1964; "Three Gray Whales Fight Off Killers in Ocean Battle," by S. A. Desick, *San Diego Union*, Jan. 10, 1966; Notes in Gilmore papers; "Killer Whale: *Orcinus orca* (Linnaeus, 1758)," by Marilyn E. Dahlheim and John E. Heyning, in *Handbook of Marine Mammals*, vol. 6; "Killer Whales," in *The Smithsonian Book of North American Mammals*, by D. E. Wilson and S. Ruff (Smithsonian Institution Press, 1999); "Pack of Killers Pounces on Two Near La Jolla," *San Diego Union*, Jan. 12, 1947; *Guardians of the Whales.*

CHAPTER 11: OREGON AND WASHINGTON: SCHOLARS OF THE GREAT MIGRATION

Insight Guides U.S. National Parks West, ed. by John Gattuso (Houghton Mifflin, 1997), and *Olympic National Park and Peninsula*, by Nicky Leach (Sierra Press, Mariposa, Calif., 1998), proved useful in describing the region. Whale entanglement figures: *Status Review of the Eastern North Pacific Stock of Gray Whales*, by D. J. Rugh et al., NOAA Technical Memorandum NMFS-AFSC-103, Aug. 1999. Author's interview with Bruce Mate, Newport, Oreg., Apr. 27, 2000. Background on tagging of whales: "Watching Habits and Habitats from Earth Satellites," by Bruce R. Mate, *Oceanus*, Spring 1989; "Ocean Movements of Radio-Tagged Gray Whales," by Bruce R. Mate and James T. Harvey, in *The Gray Whale;* "Assessing Critical Habitats of Whales by Satellite Tracking," by Bruce R. Mate, *IBI Reports*, Nov. 4, 1993; "Satellite-Monitored Radio Tracking of Cetaceans: Developments in Transmitter Package and Attachment," abstract by Bruce R. Mate and Rod Mesecar, *Forum on Wildlife Telemetry*, Snowmass, Colo., Sept. 1997.

Scammon background: *Beyond the Lagoon;* Bancroft papers. "About the Shores of Puget Sound," *OM*, Sept. 1871; "Lumbering in Washington Territory," *OM*, July 1870; "In and Around Astoria," *OM*, Dec. 1869.

Author's interview with John Calambokidis, Olympia, Wash., Apr. 26, 2000. Background: "Photographic Identification Research on Seasonal Resident Whales in Washington State," by John Calambokidis and Jennifer Quan, *Status Review . . .* (1999); "Encounter with a Humpback," by Calambokidis, in *Life on the Edge. A Guide to California's Endangered Natural Resources*, ed. by Carl G. Thelander and Margo Crabtree (Biosystems Books, Santa Cruz, Calif., 1994); "Gray Whales of Washington State: Natural History and Photographic Catalog," by Calambokidis et al., 1984; "Final Report: Gray Whale Photographic Identification in 1998," by Calambokidis et al. (Cascadia Research Collective, May 1999); "Watcher of the Whales," by Christopher Dunagan, *The Sun*, May 14, 1995; "Gray Whale Deaths in Puget Sound: A Perspective," by John Calambokidis, *Puget Sound Notes* 28: 5–7, 1992; "The Makah Gray Whale Hunt," by Calambokidis, *Sea Kayaker*, June 1999.

CHAPTER 12: THE KILL

The description of the Makah Tribe's whale hunt and its aftermath was drawn from numerous sources, including: Author's interviews in Neah Bay with Theron Parker (Apr. 29, 2000), Alberta Thompson (Apr. 29, 2000), Jonathan Paul (Apr. 30, 2000), Gary Ray (Apr. 30, 2000), Charles Claplanhoo (Apr. 30, 2000); in Olympia, Washington, with Micah McCarty (Apr. 28, 2000); in Seattle with Deborah Brosnan (May 4, 2000); at San Ignacio Lagoon with Jeff Pantukhoff (Mar. 6, 2000); and by phone with Paul Watson (Oct. 15, 2000); video footage in *Gray Whale Kill*, compiled by Fourth Corner Whale Project (www.pacoasis.com).

Press accounts: "Reviving Tradition, Tribe Kills a Whale," by Sam Howe Verhovek, *New York Times*, May 18, 1999; "Makahs Harpoon, Shoot Whale," by Lynda

V. Mapes, *Seattle Times*, May 17, 1999; "Tribe Reclaims Its Past with Harpoon Thrust," by Peggy Andersen/AP, *Boston Globe*, May 18, 1999; "Makahs Celebrate Catch, Claiming Victory amid Strong Protest of Whale Kill," by Lynda V. Mapes and Chris Solomon, *Seattle Times*, May 18, 1999; "After the Hunt, Bitter Protest and Salty Blubber," by Sam Howe Verhovek, *New York Times*, May 19, 1999; "Tradition vs. a Full-Blown PR Problem," by Eric Sorensen, *Seattle Times*, May 18, 1999; "Whales' Killing Stirs Intense Reactions," by Ross Anderson, *Seattle Times*, May 18, 1999; "Readers React Strongly to the Hunt," by Seattle Times staff, *Seattle Times*, May 18, 1999; "Significance of Whale Hunt 'Starting to Sink in Now,' " by Eric Sorensen and Chris Solomon, *Seattle Times*, May 19, 1999; "Threats, Disdain Continue to Mount," by Chris Solomon, *Seattle Times*, May 20, 1999; "Readers Write About the Makah Whale Hunt, Other Topics," *Seattle Times*, May 20, 1999; "Native Peoples from All Over Share Makah Potlatch," by Lynda V. Mapes, *Seattle Times*, May 23, 1999; "Harvest from the Sea" (op-ed), by Wayne Johnson, *New York Times*, May 21, 1999; "Sun Shines on Makah Parade, Potlatch; Protesters Mourn," by Peggy Anderson, AP, May 23, 1999; "More Letters About the Makah Whale Hunt," *Seattle Times*, May 23, 1999; "E-mails, Phone Messages Full of Threats, Invective," by Alex Tizon, *Seattle Times*, May 23, 1999.

Scammon quotes: *MM. The Ecological Indian: Myth and History*, by Shepard Krech III (W. W. Norton, 1999). Japan and IWC: "Japan Wants Some Whales off Endangered List," AP, *Los Angeles Times*, May 27, 1999; "International Whaling Police Seem Afloat in an Ocean of Inertia," by Mark Fineman, *Los Angeles Times*, June 1, 1999.

Chapter 13: Whalemen of Vancouver Island

Francis Frank quotation: "Canada's Nuu-chah-nulth Tribe Rejoices, Considers Its Own Hunt," by Alex Tizon, *Seattle Times*, May 18, 1999. Author's interview with Tom Happynook, Brentwood, B.C., May 1, 2000. Background on World Council of Whalers: "Whaling Lobby Gets Federal Cash, PR Assistance," by Sarah Schmidt, *Globe and Mail*, June 14, 1999; www.worldcouncilofwhalers.com Web site; *World Council of Whalers 1998 General Assembly Report*, Victoria, B.C., Mar. 2–4, 1998; *World Council of Whalers 1999 General Assembly Report*, Reykjavík, Iceland, Mar. 27–30, 1999. Japan and CITES 2000 meeting: "Japan Squares Up to Conservationists over Grey Whale's Status," *Nature*, Apr. 6, 2000; "So Much for Saving the Whales," by Sharon Begley and Thomas Hayden, *Newsweek*, Apr. 17, 2000. Steven Boynton background: " 'Arkansas Project' Led to Turmoil and Rifts," by staff, *Washington Post*, May 2, 1999; *The Hunting of the President*, by Joe Conason and Gene Lyons (St. Martin's Press, 2000). Background on Eugene Lapointe: *Green Backlash: Global Subversion of the Environmental Movement*, by Andrew Rowell (Routledge, 1996); "Wildlife Guardians Accused of Backing Ivory Trade," by Polly Ghazi and Geoffrey Lean, *London Observer*, Oct. 8, 1989; www.iwmc.org, IWMC/World Conservation Trust Web site; author's interviews with confidential sources at IFAW, Humane Society/USA, and Monitor Consortium, Aug. 2000. For more on the wise use movement, see *The War Against the Greens*, by David Helvarg (Sierra Club Books, 1994). Also see "Crime Against Nature: Organized Crime and the Illegal Wildlife Trade," investigative report by the Endangered Species Project (Earth Island Institute/Marine Mammal Fund, San Francisco, 1995).

Background on Vancouver Island: *Meares Island* (Friends of Clayoquot Sound/

Western Canada Wilderness Committee, 1985); "The Islands," 2000 Vacation Guide. Author's interview with Jim Darling, Tofino, B.C., May 2–3, 2000. Darling background: *Gray Whales* (Darling), *Guardians of the Whales;* "Gray Whale (*Eschrichtius robustus*) Habitat Utilization and Prey Species off Vancouver Island, B.C.," by Darling et al., *Marine Mammal Science*, Oct. 1998; "The Whale Tomorrow?" by Darling, *Ocean Realm*, Summer 1991; "Humpback Whales Win Japanese," by Darling, *Whalewatcher* 25 (2), 1991; "Whales: An Era of Discovery," by Darling, *National Geographic*, Dec. 1988; "Humpback Whales," by Douglas H. Chadwick, *National Geographic*, July 1999.

Chapter 14: Alaskan Journey: Beginnings

Background on Scammon and the Western Union Telegraph Expedition (WUTE) from numerous sources, including Smithsonian Institution Archives in Washington, D.C. (letters and papers of William H. Dall, record unit 7073; Spencer Baird; and papers relating to the WUTE, record unit 7213); *Beyond the Lagoon;* Scammon's logbooks and scrapbooks in Bancroft Library; "The (Almost) Russian-American Telegraph," by Phillip H. Ault, *American Heritage*, June 1975; Dall obituary, *Science*, Apr. 8, 1827; Dall entry in *Dictionary of Scientific Biography*, vol. 3, 1971; "An Illustrious Family Name," (Dall) by James W. Willoughby, *Alta Vista Magazine*, Feb. 13, 1994; *The First Scientific Exploration of Russian America and the Purchase of Alaska*, by James Alton James (Northwestern University Press, 1942); and *Continental Dash: The Russian-American Telegraph*, by Rosemary Neering (Horsdal & Schubart, Ganges, B.C., 1989).

Background on "Inside Passage" and Sitka: *Portrait of Alaska's Inside Passage*, by Kim Heacox (Graphic Arts Center Publ., Portland, Oreg., 1997); "Sitka, Alaska," The Official Sitka Vacation Planner; *Alaska-Yukon Handbook*, by Deke Castleman and Don Pitcher (Moon Publications, 1997); *Insight Guide: Alaska* (APA Publications, 1998); "A Land Use Struggle over a Forest Bounty," by Sam Howe Verhovek, *New York Times*, May 27, 2000.

Migration: Author's interviews with Walt Cunningham (June 17, 1999) and Jan Straley (June 26, 1999); "Gray Whale Frolics Off Sitka," by Shannon Haugland, *Daily Sentinel*, Apr. 27, 1999; "The Return of the Gray Whale," by Raymond M. Gilmore, *Scientific American* 192 (1), 1955; "Migration and Feeding of the Gray Whale," by Gordon C. Pike, *Fisheries Research Board, Canada* 19 (5), 1962; "Observations of Migrating Gray Whales at Cape St. Elias, Alaska," by Walt Cunningham and Susan Stanford, report for Alaska Department of Fish and Game, 1979.

Chapter 15: Whales in Strange Places: Kenai, Kodiak, and Unimak Pass

Author's interviews with John Hall, Walnut Creek, Calif., Apr. 19, 1999; David Rugh, Seattle, Apr. 25, 2000. E-mail to author from Michael Poole, May 2000. Scent Trail: "Investigations on the Postanal Sac of the Gray Whale," by Floyd E. Durham and John W. Beierle, and "An Analysis of the Fluid Contents in the Postanal Sac of the Gray Whale," by Beierle et al., in *Bulletin Southern California Academy of Sciences* 75, May 1976.

Kodiak Island background: *Alaska-Yukon Handbook; Insight Guide: Alaska;* "Kodiak, Alaska's Island Refuge," by John L. Eliot, *National Geographic*, Nov. 1993. Author's interview with Eric Stirrup, June 22, 1999. Whaling statistics: "Whaling Results at

Akutan (1912–1939) and Port Hobron (1926–1937), Alaska," report by Randall R. Reeves et al., submitted to Scientific Committee, IWC, June 1984. Kodiak native whaling history: Author's interview with Lydia Black, June 21, 1999; "The Konyag . . . ," trans. and ed. by Lydia T. Black, *Arctic Anthropology* 14 (2), 1977; "Whaling in the Aleutians," by Lydia T. Black, *Études/Inuit/Studies* 11 (2), 1987; "Masked Rituals of the Kodiak Archipelago," doctoral thesis by Dominique Desson (University of Alaska, Fairbanks, May 1995).

Kodiak and military: Author's interview with Kate Wynne, June 21, 1999; "Environmental Assessment for U.S. Air Force Atmospheric Interceptor Technology Program" (Department of the Air Force, Nov. 1997); "Evaluation of the Potential Impacts of Launches of the USAF Atmospheric Interceptor Technology (AIT) Test Vehicle from the Kodiak Launch Complex (KLC) on Threatened and Endangered Species of Wildlife . . . Noise Monitoring Results," by Brent S. Stewart, Hubbs–Sea-World Research Institute Technical Report 99-291; Author's interview (June 21, 1999) and E-mail communications from Carolyn Heitman, Kodiak; Author's interview with Susan Payne, June 21, 1999.

Seward Peninsula: Materials from Kenai Fjords Tours. WUTE: Scammon, Dall papers. Unimak Pass background: "Census of Gray Whales at Unimak Pass, Alaska, November–December 1977–1979," by David J. Rugh in *The Gray Whale;* "Gray Whale Census in Alaska," by Rugh, *Whalewatcher* 14 (4), 1980; "Gray Whale Migration into the Bering Sea," paper by Pauline Hessing, NOAA/OMPA Contract: NA 8IRGA 00080, Spring 1981.

CHAPTER 16: INTO THE BERING SEA

Terry Johnson: E-mail communication with author, spring 2000. "Community of Scammon Bay" Web site. WWF Bering Sea Conference, Anchorage, July 8, 1999. Bering Sea: "An Ill Tide Up North," by Eugene Linden, *Time*, Aug. 16, 1999; "On the Edge of the World and on the Brink of Losing Nature's Gifts," by John Balzar, *Los Angeles Times*, Aug. 21, 1999; "The Unbearable Capriciousness of Bering," by Robert A. Saar, *Science*, Feb. 25, 2000; "The Bering Sea Ecoregion," report by World Wildlife Fund/Beringia Conservation Program, 2000. Dutch Harbor: Author's visit, July 13, 1999; fisheries statistics from *National Fisherman*. Scammon on Pribilofs: "Fur Seals," *OM*, Nov. 1869, and "Seal Islands of Alaska," *OM*, Oct. 1870. Pribilofs background: *Seven Words for Wind*, by Sumner Macleish (Epicenter Press, Fairbanks and Seattle, 1997); "Islands of the Seals: The Pribilofs," *Alaska Geographic* 9 (3), 1982; "Aleuts Ask for Justice," by David Whitney, *Anchorage Daily News*, July 30, 1999. Author's interview with Terry Spraker, St. Paul, July 11, 1999.

Marine mammal declines: "Bering Sea Task Force—Report to Governor Tony Knowles," Mar. 1999. Gray whale strandings: "Starved Gray Whales Washing Up in Record Numbers," AP, *Anchorage Daily News*, July 15, 1999. Author's interviews with Sue Moore, Marilyn Dahlheim, Seattle, Apr. 25, 2000. Scammon quotations: *MM*, papers.

Gray whale feeding: "Side-Scan Sonar Records and Diver Observations of the Gray Whale (*Eschrichtius robustus*) Feeding Grounds," by John S. Oliver and Rikk G. Kvitek, *Biology Bulletin* 167, Aug. 1984; "Side-Scan Sonar Assessment of Gray Whale Feeding in the Bering Sea," by Kirk R. Johnson and C. Hans Nelson, *Science*, Sept. 1984; "Gray Whales (*Eschrichtius robustus*) in the Western Chukchi and East Siberian Seas," by R. V. Miller et al., *Arctic*, Mar. 1985; "The Association of Marine Birds and

Feeding Gray Whales," by Craig S. Harrison, *Condor* 81, 1979; "Gray Whales and the Structure of the Bering Sea Benthos," paper by Mary K. Nerini and John S. Oliver; 1991 Study: *Gray Whales* (Gordon, Baldridge); "Feeding of a Captive Gray Whale," by G. Carleton Ray and William E. Schevill, *Marine Fisheries Review*, Apr. 1974; "Whales and Walruses as Tillers of the Sea Floor," by C. Hans Nelson and Kirk R. Johnson, *Scientific American*, Feb. 1987; "Feeding Habits of the Gray Whale off Chukotka," by L. S. Bogoslovskaya et al., *Report of the International Whaling Commission* 31, 1981; *Gray Whales: Wandering Giants*; "Productivity of Arctic Amphipods Relative to Gray Whale Energy Requirements," by Raymond C. Highsmith and Kenneth O. Coyle, *Marine Ecology Progress Series* 83 (2 and 3), July 16, 1992; "Seabird Feeding on Benthic Amphipods Facilitated by Gray Whale Activity in Northern Bering Sea," by Jacqueline M. Grebmeier and Nancy M. Harrison, *Marine Ecology Progress Series* 80, Mar. 3, 1992; "Pelagic-Benthic Coupling on the Shelf of the Northern Bering and Chukchi Seas, II: Benthic Community Structure," by Grebmeier et al., *Marine Ecology Progress Series* 51 (3), Feb. 6, 1989; John Hall interview.

John Heyning: Author's interview, Long Beach, Calif., May 11, 2000; "Thermoregulation in the Mouths of Feeding Gray Whales," by John E. Heyning and James G. Mead, *Science*, Nov. 7, 1997; "Tongues Are Whales' Key to Warmth While Eating," by Valerie Burgher, *Los Angeles Times*, Dec. 4, 1997; "Cold Tongues, Warm Hearts," by John E. Heyning, *Terra*, Jan.–Feb. 1998; "Rescued Whale J.J. Begins Long Journey Home," by Tony Perry, *Los Angeles Times*, Apr. 1, 1998.

Chapter 17: Among the Hunters at Bering Strait

Author's interviews with Dan Richard, Luther Komenseak, Raymond Seetook, Toby Anungazuk, Jr., in Wales, June 29–July 3, 1999; with Jim Stimpfle in Nome, June 27–28, 1999.

Migration: "Recent Records of the California Grey Whale (*Eschrichtius glaucus*) Along the North Coast of Alaska," by William J. Maher, *Arctic* 12, 1960; "Behavior of Gray Whales Summering near St. Lawrence Island, Bering Sea," by Bernd Würsig et al., *Canadian Journal of Zoology* 64, 1986. The early history of the region (Libbysville, Trollope, and so on) is detailed in *The Eskimos of Bering Strait, 1650–1898*, by Dorothy Jean Ray (University of Washington Press, 1975); "Nome: 'City of the Golden Beaches,' " *Alaska Geographic* 11 (1), 1984. Whaling history and background in *Whales, Ice, and Men*, by John R. Bockstoce (University of Washington Press, 1995); *Hunting the Largest Animals*, ed. by Allen P. McCartney (Canadian Circumpolar Institute, University of Alberta, 1995); "Alaska Whales and Whaling," *Alaska Geographic* 5 (4), 1978; "Story of a Whale Hunt," by Suzanne R. Bernardi, *Alaska*, Aug. 1981; *People of the Ice Whale*, by David Boeri (E. P. Dutton, 1983); "Gray Whale Distribution and Catch by Alaskan Eskimos: A Replacement for the Bowhead Whale?" by Willman M. Marquette and Howard W. Braham, *Arctic* 35 (3), Sept. 1982; "The Hunt of Gray Whales by Alaskan Eskimos: A Preliminary Review," report by Ronn Storro-Patterson for National Oceanic and Atmospheric Administration, circa 1980. Additional background on Wales: *The Wake of the Unseen Object*, by Tom Kizzia (University of Nebraska Press, 1991). Scammon quotations: *MM*.

Chapter 18: To the Diomedes: Life on the Edge

Author's trip to Little Diomede: July 3, 1999. The whaling legend was derived from an account in *Natives of the Far North*, by Shannon Lowry (Stackpole Books, 1994).

Additional background on the Diomede Islands: *Imaging the Arctic*, ed. by J. C. H. King and Henrietta Lidchi (University of Washington Press, 1998); *Arctic Passages*, by John Bockstoce (Hearst Marine Books, New York, 1991); "People of the Tusk," by Fred Bruemmer, *Equinox*, Mar.–Apr. 1992; "Last of the Umiaks," by Fred Bruemmer, *Natural History*, Oct. 1992; "Little Diomede: Old Whaling Interviews with Peter Oscar Akhinga," June 23, 1997 (transcriptions provided to author by Akhinga); Diomede Community Web site; letters, clippings, and unpublished ms. provided to author by Sue Steinacher. For more on Arctic native whaling, see *Chasing the Bowhead*, as told by Captain Hartson H. Bodfish (Harvard University Press, 1936); *Pursuing the Whale*, by John A. Cook (Houghton Mifflin, 1926); *An Eskimo Village in the Modern World* [St. Lawrence Island], by Charles Campbell Hughes (Cornell University Press, 1960); "Ipiutak and the Arctic Whale Hunting Culture," by Helge Larsen and Froelich Rainey, *Anthropological Papers of the American Museum of Natural History* 42, 1948; *Eskimo Prehistory*, by Hans-Georg Bandi (University of Alaska Press, 1969).

Background on Alaska Eskimo Whaling Commission and IWC: "Eskimos, Yankees, and Bowheads," by Howard W. Braham, *Oceanus*, Spring 1989; "International Management of Whales and Whaling: An Historical Review of the Regulation of Commercial and Aboriginal Subsistence Whaling," by Ray Gambell, *Arctic* 46 (2), June 1993; "The Role of Eskimo Hunters, Veterinarians, and Other Biologists in Improving the Humane Aspects of the Subsistence Harvest of Bowhead Whales," by Todd M. O'Hara et al., *Journal of the American Veterinary Medical Association* 214 (8), 1999; "Chairman's Report of the Forty-ninth Annual Meeting, 20–24 October, 1997," *Report of the International Whaling Commission*, 1998. Whale landings in Wales, Diomede: Author's interviews. Killer whales and gray whales: "Observation of Killer Whale (*Orcinus orca*) Predation in the Northeastern Chukchi and Western Beaufort Seas," *Marine Mammal Science* 14(2), Apr. 1998; *Arctic Alaska and Siberia, or Eight Months with the Arctic Whalemen*, by Herbert L. Aldrich (Rand McNally, 1889). Nelson quotation: in *Whales, Ice, and Men*, p. 139. Scammon quotation: *MM*.

Chapter 19: Sakhalin Island: Last of the Western Grays

The author visited Sakhalin Island, July 17–30, 1999. 1983 sightings: "On the Korean-Okhotsk Population of Gray Whales," by S. A. Blokhin et al., *Report of the International Whaling Commission* (SC/36/PS7), 1984. Background on the island: *Sakhalin Region*, by Mikhail Vysokov (Sakhalin Book Publishing House, Yuzhno Sakhalinsk, 1998); *Sakhalin: A History*, by John J. Stephan (Clarendon Press, Oxford, 1971); *Siberia: Its Conquest and Development*, by Yuri Sëmyonov (Helicon Press, Baltimore, 1964); *The Island*, by Anton Chekhov (Washington Square Press, New York, 1967); *Letters of Anton Chekhov*, ed. Simon Karlinsky and Michael Henry Heim (Harper & Row, 1973); *In the Soviet House of Culture*, by Bruce Grant (Princeton University Press, 1995); "Sakhalin Oblast," report by Friends of the Earth Japan, 1996; "Sakhalin," by Carl Hoffman, *Islands*, Feb. 1999; "Guide to the Sakhalin Museum"; *The Russian Far East: A Business Reference Guide*, 4th ed., ed. by Elisa Miller and Alexander Karp (Russian Far East Advisory Group, Seattle, 1999); *The Russian Far East*, by Erik Azulay and Allegra Harris Azulay (Hippocrene Books, New York, 1995).

Scammon on Sakhalin: "Northern Whaling," *OM*, June 1874; "The Pacific Coast Cod-Fishery," *OM*, May 1870; logbooks. Gray whale history: *Explorations of Kamchatka . . .* , by Stepan Petrovich Krasheninnikov, trans. by E. A. P. Crownhart-

Vaughan (Oregon Historical Society, Portland, 1972); *Whale Hunting with Gun and Camera,* by Roy Chapman Andrews (D. Appleton, 1916); Andrews, "The California Gray Whale."

Oil companies and Sakhalin: Author's interview with Robert Brownell, San Diego, Jan. 17–18, 2000; "The Sakhalin-2 Project," brochure of Sakhalin Energy Investment Company Ltd.; "Sakhalin's Oil: Doing It Right," by Dan Lawn, Rick Steiner, and Jonathan Wills, Sakhalin Environment Watch and Pacific Environment and Resources Center, Nov. 1999; "Oil Spills: Lessons from Alaska for Sakhalin," report by Rick Steiner (University of Alaska, July 1999); "Whales of the Asian Pacific in Peril," by Michelle Boyd and Gary Cook, *Earth Island Journal,* Winter–Spring 1999; "Sakhalin Celebrates 'First Oil,' " by Irina Panteleyeva and Richard Thomas, *Eastern Russian Journal,* July 30, 1999; "Sakhalin's Projects," *Pacific Russia Oil and Gas Report,* Winter 1999; "Sakhalin," by David Gordon, *Biodiversity Briefings from Northern Eurasia,* ed. by Margaret Williams (Pocono Environmental Education Center and The Center for Russian Nature Conservation/Tides Center), 1998.

Western gray whale migration: "A Sighting of Gray Whale off Kochi, Southwest Japan in July 1997, with Some Notes on Its Possible Migration in Adjacent Waters of Japan," by Hidehiro Kato and Yukio Tokuhiro, *Report of the International Whaling Commission* (SC/49/AS17), 1998; "Strandings and Sightings of the Western Pacific Stock of the Gray Whale *Eschrichtius robustus* in the Chinese Coastal Waters," draft report by Qian Zhu, Institute of Oceanology, Chinese Academy of Sciences, 1998; "Sluggish Giant with a Butterfly-Tail," by S. Blokhin, *Sovietskiy Sakhalin,* July 18, 1996; "Distribution and Migration of the Western Pacific Stock of the Gray Whale," by Hideo Omura, *Scientific Reports of the Whales Research Institute* 39, 1988; "Distribution of the Gray Whale (*Eschrichtius gibbosus*) off the Coast of China," report by Wang Peilie; "Notes on a Gray Whale Found in the Ise Bay on the Pacific Coast of Japan," by Masami Furuta, *Scientific Report Whales Research Institute,* 35, 1984; "Probable Existence of the Korean Stock of the Gray Whale (*Eschrichtius robustus*)," by Robert L. Brownell, Jr., and Chan-II Chun, *Journal of Mammalogy,* May 1977; "History of Gray Whales in Japan," in *The Gray Whale; Men and Whales.* Scammon quotation: *MM.* Andrews quotation: 1914 monograph. Hainan background: Various Web sites, including Hainan Professionals Work's International Travel Service.

Gray whale studies: "Gray Whales (*Eschrichtius robustus*) off Sakhalin Island, Russia: Seasonal and Annual Patterns of Occurrence," by David W. Weller et al., *Marine Mammal Science* 15 (4), Oct. 1999; "Observations on Okhotsk-Korean Gray Whales on Their Feeding Grounds off Sakhalin Island," by R. L. Brownell et al., *Report of the International Whaling Commission,* 47, 1997, app. 4; "Observations on the Behavior of *Eschrichtius gibbosus* Erxl., 1777 on the Shelf of Northeastern Sakhalin," by E. I. Sobolevskii, *Russian Journal of Ecology* 29 (2), 1998.

CHAPTER 20: BREAKTHROUGH ACROSS TROUBLED WATERS

1988 gray whale rescue: "A Rescue That Moved the World," by Mark A. Fraker, *Oceanus* 32(1), spring 1989; *Men and Whales;* "Helping Out Putu, Siku and Kanik," by Eugene Linden, *Time,* Oct. 31, 1988; *Connection on the Ice,* by Patti H. Clayton (Temple University Press, 1998); *Freeing the Whales,* by Tom Rose (Birch Lane Press, 1989); *The Charged Border.* Author's interview with Tom Albert, Barrow, Alaska, July 6, 1999.

Chapter 21: Catastrophe in Chukotka

The author visited the Provideniya region of Chukotka, Aug. 3–10, 1999. Opening epigraph: *Inuit, Whaling, and Sustainability*, by Milton M. R. Freeman et al. (AltaMira Press, Walnut Creek, Calif., 1998); Basic background on whaling and the region was found in the following books: *Arctic Adaptations: Native Whalers and Reindeer Herders of Northern Eurasia*, by Igor Krupnik (Dartmouth College, 1993); *Hunting the Largest Animals: Fairy Tales and Myths of the Bering Strait Chukchi*, ed. by Alexander B. Dolitsky (Alaska-Siberia Research Center, Juneau, 1997); *The Eskimos of Bering Strait: Divided Twins*, by Yevgeny Yevtushenko (Viking Studio Books, 1988); *Anthropology of the North Pacific Rim*, ed. by William W. Fitzhugh and Valérie Chaussonet (Smithsonian Institution Press, 1994); *Men and Whales* ("Pirate whaling" chapter); *Ten Months Among the Tents of the Tuski*, by W. H. Hooper (John Murray, London, 1853); *Travel and Adventure in the Territory of Alaska*, by Frederick Whymper (Harper & Bros., 1869); *Tent Life in Siberia*, by George Kennan (G. P. Putnam's Sons, 1910); *Contributions to North American Ethnology*, vol. 1 (U.S. Government Printing Office, 1877); *Spirit of Siberia*, by Jill Oakes and Rick Riewe (Smithsonian Institution Press, 1998); *The Peoples of the Soviet Far East*, by Walter Kolarz (Archon Books, 1969); *On the Northwest: Arctic Alaska and Siberia* (Aldrich); *A Year with a Whaler*, by Walter Noble Burns (Macmillan, 1919); *Account of a voyage of Discovery to the North-east of Siberia, the Frozen ocean, and the North-east Sea*, by Gawrila Sarytschew (Richard Phillips, London, 1806); *The Gray Whale*. Chukchi legend: "A Whale of a Tale," by Shelley Gill, *Fly Me Away* magazine, circa 1981.

Articles and reports: "The Role of Subsistence System of Chukotka Natives in a Biological Diversification of the Bering Sea," by Nickolai I. Mymrin, report for World Wildlife Foundation, June 1999; "Role of the Eskimo Society of Chukotka in Encouraging Traditional Native Use of Wildlife Resources by Chukotka Natives and in Conducting Shore Based Observations on the Distribution of Bowhead Whales, *Balaena mysticetus*, in Coastal Waters off the South-eastern part of the Chukotka Peninsula (Russia) During 1994," report to Department of Wildlife Management, North Slope Borough; "Role of . . . During 1995"; "Report: Preservation and Development of the Subsistence Lifestyle and Traditional Use of Natural Resources by Native People (Eskimo and Chukchi in Several Coastal Communities: Inchoun, Lorino, New Chaplino, Sireniki, Enmelen) of Chukotka in the Russian Far East During 1997," Shared Beringian Heritage Program, U.S. National Park Service, Anchorage, and Department of Wildlife Management, North Slope Borough, Mar. 1999.

"Whale Alley," by Mikhail A. Chlenov and Igor I. Krupnik, *Expedition*, Winter 1984; "A Report on Russian Aboriginal Whaling," by Eleanor O'Hanlon, report by Humane Society International, 1997; "Return to Siberia," by Michael Kundu, *Sea Shepherd Log*, Spring 1998; "The History of Gray Whale Harvesting off Chukotka," by M. V. Ivashin and V. N. Mineev, *Report of the International Whaling Commission* 31, 1981; "In Russia, a Whale of a Conservation Question," by Richard C. Paddock, *Toronto Star*, Sept. 25, 1997; *Antler on the Sea: Chukchi, Yupik, and Newcomers in the Soviet North*, dissertation by Anna Marie Kerttula (University of Michigan, 1997); *Beringia Natural History Notebook Series*, National Audubon Society, Sept., 1992; "Alaska at a Cross Roads," by L. Saunders McNeill, *True North*, Spring 1999; "Sea Hunters of Sireniki," by Alexander Milovsky, *Natural History*, Jan. 1991; "An

Overview of Two Boating Accidents During Early September near the Chukotka Villages of Provideniya and Sireniki in the Russian Far East," North Slope Borough Memorandum, Sept. 29, 1995; "Siberian Eskimo," by Charles C. Hughes, chapter in *Handbook of North American Indians*; "Chukchi: Warriors and Traders of Chukotka," by S. A. Arutionov, chapter in *Crossroads of Continents* by William W. Fitzhugh and Aron Crowell (Smithsonian Institution Press, 1988); "Travels in Chukotka," by Ken Leghorn, *Juneau Empire*, Feb. 28, 1993; "U.S. and Soviet Environmentalists Join Forces Across the Bering Strait," by Frank Graham, Jr., *Audubon*, July–Aug. 1991; "Forsaken in Russia's Arctic: 9 Million Stranded Workers," by Michael R. Gordon, *New York Times*, Jan. 6, 1999; articles in *Anchorage Daily News*, Aug. 5–9, 1999, about missing Eskimo boaters. Legend: "About a Little Whale," provided author by Alexander B. Dolitsky.

Chapter 22: Northern Coda: End of the Expedition

The Dall letters and news accounts of the expedition were found in the Smithsonian archives. *Beyond the Lagoon* recounts the Bulkley charges against Scammon and Dall. The author interviewed Mikhail Vysokov in Yuzhno Sakhalinsk, with Michael Allen as translator, on July 29, 1999.

Chapter 23: Christopher Reeve and the Gray Whale

Gray whale strandings: "High Gray Whale Mortality and Low Recruitment in 1999: Potential Causes and Implications," by B. J. Le Boeuf et al., *Cetacean Research Management* (in press). Russian whaling: International Fund for Animal Welfare, 1999 annual report. Payne and Brosnan quotations from press conference, Sept. 30, 1999: video footage supplied to author by IFAW. Meeting with Mitsubishi official: Interview with Jared Blumenfeld, Yarmouthport, Mass., Aug. 2000. UNESCO report: "UN Report Spells Bad News for Mitsubishi Salt Plant," by Cat Lazaroff, *ENS*, Nov. 29, 1999. "Ecology Groups Use Gray Whale to Re-Engineer Mexican Project," by Jonathan Friedland, *Wall Street Journal*, Aug. 24, 1999. Author's interview with Kathryn Fuller, telephone, Oct. 10, 2000. World Wildlife Fund report: "Evaluating Development in Protected Areas: Decision-Making in the Vizcaíno Biosphere Reserve" (draft 3/3/99). Prince Bernhard letter: Copy supplied by Betty Aridjis. Author's interviews with Blumenfeld, Fred O'Regan, at IFAW, Yarmouthport, Nov. 1999. "How Saving Whales Advances Democracy," by Homero Aridjis and Serge Dedina, *Los Angeles Times*, Nov. 25, 1999.

The author's interview with Christopher Reeve took place at his home in New York State on Nov. 22, 1999. Background on Reeve: "New Hopes, New Dreams," by Roger Rosenblatt, *Time*, Aug. 26, 1996; " 'We Draw Strength from Each Other,' " by Liz Smith, *Good Housekeeping*, June 1996; "Local Hero," *People*, Jan. 27, 1997. "Gray Whales with Christopher Reeve," from *In the Wild*, is available on PBS Home Video.

Chapter 24: Scientific Puzzles in San Diego

Carl L. Hubbs Papers, 1927–79, at Scripps Institution of Oceanography, La Jolla, Calif. Articles by Raymond Gilmore: "Calving of the California Grays," *Pacific Discovery* 7, 1954; "The Gray Whale," *Oceans*, Jan. 1969; "The Return of the Gray Whale," *Scientific American*, Jan. 1955; "Ecology of the Gray Whales," *Environment*

Southwest, Winter 1976. Gilmore background: "R. M. Gilmore, Whale Watcher, Is Dead at 77," by Lanie Jones, *Los Angeles Times*, Jan. 2, 1984; "Raymond Maurice Gilmore, 1 January 1907–31 December 1983," *Environment Southwest*, Spring 1984. Gilmore papers at San Diego Museum of Natural History. Scammon quotation: "Pacific Sea-Coast Views III," *OM*, Mar. 1872. Whale-watching figures: "Whale Watching 2000: Worldwide Tourism Numbers, Expenditures, and Expanding Socioeconomic Benefits," report by Erich Hoyt for International Fund for Animal Welfare, Aug. 2000.

Author's interview with Jim Sumich, Jan. 18–19, 2000. Gigi background: *Gigi: A Baby Whale Borrowed for Science and Returned to the Sea*, by Eleanor Coerr and William E. Evans (Putnam, 1980). Whale sampling: *The Life History and Ecology of the Gray Whale*, by D. W. Rice and A. A. Wolman, Special Publication American Society of Mammalogists 3, 1971. Author's interview with Wayne Perryman, La Jolla, Jan. 17, 2000. "Annual Calf Production for the California Stock of Gray Whales and Environmental Correlates, 1994–2000," by Wayne L. Perryman et al., Southwest Fisheries Science Center (SC/52/AS18) 2000. See also "Diel Variation in Migration Rates of Eastern Pacific Gray Whales Measured with Thermal Imaging Sensors," by Wayne L. Perryman et al., *Marine Mammal Science* 15 (2), Apr. 1999; "ACS/LA Gray Whale Census and Behavior Study: 1998–99," paper by Alisa Schulman-Janiger, American Cetacean Society.

Chapter 25: Mysterious Evolutions

Beyond the Lagoon is a primary source on Scammon's life after the telegraph expedition. Scammon correspondence re: Cope: Spencer Baird papers at Smithsonian Archive, also Dall papers. Cope background: *The Gilded Dinosaur*, by Mark Jaffe (Crown Publishers, 2000); Internet sources. Scammon as describer: "The Fossil Whale, *Balaenoptera davidsonii* (Cope, 1872), with a Review of Other Neogene Species of *Balaenoptera* (Cetacea: Mysticeti)," by Thomas A. Deméré, *Marine Mammal Science* 2 (4), Oct. 1986.

Gray whale history: "On the Nomenclature of the Pacific Gray Whale," by William E. Schevill, *Breviora* (Museum of Comparative Zoology, Cambridge, Mass., Sept. 29, 1952). McLeod and Barnes quotations: "Gray Whales," by Samuel A. McLeod and Lawrence G. Barnes, *Terra*, Nov.–Dec. 1983. Gray whale fossil: *Terra* article and "The Fossil Record and Phyletic Relationships of Gray Whales," by Lawrence G. Barnes and Samuel A. McLeod, chapter in *The Gray Whale*. Author's interview with Barnes, Los Angeles, Dec. 2, 1999.

Whale paleontology: *Marine Mammals: Evolutionary Biology*, by Annalisa Berta and James L. Sumich (Academic Press, 1999); "The Evolutionary History of Whales and Dolphins," by R. Ewan Fordyce and Lawrence G. Barnes, *Annual Review of Earth Planetary Science* 22, 1994; "Whales, Dolphins and Porpoises: Origin and Evolution of the Cetacea," by Lawrence G. Barnes, "Mammals: Notes for a Short Course Organized by P. D. Gingerich and C. E. Badgley," ed. by T. W. Broadhead, *Studies in Geology* 8, 1984; "How the Whale Lost Its Legs and Returned to the Sea," by John Noble Wilford, *New York Times*, May 3, 1994; "Genetic Science Looks at Whale, Sees Hippo Under the Blubber," Reuters, *Boston Globe*, Aug. 31, 1999.

Atlantic gray whales: "Atlantic Gray Whales," by James G. Mead and Edward D. Mitchell, in *The Gray Whale*; *The North Atlantic Gray Whale. . . . An Historical Outline*

Based on Icelandic, Danish-Icelandic, English and Swedish Sources Dating from ca. 1000 A.D. to 1792, dissertation by Ole Lindquist, Centre for Environmental History and Policy, Universities of St. Andrews and Stirling, Scotland, Mar. 2000; *Men and Whales; Deep Atlantic*, by Richard Ellis (Alfred A. Knopf, 1998); *Sea of Slaughter*, by Farley Mowat (Atlantic Monthly Press, 1984); "A Whale of a Story," by Neil Savage, *Boston Globe*, May 23, 2000.

The author traveled to La Paz and Magdalena Bay with Larry Barnes, Feb. 1–3, 2000. See Barnes's "Summary of the Fossil Cetacean Record of Mexico," *Somemma*, Apr. 18–22, 1999.

Chapter 26: Scammon's Legacy

J. Ross Browne: *Etchings of a Whaling Cruise*, by J. Ross Browne (1846; rept., Belknap Press, Harvard University Press, 1968); *Resources of the Pacific Slope*, by J. Ross Browne (D. Appleton & Co., 1869); *J. Ross Browne: His Letters, Journals and Writings*, ed. by Lina Fergusson Browne (University of New Mexico Press, 1969); *A Western Panorama, 1849–1875: The Travels, Writings and Influence of J. Ross Browne . . .*, by David Michael Goodman (Arthur H. Clark Co., Glendale, Calif., 1966); Browne papers, Bancroft Library. See also: *Melville in the South Seas*, by Charles Roberts Anderson (Dover rpt., 1966).

The Balaenic Games; or, The Whale's Jubilee, by Mrs. Linzorn (Darton and Harvey, London, 1808), found at New Bedford Whaling Museum library. Scammon "try-works" quotation: *MM*. Scammon letters: Dall and Baird collections, Smithsonian archives. Father's letter: Scammon papers, Bancroft Library. Reviews: Scammon papers. *Louis Agassiz: Speaking for Nature*, by Paul Brooks (Houghton Mifflin, 1980); Internet sources. Swedenborg: *The Essential Swedenborg*, by Sig Synnestvedt (Swedenborg Foundation, Westchester, Pa., 1977); *Emanuel Swedenborg: Scientists and Mystic*, by Signe Toksvig (Swedenborg Foundation, 1983). Carmany letter: Scammon papers. *History of Sonoma County:* Bancroft Library. Author's interview with Scammon descendants: Sacramento, Dec. 13, 1998.

Early conservationists and Leopold quotation: *A Fierce Green Fire*, by Philip Shabecoff (Hill and Wang, 1993). M. E. Bowles quotation: Found at Kendall Whaling Museum library. *Prentice Mulford's Story*, by Prentice Mulford (F. J. Needham, New York, 1889). *The Rise of Theodore Roosevelt*, by Edmund Morris (Ballantine Books, 1988). Hemingway quotation: *By-line: Ernest Hemingway*, ed. by William White (Touchstone/Simon and Schuster, 1998). Penitent Butcher's Society: Interview with Dr. Gregg Mitman, Myopia Hunt Club, Mass., July 26, 2000.

Chapter 27: Victory at San Ignacio Lagoon

The author wrote about the saltworks decision in "David: 1 Goliath: 0," *Amicus Journal*, Summer 2000. Other articles included "In Mexico, Nature Lovers Merit a Kiss from a Whale," by Julia Preston, *New York Times*, Mar. 5, 2000; "Mexico Deep-Sixes Plan for Baja Lagoon Saltworks," by Mary Beth Sheridan and James F. Smith, *Los Angeles Times*, Mar. 3, 2000; "Environmentalists Persuade Mexico to Save the Whales," by John Ward Anderson, *Washington Post*, Mar. 3, 2000; "Green Triumph in Shades of Grey," by James Hrynyshyn, *Vancouver Sun*, Mar. 14, 2000; "Whales and Mexican Environmentalists Win One," by Susan Ferris, Cox News Service, Apr. 2, 2000; "Whales Sink Plans for Mexican Salt Plant," by Richard Chacon, *Boston Globe*,

Apr. 2, 2000; "Activists Break New Ground to Help Shake Off Saltworks Project," by James F. Smith, *Los Angeles Times*, Apr. 23, 2000.

EPILOGUE: AN UNCERTAIN FUTURE

Author's interviews on saltworks decision: Jorge Urbán, La Paz, Mar. 7, 2000; Alberto Szekely, Andrés Rozental, and Jared Blumenfeld, Yarmouthport, Mass., Aug. 2000; Homero Aridjis, New York, Aug. 9–10, 2000; James Brumm, New York, Oct. 19, 2000. Mitsubishi construction contract: "Mexico: Construction Contract Award for Planned $278,000,000 Power Plant, Mitsubishi," Newsletter, Information Access Company, Nov. 1, 1999.

Makah hunt: "Activist Prevents Tribe from Killing Gray Whale," Reuters, *Los Angeles Times*, Apr. 21, 2000; "Tribe Loses Round on Bid to Hunt Whales," by Kim Murphy, *Los Angeles Times*, *Boston Globe*, June 10, 2000; "Court Voids Approval of Makah Whaling," by Hal Bernton and Lynda V. Mapes, *Seattle Times*, June 10, 2000. Japan and whaling: "Banned Gray Whale Meat Sold in Japan: N.Z. Researchers," *Japan Times*, May 14, 2000; "Our Whale Vote Was Bought, Says Minister," by Belinda Huppatz, *Adelaide Advertiser*, July 7, 2000; "Japan, Feasting on Whale, Sniffs at 'Culinary Imperialism' of U.S.," by Calvin Sims, *New York Times*, Aug. 10, 2000; "A Reprehensible Whale Hunt," editorial, *New York Times*, Aug. 15, 2000; "Japan Widens Whale Hunt, Provoking Objections," by Andrew C. Revkin, *New York Times*, Aug. 29, 2000; "U.S. to Move Against Japan over Whales," by David E. Sanger, *New York Times*, Sept. 13, 2000.

The author visited Wayne Perryman at Piedras Blancas Point, Apr. 21, 2000; phone interview, Nov. 2, 2000. Bering Sea Problems/Climate Change: "The Unbearable Capriciousness of Bering," *Science*, Feb. 25, 2000; "Aquamarine Waters Recorded for First Time in Eastern Bering Sea," *Eos Transactions American Geophysical Union* 79 (10), Mar. 10, 1998; "Global Warming Could Be Bad News for Arctic Ozone Layer," *Nature*, Apr. 6, 2000; "Observations and Predictions of Arctic Climatic Change: Potential Effects on Marine Mammals," by Cynthia T. Tynan and Douglas P. DeMaster, *Arctic* 50 (4), Dec. 1997; "Incorporating Climate Change Effects into the Process for Evaluating Management Regimes for Aboriginal Subsistence Whaling," by Cynthia Tynan and Douglas P. DeMaster, *Report of the International Whaling Commission* 47, 1997; "Proceedings of the 1998 Science Board Symposium: The Impacts of the 1997–98 El Niño Event on the North Pacific Ocean and Its Marginal Seas," North Pacific Marine Science Organization, Mar. 1999; "Age-Old Icecap at North Pole Is Now Liquid, Scientists Find," by John Noble Wilford, *New York Times*, Aug. 19, 2000; "Open Water at Pole Is Not Surprising, Experts Say," by Wilford, *New York Times*, Aug. 29, 2000; "Through Northwest Passage in a Month, Ice-Free," by James Brooke, *New York Times*, Sept. 5, 2000; "Thinning Sea Ice Stokes Debate on Climate," by William K. Stevens, *New York Times*, Nov. 17, 1999; "Melting of Arctic Ice Tied to Global Warming," AP, *Los Angeles Times*, Dec. 3, 1999; "Research Predicts Summer Doom for Northern Icecap," by Walter Gibbs, *New York Times*, July 11, 2000; "From the Harpoon to the Heat: Climate Change and the International Whaling Commission in the Twenty-first Century," report by William C. G. Burns, Pacific Institute for Studies in Development, Environment, and Security, June 2000; "Turning Up the Heat: How Global Warming Threatens Life in the Sea," report by Amy Mathews-Amos and Ewann A. Bernston, World

Wildlife Fund/Marine Conservation Biology Institute, 1999; "Answers from the Ice Edge," report by Greenpeace Arctic Network, June 1998.

Lawrence quotation: *Studies in Classic American Literature*, by D. H. Lawrence (1922; rept., Viking Press, 1971). Jung quotation: *Psychological Reflections*, by C. G. Jung, ed. by Jolande Jacobi (Princeton University Press, 1970). Hillman quotation: *A Blue Fire*, by James Hillman, ed. by Thomas Moore (Harper & Row, 1989).

Other Sources Consulted

Whale Nation, by Heathcote Williams (Jonathan Cape, London, 1988); *A Window on Whaling in British Columbia*, by Joan Goddard (Jonah Publications, Victoria, B.C., 1997); *History of the American Whale Fishery*, by Alexander Starbuck (Castle Books, Secaucus, N.J., 1989 ed.); *Reflections of a Whale-Watcher*, by Michelle A. Gilders (Indiana University Press, 1995); *Green Alaska*, by Nancy Lord (Counterpoint, Washington, D.C., 1999); *Whale Tales*, vol. 1, collected by Peter J. Fromm (Whales Tale Press, Friday Harbor, Wash., 1996); *The Whale and His Captors . . .* , by Henry T. Cheever (Ye Galleon Press, Fairfield, Wash., 1991); *At the Hand of Man*, by Raymond Bonner (Alfred A. Knopf, 1993); *The Presence of Whales*, ed. by Frank Stewart (Alaska Northwest Books, 1995); *Coming into the Country*, by John McPhee (Noonday Press, 1976); *Ocean Warrior*, by Paul Watson (Key Porter Books, 1994); *Arctic Dreams*, by Barry Lopez (Bantam Books, 1996 ed.); *The Reader's Companion to Alaska*, ed. by Alan Ryan (Harvest/Harcourt Brace, 1997); *Ancient Land: Sacred Whale*, by Tom Lowenstein (Farrar, Straus & Giroux, 1993); *The Outer Coast*, by Richard Batman (Harvest/HBJ, 1985); *Hunting the Desert Whale*, by Erle Stanley Gardner (Companion Book Club, London, 1960); *Among Whales*, by Roger Payne (Delta, 1995); *A Whale Hunt*, by Robert Sullivan (Scribner, 2000); *Secrets of the Ocean Realm*, by Michele and Howard Hall (Carroll & Graf/Beyond Words, 1997); *Whales, Dolphins and Porpoises* (Facts on File, 1988); *Whale*, by Eric S. Grace (Laurel Glen, San Diego, 1997).

USEFUL WEB SITES

American Cetacean Society: www.acsonline.org

John Calambokidis: www.cascadiaresearch.org

Campaign Whale: www.campaign-whale.clara.net

Center for Whale Research: www.rockisland.com

Coast Watch Society: www.coastwatchsociety.org

Jim Darling: www.westcoastgraywhale.org

Serge Dedina: www.wildcoast.net

Steph Dutton/Heidi Tiura: www.sanctuarycruises.com

Gray Whales with Winston: www.geocities.com

International Fund for Animal Welfare: www.ifaw.org

International Whaling Commission: http://ourworld.compuserve.com/homepages/iwcoffice

Interspecies Communication, Inc. (Jim Nollman): www.interspecies.com

Bruce Mate: http://hmsc.orst.edu/groups/marinemammal

National Marine Mammal Laboratory homepage: http://nmm101.afsc.noaa.gov/

Natural Resources Defense Council: www.nrdc.org

Ocean Alliance: www.oceanalliance.org

Oceanic Society: www.oceanic-society.org

Christopher Reeve Paralysis Foundation: www.apacure.com

Save the whales: www.savethewhales.org (updates on Makah hunt, Japanese whaling)

Sea Shepherd Conservation Society: www.seashepherd.org

Society for Marine Mammalogy: http://pegasus.cc.ucf.edu/~smm/

Southwest Fisheries Center (Bob Brownell, Dave Weller, Wayne Perryman): http://swfsc.ucsd.edu

U.S. Citizens Against Whaling: www.usagainstwhaling.org

Whale Center of New England: www.whalecenter.org

Whaleman Foundation: www.whaleman.org

WhaleNet: http://whale.wheelock.edu

Whaling: www.whaling.com (latest news stories)

World Wildlife Fund: www.wwf.org

Acknowledgments

WRITING this book was far from a solitary endeavor, but one in which I received inspiration and guidance from so many friends, associates, and experts in marine science. I can only begin by honoring my "extended family," who made it all possible. My wife, Alice, not only read every word—several times, a labor of love beyond the call of duty—and where necessary challenged my faulty thinking, but once again tolerated my long, inattentive hours and took care of me with beauty and grace. George Peper was not only a wonderful traveling companion who contributed numerous outstanding photographs, but proved an astute and meticulous reader whose editorial suggestions strengthened the text immensely. Jessie Benton's inspiration and vision laid a foundation for the book and, as always, she gave me a beautiful home in which to create (and incredible cuisine to accompany it). Eben Given's many painstaking days spent at his drawing table, crafting fifteen marvelous maps to accompany the journey, were an immeasurable contribution.

Brian Keating's long hours working with me on the initial culling of hundreds of pictures, as well as his own photographic contributions, were deeply appreciated. Devora Wise offered her continuous enthusiasm and scrutinous eye for detail. That first pathbreaking trip to "la laguna" only happened via the wheel of David Gude, who also contributed many provocative thoughts on human-whale interactions. Alison Peper and Richie Guerin provided invaluable translating assistance. My other traveling partners on several voyages to the lagoon offered great company: Deirdre Goldfarb, Anthony Gude, Atticus Gude, Max Burnett, Eve Lyman, Valery Lyman (filmmaker), Jacqueline Lyman, Michèle DelFonso, Cal Bernhard, Gregg Walter (my acoustical educator), Heidi Keegan, Mike McTag, and son Eric. Thanks also to Arthur Young and Dr. Bryan J. Stern, who shared that first "close encounter." And, of course, Dennis O'Brien and our three canine companions, Leroy, Taylor, and Mocha.

Thanks to Daria Lyman and Darren Kew for their hospitality on my trips to D.C.; to Jeremy Greenwood for clipping all the papers; to Laura

Given, Kurt Franck, George Burnett, Gale Lyman, Mary Curtis, Jim Kweskin, Rachel Greenwood, Candy Guerin, Susan Dewan, Faith Gude, and Carol Lazar for their preliminary readings, suggestions, and moral support (and Carol Franck, Patty Glynn, Carolee Goldfarb, and Heidi Keegan for "fronting" those expenses and bills; Rhys Burman for his assistance with my computer [il]literacy, Jay Johnson for "on-call" home maintenance, secretarial assistants Dana Gerety, Lisa Thomas, and Amy Brady, and Albert White for "guarding the office" while I was away).

Ivette Arias took hours out of her busy office schedule to help with transcribing my interview tapes. Julia Olszewski, an intern provided under the auspices of Brandeis University's Professor Laura Goldin, devoted hours to gathering research and analyzing the saltworks situation. Among my writer friends, Alan and Amy Flurry helped greatly with the proposal process, where Sy Montgomery's work provided my model; Ross Gelbspan's suggestions on giving the book a more "overt spine" provided a critical turning point in the task of weaving the material together; Richard Adams Carey and John Balzar, esteemed authors both, took the time to read and offer valuable commentary on several chapters. So did Homero and Betty Aridjis, always available amid their hectic schedule, and who kindly granted permission to reprint Homero's poem, "Eye of the Whale," as the epigraph to this book. In the realm of literary agents, I couldn't imagine a more engaged and supportive one than my friend Sarah Jane Freymann. Nor have I worked with an editor as incisive, thoughtful, and enthusiastic as Denise Roy at Simon & Schuster (how she remembered those various repetitive passages, I'll never fathom). Kathryn Lassila, my editor at *The Amicus Journal*, was instrumental in article assignments that got this project off the ground.

Without the careful reading, editorial corrections and other offerings made by numerous marine scientists, my "layman's effort" would have contained many errors and misconceptions. My deepest thanks to Dr. Bruce Mate, Dr. David Rugh, Dr. David Weller, Dr. Lawrence Barnes, Dr. Steven Swartz, Dr. Marilyn Dahlheim, Dr. John Calambokidis, Dr. James Sumich, Dr. Peter Tyack, Dr. Jim Darling, Dr. Thomas Albert, Dr. Lindy Weilgart, Dr. John Heyning, Dr. Deborah Brosnan, Dr. Wayne Perryman, Dr. Roger Payne, Francisco Ollervides, and Amanda Bradford in this regard. Also to Jared Blumenfeld, Joel Reynolds, Heidi Tiura, Christopher Reeve, and Robert Kennedy, Jr., for their comments on various sections of the book. And to Richard Donner for his ongoing concern about gray whales.

Across many thousands of miles of travel, I was privileged to find a number of gracious hosts: Bob Small Sr. in Oakland, and Bob Small Jr. in Anchorage (thanks to Lincoln Brower for the entrée); Jim and Bernie Stimpfle in Nome; Dan and Ellen Richard in Wales; Dave Weller at Piltun Lagoon and in San Diego; Saunders McNeill in Anchorage; Maldo and Catarina Fischer at San Ignacio Lagoon; Margaret Williams and Michael Ross on the World Wildlife Fund's Bering Sea tour; Grant Miller in Sitka; Steph Dutton and Heidi Tiura in Monterey; Tom Langman in Washington, D.C.; John and Nancy Spencer at La Laguna; Jorge Salorio and family at Punta Abreojos; Viktor and Lyuda Rekun in Provideniya; Igor Zagrebin and Yana Rekun on my travels through Chukotka; Dmitry Listsyn, Natasha Barannikova, and Sergei Alekseenko of Sakhalin Environment Watch in Yuzhno-Sakhalinsk; Yuri Shvetsov in Okha; Robert Brownell in La Jolla; the Orion Society in Zitácuaro, Mexico; Jeff Gossett and Frank Falcinelli at Moomba's restaurant in New York; the clerks at my Anchorage Hotel "base camp," and caretakers of Baja Expeditions, Kuyima, and Campo Cortés at San Ignacio Lagoon. Thanks, too, to those who assisted in my travel planning: Dean Franck for advice on Russia; Geoffrey C. Orth with Alaskan itinerary; Roman Bratlavsky and Tandy Wallack of Circumpolar Expeditions; Gary Cook of Baikal Watch; author Mark Taplin; Vicky Malone of Kenai Fjords Tours; Debbie Chapman of Uniglobect; Michael Allen of American Business Center in Yuzhno; B. J. Chisholm; Doug Norlen and David Gordon of Pacific Environment Resources Center.

For assistance with my piecing together the story of Charles Melville Scammon, I first express my condolences to the family of the late Mildred Decker and her son Tom Decker, both of whom passed away in 1999–2000, including Edy Decker, her son Dave Decker, and relations John Decker and Lois Decker. Also, to author Lyndall Baker Landauer, and to Poole family historian Harriette Otteson.

For extremely useful background information/material on gray whales and related matters, I thank: Dr. Burney J. Le Boeuf, Dr. James T. Harvey, Dr. Cynthia Tynan, Ronn Storro-Patterson, Craig Van Note of the Monitor Consortium; Kate O'Connell of the Whale-Dolphin Conservation Society; Michael Kundu of Project Sea Wolf; Dan Spomer of the Washington Citizens' Coastal Alliance; Lisa Denham; John Bockstoce; Brad Wellman and Pam Simpson of the Fourth Corner Whale Project; Joel Reynolds, Jacob Scherr, and Ari Hershowitz of the Natural Resources Defense Council; Mark Spalding; Jennifer Ferguson-Mitchell at International Fund for Animal Welfare; Carolyn Heitman; Erin Abbott; Joan Goddard; Rick Carver; Sue Steinacher; Serge Dedina and

Emily Young; Francisco Guzmán Lazo; Jeff Pantukhoff; Susan Payne; Nancy Ricketts; Chuck and Alice Johnstone; Charley Lean; Ken Leghorn; Ana Kerttula; Sue Arnold of Australians for Animals; Emma Wilson; Eric Stirrup; Roger Harritt; Gennady Smirnov; Nathan LaBudde; Humane Society of the U.S.; Karen S. Barton; Richard Ellis; Dr. Carl Safina; Andrew Christie of Sea Shepherd. For facilitating my meeting with Christopher Reeve: Sarah Houghton. And my deepest condolences to Susan Cunningham, whose husband Walt Cunningham died in a diving accident late in 2000; an Alaskan gentleman and naturalist of his caliber will be sorely missed.

For kindly making their photographs available for consideration: Michelle Kinzel, Cathy Gremin of Tigress Productions, and Erika Hill at Pandion Enterprises. Also, to the archivists at the Bancroft Library, Smithsonian Institution, American Museum of Natural History, Scripps Institution of Oceanography, Marine Mammal Commission, San Diego Museum of Natural History, Kendall Whaling Museum, New Bedford Whaling Museum, National Maritime Museum, Makah Cultural and Research Center, Provideniya Museum, and Alaska/Siberia Research Center.

Index

Note: Page numbers in *italics* refer to illustrations.